Mechanics of Reinforced Concrete

This book provides a practical foundation in the mechanical behaviour of reinforced concrete (RC) for advanced structural engineering students and practitioners. It examines RC at multiple scales, introducing strain and slip-based material behaviour that can simulate the mechanisms controlling cracking and crushing. These concepts are then applied to beam segments, explaining member deformation, crack width, flexural strength and shear strength. The behaviour at a segment can subsequently be used to predict more complex behaviour such as moment redistribution, the behaviour of prestressed concrete members and the behaviour of members that have been strengthened using external reinforcement. The engineering principles are equally applicable to RC as well as fibre-reinforced polymer reinforcement and fibre-reinforced concretes.

Mechanics of Reinforced Concrete

Deric John Oehlers and Phillip Visintin

CRC Press
Taylor & Francis Group
Boca Raton London New York

CRC Press is an imprint of the
Taylor & Francis Group, an **informa** business

First edition published 2026
by CRC Press
2385 NW Executive Center Drive, Suite 320, Boca Raton FL 33431

and by CRC Press
4 Park Square, Milton Park, Abingdon, Oxon, OX14 4RN

CRC Press is an imprint of Taylor & Francis Group, LLC

ISBN: 9781041193234 (hbk)
ISBN: 9781041193685 (pbk)
ISBN: 9781003711407 (ebk)

DOI: 10.1201/9781003711407

Typeset in Sabon
by codeMantra

Contents

Foreword

It is my honour to introduce *Mechanics of Reinforced Concrete* by Emeritus Professor Deric Oehlers and Professor Phillip Visintin, a significant work that fundamentally transforms our understanding of reinforced concrete behaviour. Having conducted doctoral research under Professor Visintin's supervision and been exposed to their collaborative work, I have observed firsthand their exceptional ability to make sophisticated engineering mechanics both accessible and practically applicable.

The authors' fundamental aim – to explain the mechanics in simple terms – is brilliantly realised through their extensive use of visual aids, flow charts and worked examples that bring the subject to life. From the outset, this book prioritises visual explanations, using carefully selected photographs and diagrams to illustrate fundamental mechanisms, followed by iterative numerical methods that render complex behaviour genuinely approachable. By focusing systematically on what can be seen and measured in practice, the authors ensure that the mechanics remain grounded in physical reality, avoiding esoteric abstractions while embracing practical insights.

The scope is remarkably comprehensive, systematically covering the key mechanisms governing reinforced concrete behaviour – from fundamental material properties and fibre reinforced concrete, through tension stiffening and concrete softening, to sophisticated analysis under flexure and shear. The treatment extends from members with strong reinforcement bonds, as in reinforced concrete, to weak bonds in composite steel–concrete beams, profiled slabs and FRP-retrofitted members, while thoroughly examining corrosion and its impact on sustainable design insights increasingly vital in today's engineering landscape.

What distinguishes this work is its innovative and generic displacement-based approach that directly handles discontinuities – cracks, sliding and bond deterioration – without the empirical approximations characterising traditional strain-based methods. This represents a genuine paradigm shift, enabling exact solutions where conventional approaches require approximation and opening new possibilities for innovation in materials and structural systems.

This book will resonate with diverse readers. Students will appreciate clear, patient explanations that make challenging concepts accessible, while practising engineers will value insights extending beyond conventional approaches, offering fresh perspectives on complex problems. The emphasis on understanding fundamental behaviour through material properties rather than expensive testing promises to inspire innovation in emerging materials.

Through my academic experience working with these concepts, I have recognised that this work represents more than technical knowledge – it embodies an educational philosophy of learning through understanding rather than memorisation. I am deeply grateful for their dedication to advancing our field, and I am confident that this book will become an essential resource for anyone seeking to understand reinforced concrete mechanics.

Tianyu Xie
Professor, School of Civil Engineering,
Southeast University, China
Editor-in-Chief, Case Studies in Construction Materials, Elsevier
July 2025

Preface

The Euler–Bernoulli theorem that plane sections remain plane is widely accepted as fundamental to the behaviour of reinforced concrete members subjected to some degree of flexure. The corollary to this theorem is the linear strain profile, subsequently moment–curvature and consequently the predominance of strain-based methods of analysis. Unfortunately, strain-based approaches cannot cope directly with discontinuities, that is, cracking or sliding, which are unavoidable when dealing with brittle materials like concrete and dominate RC behaviour; this is overcome by using 'fudge factors' or approximations such as smeared cracks, empirically derived hinge lengths or semi-empirical effective flexural rigidities. That is, despite more than one hundred and seventy years of development and application, reinforced concrete design is still dominated by empirical, semi-mechanical and semi-empirical approximations. The Euler–Bernoulli displacement approach provides a full mechanical solution.

To overcome this problem with discontinuities, this book does not start with the strain-based Euler–Bernoulli corollary but with the displacement-based Euler–Bernoulli theorem, with the advantage that the subsequent moment–rotation relationships can simulate discontinuities directly. It must be emphasised that this displacement-based moment–rotation approach, when applied to segments bounded by vertical sides, gives exactly the same results as the strain-based moment–curvature approach when there are no discontinuities but has the added advantage of coping with discontinuities directly without the need for empirical approximations. It is also shown that the same moment–rotation approach can be applied to segments with inclined cracks to predict shear failure based on non-linear distributions of strain. An added advantage of the Euler–Bernoulli displacement-based approach is that it can cope directly with both flexural behaviour and shear behaviour, as they are fundamentally the same mechanism and do not require separate forms of analysis. Furthermore, it is generic and can cope directly with any reinforcement, such as steel reinforcing bars and adhesively bonded fibre reinforced polymer plates, any type of concrete including those with steel or polymer fibres and any type of member axial load including pre-tensioned or post-tensioned members with bonded or unbonded cables.

This book is primarily a teaching book and uses worked examples and flow charts to aid in the explanation of reinforced concrete mechanical behaviours. We have first used simple numerical approaches that are easy to follow and can be applied to any material, whether linear or non-linear. After explaining the mechanics through these often iterative approaches, we follow with closed form solutions where possible. We have deliberately tried to keep the mathematics as simple as possible. Understanding is further enhanced by carefully chosen worked examples, even for what may be perceived as the most trivial of problems, to provide continuity in the development of understanding. We suggest that this book could be used as a preliminary undergraduate course to give students a deep understanding of the mechanics that govern reinforced concrete behaviour before they are introduced in subsequent courses to codes of practice whose main purpose is to provide a safe design.

This book has also been written to help researchers. Having understood the fundamental behaviours, we point out where further research is needed, where further research can be done, and we also provide flow charts to help researchers develop their own models. For example, for researchers who enjoy specialising in mechanics, in the formation of RC hinges required in the collapse of a structure, the hinges developed in the current model require a small constant moment region. Developing this hinge to allow for variable moments within it may be a useful improvement, but whether it adds significantly to the accuracy needs to be determined. For researchers who prefer the empirical approach, such as in the important areas of fatigue or corrosion, it now becomes clear what material testing is required to quantify long-term effects. These properties can then be used to determine elemental, segmental and member behaviour. For those focused on the sustainability of structures, we show how a deep understanding of the behaviour of reinforced concrete, from the material to the structural level, is critical for making informed decisions on material selection and structural form. Indeed, we demonstrate that trying to make decisions to improve the sustainability of structures in the absence of a deep understanding of behaviour can lead to perverse outcomes, in which more harm is done than if environmental impact were entirely ignored.

Finally, and most importantly for engineers, this book can help in the development of new products and in finding new solutions to engineering problems such as sustainability. For those developing new products, such as fibre concrete, it is made clear what material tests need to be conducted, which can then be used in the available flow charts to determine the effect of the new materials on member behaviour. Should these effects not be satisfactory, then the material characteristics can be changed until the theoretical member behaviour is satisfactory. This will save vast amounts in member testing, as only material testing needs to be done, thus expediting the development of new materials. For engineers, this displacement-based approach provides an additional valuable tool in design to help solve particular problems that fall outside the scope of codes. The displacement-based approach can be used to give an accurate assessment of the performance and strength of an existing structure and the effect of remedial action if required. Furthermore, it can be used to assess the use of new products in new structures.

Acknowledgements

We, the authors, would like to thank the Australian Research Council for funding much of the research referred to in this book through Discovery Grants. We would also like to thank the China Scholarship Council for supporting the PhD students below from China, which has not only helped in the development of this approach but also enhanced research collaboration between China and Australia. We would also like to thank The University of Adelaide for fully supporting this research and the writing of this book. I, Deric Oehlers, would like to thank the following PhD students of mine, in alphabetical order, for their outstanding contributions in developing major advancements in the displacement-based theory: Kaldoon Al-Mhawish, M.S. Mohamed Ali, Yongjian Chen, Qian Feng, Xinkai Hao, Matthew Haskett, Daniel Knight, Irene Liu, Wade Lucas, Rahim Muhamad, Rudi Seracino, Alexander Sturm, Phillip Visintin and Tao Zhang.

We would like to thank the British Institution of Civil Engineers (ICE) for awarding us and our colleagues both the Telford Gold Medal in 2012 for our paper *A Generic Unified Reinforced Concrete Model* and the Frederick Palmer Prize in 2015 for our paper *Simulating Reinforced Concrete Members. Part 1: Partial Interaction Properties*, both of which were published in the Proceedings of the Institution of Civil Engineers – Structures and Buildings. These prizes were an incentive to continue and develop the new approach expounded in this book.

There are two types of research: that which improves current approaches by extending and refining them, which is very important, and that which produces radically new approaches to provide a better understanding. The latter is much more difficult to publish as it goes against the grain. Much of our research falls into this latter category, and much has been published by the American Society of Civil Engineers' (ASCE) *Journal of Structural Engineering* and *Journal of Composites for Construction*. We would like to thank the ASCE for having a review system that allows radically new approaches to be published, and this, of course, also applies to the ICE.

We would also like to thank Dr. Alexandre B. Sturm and Dr. Lloyd Dadd for their helpful contributions to this book, as well as Jon Ayoub and Mark Peacock for their photographs of test specimens. Finally, we would like to thank our wives, Julie and Bree, for their patience and encouragement.

Author Biography

Deric John Oehlers is an Emeritus Professor of Structural Engineering at the Adelaide University of Adelaide, Australia. This is the culmination of a long career, which includes the award of the Telford Gold Medal from the UK's Institution of Civil Engineers. His other works include *Elementary Behaviour of Composite Steel and Concrete Structural Members*, also published by Taylor & Francis.

Phillip Visintin is a Professor of Structural Engineering at The University of Adelaide University, Australia. He is a Fellow of Engineers Australia, Chartered in Structural Engineering and Engineering Leadership and Management.

Part I

Introduction

Part 1 of this book provides a qualitative visual description of all the mechanisms that control the behaviour of reinforced concrete members, from the microscopic to the macroscopic, and their progressive interaction. It is crucial that the reader can visualise what is happening before quantifying the behaviour in the ensuing parts of the book.

Part 2 covers all the material properties at the engineering and meso-scale level that need to be determined experimentally. These consist of the standard strain-based full-interaction properties in Chapter 2 for the reinforcement and concrete, which include the long-term effects of creep and shrinkage. Additionally, it covers the slip-based partial-interaction (PI) material properties in Chapter 3, which include: the shear friction properties of concrete; the PI bond-slip properties between the reinforcement and adjacent concrete; and the PI tensile properties of fibre concrete.

These meso-scale properties are then used in Part 3 to determine the PI mechanical behaviour of small elements of the reinforced concrete member. Chapter 4 discusses tension stiffening, and Chapter 5 addresses size-dependent concrete softening for actively and passively confined concrete. These elemental behaviours are then used in Part 4 to determine the member behaviour over a short length, which is a segment of the member, not only for flexure in Chapter 6 but also for shear in Chapter 7.

The behaviours within a segment, primarily moment–curvature, are then used in Part 5 to determine the member behaviour. Chapter 8 covers the flexural analysis of RC members with strong bond, from serviceability to moment redistribution. Moment redistribution and collapse are dealt with from first principles, which highlights the importance of ductile hinges. Chapter 9 covers both post-tensioned and pre-tensioned members, which are members with external axial loads. Chapter 10 discusses profiled slabs and composite steel and concrete beams, which are members with a relatively weak interface bond between the steel and concrete components. Chapter 11 then considers FRP retrofitting and methods to ensure gradual, as opposed to catastrophic, failure.

Having covered member behaviour, Part 6 examines the long-term design of members. This is approached in two ways. Firstly, in Chapter 12, design for durability is considered by quantifying the influence of reinforcement corrosion on behaviour at both the serviceability and ultimate limit state. Importantly, it is demonstrated how the application of the segmental approach can assist with assessing the performance and residual capacity of an existing structure. Finally, in Chapter 13, the mechanics of sustainable design are explored by explaining key factors that are important to the

DOI: 10.1201/9781003711407-1

structural engineer when quantifying the whole-of-life environmental impact through a life-cycle assessment.

The book concludes with an epilogue in Part 7, which discusses the approach used and the sequence of research and development that has led to our current understanding.

At the end of each chapter, there is a list of 'notations and abbreviations', followed by a list of 'references and development'. Papers and books are directly referenced in the text in the usual way. Additionally, a glossary of papers that are not referenced directly in the text but are important as they show the development of the mechanics is included, which may be of use as an additional resource to develop understanding.

Chapter 1

A visualisation of reinforced concrete behaviour

This chapter gives a qualitative description of the numerous mechanisms that control the behaviour of reinforced concrete (RC) members (Oehlers 2010; Oehlers et al. 2012, 2014a,b, 2017; Oehlers and Visintin 2021). These mechanisms occur from the micro-scale, that is the material properties, and progressively interact up to the macro-scale, governing the mechanical behaviour of members. Students and engineers should be able to visualise these mechanical behaviours and understand how they function before designing members according to codes of practice or, if a more thorough analysis is required, by using the procedures outlined in this book.

There have been two significant recent advancements in concrete development: (i) fibre reinforcement in the concrete and (ii) the use of non-metallic (*FRP*) reinforcement. The inclusion of steel or synthetic fibres into concrete has been shown empirically to have enormous benefits in providing increased member stiffness, strength, ductility and durability. Furthermore, the use of non-metallic reinforcement can significantly improve the service life of structures by either strengthening existing elements or by utilising its non-corrosive nature in the construction of new elements. Importantly, the introduction of non-metallic reinforcement has also removed the reliance on material ductility to provide member ductility since, as will be shown in this book, this can be provided through design, allowing for concrete softening. Hence, this book deals with concrete with any type and amount of fibres from zero (referred to as conventional concrete) up to the maximum amount that can be incorporated (referred to as fibre reinforced concrete (FRC)) and with any type of *reinforcement*, including *FRP* reinforcement.

In this chapter, we cover all the major mechanisms that control the physical behaviour of RC. This is because we believe it is necessary to demonstrate the interrelationship between these mechanisms, which would not be possible if each were treated in isolation. Students, researchers and engineers may find this chapter challenging, and we apologise for this; however, it is necessary to show these linkages to visualise RC behaviour effectively. To aid understanding, these mechanisms will be further explained in ensuing chapters, along with examples demonstrating their application in quantifying the strength and serviceability behaviour of RC. The reader may therefore wish to return to this chapter after reading other sections of the book to develop a more comprehensive understanding of RC behaviour. To facilitate this, when new mechanisms are introduced, the reader will also be referred to chapters where they are discussed in greater detail.

DOI: 10.1201/9781003711407-2

1.1 PARTIAL INTERACTION

The terms partial interaction (*PI*), full interaction (*FI*) and no interaction (*NI*) are used throughout this book. These terms are not complex; they simply refer to the ability of a RC component to slide across an interface. PI refers to a local mechanism where any degree of sliding can occur, from a small to a large amount. The two extremities of *PI* are *NI* and *FI*. *NI* refers to the case where there is no interface shear resistance or restriction to sliding. In contrast, at the other extremity, *FI* refers to an infinite resistance or stiffness against sliding, thereby preventing sliding.

The major sliding mechanisms that can occur in *RC* beams are shown in the photograph in Figure 1.1a, showing a constant moment region to the left of the applied load and a constant shear region to the right. These cracks are illustrated in Figure 1.1b, in which the small segment of the beam is subjected to an axial load *N*, moment *M* and transverse shear force *V*; the deformations shown are considerably exaggerated. Consider Point *A* in Figure 1.1b, at the interface between the reinforcing bar and the adjacent concrete and where the reinforcement touches the concrete. First, consider the case when the flexural crack shown has not formed, that is the case prior to flexural cracking. At Point *A*, let the longitudinal strain in the reinforcing bar, that is the strain in the reinforcing bar in the direction of the length of the bar, be ε_r and the longitudinal strain in the concrete, that is the strain in the concrete in line with the direction of the reinforcing bar be ε_c. Prior to the flexural crack forming, there is no slip between the reinforcing bar and the adjacent concrete, indicating FI. Consequently, the FI strain in the reinforcing bar ε_{rFI} is equal to that in the concrete ε_{cFI}.

Now, let the flexural crack form, such that Point *A* in Figure 1.1b is adjacent to the right hand face of the crack as shown. Let the crack width at the level of the reinforcement be $2S_r$ also shown. To accommodate this crack width, the reinforcing bar at both crack faces must, by symmetry, slip S_r, which signifies PI. This slip does not occur over zero length but accumulates along the bar. This zone, B_L–B_R, where slip occurs, is labelled the 'reinforcement sliding zone'. Hence, this is a PI zone, beyond which FI occurs. That is beyond B_L–B_R, ε_{rFI} equals ε_{cFI} and interface slip is zero. Revisiting Point *A*, where PI occurs, the longitudinal strain in the concrete at the crack face ε_{cPI} is zero as the concrete is adjacent to the crack face where there is no lomgitudinal stress. Hence,

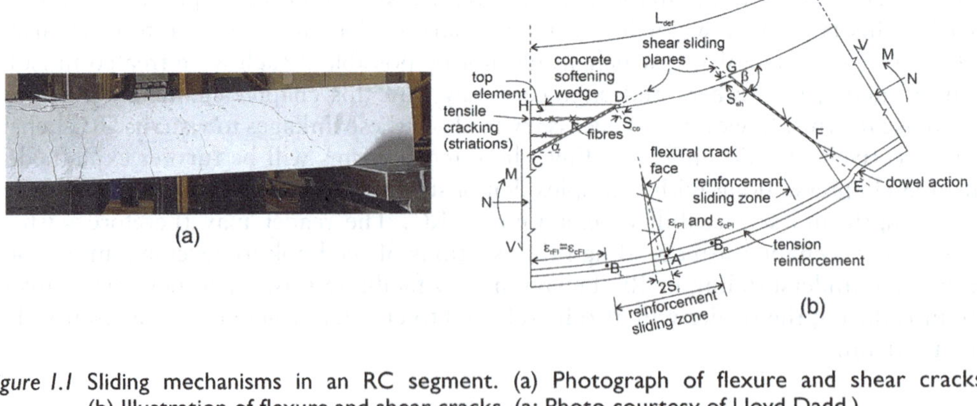

Figure 1.1 Sliding mechanisms in an RC segment. (a) Photograph of flexure and shear cracks. (b) Illustration of flexure and shear cracks. (a: Photo courtesy of Lloyd Dadd.)

the longitudinal strain in the concrete at Point *A* has reduced from ε_{cFI} to ε_{cPI}, which is zero. Furthermore, upon cracking, the longitudinal force in the reinforcement will increase to offset the loss of longitudinal stress in the concrete along the crack face. Hence, the longitudinal strain in the reinforcement at Point *A* will increase to ε_{rPI} such that $\varepsilon_{rPI} > \varepsilon_{rFI}$.

The difference in longitudinal strain between the reinforcing bar and the adjacent concrete ($\varepsilon_r - \varepsilon_c$) is referred to as the slip–strain. The integration, or accumulation, of this slip–strain over the 'reinforcement sliding zone' shown in Figure 1.1 determines the variation in slip along this zone. This mechanism will be fully explained in simple terms in the ensuing simple iterative numerical approaches. Hence, when the slip–strain $\varepsilon_{r\_}\varepsilon_c$ is zero, *FI* occurs, meaning there is no slip. Conversely, when the slip–strain is not zero, *PI* occurs, indicating interface slip. *NI* is a special case of *PI* when there is no interface bond. The slip between the reinforcing bar and the adjacent concrete at the crack face in Figure 1.1 is S_r. If the bond between the reinforcing bar and the adjacent concrete were infinitely stiff, there would be no slip, meaning the slip–strain would be zero and *FI* would occur, even though the flexural crack had developed. As this scenario of infinite bond stiffness is impossible, the crack width ($2S_r$) and the slip (S_r) at the crack face will always be finite, thus ensuring *PI*.

Transverse shear failure is another sliding mechanism, and thus another *PI* mechanism. This shear failure mechanism occurs along an inclined plane, such as the shear sliding plane *E–F–G* in Figure 1.1b. The angle of plane β often, but not always, corresponds to the weakest sliding plane. Unlike the flexural *PI* mechanism within the region B_L–B_R, where slip accumulates from slip–strain, transverse shear failure involves a rigid body displacement of S_{sh}, where one side of the interface slides relative to the other, as shown. The inclined shear sliding plane *E–F–G* crosses a compression region *F–G* and a tension region *E–F* within which the crack is inclined due to flexure and shear. If a reinforcing bar crosses this tension region as shown, it is also subject to all the *PI* mechanisms previously described for the 'reinforcement sliding zone'. In Figure 1.1b, the shear sliding plane has to resist a component of the transverse shear force *V*. Resistance is largely provided by the shear capacity of the uncracked compression region *F–G*, and some from the tension region *E–F*, as it is initially cracked. Once rigid body sliding S_{sh} occurs (that is a further *PI* mechanism occurs), the shear is resisted by shear sliding in both the compression and tension regions and, furthermore, by dowel action of the tension reinforcement as shown, introducing yet another *PI* mechanism.

Concrete crushing is associated with the formation of the concrete wedge *H–C–D* (Figure 1.1b), which is inclined at α, where the value of α is reasonably constant at about 26° (Chen et al 2015). Concrete in compression softens by slip along the edge *C–D*. The slip, denoted S_{co} at the top of the wedge, initiates where compressive strains are largest. This movement can cause tensile cracking or striations within the wedge, as depicted, and the element continues to slide as a rigid body. As compressive stresses in the element below continue to increase, they eventually induce sliding, repeating the procedure for subsequent elements. Sometimes the wedge fails as a single element or rigid body, that is horizontal tensile cracks do not occur. Hence, this softening mechanism is akin to a rigid body displacement and constitutes another significant *PI* mechanism.

Fibres, shown as crosses in Figure 1.1b, affect all the *PI* mechanisms or zones. Steel fibres are roughly half a millimetre thick or less and only a few centimetres long. In general, they act as small, unanchored or poorly anchored reinforcing bars that cross a sliding zone and, hence, are governed by the *PI* mechanisms already described for reinforcing bars. Fibres resist the sliding of the concrete softening wedge along *C–D* and help the wedge to behave monolithically by traversing the tensile cracks or striations within it. As such, they can increase concrete strength; however, more importantly, fibres increase concrete ductility, allowing for much greater plastic hinge rotations and, consequently, enhancing the ability to redistribute moment and resist collapse. Around flexural cracks, fibres not only enhance the flexural capacity by providing tensile stresses across the crack but also enhance the bond between the reinforcing bar and the adjacent concrete, thereby reducing crack widths and consequently enhancing durability. In the shear failure zone, fibres help resist sliding due to shear failure, thereby enhancing shear capacity and also making shear failure much more ductile. Tiny fibres, acting at the meso-scale, may at first glance appear insignificant, but they provide a profound beneficial effect on the global (macro-scale) behaviour of *RC* members.

1.2 EULER–BERNOULLI

The Euler–Bernoulli theorem, which states that plane sections remain plane, governs much of structural analysis and applies to any member material or member shape. In very simple terms, this can be visualised by considering a constant moment *M* applied to a segment of length $2L_{def}$, as in Figure 1.2b. When this occurs, any line on the surface, such as *A–A*, remains a line (e.g. *B–B*) after the moment is applied. Hence, if the ends of the segment are subjected to a rotation θ induced by a moment *M* at each end, the displacement *d* at the ends of the segment δ remains linear. This is shown in the displacement profile in Figure 1.2c, where δ_t is the displacement at the top surface and d_{NA} is the depth to the neutral axis. By symmetry in Figure 1.2b, θ represents the rotation over half the length of the segment (L_{def}) and is equal to δ_t/d_{NA} radians. It can be seen that this displacement has been applied to a three-dimensional structure with the dimensions shown in Figure 1.2a and b. Furthermore, the Euler–Bernoulli theorem can be characterised by the relationship M/θ.

The displacements shown in Figure 1.2c can be divided by the total length over which this displacement accumulates, that is by L_{def}, to yield the strain distribution profile ε in Figure 1.2d. The curvature χ shown is, therefore, the rotation θ divided by L_{def},

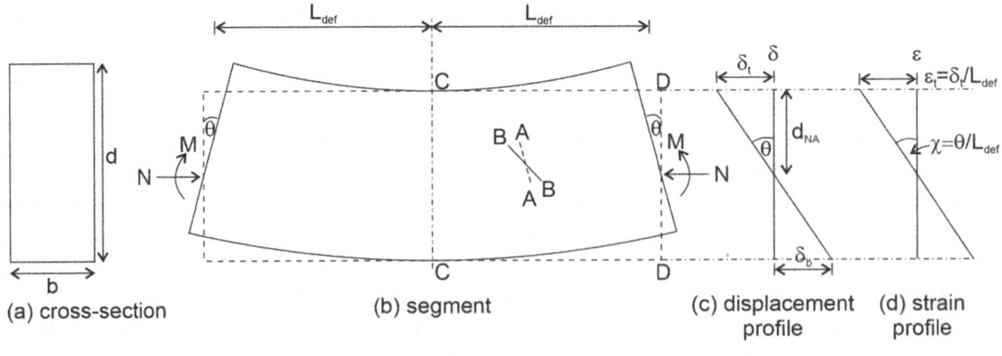

Figure 1.2 Euler–Bernoulli deformation (a) member cross-section; (b) segment of member; (c) displacement profile δ; (d) strain profile ε.

which is the rotation per unit length. This is the corollary of the Euler–Bernoulli theorem: a linear strain profile, characterised by M/χ, which is strain-based. This strain profile applies to the cross-section in Figure 1.2a and at any position, such as C–C and D–D in Figure 1.2b, and is therefore two-dimensional. From elementary mechanics, M divided by χ is the flexural rigidity, that is the flexural stiffness of the member, EI, where E is the modulus of the material and I is the second moment of area of the cross-section. The flexural rigidity EI is a very useful parameter, as it can be used directly in frame analyses. It can be seen that the flexural rigidity approach is strain-based and two-dimensional.

Strain-based approaches cannot directly cope with discontinuities or slips that occur in RC, such as those shown in Figure 1.1. Nor can they easily cope with inclined failure planes. In contrast, displacement-based approaches, as shown in Figure 1.2, can directly address these discontinuities and also inclined failure planes. This makes starting with the displacement approach preferable, as it can directly account for slip, that is PI (Oehlers et al. 2011). The discontinuities shown in Figure 1.1 can be directly incorporated into the Euler–Bernoulli displacement approach (Figure 1.2b and c). This can then be converted to the strain-based approach (Figure 1.2d) to yield real, mathematically exact flexural rigidities (not approximations) for use in flexural analyses. Similarly, the displacement-based approach can provide real or mathematically exact equivalent properties (not approximations) for strain-based analyses, such as those involving smeared properties and tensile softening concrete properties.

Worked Example WE-1.1: M/θ and M/χ

Let us consider the following simple analysis of a beam to consolidate understanding of the relationship between the Euler–Bernoulli M/θ relationship and its corollary, the M/χ relationship. The uncracked, unreinforced beam in Figure 1.2 has a depth d of 480 mm, a width b of 240 mm, a half-length L_{def} of 1000 mm, a concrete modulus E_c of 30,000 MPa and a concrete tensile strength σ_{ct} of 4 MPa.

From elementary mechanics, the second moment of area of the beam I is $bd^3/12$, which, upon inserting its properties, gives $2212 \times 10^6 mm^4$. Consequently, the flexural rigidity of the beam E_cI is 6.636×10^{13} Nmm^2. From elementary elastic beam theory, the moment M is given by $\sigma I/y$, where σ is the longitudinal stress at the distance y from the neutral axis. At the bottom fibre of the beam in Figure 1.2, y is $d/2$, that is 240 mm. Hence, the moment M to cause the bottom fibre of the beam to crack, that is to reach a stress σ_{ct} of 4 MPa, is 36.86 kNm. Furthermore, from elementary mechanics, curvature χ is M/EI. Therefore, the curvature at the onset of tensile cracking is 5.556×10^{-7}/mm. Curvature is the rotation of the beam per unit length; in this case, it is the rotation per millimetre length. Hence, the rotation θ of the beam in Figure 1.2 over the length L_{def} of 1000 mm is χL_{def}, that is 5.556×10^{-4} radians. Note θ in Figure 1.2b and c is measured in radians, where π radians equals 180°.

The deformation of the right-hand side (RHS) of the beam in Figure 1.2b is shown in Figure 1.2c. The longitudinal displacement of the top fibre δ_t is equal to $d_{NA}\theta$. The beam under consideration is symmetrical, as shown in Figure 1.2a, such that d_{NA} in Figure 1.2c is half the depth ($d/2$), which is 240 mm. Hence, the contraction at the top of the beam δ_t will equal the extension at the bottom of the beam δ_b. Hence, when the moment causing cracking is applied, the magnitude of δ_t equals that of δ_b, both of which equal $d\theta/2$, that is 0.1333 mm.

Conversely, when an Euler–Bernoulli deformation is applied to the ends of the beam in Figure 1.2b, such that in Figure 1.2c, δ_t contracts by 0.1333 mm and δ_b expands by 0.1333 mm, the rotation θ is 5.556×10^{-4} radians and the curvature throughout the beam is 5.556×10^{-7}/mm. Furthermore, the moment is 36.86 kNm, and the stress at the top and bottom fibres is 4 MPa (compression) and 4 MPa (tension), respectively. It needs to be emphasised that for an elastic, uncracked section, both approaches – the M/θ and M/χ approaches – yield not an approximate, but an exact, solution. It will, however, be shown that should cracking occur, the M/θ approach provides a direct solution, whereas M/χ does not.

1.3 *PI* CONCRETE MATERIAL PROPERTIES

The sliding mechanisms illustrated in Figure 1.1b are controlled by two PI concrete material properties. These properties quantify the resistance to interface sliding and must be measured or determined experimentally. Along both the shear sliding plane *C–D* (adjacent to the concrete softening wedge) and the transverse shear sliding plane *E–F–G* (associated with shear failure), the *PI* shear friction (SF) material properties quantify the relationship between the interface shear resistance and interface slip. In the reinforcement sliding zone of the tensile reinforcement, such as B_L–B_R, the *PI* bond-slip properties quantify the relationship between the interface resistance and interface slip.

1.3.1 Shear friction (*SF*)

An element of initially uncracked concrete is subjected to both a constant compressive confining stress σ_N, as shown in Figure 1.3a, and a gradually increasing shear stress τ_{SF}. As the shear stress τ_{SF} increases, a shear sliding crack will eventually form at a shear stress of V_u. If the aggregate is strong compared with the mortar, the shear sliding crack will circumvent the aggregate particles, as shown along *A–A–A*. Any further increase in the shear displacement will cause the slip Δ_{SF}, shown in Figure 1.3b. Because of the unevenness of the shear sliding plane, this slip will cause the separation or widening of the crack w_{SF}. This *PI* SF mechanism, often referred to as aggregate interlock, is idealised in Figure 1.3c, showing the interaction along a shear sliding plane between the shear stress τ_{SF}, normal stress σ_N, interface slip Δ_{SF} and effective crack width w_{SF}.

The SF properties are presented in Figure 1.4, where the *SF* material properties for specific confinements σ_N are plotted. For example, consider a concrete element subjected to a compressive confinement of σ_{N2}. From Figure 1.4, a shear stress of V_{u2} will cause a shear sliding crack. Subsequently, if a shear stress τ_{SF2} is applied, this will cause an interface slip of Δ_{SF2} and a crack widening of w_{SF2}. Conversely, if the crack width or slip can be measured, the shear stress can then be determined.

It should be noted that these *SF* properties are not esoteric but govern concrete softening and transverse shear failure, as illustrated in Figure 1.1b, and are therefore essential in the design of *RC* members. Hence, when developing new forms of concrete, such as those utilising alternative aggregates (e.g. construction and demolition waste or crumb rubber) or alternative binders (e.g. geopolymers), the developer must ensure that the SF properties are adequate. It can be inferred from the mechanism in Figure 1.3b that fibres restraining the shear sliding crack will enhance the resistance to slip and hence improve the SF properties. It has already been illustrated in Figure 1.3a that the ideal

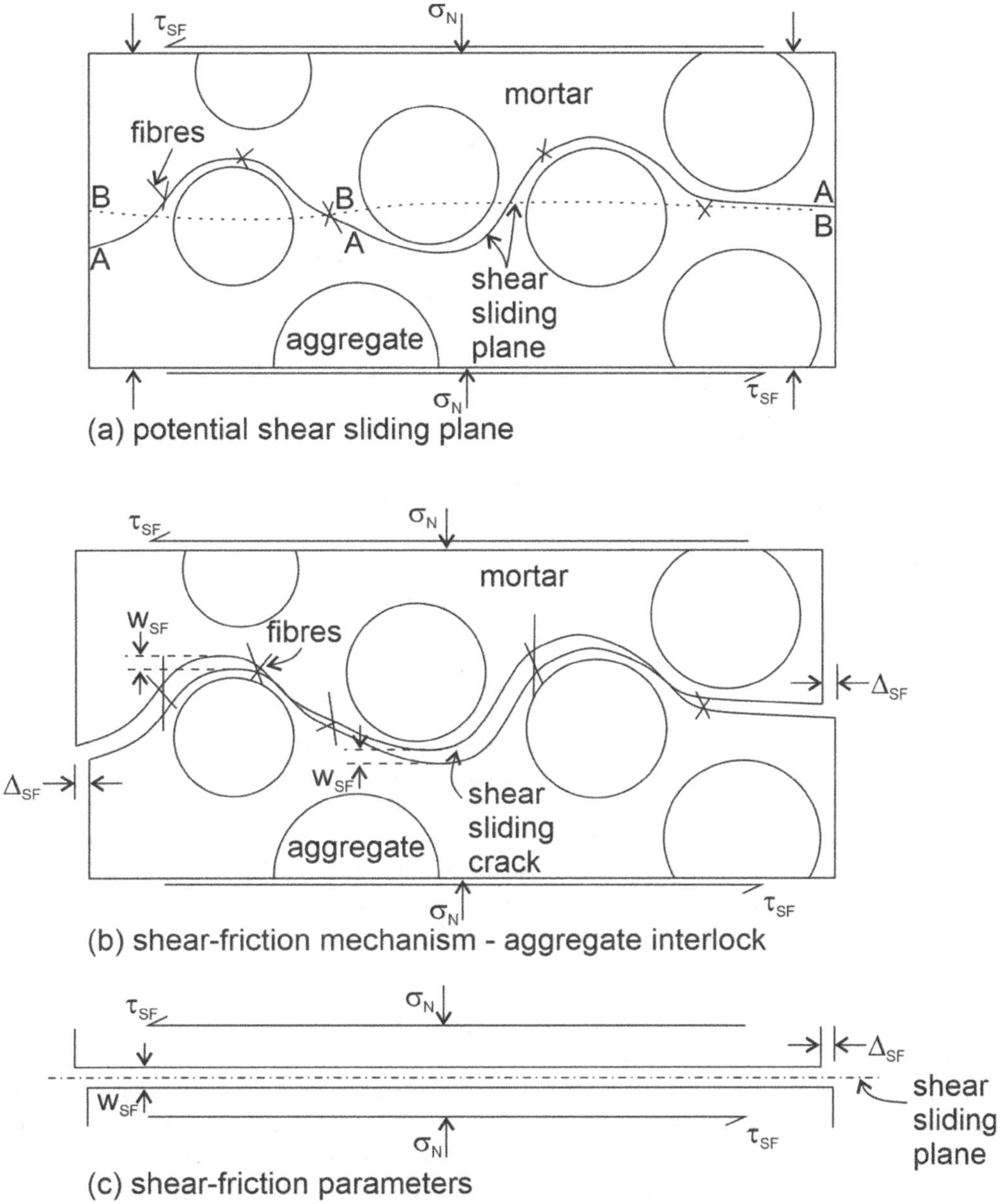

Figure 1.3 PI shear friction (a) potential shear sliding plane; (b) shear friction mechanicsm showing aggregate interlock after sliding; (c) shear friction parameters that control SF behaviour.

solution is to have strong aggregate, which causes the shear sliding plane to circumvent the aggregate, thereby significantly improving aggregate interlock and consequently the SF properties. Hence, the reader might infer that the use of recycled aggregate concrete or sintered pulverised fuel ash aggregate, which tend to produce an almost flat shear sliding plane (as shown along *B–B–B* in Figure 1.3a) should be avoided. This may not be the case, as the developer can now add fibres to improve the *SF* properties.

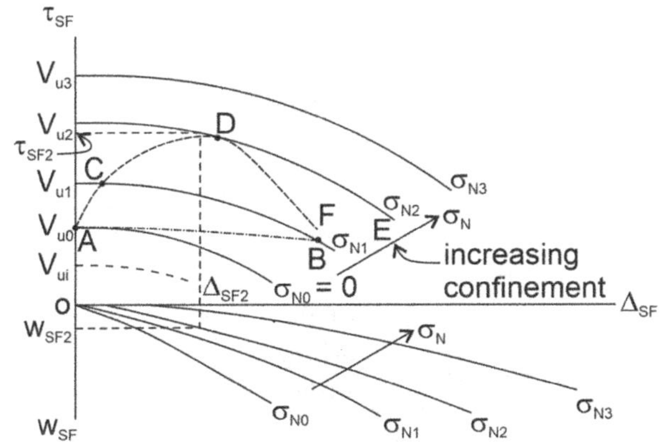

Figure 1.4 Shear friction material properties.

It should be noted that the *SF* properties in Figure 1.4 are already frequently used in design codes for the shear strength of initially cracked shear planes and are given in the following form (Oehlers and Bradford 1995):

$$v_u = 2.0f_{ct} + 0.8pf_y + 0.8\sigma_{nf} \tag{1.1}$$

Equation 1.1 is for uncracked shear planes, where v_u is the shear strength, p is the area of transverse reinforcement per unit area crossing the shear plane with a yield strength of f_y and σ_{nf} is the normal stress to the shear plane. This equation gives the shear capacity at zero slip (V_u) in Figure 1.4, where σ_N in Figure 1.4 is ($pf_y+\sigma_{nf}$). Hence, the benefits of the *SF* properties in Figure 1.4 are in quantifying the softening branch – the reduction in shear strength with slip – and also in quantifying crack widening w_{SF} due to aggregate interlock. Hence, the *SF* properties in Figure 1.4 are not unique to design but represent an extension of existing design principles.

1.3.2 Bond slip

Bond slip is the *PI* material property at the interface between the reinforcement and the adjacent concrete; it quantifies the interaction between the longitudinal shear stress and longitudinal slip at this interface. Bond slip is often determined from tests of short lengths of reinforcing bars embedded in concrete, as shown schematically in Figure 1.5a and b, where τ_B is the longitudinal shear stress and δ_B is the longitudinal slip at the interface. A short length is chosen to minimise variation of bond stress or interface slip along the embedded length.

Shrinkage of the concrete surrounding the reinforcing bar in Figure 1.5a induces the compressive radial stresses σ_N, as shown. Furthermore, interface slip in Figure 1.5b causes the concrete to ride over the reinforcement ribs, such as occurs with aggregate interlock in Figure 1.3b, which further increases the radial compressive stress σ_N in Figure 1.5a. As with the SF properties in Figure 1.4, τ_B increases with σ_N. These radial stresses in Figure 1.5a induce circumferential stresses around the reinforcing bar, which can potentially induce splitting or tearing of the concrete, as shown in Figure 1.5a.

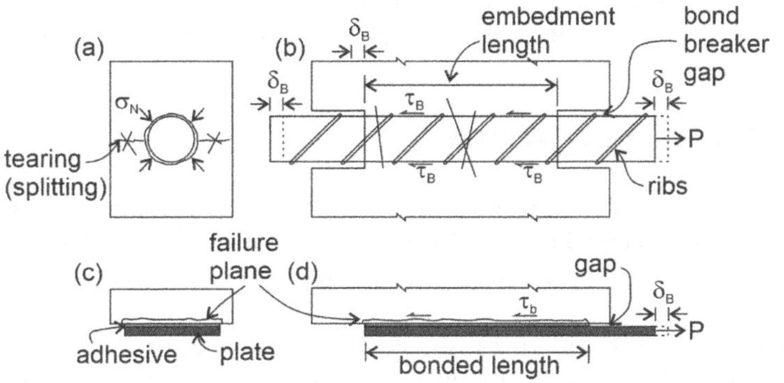

Figure 1.5 Bond slip (a) tension stiffening cross-section of encased reinforcement; (b) embedment length for internal reinforcement; (c) tension stiffening cross-section of externally plated member (d) bonded length for externally bonded reinforcement.

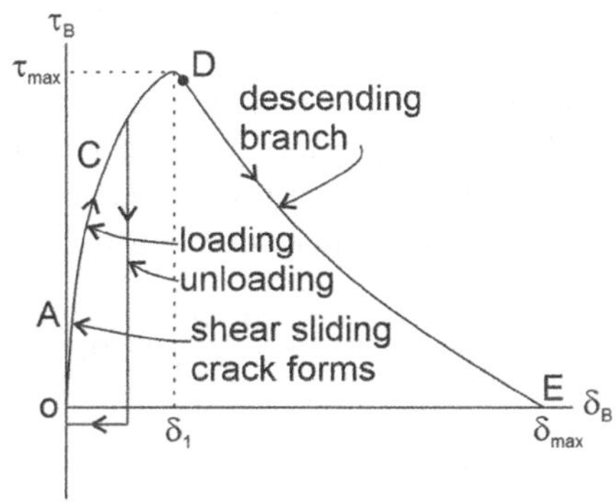

Figure 1.6 Bond-slip material properties.

Whether the concrete cover splits depends on the strength of the concrete, the concrete cover to the bar, the proximity of other reinforcing bars and the presence of stirrups. Hence, the variables that affect the bond-slip mechanism are numerous and complex.

Bond-slip test results are often represented as in Figure 1.6, where τ_{max} is the maximum shear stress at slip δ_1 and δ_{max} is the slip beyond which the shear strength can be assumed to be zero. The bond-slip mechanism in Figure 1.5a and b is akin to the SF mechanism in Figure 1.3a and b. One major difference is that the normal stresses in the SF material properties depicted in Figure 1.4 are active stresses; that is they are applied and remain constant during testing. Conversely, in the bond-slip properties shown in Figure 1.6, the confining stresses σ_N are passive; that is they change with slip. Quantifying reinforcement bond-slip properties from SF material properties is an area ripe for further research.

The bond-slip mechanism in Figure 1.5a and b applies to reinforcing bars of any type and shape, including FRP bars. Furthermore, it applies to plates adhesively bonded to the surface of concrete as depicted in Figure 1.5c and d, and to profiled sheets in composite slabs. It is also worth noting that fibres enhance performance across tearing cracks (as in Figure 1.5a) and restrain separation across the sliding plane (as in Figure 1.5b).

However, they appear to have little beneficial effect on plates bonded to the surface (Figure 1.5d), as it would be difficult for fibres to bridge the failure plane.

1.4 *PI* ELEMENTAL MECHANICS

Within a segment or small length of a member, there are elements or small regions in which PI mechanisms occur. Figure 1.7 shows the failure zones of two RC beams tested to collapse; these zones occurred in a constant moment region where there is no transverse shear.

Figure 1.7 reveals several features: the concrete crushing or softening wedge at the top of the beam, visible in both the upper and lower photographs of the sides of the beam, which has been illustrated in Figure 1.1; the vertical or near-vertical flexural cracks, also apparent in both photographs and exhibiting reasonably uniform spacing; and, in both photographs, about two-thirds down some flexural cracks, at the level of the tension reinforcement, the near-horizontal cracks emanating from them, which are the tearing cracks illustrated in Figure 1.5a. Figure 1.7 also shows the tensile cracks, or striations, within the wedge, as also illustrated in Figure 1.1. All of these features are governed by *PI* mechanics, as will be explained later in this chapter.

Figure 1.8 shows the development of cracks leading to what is commonly referred to as shear failure. Figure 1.8a shows the crack formation just prior to shear sliding failure, and Figure 1.8b shows the state just after sliding failure. In Figure 1.8a, the

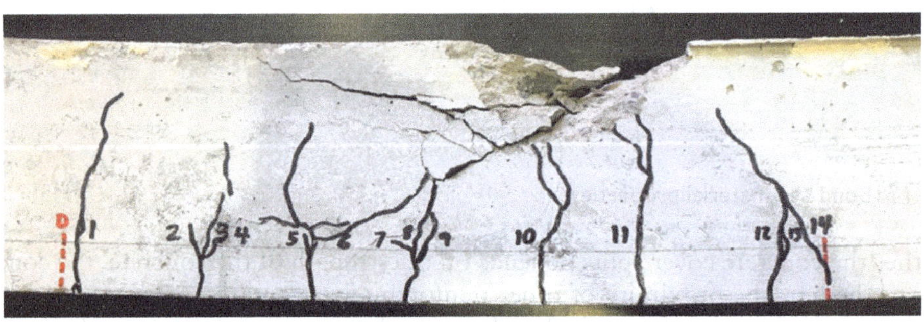

Figure 1.7 Flexural failure of *RC* beams. (Photos courtesy of Lloyd Dadd.)

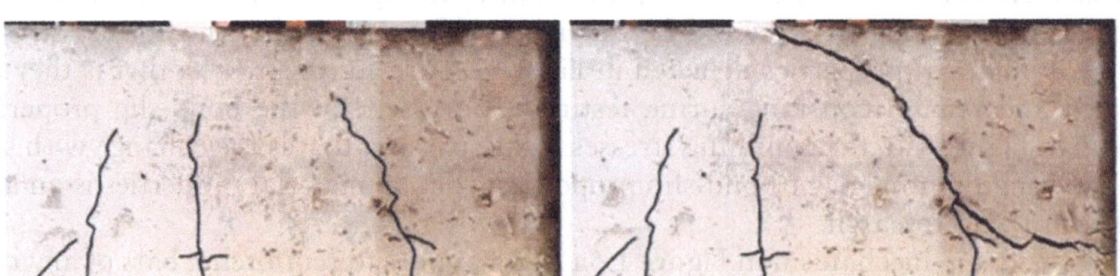

(a) Prior to sliding (b) After sliding failure

Figure 1.8 Transverse shear failure. (Reproduced from Oehlers and Visintin 2021.) (a) beam segment prior to shear sliding; (b) beam segment post sliding failure.

two flexural cracks beneath the applied load are nearly vertical because the transverse shear reduces rapidly under the applied load. Furthermore, horizontal tearing cracks are also visible in Figure 1.8a. In Figure 1.8a, the crack to the right is slightly deeper (i.e. more extended) than the two flexural cracks to the left and is inclined. Hence, the shear and flexural forces in this region have both extended and inclined the crack. This is, therefore, a flexure-shear crack. Figure 1.8b shows that shear failure occurs when the shear sliding crack extends into the compression region at the top of the beam, that is sliding has occurred along a shear sliding plane. Notably, the large horizontal crack to the right of the shear sliding crack, at the level of the tension reinforcement, is due to the dowel action of the reinforcement (illustrated in Figure 1.1) tearing the adjacent concrete, which can considerably reduce the dowel force. All of these features are also elements in which the *PI* mechanics will be described in this chapter.

Examples of the above elemental PI regions are summarised pictorially in Figure 1.9 and will be described in mechanical terms in the ensuing section. Tension stiffening, labelled (1), applies not only to the longitudinal tension reinforcement but also to the stirrups. *FRC* concrete tension is labelled (2) and applies to flexural cracks as well as along the shear sliding plane. Concrete softening is labelled (3). It applies in the flexural analysis, but it can also apply along the shear sliding plane if flexure is large.

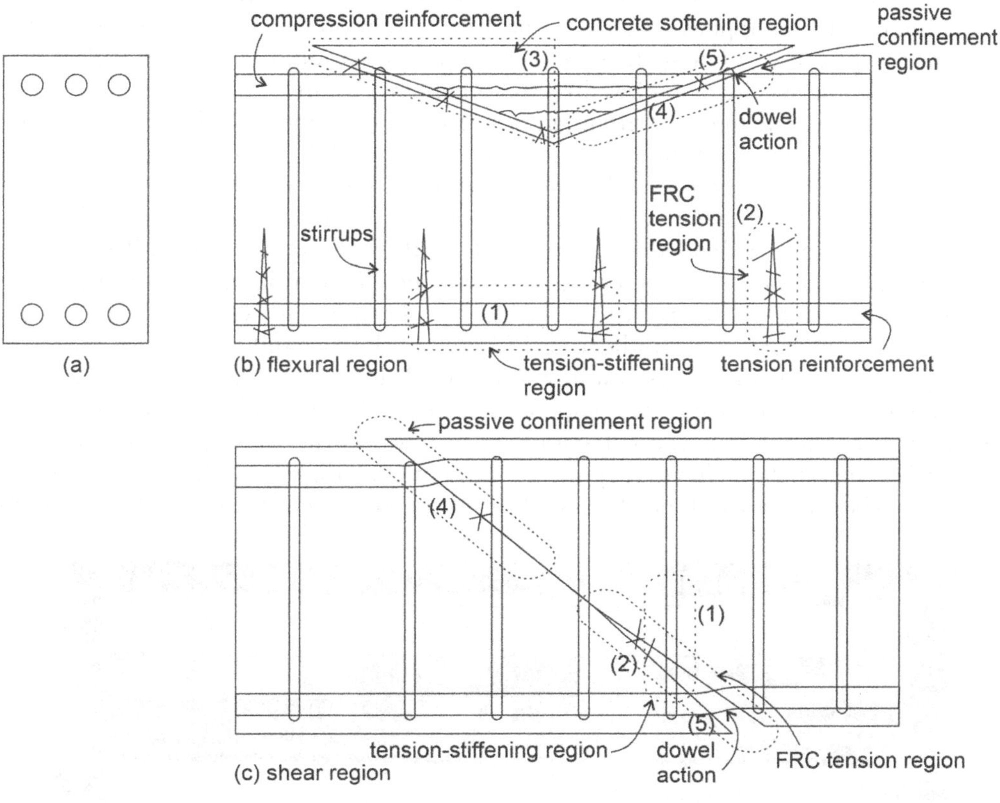

Figure 1.9 Partial-interaction elemental regions (a) beam cross section; (b) PI mechanisms in flexural regions which includes (1) tension stiffening region, (2) FRC tension region, (3) concrete softening region, (4) passive confinement region, (5) dowel action; (c) PI mechanisms in shear regions.

The passive confinement mechanism is labelled (4) and gives ductility not just to the softening region but also to the compression region of the shear sliding plane. Finally, dowel action, labelled (5), applies to both the concrete softening region and the shear sliding region. This Section 1.4 explains in qualitative terms the *PI* mechanics of each of these elements. Their interaction (i.e. the behaviour of a segment) will be addressed in the ensuing Section 1.5.

1.4.1 Tension stiffening mechanism

Figure 1.10 shows two RC specimens. Each consists of a longitudinal reinforcing bar (cross-sectional area A_r, modulus E_r) embedded in a concrete rectangular cross-section (area A_c, modulus E_c), with the reinforcing bar protruding at the ends. These specimens were pulled from their ends to produce the pattern of cracks shown. The concrete bonded to the reinforcement increases the effective axial stiffness of the tension reinforcing bar, and this PI mechanism is referred to as tension stiffening. The tension stiffening mechanism not only accounts for the increase in effective axial stiffness but also the formation, spacing and widths of the cracks shown. Consider a beam in Figure 1.7. The small portion of the beam at the level of the tension reinforcement can be considered to act as a tension stiffening prism, as in Figure 1.10. Hence, the analysis of tension stiffening in a prism applies directly to the tension reinforcing bars in a beam, that is the effective increase in bar stiffness and the crack formation, spacing and widths in a tension stiffening specimen applies directly to bars in a beam.

To further illustrate tension stiffening, test results from a tension stiffening specimen (as in Figure 1.10) are illustrated in Figure 1.11. In Figure 1.11a, the ordinate represents the force *P* applied to the bar ends, and the abscissa represents the average specimen strain ε_{sp}. This average strain, also referred to as an effective strain, is the total extension of the specimen δ divided by its length *L*. Path *O–D* has an axial stiffness E_rA_r. If there was no bond (*NI*) between the concrete and the reinforcing bar (meaning the bar took all the force), this path O-D would be followed. Path *O–A* has a combined axial stiffness of $E_rA_r+E_cA_c$, indicating *FI* between the bar and the concrete. As *P* is gradually increased, *FI* occurs initially along *O–B* until the interface bond deteriorates sufficiently to allow interface slip. Subsequently, *PI* occurs along *B–C*, and as the bond further deteriorates, the behaviour tends towards the *NI* path *O–D*. Another way of looking at the behaviour is shown in Figure 1.11b, where the ordinate represents the axial stiffness and the abscissa the effective specimen strain. The loading path

Figure 1.10 Tension stiffening specimens. (Photos courtesy of Lloyd Dadd.)

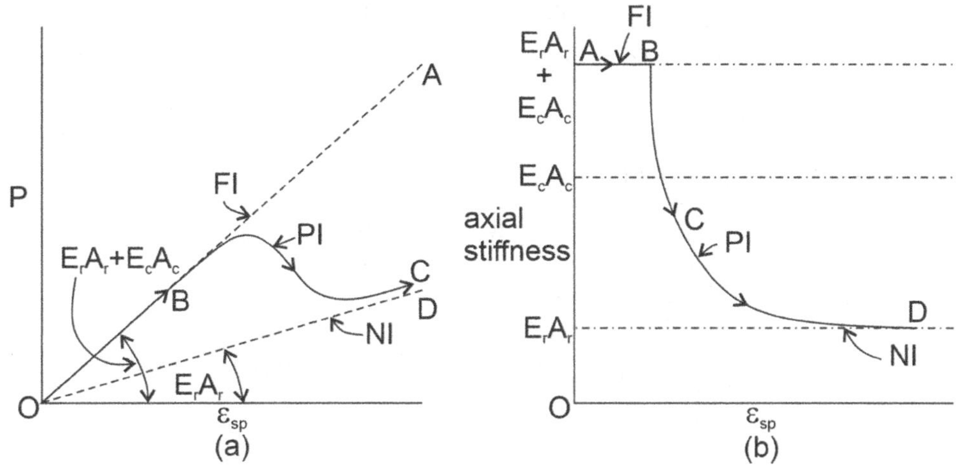

Figure 1.11 Tension stiffening test results (a) reinforcing load P versus average strain response ε_{sp}; (b) axial stiffness *EA* versus average strain response.

follows *A–B–C–D*. Initially, along *A–B*, there is *FI* at the combined axial stiffness of $E_rA_r+E_cA_c$. At Point *B*, the bond begins to deteriorate sufficiently to cause interface slip, resulting in *PI* and a reduction in axial stiffness along *B–C*. Further reduction in bond causes the axial stiffness to tend towards that corresponding to *NI*, as shown.

The tension stiffening test in Figure 1.10 is simulated in Figure 1.12. In Figure 1.12b, the specimen comprises a single bar with a cross-sectional area of A_r and an interface contact length L_{per} along its perimeter, encased in concrete with a cross-sectional area of A_c. If the bottom of the prism is considered to be the tension face of the beam, then d_r is the distance from the centroid of the bar to the tension face, such that the height of the prism is $2d_r$. The width of the tension face *b* is the portion of the tension face width of the beam associated with that bar alone. Alternatively, A_r may represent the total cross-sectional area of the tension reinforcement; L_{per} the sum of the perimeter lengths of the tension reinforcement; *b* the total width of the tension face of the beam; and d_r the distance of the centroid of the tension reinforcement from the tension face of the beam; such that the area A_c is $2d_r$ multiplied by *b* minus A_r.

Let us assume that a specimen in Figure 1.10 is pulled until the first or initial crack forms. This is represented in Figure 1.12a, where the left-hand side (*LHS*) face of the specimen corresponds to the *RHS* face of the initial crack. The bar force is *P*, and the slip of the bar relative to the RHS face of the initial crack is Δ. At the opposite face of the crack on the LHS crack face, by symmetry the bar would also have slipped Δ such that 2Δ is the width of the crack *w*. The relationship between *P* and Δ, which is equivalent to the relationship between *P* and *w*/2, depends on the bond-slip properties, such as τ_B/δ_B shown in Figure 1.6. If considering a single bar in the prism, the bond-slip properties appropriate to that bar and its cover should be used. If considering a group of bars, which may include bundled bars, then bond-slip properties appropriate to that group of bars are required.

As *P* in Figure 1.12a is increased, the slip δ_B, shear stress τ_B and concrete axial stress σ_c vary along the length of the bar as shown by Variations C in Figure 1.12c–e. A crack

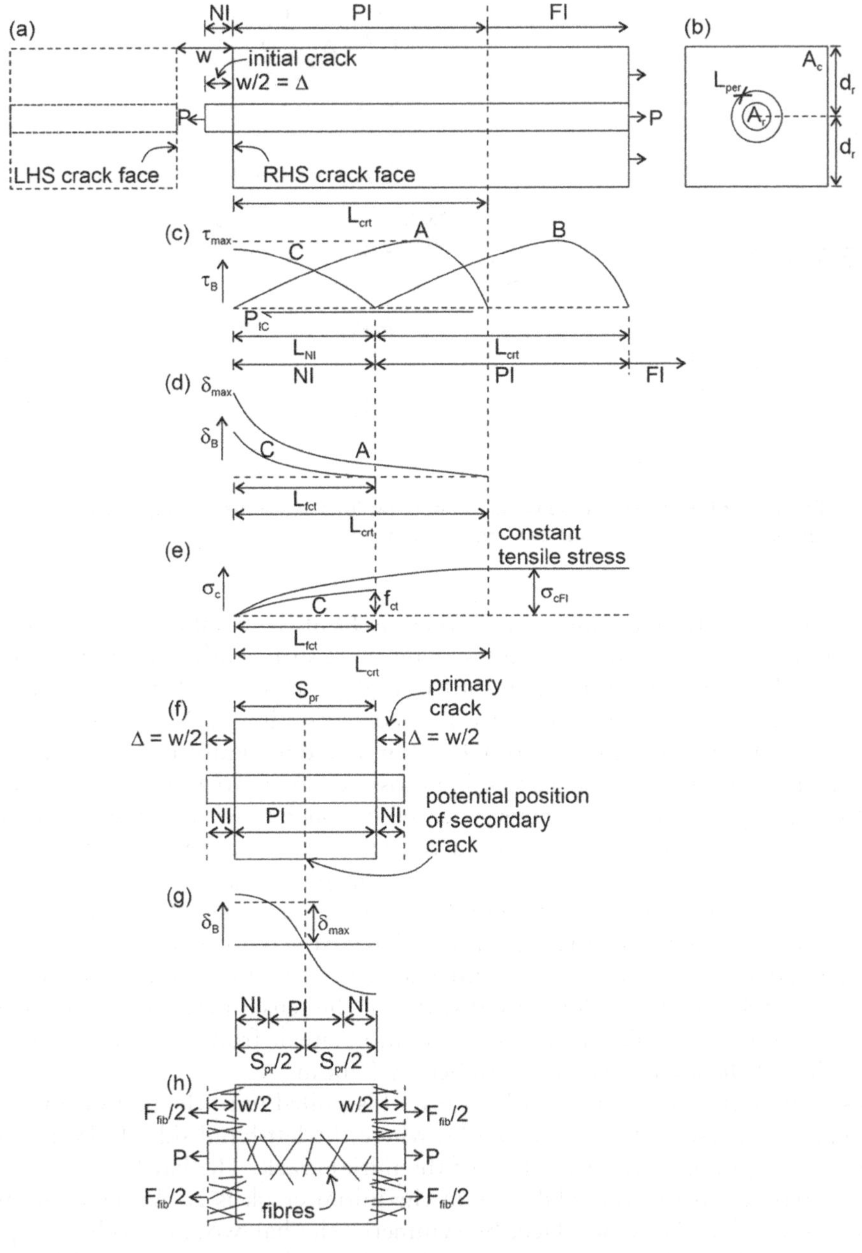

Figure 1.12 *PI* tension stiffening mechanism (a) infinite tension-stiffening prism elevation; (b) tension-stiffening cross section; (c) variation of bond stress τ distribution along prism; (d) variation of interface slip δ along prism; (e) variation of concrete stress σ_c along prism; (f) segment between primary cracks; (g) distribution of slip δ between primary cracks; (h) primary crack segment for FRC.

occurs once the concrete axial stress (Figure 1.12e) reaches the tensile strength of the concrete f_{ct}. The resulting distance L_{fct} from the initial crack defines the primary crack spacing S_{pr}. If cracking does not occur, then both the maximum shear stress τ_{max} and the slip δ_{max} (as illustrated in Figure 1.6) are attained. This is depicted by the distributions labelled *A* in Figure 1.12c and d, which extend over a bar length L_{crt}. To the right of this length is the *NI* region, where both slip and slip–strain are zero. When both δ_{max} and τ_{max} are reached, corresponding to Distributions *A*, *P* represents the maximum force the bar can sustain, referred to as the intermediate crack (*IC*) debonding resistance P_{IC}. The stress in the bar at the crack face σ_{IC} is given by P_{IC}/A_r. Consequently, the strain at the crack face ε_{IC} is equal to σ_{IC}/E_r, where E_r is the modulus of the reinforcing bar. The bar now exhibits three distinct interaction regions, as shown in Figure 1.12a; that is the *NI* region within the crack faces; the *PI* region, which extends over a distance L_{crt}; and the *FI* region, which extends beyond the *PI* region.

As the bar is gradually pulled or displaced further to the left in Figure 1.12a, and once P_{IC} is reached, the force in the bar *P* remains at P_{IC} while the bond stress distribution gradually moves to the right. As shown in Figure 1.12c, the shear distribution has moved as a block by L_{NI} to Distribution *B*, where there is *FI* to the right and as before both the slip and slip–strain remain zero. Consequently, the *NI* region has extended by L_{NI}, and as within this region there is NI across the interface, the strain in the bar remains at ε_{IC}. Hence, the bar at the *RHS* crack face has slipped a further $\varepsilon_{IC}L_{NI}$. This represents the increase in the half crack width, meaning the crack width has increased by $2\varepsilon_{IC}L_{NI}$. The interaction regions have now moved as shown in Figure 1.12c, resulting in an increased *NI* region, the same *PI* region, and a reduced *FI* region. Consequently, the length of bar in which slip occurs has increased, and the *PI* region has extended. This procedure can continue until debonding reaches the end of the bar, at which point the bar is fully debonded. This is referred to as the global debonding mechanism, since it occurs along the entire length of the member. The debonded region is where the slip is greater than δ_{max}. If the end of the bar in Figure 1.12a is less than L_{crt} from the crack face, then P_{IC} cannot be attained, and the bar will prematurely debond or slip out at a lower load.

When *P* in Figure 1.12a reaches P_{IC} (assuming premature cracking due to reaching f_{ct} does not occur), the following conditions are met: the shear stress Distribution A (Figure 1.12c) is established over the *PI* region of length L_{crt}; the slip is a maximum at the crack face (as shown in Figure 1.12d) and reduces to zero at L_{crt}, where the *FI* region begins; and the axial stress in the concrete (Figure 1.12e) reaches a maximum value of σ_{cFI} at L_{crt}. Beyond the distance L_{crt}, the concrete stress remains constant, as shown. Hence, cracking is unlikely to occur within the *PI* region (i.e. within L_{crt}) and is more likely to occur somewhere within the *FI* region. If cracking occurs at L_{crt}, then the procedure is repeated from this crack, initiating the next crack at a distance S_{pr} (equal to L_{crt}). All such cracks with spacing S_{pr} are referred to as primary cracks.

As the crack width increases, such as in Variations *C* in Figure 1.12c–e, the stress in the concrete builds up. Should it reach its tensile capacity f_{ct} at a distance L_{fct}, as shown in Figure 1.12e, cracking would occur at that position, which is less than L_{crit}. Hence, the primary crack spacing S_{pr} is now L_{fct}, and the force to cause the primary crack, P_{fct}, is $A_c f_{ct}$, which is less than P_{IC}. Hence, when premature cracking occurs at L_{fct}, L_{crit} can be considered the maximum crack spacing. Bearing in mind that the tension stiffening region is part of the beam in Figure 1.7 that is subjected to flexure, cracking can be caused when the moment M_{crFI} that causes a flexural crack in the *FI* region of the beam – which can be determined from a simple *FI* elastic analysis – is reached. It is probably better to use M_{crFI} to determine the next primary crack in a beam than to use P (as shown in Figure 1.12) that induces f_{ct}, but this is debatable. The uniformity of the primary crack spacing can be seen in the upper prism in Figure 1.10 and in the cracks on the tension face in Figure 1.7.

Once the primary cracks, at a minimum spacing S_{pr}, have occurred, the configuration shown in Figure 1.12f, with the slip distribution in Figure 1.12g, is observed between two primary cracks. When the primary cracks first form, two interaction regions are now present, as shown in Figure 1.12f: the *NI* region within the crack width and the *PI* region between the opposite crack faces. It could also be argued that *FI* exists at mid-length where the slip is always zero due to symmetry. On further loading, and where the slip exceeds δ_{max} (as depicted in Figure 1.12g), the *NI* region extends as shown. Eventually, should the bond slip be strong and stiff enough, cracking will occur at mid-length where the concrete stresses are largest. These cracks are referred to as secondary cracks. Hence, the crack spacing is now $S_{pr}/2$, and two examples of these cracks can be seen in the lower prism in Figure 1.10. When the cracks are $S_{pr}/2$ apart, the same procedure can be repeated to produce cracks at $S_{pr}/4$ (i.e. tertiary cracks), and so on, depending on the bar force and bond strength and stiffness. This development of cracks is often referred to as local debonding. Unlike global debonding, which develops over a wide length of the beam as depicted in Figure 1.12, the local debonding mechanism is restricted to zones between adjacent cracks and does not cause the bar to pull out of the member.

It will be shown that the flexural rigidity *EI* of the member is independent of the bond stiffness. Furthermore, if the bond is very weak, only one very large crack may form. Increasing the bond stiffness reduces the crack widths but increases the number of cracks. Consequently, increasing the bond stiffness has no effect on the member deflection while there is only local debonding (as in Figure 1.12f–h); however, it does reduce crack widths, despite leading to more cracks. Adding fibres (as in Figure 1.12h) increases the bond strength and stiffness by providing confinement across the bond-slip interface, subsequently reducing the crack spacing and increasing the number of cracks. The fibres also provide tensile stresses across the crack faces, such that, during the development of primary cracks, the concrete stress (Figure 1.12e) is no longer zero at the crack face. This tends to reduce the primary crack spacing S_{pr}. Furthermore, during the development of secondary cracks (Figure 1.12h), the stresses due to the fibres cause them to occur earlier, thereby increasing the likelihood of tertiary cracks. This means that the addition of fibres makes it much more likely to result in closely spaced, fine cracks, which are more ideal for durability design than fewer, wide cracks.

The tension stiffening mechanism described in Figure 1.12 applies to any type of reinforcement that crosses a crack. Hence, it applies to reinforcement that crosses flexural cracks (i.e. the 'tension stiffening region' in Figure 1.9b) and to stirrups that cross shear sliding planes such as the 'tension stiffening region' in Figure 1.9c. It also applies to plates bonded to the surface of RC members (as in Figure 1.5c and d). Furthermore, the reinforcement can be of any material such as steel or *FRP*.

It is worth noting at this stage, and it will be shown in the ensuing chapters, that increasing the bond stiffness (τ_B/δ_B Figure 1.6) does not increase the axial stiffness in

Figure 1.11. Increasing the bond stiffness simply reduces the crack spacing, resulting in more numerous but narrower cracks, which is beneficial for durability. It also needs to be emphasised that it is only when there is interface slip in the tension stiffening mechanism in Figure 1.12 that there is a loss of axial stiffness in Figure 1.11. For example, it is known from tests that cyclic loads can cause a deterioration in the bond. However, it is suggested that most *RC* structures have a surplus of bond strength, such that the axial stiffness only starts to reduce when there is sufficient deterioration to cause interface slip.

It can be seen that even though the mechanics controlling cracking (which yield L_{fct}, L_{crt} and M_{crFl}) are understood, crack widths and crack spacing are highly variable and complex phenomena. A very crude way of visualising crack development is that the total width of cracks in a beam at a specific load is constant. The development of secondary or tertiary cracks simply reduces the individual crack width by creating more cracks. Similarly, improving the bond also reduces crack widths by increasing the number of cracks.

1.4.2 FRC concrete tensile mechanism

Figure 1.13 shows the crack faces of an *FRC* specimen tested in direct tension to failure. The steel fibres were mixed into the concrete randomly, resulting in a random alignment and distribution. On tensile cracking, the fibres align with the direction of pull,

Figure 1.13 FRC concrete tensile test. (Photos courtesy of Lloyd Dadd.)

as can be seen. The axial tensile force divided by the cross-sectional area is the effective concrete tensile stress after cracking, $\sigma_{ct\text{-}fib}$. In the example shown here, the steel fibres pull out of the concrete, providing ductility to the tensile failure, which is very useful. Some steel fibres are produced with anchors at their ends, allowing them to fracture and produce a larger $\sigma_{ct\text{-}fib}$. This, however, results in a brittle failure, which is undesirable as the fibres may fracture at low crack widths before significant stress occurs in the tension reinforcement. Synthetic fibres tend to rupture because they have excellent bond and relatively low tensile strengths.

The steel fibres in Figure 1.13 are miniature reinforcing bars crossing a crack and as such are governed by the tension stiffening mechanism already explained with the help of Figure 1.12. In theory, it would be possible to test one fibre to determine its bond-slip characteristics. It would then be possible to account for random orientation and embedment across a tensile crack, using tension stiffening theory to determine the response of the fibre tensile strength ($\sigma_{ct\text{-}fib}$) with crack width. Alternatively, a tensile specimen can be tested as in Figure 1.13 to determine the material properties as illustrated in Figure 1.14.

The longitudinal stress/strain relationship at the start of testing is shown as *O*–*A* in Figure 1.14a. Fibres have little effect on either the modulus E_c or the tensile strength at first cracking, f_{ct}, although if there is a significant amount of shrinkage, fibres can restrain it internally, leading to micro-cracking and reduced tensile strengths. Once the concrete starts to crack in tension at Point *A* (corresponding to f_{ct}), the fibres begin to act. These fibres act as minute tension reinforcing bars; thus, by applying the tension stiffening mechanism of Figure 1.12, they can, if present in sufficient numbers, produce very small crack spacings, S_{cr}. This is referred to as micro-cracking and occurs beyond Point *A* in Figure 1.14a. If there is a large quantity of fibres, the strength may increase along *A*–*C*; if only a small quantity, it may decrease along *A*–*D*. Micro-cracking can be accounted for by using the effective modulus $E_{ct\text{-}micro}$ as shown. At some stage, the

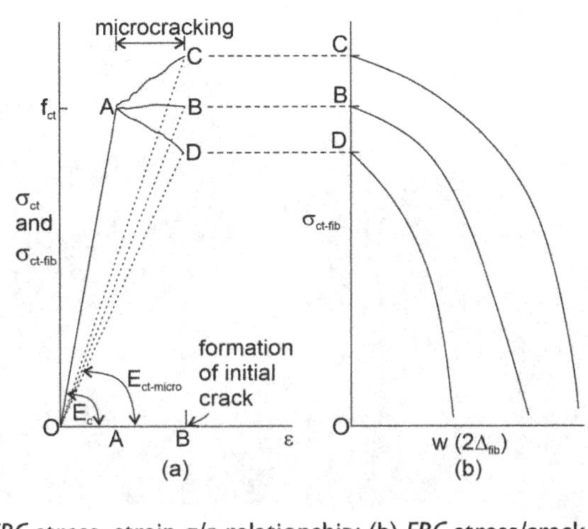

Figure 1.14 (a) Typical *FRC* stress–strain σ/ε relationship; (b) *FRC* stress/crack width σ/w relationship.

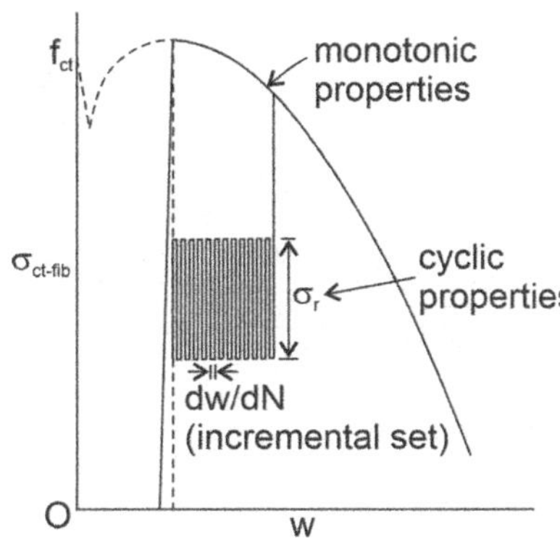

Figure 1.15 Fatigue properties of *FRC* concrete in tension.

single weakest micro-crack starts to fail or localise, and consequently widens, which is analogous to the initial crack in Figure 1.12a. When this occurs, strain no longer quantifies the behaviour; consequently, the analysis transitions to Figure 1.14b, which illustrates the fibre stress against the crack width w. It should be noted that the crack width w in Figure 1.14b corresponds to the 2Δ in Figure 1.12, allowing the information in Figure 1.14 to be utilised in a tension stiffening analysis, such as that shown in Figure 1.12h. In this context, the concrete will have a modulus of $E_{ct\text{-}micro}$ and the tensile stress across a crack is given by $\sigma_{ct\text{-}fib}$, which depends on the crack width.

It is worth noting that the monotonic properties presented in Figure 1.14b change under cyclic and fatigue loads, as illustrated in Figure 1.15. Each cyclic range of stress σ_r induces an incremental set, that is an increase in width per cycle dw/dN, as shown. On reloading, the monotonic path is followed. Hence, the crack width can be used as a reliable indicator of fatigue damage.

1.4.3 Concrete softening

It will be shown that concrete softening under compression is not a strain-based material property, such as the concrete modulus E_c, but rather a *PI* mechanism that can be quantified through the use of *SF* material properties.

1.4.3.1 Concrete softening mechanism

Figure 1.16a shows a longitudinal specimen softening under compression through slip along a sliding plane and Figure 1.16b shows the sliding plane enlarged, in which separation due to aggregate interlock occurs. When a long or slender concrete element is subjected to compression and is laterally unrestrained, as depicted in Figure 1.16a, it fails or softens through the formation of an inclined sliding plane at the angle α shown, typically about 26° (Chen et al 2015). This phenomenon can be observed in the axially loaded cylinder shown in Figure 1.17 and along the inclined sliding planes that form the

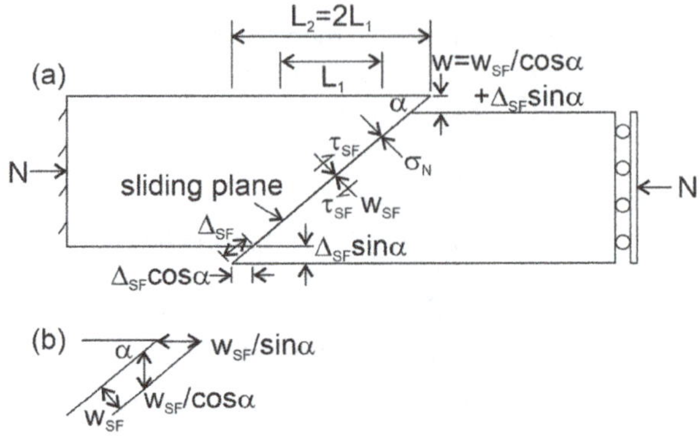

Figure 1.16 *PI* concrete softening mechanism: (a) shear sliding plane; (b) crack opening w along shear sliding plane.

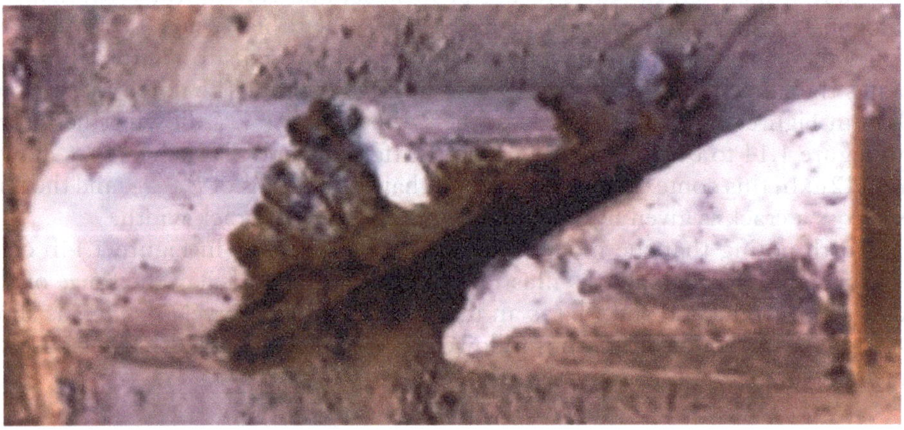

Figure 1.17 Concrete softening test specimen. (Reproduced from Oehlers and Visintin 2021.)

wedges in Figure 1.7. The behaviour of the sliding plane, as illustrated in Figure 1.16, is controlled by the *SF* properties detailed in Figure 1.4, where the interface slip Δ_{SF} in Figure 1.16a depends on the normal stress σ_N and shear stress τ_{SF} along the sliding plane.

From the resolution of displacements shown in Figure 1.16a, the specimen contracts longitudinally by Δ_{SF} cos α due to sliding and expands in the longitudinal direction in Figure 1.16b by w_{SF}/sin α, due to aggregate interlock. The total contraction, calculated as Δ_{SF} cos α less w_{SF}/sin α (where w_{SF} is due to aggregate interlock widening, as per Figure 1.4), when divided by the length of a specific zone, say L_1 (Figure 1.16a), yields the effective compressive axial strain due to sliding for that zone of length L_1. It follows that if the length of the zone is doubled to $2L_1$, the effective strain halves. This size effect is often associated with concrete in compression, and is illustrated in Figure 1.18 for laterally unconfined concrete. It is worth noting that the aggregate interlock component w_{SF}/sin α is at least an order of magnitude smaller than the sliding component Δ_{SF} cos α, and is consequently often ignored in practice. It is also worth noting that the

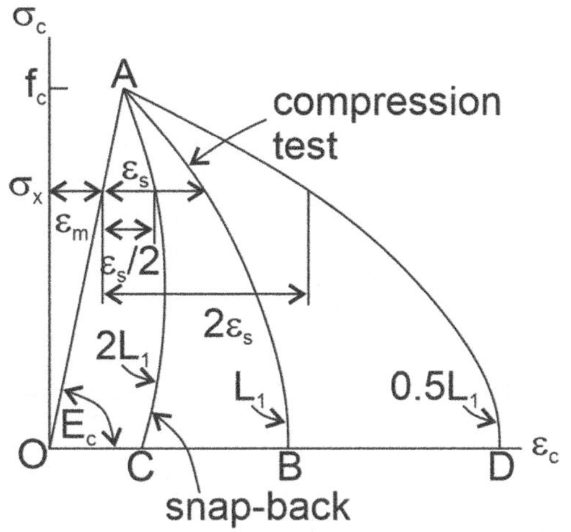

Figure 1.18 PI concrete softening size effect for unconfined concrete.

lateral expansion w (Figure 1. 16a), when divided by the depth of the member is the Poisson effect during softening. Hence, the Poisson ratio is not only a material property but also a mechanics property.

Let O-A-B in Fig. 1.18 be the stress/strain properties obtained from testing a concrete cylinder of length L_1. The ascending branch $O–A$ is depicted as linear with modulus E_c, although its actual shape and magnitude are determined experimentally. In the descending, or concrete softening, branch $A–B$, the material strain remains constant, with an additional effective strain arising from sliding, as previously described. At a stress level σ_x, the material strain in the ascending branch is ε_m and the softening strain due to sliding is ε_s. If a cylinder of twice the length (i.e. $2L_1$) were tested, given that there is still only one sliding plane, the effective strain due to sliding would be exactly half, as shown along $A–C$, and snap-back behaviour might even be observed as shown. Furthermore, the cylinder's behaviour becomes brittle. Conversely, if a cylinder of half the length were tested, it would exhibit more ductile behaviour, as shown along $A–D$. Hence, if the effective compressive stress/strain relationship for a specific zone length is known – whether determined experimentally or, as will be shown, theoretically from the *SF* material properties – then the stress/strain relationship for any other zone length can be constructed. Furthermore, the *SF* material properties can be used to construct the stress/strain relationship for actively confined concrete, a relationship that is very difficult to determine experimentally. It should be remembered that these size-dependent stress/strain relationships are merely a convenient means of accounting for sliding in a strain-based analysis. They do not represent an approximation stemming from a lack of understanding of the underlying mechanics.

1.4.3.2 Concrete compressions tests

It has been shown, both experimentally and theoretically, that concrete softens by sliding along a plane at about α equal to 26° (Figure 1.16). This causes lateral movement, manifested as the aggregate component $w_{SF}/\cos\alpha$ plus the sliding component $\Delta_{SF}\ \sin\alpha$.

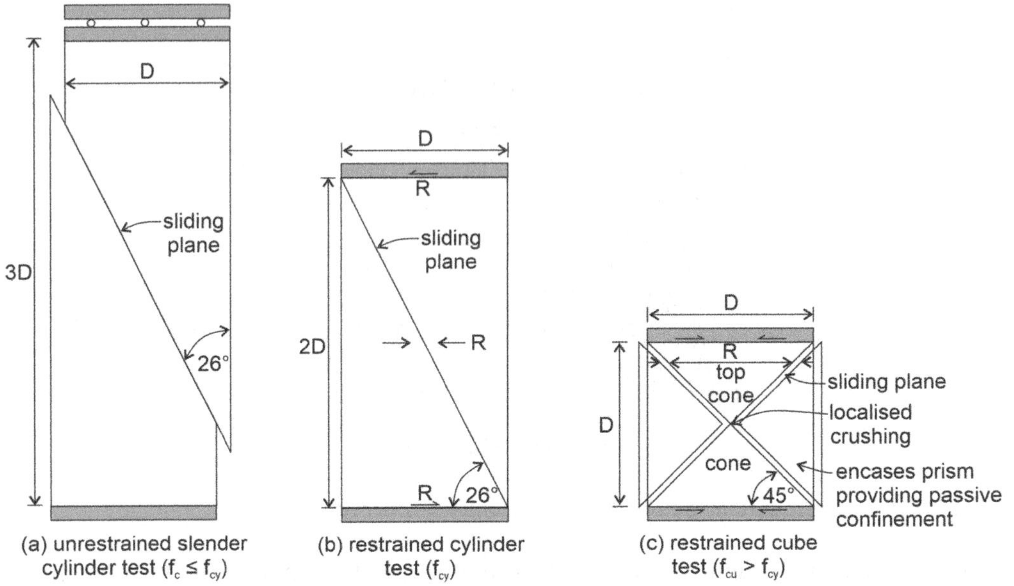

Figure 1.19 Concrete compression tests (a) unrestrained slender cylinder tests such that the concrete strength f_c is less than in a cylinder test f_{cy}; (b) standard restrained cylinder test in which concrete strength is denoted by f_{cy}; (c) standard restrained cube test in which concrete strength $f_{cu}>f_{cy}$.

If this natural lateral movement is restricted, this induces transverse passive confinement, thereby enhancing the concrete's capacity and ductility. Figure 1.19 illustrates some typical tests for experimentally determining the compressive strength and behaviour of concrete. Figure 1.23a represents a specimen that is long enough to encompass the natural sliding plane at 26° and is not artificially restrained laterally. This yields both the true, or fundamental, compressive strength f_c and the falling branch, as depicted in Figure 1.18; however, because these specimens are slender, the falling branch is steep and challenging to measure experimentally.

Concrete cubes (Figure 1.19) are often used in codes to determine the compressive strength f_{cu}. The problem with cubes is that this system does not allow for the natural failure mode, as illustrated in Figure 1.19a, which involves unhindered rigid body sliding. The short length of the cube (Figure 1.19c)forces the sliding plane to occur at 45°, thereby enhancing the strength, as the natural angle is 26°. Furthermore, the restraint R from the platens (Figure 1.19c) fixes the position of the top and bottom cones or pyramids once they form. Upon further contraction, localised material crushing occurs. The resulting shortening of the cube is then passively restrained by the remainder of the concrete encircling the upper and lower pyramids, further enhancing the strength. Sliding now occurs on the surface of the pyramids. This mechanism is referred to as a circumferential wedge failure, as the sliding plane encircles the cone or pyramid. The strength f_{cu} is generally about 20% greater than that observed in cylinder tests (Figure 1.19b). Hence cube tests do not simulate the natural sliding plane angle and induce passive confinement to further increase the strength. A cylinder test is illustrated in Figure 1.19b. This configuration can just accommodate the natural sliding plane, but the ends are restrained by the plates, inducing some passive confinement R across the sliding plane, which enhances the strength. Because of this lateral restraint, these cylinders sometimes fail by the circumferential wedge mechanism (Figure 1.19c) and sometimes by sliding (Figure 1.19a). The cylinder strength f_{cy} is less than the cube strength f_{cu} but greater than or equal to the laterally unrestrained strength f_c (Figure 1.19a). The addition of fibres can enhance both strength and ductility substantially. The stress/strain relationship can also be determined directly from the SF material properties of the fibre concrete.

1.4.4 Passive confinement mechanism

Figure 1.20 (Chen 2015) is an example of the passive confinement mechanism. The tube (Figure 1.20a) was first filled with concrete; then, only the concrete within the tube was compressed to failure. Subsequently the concrete was extracted and is shown in Figure 1.20b. It can be seen that failure was caused by sliding, which resulted in distortion of the top half of the tube (Figure 1.20a). The gradual increase in resistance of the tube to sliding represents the passive resistance, which can increase strength

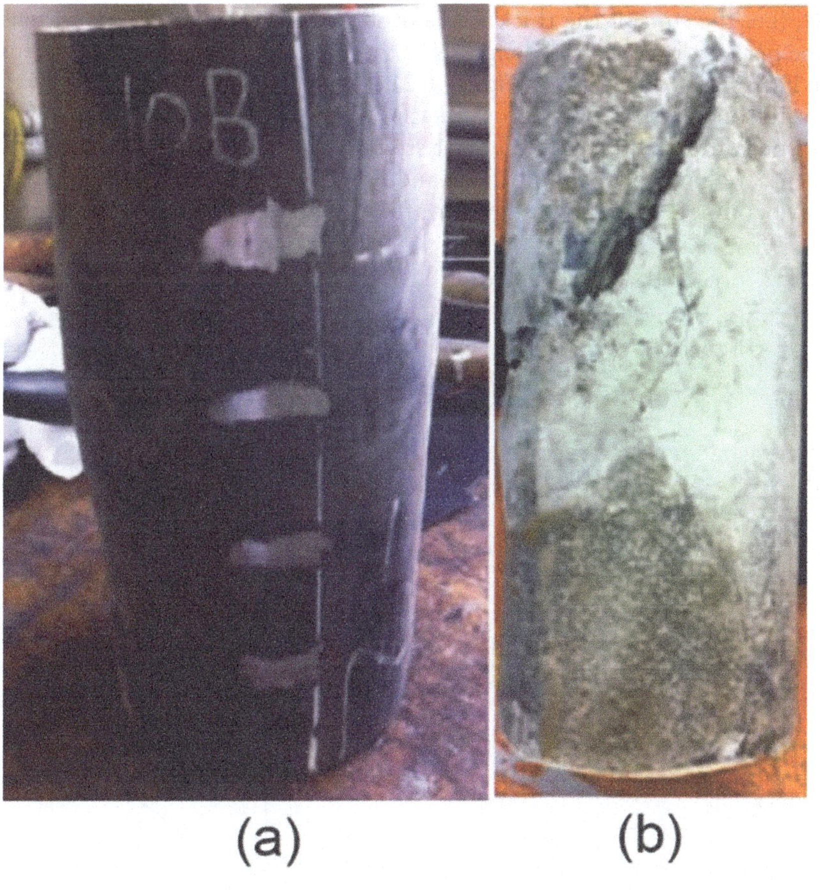

Figure 1.20 Passively confined concrete. (Reproduced from Oehlers and Visintin 2021.) (a) deformation of a steel tube passively confined concrete cylinder; (b) steel tube removed to show deformation of the encased concrete.

and certainly increases ductility. To incorporate the benefits of passive confinement in our analyses, we need to derive the stress–strain relationship of the passively confined concrete.

We first consider the tube filled with concrete in Figure 1.21a and b, and, for ease of explanation, assume that there is no bond between the concrete and the tube. As explained in Figure 1.16a and b, sliding along a shear plane causes the specimen to widen laterally and contract longitudinally through two components: aggregate interlock and sliding along the inclined plane. The aggregate interlock component is usually an order of magnitude smaller than that due to sliding; hence, we will only consider sliding along the inclined plane here, although it is very easy to include the effects of aggregate interlock in the analysis.

Let us assume that Points *A* and *B* in Figure 1.21b first coincide on opposite faces of the sliding plane. Slip Δ_{SF} along the sliding plane in Figure 1.21b causes these points to move apart by Δ_{SF}, as shown such that vertical or lateral component of the slip $\Delta_{SF} \sin \alpha$ causes the tube to widen. Hence, the lateral strain in the tube ε_{tube} is equal to $\Delta_{SF} \sin \alpha / D$, where *D* is the width of the tube. This tensile strain causes a tensile stress in the tube, $\sigma_{tube} = E_{tube}\varepsilon_{tube}$, where E_{tube} is the modulus of the tube. If the total thickness of the tube is $2t$, (i.e. the sum of the thicknesses on both sides), then the lateral tensile force in the tube per unit longitudinal length is $2tE_{tube}\varepsilon_{tube}$. This is shown as the tensile force $P_{conf}/2$ on both sides of the tube in Figure 1.21a. For equilibrium, there must be an equal and opposite compressive force across the sliding plane in the concrete, shown as P_{conf}, which can be converted to a confinement stress σ_{conf}. It should be noted that in passive confinement, we have local equilibrium, and as such, these forces (both the tensile and compressive components of P_{conf}) do not affect global equilibrium. The horizontal or

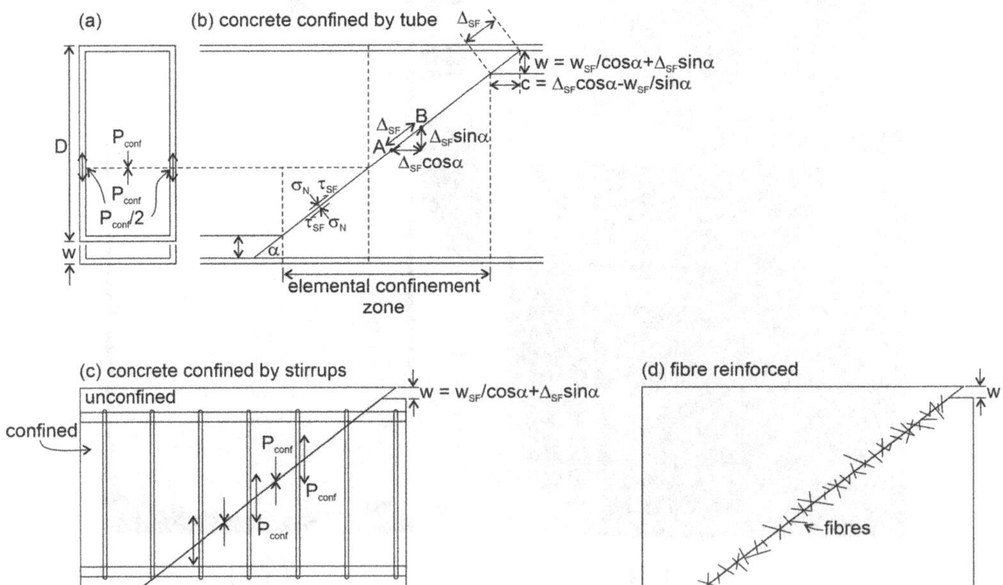

Figure 1.21 *PI* passive confinement mechanism (a) cross-section of concrete confined by a rectangular tube; (b) elevation of concrete confined by rectangular tube; (c) concrete confined by stirrups; (d) concrete along sliding plane passively confined by fibres.

longitudinal component of sliding $\Delta_{SF}\cos\alpha$ causes the concrete to compress or shorten. This, divided by the length L of the element being considered, gives the axial compressive strain due to sliding ε_{SL}. Hence, we can derive the axial strain ε_{SL} for a specific lateral confinement, σ_{conf}. However, the axial stress due to the axial force N when this occurs remains to be determined.

Figure 1.22 shows a series of stress–strain relationships for actively confined concrete, which can be obtained from actively confined experimental tests or derived from *SF* properties. First, consider the active confinement curve *B–G* where the lateral confinement is σ_{conf1}. From Figure 1.21a, we can derive the lateral strain in the tube from which P_{conf1} induces a confining stress σ_{conf1}. From this lateral strain in the tube, we can derive the widening of the tube w_1 in Figure 1.21b. From this widening of the tube w_1, we can derive the interface slip Δ_{SF1} using the geometries in Figure 1.21b. Knowing Δ_{SF1} and using the geometries in Figure 1.21b, we can derive the longitudinal component of slip ($\Delta_{SF1}\cos\alpha$) and, consequently, the effective longitudinal strain ε_{SL1} due to sliding ($\Delta_{SF1}\cos\alpha/L$). The determination of Point I in Figure 1.22 then involves fitting ε_{SL1} between *O–A–B* and *B–G*. The same approach can be used to determine Point *J* when the lateral confinement is σ_{conf2}. Should, for example, σ_{conf2} occur when the tube has started to yield in the transverse direction (i.e. the lateral confinement is constant with increasing slip), then the passive stress/strain will follow the path *J–H* until it strain hardens. Thus, the passive stress/strain can be constructed from the active stress/strain for any specific element length and member cross-section.

Lateral reinforcement or stirrups also provide passive confinement, as shown in Figure 1.21c. In this case, the tensile force in the stirrups is determined from the tension stiffening mechanism in Figure 1.12a–e, where w in Figure 1.21c is the crack width crossed by the reinforcement. Once again, the tensile force in the stirrups (P_{conf}) is balanced by an equal compressive confining force across the sliding plane (P_{conf}), and these forces do not affect global equilibrium. It can be seen that the stirrups not only increase the shear capacity but also increase the confinement to concrete softening making it more ductile. Hence, placing stirrups where they are not required as transverse shear reinforcement is beneficial as this improves the flexural ductility. This is commonly done in the high moment regions of columns where ductility is essential for overall structural performance under earthquake loading. Furthermore, it can be seen

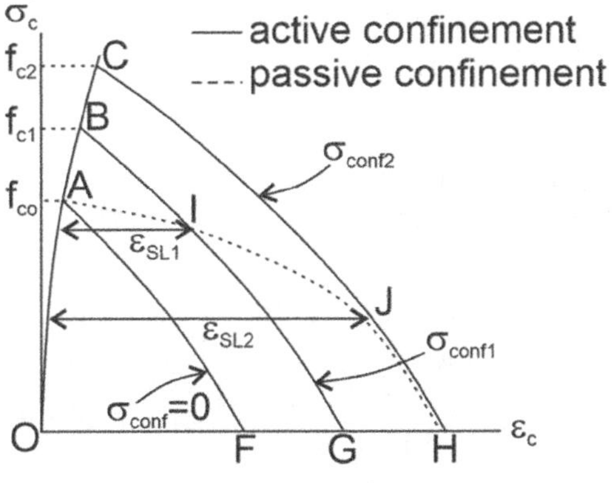

Figure 1.22 Passive confinement stress–strain from actively confined concrete.

in Figure 1.21c that this approach can distinguish between the confined region and the unconfined region, both of which exhibit distinctly different stress/strain relationships.

It is also worth noting that this displacement-based approach can easily accommodate concrete shrinkage in tubes, as shrinkage simply increases the displacement w (Figure 1.21b), which induces confinement. In contrast, it can accommodate the benefits of concrete expansion in tubes, where the concrete starts in a confined state. Thus, passive concrete confinement requires two elemental mechanisms: tension stiffening mechanics and shear sliding mechanics, which occur in both concrete softening and shear failure.

1.4.5 Reinforcement dowel action

To understand the dowel action of reinforcing bars in *RC* members, an analogy can be drawn with stud shear connectors in composite steel and concrete beams, a topic of extensive research (Oehlers and Bradford 1995, 1999). Reinforcing bars crossing a crack plane (i.e. a sliding plane), as shown in Figure 1.1, will distort as depicted in Figure 1.23. This distortion induces flexure M_{dow}, shear F_{dow} and axial load N_{dow} at the base. The axial load N_{dow} is primarily due to the bar's tendency to pull out of the concrete, similar to tension stiffening effects. The bar of diameter D pushes against the concrete. Because of the distortion of the bar, the bearing zone is concentrated at the base as shown. For a 19 mm diameter bar, the bearing force F_{dow} is concentrated over the surface of the bar in an area about the size of a thumbnail. The bearing force P_{dow} causing dowel fracture in the failure zone can be as much as ten tonnes. This is a huge force acting over a very small volume of concrete, and it can be sustained provided the concrete does not split or fail in tension due to the dispersal of this concentrated force into the concrete slab.

Dowel action is a *PI* mechanism and thus requires slip to occur. The dowel *PI* bond-slip behaviour is often idealised as shown in Figure 1.24. The dowel force F_{dow} reaches its maximum value at a slip of about $0.1D$ and maintains this force up to a slip of about $0.4D$ (i.e. it is ductile), after which rapid fracture occurs. The dowel strength at fracture P_{dow} has been quantified empirically as follows, and depends to a large extent on the bar's fracture strength (calculated as $f_u \pi D^2/4$, where f_u is the material fracture strength).

$$P_{dow} = \left(4.3 - \frac{1.1}{\sqrt{n}}\right) A_{sh} f_u^{0.65} f_c^{0.35} \left(\frac{E_c}{E_s}\right)^{0.40} \tag{1.2}$$

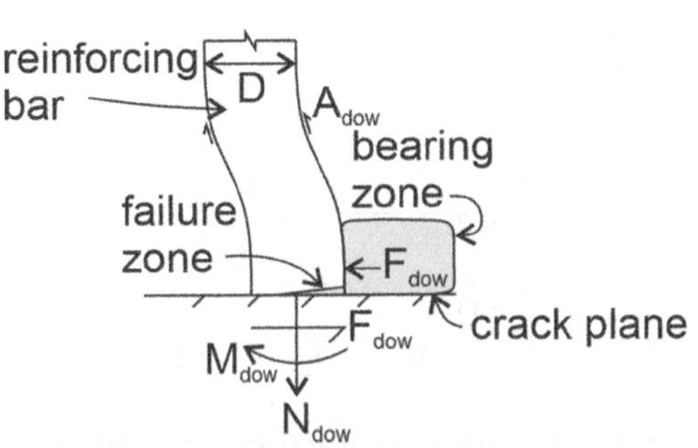

Figure 1.23 Dowel mechanism.

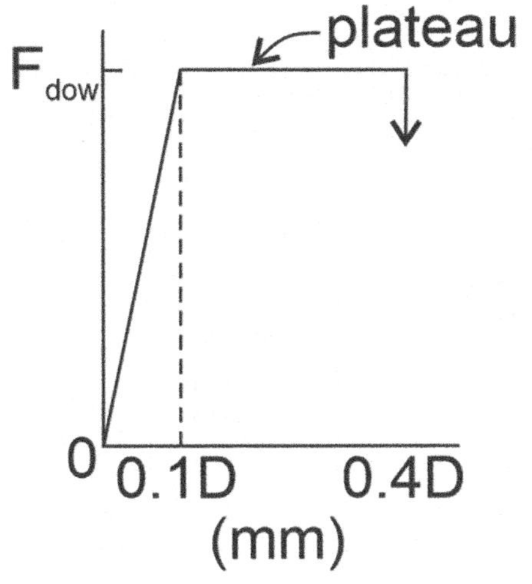

Figure 1.24 Bond slip of mechanical shear connector.

where A_{sh} is the cross-sectional area of the stud or bar, P_{dow} is the 5% characteristic strength and n is the number of bars in a group. Figure 1.24 showed that dowel fracture can be ductile; thus, if n bars fail as a group, the characteristic value increases as shown in the first term. Hence, if n tends to infinity in Eq. 1.2, the mean strength is obtained; conversely, when n equals unity, the characteristic strength of an individual is given. It should be noted that there are numerous empirically derived equations for predicting the dowel strength, many of which are dimensionally incorrect. The above equation has been presented because it is dimensionally correct.

It can be seen from Eq. 1.2 that the dowel contribution to resisting shear sliding can be significant. As a mechanical shear connection, it requires relatively large slips to activate its shear capacity, and importantly, it is ductile. All the information required to incorporate dowel action into the shear sliding resistance is readily available. Unfortunately, the concrete into which the dowel force is dispersed may not be able to resist this concentrated dowel force.

The dispersal of the concentrated dowel force into the surrounding concrete is illustrated in Figure 1.25. To understand the problem, consider the loaded strip of length h and width D in Figure 1.25b, applied to a prism of width b. The patch load is positioned $b/2$ from the free surface of the beam that is the cover of $c+D/2$. Let us first consider the situation when the strip is applied to the surface of the slab as in Figure 1.25a. The dispersal of the dowel force F_{dow} into the slab causes the lateral stress distribution shown in Figure 1.25a. This distribution is initially compressive, resulting in a transverse compressive force of magnitude P_{tr} then a lateral transverse tensile stress producing a tensile force of the same magnitude P_{tr} to maintain transverse equilibrium. These are local stress distributions that do not affect the global distribution of forces and are therefore much akin to the passive concrete confinement already discussed. The lateral compressive force P_{tr} is very beneficial as it confines the bearing zone adjacent to the patch load, thereby enhancing its ability to resist the bearing stresses in Figure 1.23.

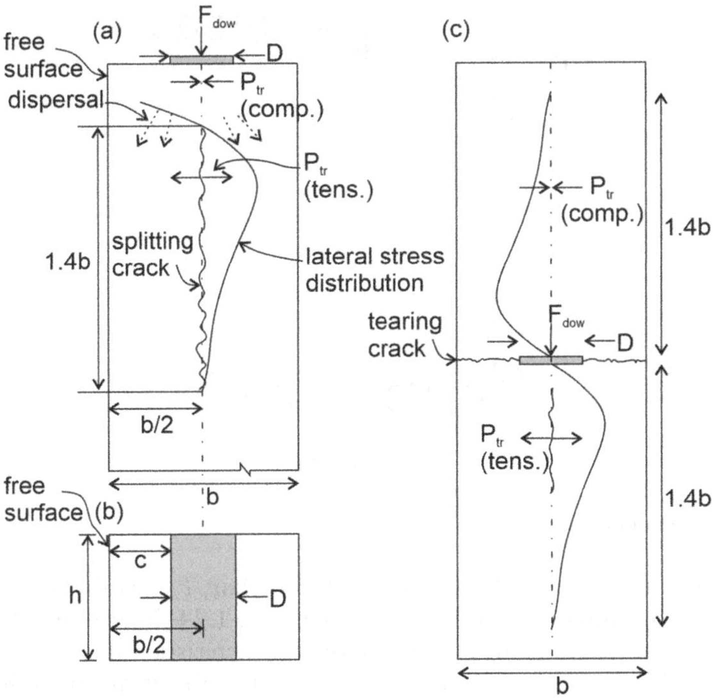

Figure 1.25 Splitting and tearing: (a) elevation of patch load applied externally to prism; (b) plan view of prism with external patch load; (c) internally applied patch load.

The transverse tensile force P_{tr} in Figure 1.25a acts over a region of about 1.4b long as shown. From mechanics, the lateral force P_{tr} is given by Eq. 1.3, and is a function of the concentrated load F_{dow} and the bar diameter to slab width ratio D/b.

$$P_{tr} = \frac{F_{dow}}{\pi}\left(1 - \frac{D}{b}\right)^2 \tag{1.3}$$

For bars such as stud shear connectors, the effective height of the bar to cause splitting h in Figure 1.25b is about 1.8D. The lateral force to cause splitting is then given by the following equation.

$$P_{split} = \frac{1.08bDf_{ct}\pi}{\left(1 - \frac{D}{b}\right)^2} \tag{1.4}$$

where for a given D/b, it is highly dependent on the width of the slab b (equal to $2c+D$), and thus highly dependent on the cover c. The cover to stirrups is frequently very small, such that P_{split} is much less than the dowel force required to fracture the dowel (P_{dow}). Consequently, the dowel strength is probably severely limited by the formation of splitting cracks.

When a bar acts within a slab, as in Figure 1.25c, the lateral stress distributions change as shown, but their magnitudes and distributions are very similar to the tensile distribution in Figure 1.25a. Therefore, Eqs. 1.3 and 1.4 still apply, and the likelihood of splitting remains the same. Should F_{dow} in Figure 1.25c be placed closer to a free end than the 1.4b from the bottom of the prism, then the tensile force P_{tr} remains the same but is concentrated over a smaller length; consequently, the peak lateral tensile stress will increase, causing an increased likelihood of splitting. The concentrated force in Figure 1.25c can also cause the tearing cracks shown, which is generally not important in composite steel and concrete beams and, therefore, has not been quantified in relation to the action of stud shear connectors. From the above discussion, it appears that any dowel action that might occur across a shear sliding plane may be severely compromised by concrete splitting and by tearing, which is visible in Figure 1.8b as the horizontal crack on the RHS.

When a bar crosses a sliding plane, it has a maximum axial capacity to resist dowel action, which we will call $P_{dow\text{-}max}$. Similarly, when it resists sliding shear by passive concrete confinement, which entails tension stiffening, it has a maximum axial capacity, $P_{pass\text{-}max}$. It is evident that when a bar resists both dowel action and passive confinement, it cannot resist the sum of both $P_{dow\text{-}max}$ and $P_{pass\text{-}max}$; that is one action will reduce the capacity to resist the other. If this interaction could be determined either experimentally or theoretically, both mechanisms could then be used in quantifying the resistance to shear sliding. In this book, we ignore dowel action because we believe that the dowel capacity is severely reduced by concrete tearing and concrete splitting and is only reached at large slips, which are at least an order of magnitude greater than those associated with shear friction. Instead, we assume the bar is subjected only to tension stiffening across a sliding plane and, as dowel action is ignored, all of its capacity can be utilised up to $P_{pass\text{-}max}$. It should be emphasised that the *PI* mechanisms postulated in this book still apply regardless of whether dowel action is considered or not.

Worked Example WE-1.2: Dowel resistance

The sliding plane on the RHS of Figure 1.8b is idealised in Figure 1.26a. As the sliding plane in Figure 1.8b is nearly vertical, for ease of analysis we assume β in Figure 1.26a tends to 90°. A reinforcing bar is shown at mid-depth, with a diameter D of 16 mm, a strength f_u of 400 MPa and a modulus E_s of 210 GPa. On shear sliding, the *RHS* moves upwards relative to the *LHS*, causing the bar to distort as shown. The bar on the RHS bears against the concrete in the zone above the bar, as shown, and the converse occurs on the LHS, such that equilibrium is maintained; that is the bearing forces generated, F_{dow}, are of equal magnitude and opposite direction.

The *RHS* of the beam in Figure 1.26a is shown in Figure 1.26b (note that the *LHS* is simply a mirror image of the *RHS*). Let the depth of the beam d be 500 mm, the cover to the bar c be 30 mm, the tensile strength of the concrete f_{ct} be 4 MPa, its compressive strength 30 MPa and its modulus 30 GPa. From the dowel distortion in Figure 1.26a, the bearing zone above the bar in Figure 1.26b experiences transverse tension, while below the bar, transverse compression occurs, as shown. Hence the zone above the bar is susceptible to splitting.

The maximum mean dowel strength of the bar from Eq. 1.2, P_{dow}, is 64.1 kN. The effective width of the prism that resists the dowel force b is equal to $D+2c$ in Figure 1.25b, which is 86 mm. Applying Eq. 1.4, the resistance of the slab to splitting P_{split} is 28.2 kN. Since $P_{split} \ll P_{dow}$, the dowel force is limited to the splitting resistance of 28.2 kN. The splitting zone has a length of 1.4b, as shown in

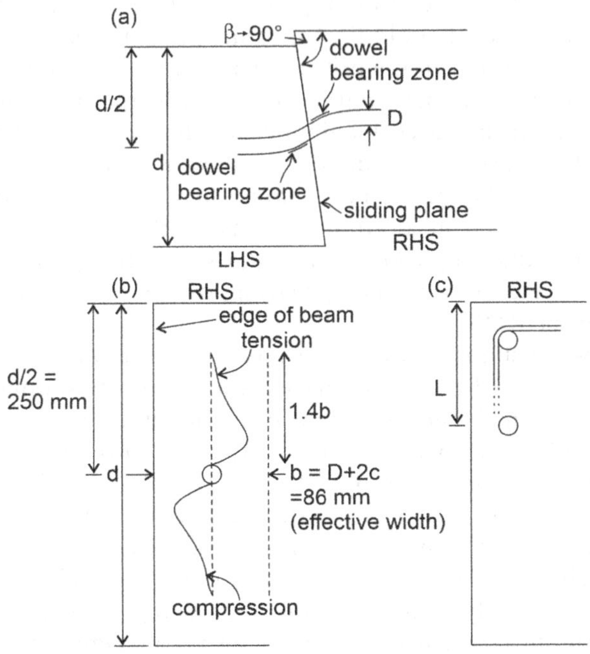

Figure 1.26 Dowel resistance (a) elevation of beam showing dowel deformation and resistance across sliding plane; (b) cross-section of sliding plane showing splitting forces required to resist the dowel forces; (c) beneficial effect of stirrups in resisting the dowel force.

Figure 1.25, which for this example is 1.4×86 mm, or 120 mm. Hence, as long as the distance L of the bar in Figure 1.26c is greater than 120 mm, the splitting resistance remains at 28.2 kN. Should the distance be less than 120 mm, the splitting resistance reduces, thereby reducing the maximum dowel force. However, should a stirrup be present, as with the upper bar in Figure 1.26c, then the stirrup itself can resist the dowel force. This provides yet another reason for using stirrups, even when they are not required to resist transverse shear. It is also worth noting that fibres crossing the sliding plane may increase f_{ct} and, thereby, allow a greater splitting force; however, equally importantly, if not more so, they allow the splitting force to be maintained, thus allowing for ductility. It is also worth noting that the slips associated with dowel action, as illustrated in Figure 1.24, are at least an order of magnitude greater than those associated with SF in Figure 1.4. As the latter system reduces with increasing slip, it is unlikely that the two systems of transferring shear will act together, and certainly not prior to interface slip.

1.4.6 FRC concrete SF mechanism

The fibres crossing the sliding plane, as shown in Figure 1.21d, act as tiny reinforcing bars. The fibres that go into tension produce passive confinement, as they act like the stirrups in Figure 1.21c; that is they obey the tension stiffening rules. Those fibres that go into compression exhibit dowel action, which may be significant because, being so small, they are unlikely to cause concrete splitting or tearing. Because of the complexity of the behaviour, the SF material properties, as shown in Figure 1.4, are often determined experimentally.

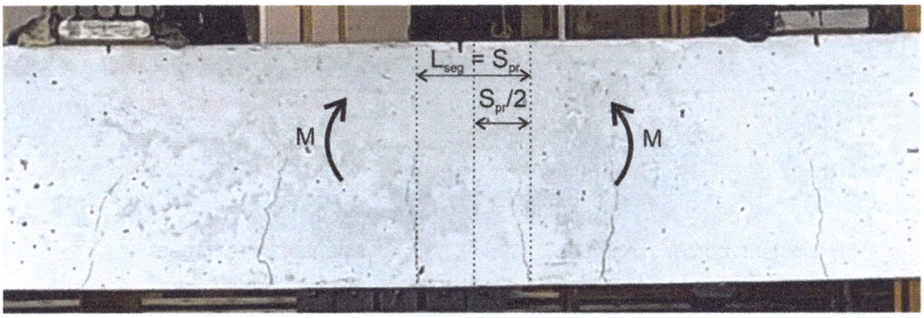

(a) serviceability

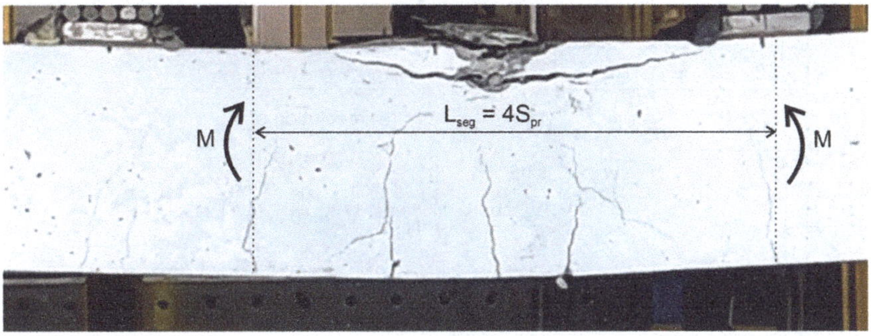

(b) ultimate

Figure 1.27 *RC* segments (Photos courtesy of Lloyd Dadd.) (a) crack patterns at serviceability limit state; (b) crack patterns at ultimate limit state.

1.5 *PI* SEGMENTAL MECHANICS

The *PI* elemental mechanical behaviours described above apply to small regions within the depth of a member. These can now be combined to simulate the *PI* behaviour of a segment – a short length of a member extending over its full depth and subjected to any combination of moment, shear and axial load, as shown in Figure 1.27a. Here, the segment acts over a primary crack spacing S_{pr} and the cracking is at serviceability. Figure 1.27b shows a segment in an *RC* beam with fibres, where the cracking is at ultimate. The segment length L_{seg} now acts over $4S_{pr}$ to accommodate the softening wedge. It can be seen that the segment length L_{seg} is usually governed by the primary crack spacing S_{pr}, or by minimum multiples of S_{pr} if a softening wedge is to be included to allow for the formation of a hinge. It is important to realise that the segment length does not affect the overall result, and the same M/χ will be obtained whether L_{seg} in Figure 1.27a is S_{pr} or $3S_{pr}$. However, it is important to keep the segment length L_{seg} as short as possible because the analysis requires a constant moment region; the presence of a moment gradient introduces an error that is diminished by making the length as small as possible.

1.5.1 Segmental stress resultants

The two types of segments used to quantify all the behaviours of an *RC* member are shown on the beam in Figure 1.28a. This beam has a depth d_{seg} and width b_{seg}, and is

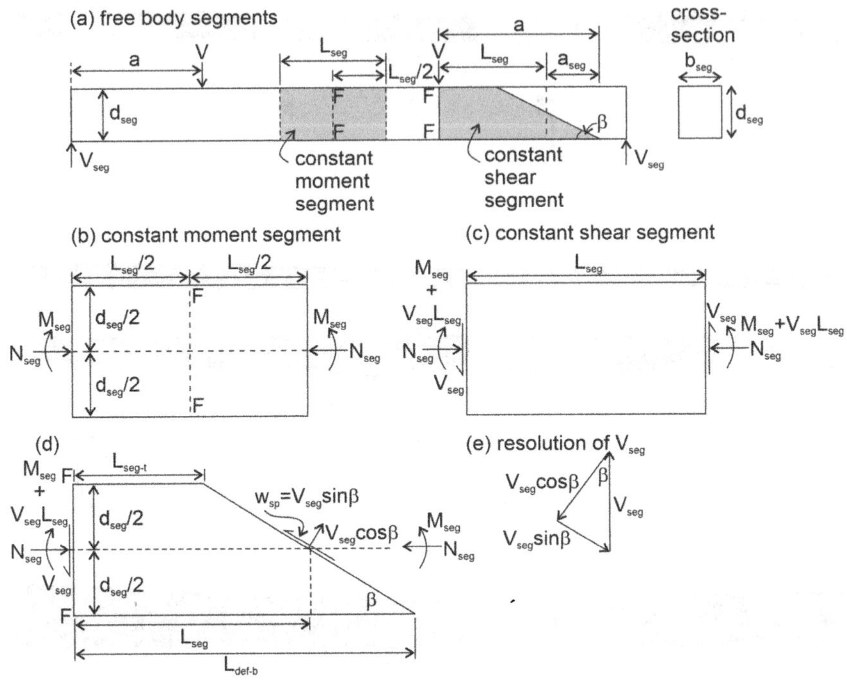

Figure 1.28 Segmental stress resultants (a) free-body segments for both flexure and shear; (b) a segment with vertical sides in a constant moment region; (c) a segment with vertical sides in a constant shear region; (d) to allow for inclined cracks associated with transverse shear, a segment with an inclined side in a constant shear region;(e) resolution of the transverse shear force V_{seg}.

symmetrically loaded with shear spans a, such that a constant moment region exists at its centre, with constant shear regions on either side. Flexural behaviour is quantified from the rectangular segment in the constant moment region, where there is no transverse shear. Shear, that is shear flexure behaviour, is quantified using the trapezoidal segment in the constant shear region. As shear failure occurs along inclined planes, the right-hand face of the segment is inclined at an angle β. A convenient distance, a_{seg}, from the end of the shear span (of length a) to this failure plane is also shown. Furthermore, the shear segment has been located adjacent to the applied load, such that it encompasses the region of high moment and as such encompasses all the maximum stress resultants in the shear span.

The segment in the constant moment region is shown in Figure 1.28b. The moments at the ends are M_{seg}, the transverse shear is zero in a constant moment region and the segment is subjected to an arbitrary axial load N_{seg} which could be zero. These stress resultants or design actions are externally applied to the segment, which is a free body and consequently must be in equilibrium. It can be seen that this free body in Figure 1.28b is in rotational equilibrium because the end moments M_{seg} are equal and opposite, and in horizontal translational equilibrium because the axial forces N_{seg} are also equal and opposite. Crucially, the centreline in Figure 1.28b, labelled *F–F*, is fixed horizontally due to symmetry. Hence, any horizontal deformation of the right-hand face due to M_{seg} and N_{seg}, when divided by $L_{seg}/2$, represents the resulting strain. The same can be said of the left-hand face.

The rectangular constant moment segment from Figure 1.28b is moved to the constant shear region in Figure 1.28a and shown in Figure 1.28c, with shear forces V_{seg} added. The shear forces cause an anti-clockwise couple of $V_{seg}L_{seg}$; therefore, to maintain rotational equilibrium, the moment on the LHS has been increased by $V_{seg}L_{seg}$, as shown. To incorporate an inclined shear sliding plane, as observed in practice (Figure 1.8b), the right-hand face has been rotated by angle β, as shown in Figure 1.28d. The shear force V_{seg} has been resolved into its components along the sliding plane in Figure 1.28e. These components have been added to the free body diagram in Figure 1.28d, where the shear component due to the transverse shear V_{seg}, shown as W_{sp}, is $V_{seg}\sin \beta$, and its component normal or perpendicular to the sliding plane is $V_{seg}\cos \beta$. For analysis, the constant shear segment (also shown in Figure 1.28a) can be analysed as a simply supported beam of length $2a$, where the left hand side, that is to the left of F-F in Figure 1.28d is the mirror image of the beam to the right of F-F. Very importantly, the left hand face in Figure 1.28d, labelled F-F, can be assumed to be fixed horizontally due to symmetry. Consequently, any horizontal deformation of the sloping side due to the applied stress resultants, when divided by the appropriate L_{def} (the distance from F - F, which varies from $L_{def\text{-}t}$ to $L_{def\text{-}b}$), represents the resulting strain caused by the horizontal deformation.

The free bodies in Figure 1.28 are generic, applying to any combination of moment, shear and axial load, and to any inclination of the sliding plane or inclined plane β. These free bodies can be used for shear flexure analyses to determine the so called shear strength. The flexural analysis shown in Figure 1.28b, within the constant moment region of Figure 1.28a, corresponds to the free body in Figure 1.28d when the transverse shear is zero and the sliding plane is vertical (β=90°). Hence, flexural analysis is a special case of the generic shear analysis, and being the simplest, it will be considered first and in detail.

1.5.2 Flexural segmental analysis

The segment shown in Figure 1.28a in the constant moment region, also shown in Figure 1.28b, is used to derive the flexural properties. Because the segment is symmetrically loaded, only half of it needs to be considered, as shown in Figure 1.29a. For ease of explanation, we shall assume that the length of the segment is half the primary crack spacing determined from a tension stiffening analysis (see Figure 1.12) and that the segment is large enough to encompass any potential softening wedge. We shall assume that the *RC* segment includes both tension and compression reinforcement and that the concrete contains fibres.

An Euler–Bernoulli deformation is applied to the *RHS* of the segment in Figure 1.29a to induce a rotation θ, as depicted by the displacement profile δ in Figure 1.29b. The displacement shown in Figure 1.29b, when divided by the length that deforms (L_{def}), yields the strain profile presented in Figure 1.29c. Depending on whether the material properties are strain-based or PI-based, either the strain profile in Figure 1.29c or the displacement profile in Figure 1.29b will yield the stress distribution shown in Figure 1.29d. From these, the element forces shown in Figure 1.29e can be determined. These forces are in the axial direction of the member which is in the same direction as

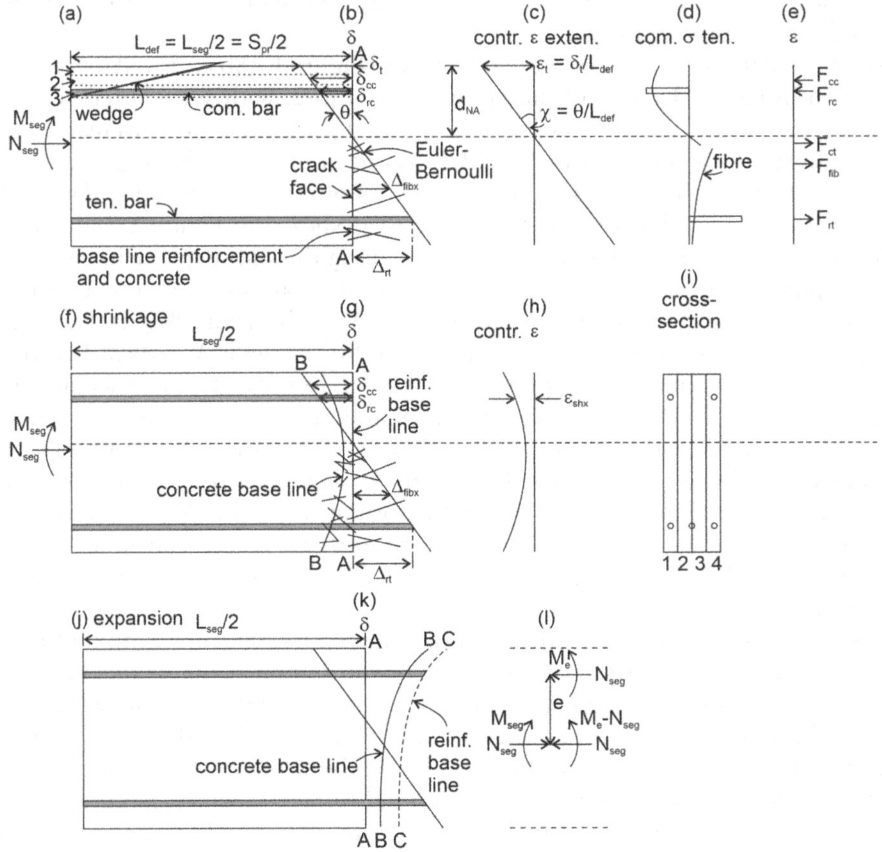

Figure 1.29 Flexural segmental analysis (a) segment that includes concrete softening and flexural cracking; (b) Euler-Bernoulli displacement δ at segment end; (c) contracting and extending axial strains ε due to Euler-Bernoulli displacement; (d) compressive and tensile axial stresses σ due to Euler-Bernoulli displacent; (e) resulting compressive and tensile axial forces F from the distribution of axial stresses σ; (f) segment that allows for shinkage and fibre concrete and an introduction to base-line theory; (g) Euler-Bernoulli displacement δ for segment subjected to shrinkage; (h) axial strain profile ε for a segment with shrinkage; (i) segments used to allow for variations in shrinkage over the beam width; (j) segment that allows for axial expansion; (k) Euler-Bernoulli deformation of segment subjected to expansion; (l) importance of taking moments at the level of the imposed axial load.

any applied axial forces (N_{seg}) on the member. Before determining the forces induced by the deformation, the material properties must be known. If this deformation produces neither flexural cracking nor concrete softening (i.e. it does not result in PI), then the material moduli E can be used to determine the effect of the deformation. If the moduli are constant, then this analysis will yield precisely the same results – not approximately, but exactly – as a conventional linear elastic analysis at a cross-section.

Let us assume that we want to determine the effects of *PI*, that is the effects of concrete softening and flexural cracking in the segment shown in Figure 1.29a. We assume we have all the strain-based material properties. For the concrete, we need the size-dependent stress–strain relationship as depicted and explained in Figure 1.18.

For example, it could be *O–A–D* for unconfined concrete (σ_{conf}=0), with an element length of $0.5L_1$, where $0.5L_1 = L_{def}$ in Figure 1.29a), which equals half the primary crack spacing S_{pr} from the tension stiffening properties in Figure 1.12f. Alternatively, if we wish to include the benefits of confinement, it could be, for example, *O–A–I–J–H* in Figure 1.22.

The top of the concrete wedge in Figure 1.29a has been sliced into three elements, labelled (1) to (3), spanning the depth of the wedge. Each element has the same length L_{def}. At first glance, the reader might think that, because the wedge slices each element at a different location, they would be controlled by a different size-dependent stress–strain relationship, as shown in Figure 1.18. This is not the case. As each element in Figure 1.29a has the same length L_{def}, it will have the same rate of deformation due to the material properties *O–A* shown in Figure 1.18. Furthermore, as each has the same shape of the sliding plane, albeit in different positions, the falling branch, such as *A–D* in Figure 1.18, will be the same. The topmost Element (1) in Figure 1.29a is subjected to the highest strains, as shown in Figure 1.29c, and hence will reach its peak stress first; that is it will achieve V_{u0} in Figure 1.4 first. Hence, the top element in Figure 1.29a will start sliding first, as sketched. This sliding will be restrained by the adjacent uncracked concrete in the potential wedge. This may cause the horizontal striations in the wedge, as illustrated in Figure 1.1 and evident in the numerous striations in the photograph in Figure 1.7b, or the tensile capacity of the concrete may be sufficient to allow only a few striations, as shown in Figure 1.7a. In conclusion, the same stress–strain relationship can be used for all the elements, bearing in mind that the strains in each element vary.

For the tension reinforcing bars in Figure 1.29a, which are shown in a cracked region, the force in the bar depends on the half crack width Δ_{rt} (Figure 1.29b). The relationship between the force in the bar *P* and Δ_{rt} is a tension stiffening property, with primary cracks as shown in Figure 1.12f. The force in the fibres crossing the flexural crack in Figure 1.29b depends on the half crack width Δ_{fibx}, which varies linearly over the depth of the flexural crack. Its properties can be determined from material testing, as shown in Figure 1.14.

We now have all the material properties for the analysis in Figure 1.29a–e. We need to find the rotation θ and the depth of the neutral axis d_{NA} (Figure 1.29b) that achieve the desired M_{seg} and N_{seg}. All deformations in Figure 1.29b are measured from the end of the segment labelled A-A, which is referred to as the base-line because it corresponds to the position of the segment side when its components are unstressed.

Let us first fix θ and d_{NA}. The ratio θ/L_{def} yields the curvature χ (Figure 1.29c) and the contraction of the top fibre δ_t (Figure 1.29b) divided by L_{def} provides the strain in the top fibre ε_t in Figure 1.29c. In the uncracked region, above the flexural crack, strain-based material moduli or properties, such as the effective stress/strain that accounts for softening, can be used to determine the stress distribution, as shown in Figure 1.29d. From this stress distribution, the magnitude and position of the resultant compressive force in the concrete (F_{cc}) (Figure 1.29e), the resultant compressive force in the compression reinforcement (F_{rc}) and the resultant tensile force in the uncracked concrete in tension (F_{ct}) can be determined; however, the latter is often neglected in practice. In the remaining cracked region, the displacements in Figure 1.29b determine the slip at the crack face Δ_{rt} and, consequently, the magnitude of F_{rt} (Figure 1.29e). From the variation of the half crack width Δ_{fib} (Figure 1.29b), the magnitude and position of the resultant fibre force F_{fib} (Figure 1.29e) can also be determined.

The objective is to find d_{NA} and θ (Figure 1.29a) that yield the target N_{seg} and M_{seg}. This can be solved through a simple iterative numerical approach or by deriving closed-form solutions. For example should N_{seg} not exist then for a fixed value of rotation θ_1 (and consequently curvature χ_{seg1}), an initial guess for d_{NA1x} is made. Summing the axial forces (Figure 1.29e) gives the first estimate for N_{seg}, denoted N_{seg1}. If N_{seg1} is not zero, then the depth of the neutral axis must be varied until the axial forces sum to zero. Once the axial force is zero, moments can be taken at any level to derive the moment M_{seg1} for that specific imposed rotation θ_1 and consequently curvature χ_{seg1}.

When an axial load N_{seg} is imposed. The same procedure as in the previous paragraph can be followed, where the resulting axial force must now sum to N_{seg}. However, the moment must now be taken at the level of the imposed axial load N_{seg} to determine the true M_{seg}. To illustrate this simple but important concept, let the moment of the forces in Figure 1.29e be taken at an eccentricity e relative to the level of the imposed axial load N_{seg}. The result is shown in Figure 1.29*l*, where at eccentricity e, the stress resultants are N_{seg} and M_e. Transposing the axial force N_{seg} to the level of the imposed axial force (i.e. by dropping the level by e) causes an anti-clockwise moment $N_{seg}e$ such that the moment M_e reduces by $N_{seg}e$, as shown. It should be remembered that all combinations of the stress resultants in Figure 1.29*l* (i.e. the combination of M_e and N_{seg} at any eccentricity e) will produce exactly the same strains, stresses and forces as in Figure 1.29c–e, but the true value is that corresponding to an eccentricity e of zero. Another way of looking at this concept is that when the eccentricity is zero, the axial force within the member is in line with the axial force applied to the member.

Allowing for shrinkage in the analysis is straightforward. For ease of design, it is normal practice to assume that the longitudinal shrinkage strain ε_{sh} is constant over the depth and width of the member. To show the versatility of this segmental approach, we assume that the longitudinal shrinkage strain varies over the depth, as shown by ε_{shx} in Figure 1.29h, as it does in practice. Multiplying this shrinkage strain by the length over which it acts ($L_{seg}/2$) gives the longitudinal contraction of the segment, which represents the horizontal distance between the original position *A*–*A* in Figure 1.29g and the curve *B*–*B*. The end of the concrete component of the segment after shrinkage contraction is now the curve *B*–*B*. This is the position of the concrete if it were totally free and able to contract due to shrinkage, unrestrained and without developing stress. This is, therefore, the base-line for the concrete; that is any displacement from this position will induce stress in the concrete. If we assume that the reinforcement does not shrink (which is the usual case), the base-line for the reinforcement is still *A*–*A* and any displacement of the reinforcement from this line will induce stress. The analysis follows the same procedure outlined in Figure 1.29a–e, but with new base-lines from which the displacements are measured.

It is interesting to note that shrinkage causes Δ_{fibx} in Figure 1.29g to increase because the base-line moves from *A*–*A* to *B*–*B* due to shrinkage. Hence, fibres are beneficial in suppressing crack widening due to shrinkage.

The same procedure can be used to allow for variations in shrinkage within the width of the member in Figure 1.29i, as occurs in practice. This can be done by assigning a different shrinkage strain profile to each of the four elements labelled (1–4), that is varying the shrinkage strain shown in Figure 1.29h for each element. However, the same displacement as in Figure 1.29g is applied to all four elements. The segmental approach can also readily cope with expansion, as shown in Figure 1.29j, which could be due to thermal gradients or concrete designed to expand. Needless to say, it can cope with any combination, for example, of shrinkage, contraction and thermal expansion. It is simply a matter of determining the base-line for each component, such as *A–A*, *B–B* or *C–C*.

The segmental moment/rotation approach gives the behaviour of a potential hinge region directly. By dividing the rotation by L_{def}, the result can be easily converted to yield the moment/curvature, the slope of which is the flexural rigidity, an important parameter in design. In mechanical terms, this is not a crude approximation but an exact solution. It can be seen that the segmental approach directly simulates what is observed or measured in practice. It should also be remembered that the segmental approach allows for local debonding in the *PI* region, as shown in Figure 1.12f, and provides the properties along the whole length of the beam. Members are usually designed only for local debonding; that is to say, they are designed so that global debonding does not occur. Global debonding, as shown in Figure 1.12c, is dealt with by considering the whole length of the member rather than just a segment.

1.5.3 Shear flexure segmental analysis

It has been explained in Section 1.5.1 that there is a close link between flexural behaviour and shear flexure behaviour because the segment used for the shear flexure analysis in Figure 1.28a is generic and the segment used in the flexural analysis is simply one case of this generic model.

Let us take this analogy between flexural behaviour and shear flexure behaviour one step further. In general, the flexural segmental approach in Figure 1.29 shows the following three stages in the flexural behaviour of RC members:

1. Formation of flexural cracks (or the moment $M_{cr\text{-}ini}$ that causes them), initiating in an FI region where the cracks are vertical. Consequently, conventional strain-based analyses at a cross-section can be used to determine the onset of cracking.
2. Onset of concrete crushing, often corresponding to the moment capacity M_{cap}. Unless the member has very little reinforcement $M_{cap} >> M_{cr\text{-}ini}$.
3. The rotational capacity, that is the moment ductility.

Now let us consider shear flexure behaviour. As previously explained using Figure 1.8, shear behaviour is characterised by the formation of an inclined shear flexure crack. In this case, the shear flexure crack formed early and then, following an increase in load, the specimen failed along a sliding plane; the failure plane can be considered bilinear. This stable failure is not always the case. In Figure 1.30, the shear flexure crack formed suddenly from a very small initial crack, leading to immediate failure, which could be considered unstable. The failure plane in the compression region being more shallow than that in the tension region, the failure plane can be considered bilinear. As a matter of interest, this figure also shows the formation of flexural secondary cracks described in Figure 1.12f. In the lower beam in Figure 1.30, the shear flexure crack formed through existing flexural cracks that is the shear-flexure crack formed independently of the flexural cracks. This failure plane was also bilinear, being slightly shallower in the compression region, though not as noticeable as in the other beam.

Figure 1.30 Formation of shear flexure crack. (Reproduced from Dadd et al. 2025.)

From these observations, it is suggested that there are also three stages in shear flexure behaviour as follows:

1. The shear to form a shear flexure crack ($V_{cr\text{-}ini}$) or the shear to cause the start of the shear flexure crack occurs in an FI region and in which the cracks are inclined.
2. The shear to cause the onset of concrete sliding in the compression region (V_{sl}), which is often associated with concrete crushing, can be greater or less than $V_{cr\text{-}ini}$, that is of the same order of magnitude.
3. The sliding behaviour, that is the ability to maintain shear whilst sliding, is the shear ductility.

Unlike flexural failure, where in general $M_{cap} >> M_{cr\text{-}ini}$, in shear failure V_{sl} and $V_{cr\text{-}ini}$ are of the same order of magnitude, and thus, both can govern the shear capacity. To understand their interaction, typical variations in shear strength with β, the angle of the inclined plane, are shown in Figure 1.31. The analysis at β=90° is shown in Figure 1.31, where the shear capacity to cause sliding (V_{sl}) is much greater than the shear to cause initial cracking ($V_{cr\text{-}ini}$). As β is reduced, V_{sl} reduces and $V_{cr\text{-}ini}$ increases, as shown, and they intersect at V_{cap} at the crack inclination β_{cap}. Even though we understand the mechanics that govern crack development, some of which are explained in Figure 1.12, there are so many variables that cracking and crack spacing are at best a semi-random phenomenon. Should a crack occur at β_{stb}, then the initial crack would occur first at Point *A* and then fail at Point *B*. This is referred to as a stable failure, as it provides ample warning, as illustrated in Figure 1.8. In contrast, should a crack occur at β_{unstb}, then an initial crack would occur at Point *D*; as this shear is greater than V_{sl} at Point C, it would fail immediately. This is referred to as unstable or catastrophic failure, as there is no warning, and it is illustrated in Figure 1.30a. A safe design is therefore the lower bound value at the intercept, which is V_{cap}. It can be seen that the variation of

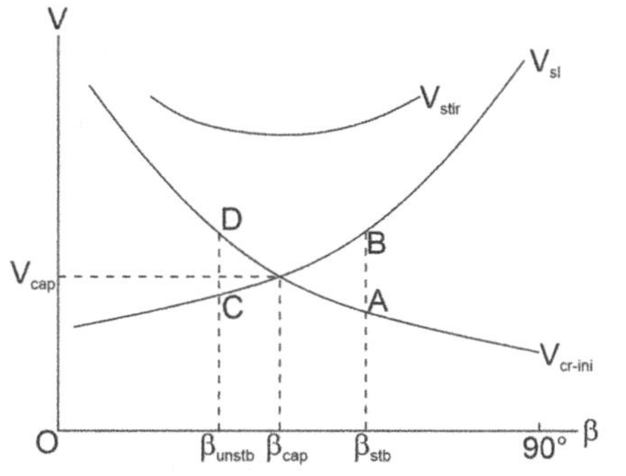

Figure 1.31 Interaction between $V_{cr\text{-}ini}$ and V_{sl}.

both capacities (V_{sl} and $V_{cr\text{-}ini}$) has to be determined to establish the safe design capacity, V_{cap}. It is worth noting that beams with stirrups have a sliding capacity V_{stir} which is well above $V_{cr\text{-}ini}$; therefore, $V_{cr\text{-}ini}$ is immaterial.

1.6 *PI* MEMBER MECHANICS

1.6.1 Members with local debonding only

The PI segmental analysis for flexure in Section 1.5.2 and Figure 1.27 provides all the local mechanical properties required to analyse an *RC* member when only local debonding, such as that illustrated for the primary cracks in Figure 1.12f, has occurred. The segmental analysis provides the flexural behaviour, including crack spacings and widths. Hence, everything a designer requires to analyse an *RC* member at both the serviceability and ultimate limit states is available. This includes members with axial loads, such as columns or prestressed members.

An encastre *RC* beam is shown in Figure 1.32a. It is noteworthy that symmetrical collapse of the beam requires the formation of the four hinges labelled *A*, *B*, *C* and *D*, rather than the three hinges typically required by conventional analyses, because each hinge is taken to be of the same shape. The flexural segmental analysis quantifies the moment–rotation properties of the potential hinges and the variation of the flexural rigidities *EI* between the hinges. Hence, moment redistribution, as illustrated in Figure 1.32b, can be simulated; for example, by determining whether the rotation capacity of the hogging (*–ve*) hinge at *A* is sufficient to allow the moment capacity of the sagging (*+ve*) hinge at *B* to be achieved. Collapse, as shown in Figure 1.32c, can also be simulated; in this scenario, the objective is to determine the deflection, which indicates the energy absorption capacity, when the rotation capacity of one of the hinges is first reached.

1.6.2 Members with global debonding

Figure 1.12c, illustrating *IC* debonding, is an example of one form of global debonding that occurs over the length of a member. Conventional RC beams are designed so that global debonding does not occur. However, global debonding can occur in

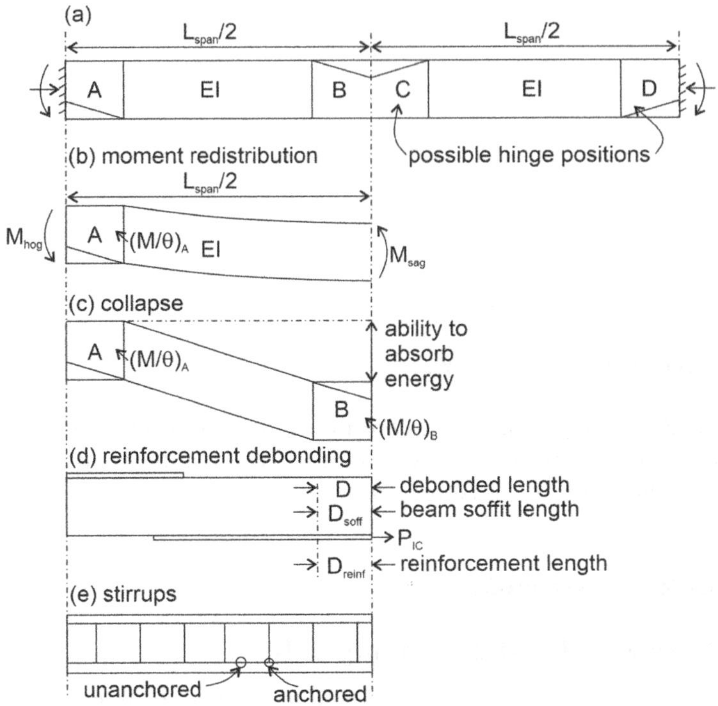

Figure 1.32 Partial-interaction member analysis (a) encastre (built-in) beam showing potential hinge positions; (b) length of beam subjected to moment redistribution; (c) beam collapse; (d) reinforcement global debonding; (e) provision of stirrups as mechanical anchors to inhibit global debonding.

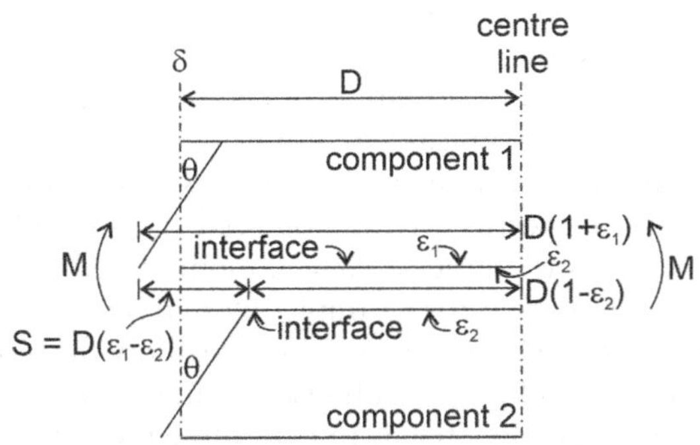

Figure 1.33 Global debonding through slip–strain.

non-conventional members such as *RC* beams with externally bonded *FRP* plates. Global debonding occurs when slip accumulates across an interface between components or segments along the length of a member.

To understand the global debonding mechanism, consider a short segment of a member, of half-length *D*, as shown in Figure 1.33. This segment consists of an upper Component 1 and a lower Component 2. For ease of explanation, let us assume that the member is subjected to a constant moment *M*. Furthermore, the centreline shown is the axis of symmetry for this symmetrically loaded segment of half-length *D*, such that its RHS is a mirror image of its LHS. We will also assume that there is no vertical separation between the two components, ensuring that the rotations induced at the ends, θ, are the same. We are in effect doing a flexural segmental analysis as in Figure 1.29.

Defining an extending strain as positive (+ve) and a contracting strain as negative (–ve). Let the strain in the bottom fibre of Component 1 in Figure 1.33 be a tensile strain $+\varepsilon_1$ and the strain at the top fibre in Component 2 be a compressive strain $-\varepsilon_2$. It should be noted that, for clarity, the components have been separated vertically in the illustration, creating a notional gap at the interface which is absent in practice. At the interface, the length of Component 1 will increase from *D* to $D(1+\varepsilon_1)$ and the length of Component 2 will reduce from *D* to $D(1 - - \varepsilon_2)$ that is $D(1+\varepsilon_2)$. Consequently, the slip *S* at the interface at the end of the beam which is the difference between these lengths, is $D(\varepsilon_1 - \varepsilon_2)$, which corresponds to *D* multiplied by the slip–strain $\varepsilon_1-\varepsilon_2$, as introduced in Section 1.1. The sum of these slips along the beam, which in structural mechanics terms is the integration of the slip strain, gives the variation of slip along the member which is resisted by the bond slip between the components, as shown in Figure 1.6, and potentially by mechanical shear connectors, as illustrated in Figure 1.24.

Now consider an element of an *FRP* or steel plated beam, as in Figure 1.32d, of length *D*, as depicted in Figure 1.34. For clarity, the plate is shown slightly separated from the *RC* beam; however, in practice, they are in contact at the interface. This is an example of global debonding, as illustrated in Figure 1.12c, occurring when the force in the reinforcement reaches its peak value, termed the *IC* debonding resistance P_{IC}. Solving this problem requires consideration of not only the local segmental properties but also their interaction. Figure 1.34 illustrates the development of global debonding. Once the force in the plate attains its *IC* debonding resistance P_{IC} at a strain ε_{IC} (as shown), any subsequent increase in the applied beam load instigates debonding along the member as illustrated in Figure 1.12c.

Let us assume that the plate has debonded over the length *D* (Figure 1.32d), as shown in Figure 1.34. The *RC* member over the length *D* is now acting as an externally post-tensioned *RC* member with a prestressing force of P_{IC}. The flexural segmental analysis (Figure 1.29), modified with an axial load P_{IC}, can be used to determine the length of the soffit of the beam D_{soff} (Figure 1.34) for the variation of the applied moments acting over length *D*. The length D_{soff} will be slightly larger than *D* due to flexure. Should the force in the plate remain at the *IC* debonding force P_{IC}, then the strain in the plate over the length *D* remains constant at the *IC* debonding strain ε_{IC}.

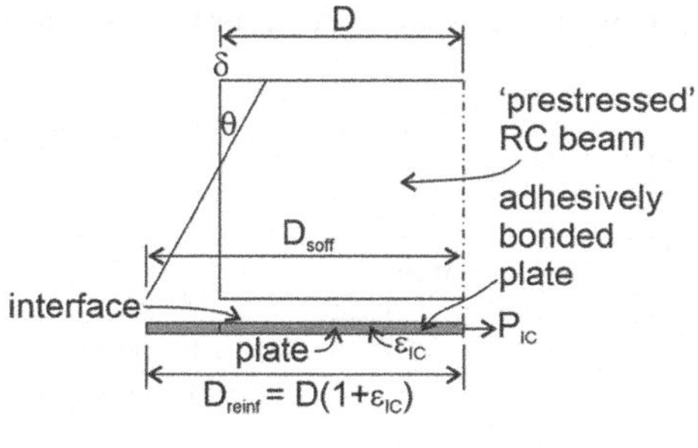

Figure 1.34 Stable and unstable global debonding of external plate.

The length of the debonded plate D_{reinf} is therefore $D(1+\varepsilon_{IC})$. For compatibility, D_{soff} must equal D_{reinf}. Should D_{reinf} be more than D_{soff}, this indicates that the strain in the plate is less than ε_{IC}; that is the applied moment to the beam has to be increased to achieve further global debonding. This is stable debonding. Should D_{reinf} be less than D_{soff}, this indicates that the applied moment has to be reduced to achieve compatibility; or, if the applied moment is maintained, debonding continues along the beam until compatibility is achieved. This is unstable debonding. Should there be no position along the beam where compatibility is achieved, the plate debonds completely.

The above global debonding mechanism, applied to an externally plated *RC* beam, also applies to the behaviour of tension reinforcement embedded in concrete, as shown in Figure 1.32e. When there are no stirrups, the force in the tension reinforcement is restricted to the *IC* debonding resistance P_{IC} of the bar, which depends on the interface bond-slip material property. It does not matter how long the bar is, this *IC* debonding force cannot be increased, assuming that at δ_{max} (Figure 1.6), τ_B is zero, which implies negligible friction, as for a smooth bar. The force can, however, be increased by providing stirrups that anchor or mechanically restrain the bars at discrete positions (as shown in Figure 1.32e), as these can behave as mechanical anchors (as in Figure 1.24) and can be included in a global analysis. This global analysis also applies to composite steel and concrete members such as I-beams that support slabs and are connected using stud shear connectors (Oehlers and Bradford 1995, 1999). It can also be applied to profiled slabs.

The tools are now available to study *RC* members subjected to local and global debonding. They can be used to study the effects of time-dependent deterioration, such as that due to corrosion and fatigue. This deterioration may be local, such as a change in a material property, but its effect on member behaviour can now be studied without the need for member testing.

Worked Example WE-1.3: Global debonding

A one-way spanning, simply supported slab is to be strengthened with externally bonded plates at 300 mm spacing. A portion of the slab is shown in Figure 1.35. The plate must have an *IC* debonding resistance of 80 kN. This can be determined experimentally, as shown in Figure 1.5c and d (provided the plate length in Figure 1.5d is sufficient to achieve the *IC* debonding resistance), or theoretically, through analyses such as that in Figure 1.12c, if the bond-slip properties are known. These methods also yield the *IC* debonding strain ε_{IC} required in Figure 1.35. The *IC* debonding strain is a variable, as simply reducing the thickness of the plate (t) increases the *IC* debonding strain. It will be assumed that the flexural rigidity of the cracked unplated section is know, as it can be accurately quantified from *PI* analyses

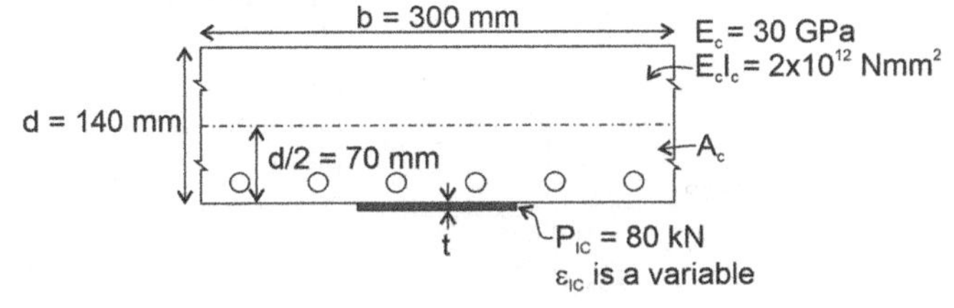

Figure 1.35 Slab properties.

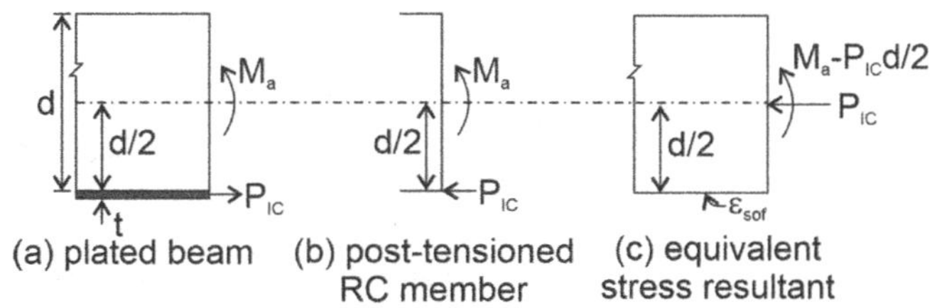

Figure 1.36 **Stress resultants (a) stress resultants acting on plated beam at commencement of IC debonding; (b) stress resultant due to post tensioning behaviour on debonding; (c) converting the stress resultants to equivalent stress resultants.**

in Chapter 8. The slab spans 4 m and is subjected to a line load at mid-span that induces a maximum applied moment M_a of 15 kNm.

The stress resultants acting on the slab at the commencement of *IC* debonding are shown in Figure 1.36. At the commencement of debonding, the force in the plate is $P_{IC.}$ Over the debonded region, the force in the plate remains constant at P_{IC}. Consequently, the slab no longer functions as a composite slab (i.e. a slab with reinforcing bars and a plate), but rather as a post-tensioned member where the plate supplies the post-tensioning force. Therefore, the forces acting on the *RC* component are those depicted in Figure 1.36b. As the plate thickness *t* is usually much smaller than the slab depth *d*, the post-tensioning force P_{IC} has been applied at the soffit of the *RC* slab, as shown, to simplify the explanation.

For ease of analysis, the stress resultants in Figure 1.36b have been converted to the equivalent stress resultants in Figure 1.36c At mid-depth, there is now an axial compressive load P_{IC} and a total moment equal to the applied moment M_a less the moment due to the eccentricity of P_{IC} (i.e. $P_{IC}d/2$). The variation of the soffit strain ε_{sof} must now be determined. From elementary structural mechanics i.e. $\sigma = My/I + P/A$, the strain in the soffit is given by

$$\varepsilon_{sof} = \frac{My}{EI} - \frac{P}{EA} = \frac{M_a \frac{d}{2}}{E_c I_c} - \frac{P_{IC} \frac{d^2}{4}}{E_c I_c} - \frac{P_{IC}}{E_c A_c} \tag{1.5}$$

The first term $(M_a d/2)/E_c I_c$ is the strain in the soffit due to the applied moment $\varepsilon_{a\text{-}m}$. The second term is the reduction in strain due to the eccentricity of the post-tensioning load $\varepsilon_{p\text{-}m}$. The third term $(P_{IC})/E_c A_c$ is the reduction in tensile soffit strain due to the post-tensioning compressive load $\varepsilon_{a\text{-}a}$.

The applied moment is assumed to vary from 15 kNm at mid-span to zero at 2 m from mid-span, as shown in Figure 1.37. The results of applying Eq. 1.5 are also shown in Figure 1.37. In this case, the abscissa is the extent of debonding from mid-span. The ordinate represents the strains, considering the linear variation of the applied moment shown. Applying Eq. 1.5, the maximum soffit strain due to the applied moment is $\varepsilon_{a\text{-}m} = 525 \times 10^{-6}$, which reduces linearly to zero at the support, 2 m from mid-span, as shown. The reduction in the soffit strain $\varepsilon_{p\text{-}m}$ due to the eccentricity of the post-tensioning force P_{IC} is constant at 196×10^{-6}, while the reduction in the soffit strain due to $\varepsilon_{a\text{-}a}$ is also constant at 64×10^{-6}. Hence, the variation in the soffit strain ε_{sof} is given by the line *A*–*B*, reaching a maximum value of 265×10^{-6}.

The variation of the soffit strain *A*–*C*–*B* in Figure 1.37 can be used to calculate the extension of the soffit from mid-span. For example, if the debonded region

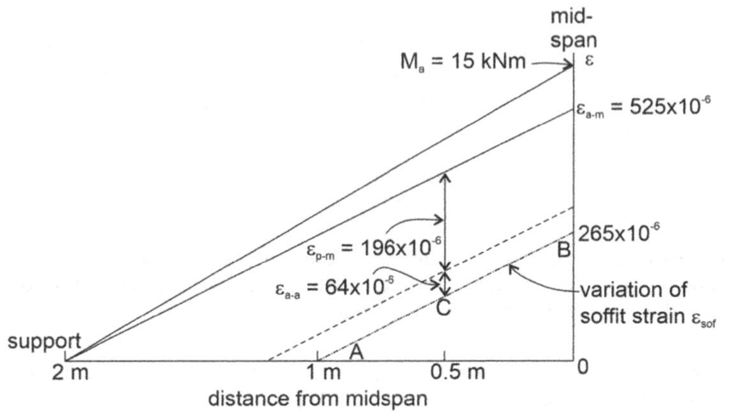

Figure 1.37 Variation of soffit strains.

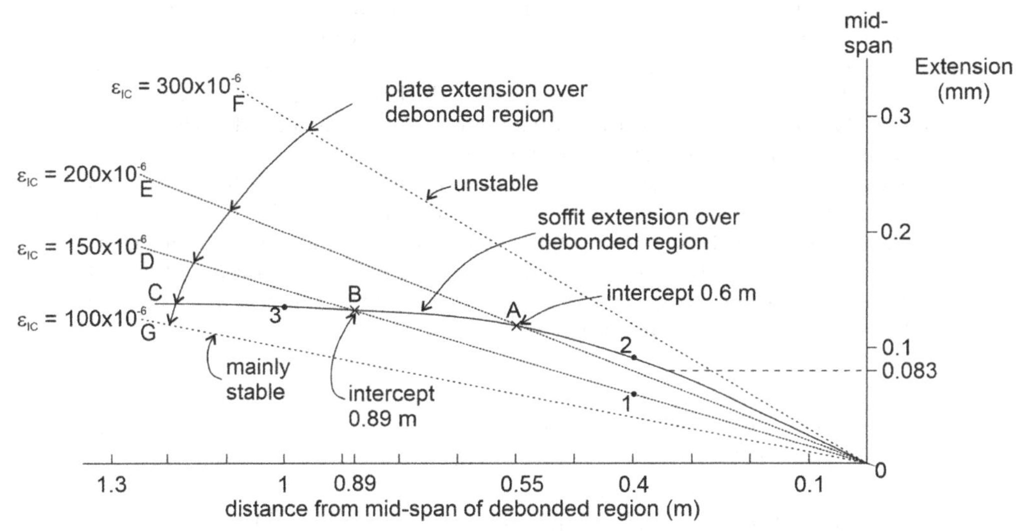

Figure 1.38 Stable and unstable debonding.

extends 0.5 m, the variation of strain from C to B can be used to determine the extension. In this case, as the variation is linear, it is simply the average strain from C to B multiplied by the length 0.5 m. These calculations are used to plot the 'soffit extension over debonded region' (line O–A–B–C in Figure 1.38). For example, when the debonded region extends to 0.4 m, the extension is given by Point 1 at about 0.083 mm. The critical question is whether this amount of debonding is stable or unstable; specifically, whether it can actually occur.

Assume we have designed the plate, primarily by varying its thickness t, such that the IC debonding strain ε_{IC} remains at 150×10^{-6}. As the strain in the plate over the debonded region is constant, the extension is simply the product of the debonded length and ε_{IC}. Hence, the extension varies linearly, as shown by O–D in Figure 1.38. The variation O–D represents the extension of the plate, while line O–A–C represents the extension of the adjacent soffit. These extensions need to be compatible, which, in this case, means they must be equal. At 0.4 m, the extension of the plate is shown at Point 1, whereas the extension of the soffit is larger, shown at Point 2. These points are therefore not compatible. The only way to make them compatible is to reduce the applied moment, since this does not affect the extension of the plate. Therefore, when the applied load is maintained, the debonded region will continue to extend in an attempt to achieve compatibility and the behaviour is therefore unstable. In this case, debonding will continue to Point *B* at about 0.89 m, where the extensions are identical; that is compatibility is achieved. Upon further extending the debonded region to 1 m, at Point 3, the soffit strain is now less than the plate strain at Point 4; therefore, the only way to achieve compatibility is to increase the applied load, resulting in stable debonding.

Designing the plate such that P_{IC} remains the same but ε_{IC} increases to 200×10^{-6} yields line *O–E* in Figure 1.38. In this case, the debonded length stabilises at Point *A* at 0.55 m. To design for stable debonding, it is necessary to increase the *IC* debonding strain further, for example, to 300×10^{-6} along line *O–F*. Stable debonding can be defined as occurring when the debonded region increases only with an increase in the applied load. In contrast, if there were a very low *IC* debonding strain of 100×10^{-6} along line *O–G*, then unstable debonding would occur over most of the span, rendering plating ineffective. Unstable debonding can be defined as occurring when the debonded region continues to increase without an increase in applied load. It should be noted that this analysis deals with *IC* debonding, which starts at a position of maximum moment and progresses towards regions of lower moment. Plates can also start debonding from their ends due to transverse shear; this can be designed against or simply prevented by bolting the ends (Oehlers and Seracino 2004). These end bolts have no effect on IC debonding.

1.7 *PI* METHODS OF ANALYSIS

Throughout this book, the mechanics of *PI* behaviour is explained using simple iterative analysis techniques. To assist the reader, algorithms, presented as flowcharts, are provided for many of these techniques. Within these simple iterative analysis techniques, the mechanics is fully explained, enabling the reader to visualise and understand the underlying processes. An iterative procedure is employed to accommodate any non-linearity in the material properties. The iterative analysis procedure often employs a shooting method, in which a parameter (e.g. the slip at a specific point in the member) is initially estimated. The analysis then proceeds along the member to a known boundary condition; for instance, where the slip is zero due to symmetry, or where, in the presence of *FI*, both the slip and slip-strain are zero. If this condition is not met, the initial estimate is iteratively refined until the boundary condition is satisfied. Adaptations of the shooting method are also used where, for example, the curvature and neutral axis depth are varied (or iteratively adjusted, as shown in Figure 1.29) until the desired stress resultants are obtained. Hence, the reader can use the mechanics to derive a solution, in contrast to a finite element analysis, which is treated by some users as a 'black box' providing solutions with little or no understanding of the underlying mechanics.

Having fully explained the mechanics through iterative non-linear analysis, enabling the reader to visualise and understand the underlying processes, the mechanics derived from this simple iterative analysis procedure is sometimes used as the basis for deriving closed-form solutions. Invariably, simplifications must be made to the mechanics and material properties to derive closed-form solutions, which typically results in a loss of accuracy. Design is not solely a matter of ensuring strength and serviceability; other considerations must be taken into account. Currently, environmental sustainability, such as minimising global warming impact, is of great importance.

1.8 INTRODUCTION TO REMAINDER OF BOOK

All the fundamental mechanisms and material properties governing the physical behaviour of RC have now been explained. In particular, this chapter has described the interaction and interplay between these mechanisms and properties. In the remainder of the book, these mechanisms and properties will be revisited in greater detail to further consolidate the reader's understanding.

The material properties of *RC* members can be categorised into *FI* properties, based on strain, and *PI* properties, which allow for interface slip. Chapter 2 covers *FI* strain-based material properties sufficiently for use only in the remainder of the book, whereas Chapter 3 covers slip-based material in more depth, from testing to quantification. We now move from material properties to mechanical elements within *RC* members, including tension stiffening (Chapter 4) and concrete compressive softening (Chapter 5). Having covered the mechanics of small elements, we move on to beam segments to quantify flexural behaviour (Chapter 6) and the effect of transverse shear on flexure (Chapter 7).

We now have sufficient information to analyse flexural members. Chapter 8 deals with flexural members in which the reinforcement has strong bond, such as those with reinforcing bars, where only local debonding occurs. Chapter 8 covers analyses from serviceability to the ultimate limit, including moment redistribution; Chapter 9 deals with prestressed members. We then move on to members where the reinforcement bond is weak, including profiled slabs and composite steel and concrete beams (Chapter 10), and FRP-retrofitted RC members (Chapter 11). All *RC* structures need to be designed for sustainability. Chapter 12 covers the impact of corrosion, and Chapter 13 addresses sustainability.

We conclude with an Epilogue that summarises the research approaches employed in developing the mechanics for safe design.

1.9 NOTATION AND ABBREVIATIONS

A_c	cross-sectional area of concrete in tension stiffening prism
A_{dow}	axial shear force on dowel
A_r	cross-sectional area of reinforcement in tension stiffening prism
a	shear span of beam
a_{seg}	shear span to inclined face of segment
b	width or breadth of member
b_{seg}	width of segment
C	longitudinal contraction due to sliding; coefficient; confinement force
c	side cover to concentrated force
conc	concrete

com	compression
$contr$	contraction
D	width or diameter of concrete prism or dowel; debonded length of beam; short length of the beam
D_{reinf}	debonded length of reinforcement
D_{soff}	debonded length of soffit of beam
d	depth of member; depth
d_{NA}	depth to neutral axis from top fibre
d_r	depth from centroid of A_r to tension face
d_{seg}	depth of segment
dw/dN	increase in crack width per cycle of load
E	modulus
E_c	concrete modulus
$E_{ct\text{-}micro}$	effective tensile concrete modulus after micro-cracking
E_r	reinforcement modulus
EI	flexural rigidity
e	eccentricity
$exten$	extension
F	force profile
FI	full interaction
FRC	fibre-reinforced concrete
FRP	fibre-reinforced polymer
F_{cc}	force in concrete in compression
F_{ct}	force in concrete in tension
F_{dow}	dowel force; concentrated force induced by dowel
F_{fib}	tensile force in fibres bridging a crack
F_{rc}	force in reinforcement in compression
F_{rt}	force in reinforcement in tension
f_c	compressive strength of unrestrained concrete
f_{co}	unconfined cylinder compressive strength
f_{ct}	tensile strength of concrete just prior to cracking
f_{cu}	compressive cube strength of concrete
f_{cy}	compressive cylinder strength of concrete
f_u	fracture strength of steel dowel
f_y	yield capacity
h	height of concentrated load
I	second moment of area
L	length of concrete softening region; length
LHS	left-hand side
L_{crt}	minimum length required to develop P_{IC}
L_{def}	length of segment subjected to deformation
$L_{def\text{-}b}$	length of segment at bottom
$L_{def\text{-}t}$	length of segment at the top
L_{fct}	length in tension stiffening analysis to develop a crack
L_{NI}	debonded length
L_{per}	length of perimeter between reinforcement and adjacent concrete
L_{seg}	length of rectangular segment

L_{span}	beam span
M	externally applied moment; moment
M_a	externally applied moment
M_{cap}	moment capacity
M_{crFI}	moment to cause flexural cracking in FI region
$M_{cr\text{-}ini}$	moment to cause initial crack
M_{dow}	moment induced by dowel action
Me	moment at eccentricity e
M_{hog}	hogging moment; –ve moment
M_{sag}	sagging moment; +ve moment
M_{seg}	externally applied moment to segment; $V_{seg}a_{seg}$
M/θ	moment–rotation
N	externally applied axial force in the longitudinal direction of member
N_{dow}	axial force induced by dowel action
N_{seg}	externally applied axial force on segment
NI	no interaction
P	force along longitudinal axis of reinforcement or prism
PI	partial interaction
P_{conf}	passive confinement force
P_{dow}	dowel strength at fracture
P_{fct}	reinforcement force to cause primary crack governed by f_{ct}; A_cf_{ct}
P_{IC}	intermediate crack debonding resistance
P_{tr}	transverse force induced by splitting action
P_{split}	dowel force to cause splitting
p	area of transverse reinforcement per unit area of shear plane
R	lateral or transverse restraint
RC	reinforced concrete
RHS	right-hand side
SF	shear friction
S_{co}	maximum slip along concrete softening sliding plane
S_{cr}	crack spacing
S_{pr}	primary crack spacing
S_r	reinforcement slip relative to crack face
S_{sh}	rigid body slip along shear sliding plane
t	tension; tube thickness
ten	tension
V	externally applied transverse shear force in the transverse direction of member
V_{cap}	shear capacity with zero stirrups
$V_{cr\text{-}ini}$	V to cause the initial shear crack
V_{seg}	externally applied shear force to segment
V_{sl}	sliding shear capacity without stirrups
V_{stir}	shear capacity of beam with stirrups
V_u	shear capacity of initially uncracked sliding plane; shear at the onset of sliding
V_{uO}	V_u of an unconfined sliding plane
v_u	shear capacity in design codes
w	width of crack; 2Δ; effective width of crack
w_{SF}	width of crack measured normal to sliding plane governed by SF properties

α	angle of concrete softening sliding plane from longitudinal direction (about 26°)
β	angle of shear sliding plane from longitudinal direction
β_{cap}	β at V_{cap}
β_{stb}	β when stable shear failure occurs
β_{unstb}	β when unstable shear failure occurs
χ	curvature; θ/L_{def}
Δ	slip of reinforcement relative to crack face
Δ_{fib}	displacement of fibres in tension that cause $\sigma_{ct\text{-}fib}$
Δ_{rc}	Δ of compression reinforcement
Δ_{rt}	Δ of tension reinforcement
Δ_{SF}	slip across sliding plane measured in sliding plane direction and governed by *SF* properties
δ	displacement profile; displacement
δ_B	bond slip; slip between reinforcement and adjacent concrete
δ_b	displacement of bottom fibre
δ_{cc}	displacement of concrete in compression that causes stress
δ_{max}	δ_B when τ_B tends to zero
δ_{rc}	displacement of reinforcement in compression that causes stress
δ_t	displacement at top fibre
δ_1	bond slip at τ_{max}
ε	strain profile; strain
$\varepsilon_{a\text{-}a}$	reduction in soffit strain
$\varepsilon_{a\text{-}m}$	strain in beam soffit due to applied moment
ε_c	concrete strain
ε_{cFI}	concrete strain in *FI* region adjacent to and in the same direction as ε_{rFI}
ε_{cPI}	concrete strain in *PI* region adjacent to and in the same direction as ε_{rPI}
ε_{IC}	longitudinal strain in plate when plate load is P_{IC}
ε_m	material strain
$\varepsilon_{p\text{-}m}$	reduction in strain due to eccentricity of post-tensioning force
ε_r	reinforcement strain
ε_{rFI}	reinforcement strain in the *FI* region
ε_{rPI}	reinforcement strain in the *PI* region
ε_s	effective compressive strain due to sliding
ε_{sh}	shrinkage strain
ε_{sl}	longitudinal strain due to sliding
ε_{sof}	soffit strain
ε_{sp}	average specimen strain; effective strain
ε_t	top fibre strain
θ	Euler–Bernoulli rotation in radians
σ	stress profile; stress
σ_c	axial compressive stress
σ_{cFI}	stress in concrete in the *FI* region
σ_{conf}	lateral confinement
σct	tensile stress in concrete
$\sigma_{ct\text{-}fib}$	equivalent tensile stress in concrete after cracking due to fibres
σ_N	stress normal to the sliding plane or potential sliding plane; confining stress

σ_r range of cyclic stress
τ shear stress across interface between reinforcement and adjacent concrete; shear stress profile
τ_B shear stress between reinforcement and adjacent concrete
τ_{max} maximum value of τ_B; bond slip shear capacity
τ_{SF} shear stress along the sliding plane or potential sliding plane governed by *SF* properties

REFERENCES AND DEVELOPMENT

Chen, Y., 2015. *Quantifying the compressive ductility of concrete in RC members through shear friction mechanics* (Doctoral dissertation, Ph. D. thesis, School of Civil, Environmental and Mining Engineering, Univ. of Adelaide).

Chen, Y., Visintin, P. and Oehlers, D.J., 2015. Concrete shear-friction material properties: derivation from actively confined compression cylinder tests. *Advances in Structural Engineering*, *18*(8), 1173–1185.

Dadd, L., Bennett, B., Xie, T. and Visintin, P., 2025. Shear behaviour and environmental impact of single and multi-generation recycled aggregate concrete beams. *Engineering Structures*, *322*, 119200.

Oehlers, D., Haskett, M., Mohamed Sadakkathulla, M., Lucas, W. and Muhamad, R., 2011. Our obsession with curvature in RC beam modelling. *Advances in Structural Engineering*, *14*(3), 391–404.

Oehlers, D.J., 2010. The hunt for the elusive concept. *Advances in Structural Engineering*, *13*(5), 755–772.

Oehlers, D.J. and Bradford, M.A., 1995. *Composite Steel and Concrete Structural Members: Fundamental Behaviour*. Oxford: Pergamon.

Oehlers, D.J. and Bradford, M.A., 1999. *Elementary Behaviour of Composite Steel and Concrete Members: Fundamental Behaviour*. Oxford: Butterworth Heinemann.

Oehlers, D.J. and Seracino, R., 2004. *Design of FRP and Steel Plated RC Structures*. Amsterdam, The Netherlands: Elsevier.

Oehlers, D.J. and Visintin, P., 2021. A displacement based approach for the analysis, development, and design of reinforced concrete. *Structural Concrete*, *22*, 2105–2125.

Oehlers, D.J., Ali, M.M., Griffith, M.C., Haskett, M. and Lucas, W., 2012. A generic unified reinforced concrete model. *Proceedings of the Institution of Civil Engineers-Structures and Buildings*, *165*(1), 27–49.

Oehlers, D.J., Visintin, P., Chen, J.F. and Ibell, T., 2014b. Simulating RC members part 2: displacement based analysis. *ICE Structures and Buildings*, *167*(12), 718–727.

Oehlers, D.J., Visintin, P., Chen, J.F., and Ibell, T., 2014a. Simulating RC members part 1: partial interaction properties. *ICE Structures and Buildings*, *167*(11), 646–653.

Oehlers, D.J., Visintin, P., Chen, J.F., Seracino, R., Wu, Y.F., and Lucas, W., 2017 Reinforced concrete behaviour, research, development and design through partial interaction mechanics. *ASCE Journal of Structural Engineering*, *143*(7), 02517002.

Part 2

Engineering and mesoscale material properties

In this section of the book, we deal with the engineering and mesoscale properties of the material, specifically stress, strain, creep, shrinkage and interface slip. We do not cover particle mechanics, which is outside the scope of this book; instead, we focus on particle properties. Strain-based properties, such as modulus, creep and shrinkage, are discussed in Chapter 2 for use in worked examples. Chapter 3 addresses slip-based properties, beginning with the partial-interaction (PI) shear friction material properties across a potential sliding plane. These properties are crucial in determining the shear capacity of a member, as well as in compressive concrete softening and confinement. This is followed by a discussion of PI bond-slip properties, not just for steel reinforcement but also for FRP reinforcement. These properties are important in quantifying flexural crack spacings and widths, which control the serviceability flexural rigidity of a member and debonding at the ultimate limit state. Finally, we examine the PI behaviour of fibre concrete under tension, addressing the fibre concrete tension mechanism after cracking when fibres are present in the concrete and demonstrating how to manage microcracking and the falling branch when fibres pull out, resulting in debonding.

DOI: 10.1201/9781003711407-3

Chapter 2

Strain-based material properties

This is the only chapter that does not deal specifically with a mechanism that controls the behaviour of *RC* members. There is an extensive body of excellent research and applications on strain-based material properties. The aim of this chapter is not to provide a comprehensive overview, but rather to offer a basic understanding of strain-based properties sufficient: for their inclusion in developing the numerical and closed-form solutions of the mechanical models in the ensuing chapters; for enhancing the reader's ability to visualise these mechanisms; and for familiarising them with fundamental strain-based properties. Some concepts covered are very basic; we apologise in advance if their inclusion seems redundant to some readers. Readers may initially skip or glance through this chapter to grasp the main points, referring back to it in detail only when required in other chapters. The same cannot be said for Chapter 3 on slip-based properties, as these are based on shear friction properties, which themselves constitute a mechanism governing the *PI* behaviour of reinforced concrete members. As material properties are frequently updated with the development of new tests and materials, it is also worth noting that the properties given in this chapter should not be taken as definitive, but only as examples for illustrating particular analyses or behaviours. In contrast, the concepts in the remainder of the book, which are based on fundamental mechanics, remain virtually unchanged. It should be noted that many of the equations given in this chapter have been determined empirically and may be dimensionally incorrect; for such equations, the units are in N and mm.

It was explained in Chapter 1 that in reinforced concrete members there are two distinct types of material properties to consider the strain-based material properties associated with *FI* behaviour; and the slip-based material properties associated with *PI* behaviour. This chapter provides a qualitative and quantitative description of the strain-based material properties of concrete and reinforcement (both steel and fibre-reinforced polymer (*FRP*)) and how these properties change over time. Wherever possible, the descriptions of material behaviour are explained in terms most appropriate for structural engineers and hence with a focus on describing behaviour at the engineering scale. Much emphasis is placed on creep and shrinkage.

Globally, concrete is the 'most in-demand' material, second only to water; this immense demand for concrete results from its versatility, low cost and low barrier to entry associated with concrete production and construction. That is, the materials required to produce concrete are widely available, and while the production of high-quality and high-performance concrete is a scientific endeavour, lower-quality

 DOI: 10.1201/9781003711407-4

concretes suitable for many applications can be produced with little underlying knowledge or skill. It is, however, important to note that unlike other materials used in construction, concrete is highly non-homogeneous, as it can be produced from a diverse range of source materials and with a diverse range of material properties. Beyond the typical material properties of strength and stiffness, the behaviour of concrete is made more complex by its expansion and contraction with time and exposure to differing environmental conditions, not to mention its development of cracks, which, as shown throughout this book, impact virtually every aspect of behaviour.

The strain-based properties of concrete are extremely complex and could themselves be the topic of a book. To assist the reader, this chapter summarises those properties considered most important in the context of this book as a whole; other properties relevant only to a specific subsection of the book will be introduced as they arise. Concrete classification is first explained in Section 2.1. This is then followed by the components of concrete in Section 2.2. Having described concrete as a material, its mechanical properties are then given in the ensuing sections, beginning with those that occur under an instantaneous load. The compressive properties are given in Section 2.3, followed by the tensile properties in Section 2.4. The impact of time on concrete material properties is then considered in Section 2.5, starting with concrete shrinkage and thermal strains, which occur in the absence of applied loads, and then concrete creep, which occurs as a result of a sustained load. The chapter finishes with the strain-based properties of steel and *FRP* reinforcement.

2.1 CONCRETE CLASSIFICATIONS

In its broadest sense, concrete is an artificial stone consisting of a skeletal structure of fine and coarse aggregates cemented together by a mortar phase, itself consisting of a fine aggregate and a binder (Figure 2.1). The art of concrete mix design, therefore, involves specifying the properties and quantities of the coarse and fine aggregates (also known as fillers) and the binder to achieve a desired set of performance criteria. These performance criteria have traditionally related to a minimum set of mechanical and durability properties that must be met to ensure adequate structural performance, while minimising the cost to the concrete producer. More recently, concrete mix design has evolved such that minimising impact on the environment is now also important. This will be discussed in detail in Part 6 of the book.

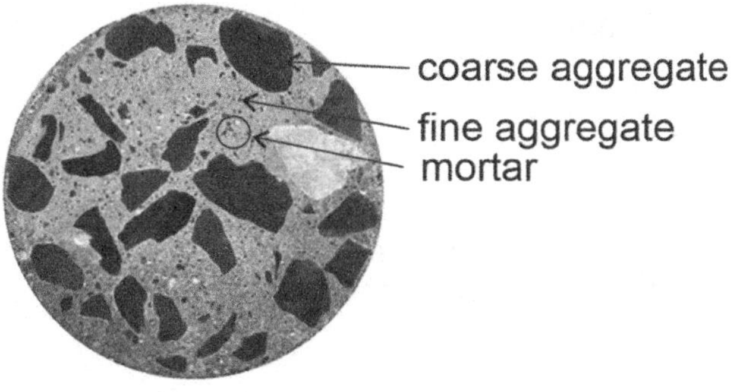

Figure 2.1 The skeletal structure of concrete. (Photo courtesy of Lloyd Dadd.)

Concrete is generally classified on the basis of the compressive strength (f_c) of the set or hardened concrete and the workability of the unset or fluid concrete. In special cases, additional considerations such as the unit weight or the addition of fibre reinforcement may be introduced into the overall classification and will be briefly discussed here.

2.1.1 Concrete compressive strength

The most basic classification of concrete is its compressive strength. Concrete is generally classified into the following categories:

- Normal-strength concrete (*NSC*) typically ranges in compressive strength from as low as 5 MPa, for non-structural components like paths and kerbs, up to 50 MPa for structural components.
- High-strength concrete (*HSC*) generally encompasses concrete strengths greater than 50 MPa and up to approximately 100 MPa. HSC is generally applied in high-rise structures, particularly where its high compressive strength can be used to minimise section size, for example for columns.
- Ultra-high-performance concrete (*UHPC*), also known as reactive powder concrete, is a relatively new class of concrete and is commonly classified as having compressive strengths of 120–200 MPa, but arguably, its very high durability is more important.

Concrete compressive strength is determined by subjecting either a cylindrical or cubic specimen to monotonically increasing load until failure as shown in Figure 2.2. Although the choice of specimen shape and size may initially appear to be unimportant,

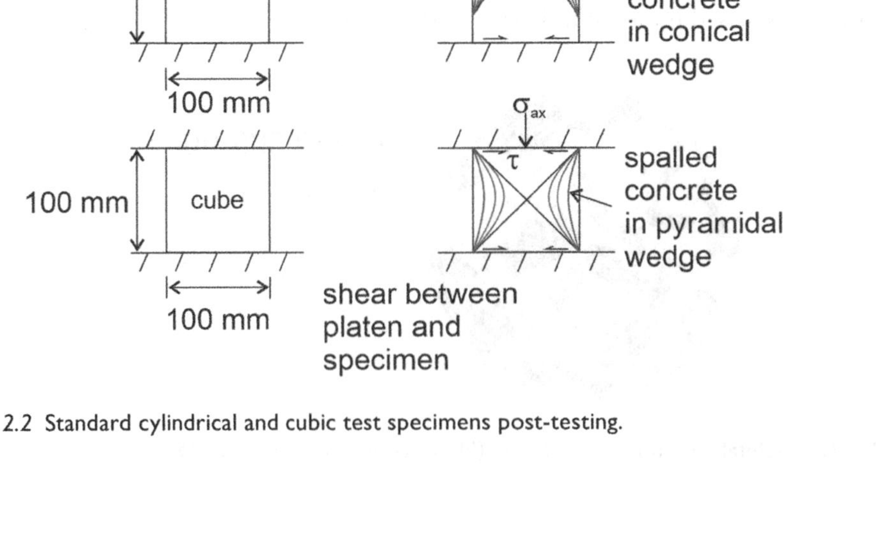

Figure 2.2 Standard cylindrical and cubic test specimens post-testing.

it nevertheless has a significant impact on the measured material properties and is, therefore, worthy of further consideration.

Figure 2.2 shows standard cylindrical (100 mm diameter×200 mm height) and standard cubic (150 mm) specimens after testing. Crucially, the results obtained from these two test specimens will not be the same even when accounting for their differing cross-sectional areas. This is because the restraint (friction) that occurs between the loading platens and the loaded surfaces of the specimen creates a shear force (τ) which confines the specimen. This lateral expansion occurs throughout the entire loading process and, before the sliding of wedges, is entirely due to Poisson's effect, where the specimen expands laterally as it contracts axially.

Importantly, the 2:1 aspect ratio of the cylindrical specimen in Figure 2.2 minimises the confinement provided to the specimen in the central region, which is most likely to fail. Thus, it can be expected that the compressive strength obtained from cylindrical tests most closely approximates the true uniaxial compressive strength. If test results are only available from cubic test specimens, then an approximate conversion factor of 0.8 can be applied. It should, however, be noted that this is only an approximation and sufficient data is not available to confirm its accuracy for all types of concrete and across all concrete strengths.

2.1.2 Concrete workability

Critically, the required concrete compressive strength will only be achieved if the concrete is adequately placed and compacted. That is, for a design to achieve the concrete material properties assumed within it, it is essential that the concrete is sufficiently workable. The workability of concrete cannot be easily measured directly; instead, it is approximated, most commonly using a slump test. In the slump test, a frustum is placed on a flat level surface and filled with concrete which is properly compacted by rodding. Immediately after compaction, the frustum is lifted and the change in height or 'slump' of the concrete is measured relative to its original height, as shown in Figure 2.3a.

Figure 2.3 Concrete workability: (a) slump test, and (b) flow table test. (Photos courtesy of Lloyd Dadd.)

Generally, the concrete contractor will specify workability based on the application, and national standards such as ACI 211–91 (2002) provide guidance on typical concrete slump measures suitable for different applications. For example, for typical reinforced concrete beams, walls and columns, it is recommended that a concrete slump of between 25 and 100 mm be used to allow it to adequately flow and be compacted around reinforcement. Conversely, for mass concrete without significant reinforcement, the suggested slump is between 25and 50 mm.

Special classes of concrete can also be specified on the basis of workability. No-slump concrete, as the name suggests, has very little to no slump (0–25 mm) and is placed with the assistance of vibrating plates and rollers. No-slump concrete is most commonly used in precast applications or in the construction of rigid pavements where the ability to increase the rate of production by rapidly removing formwork is beneficial.

On the other end of the spectrum, 'self-compacting concrete' flows and compacts under its own weight and does not require additional vibration. Self-compacting concrete is, therefore, most commonly applied in large pours such as slabs in commercial and industrial buildings, where its self-compacting nature can improve constructability. For highly flowable concretes, such as self-compacting concrete, the slump test does not yield an adequate measure of workability. Instead, the flow table test is commonly used. In this test, a frustum of concrete is compacted in the centre of a hinged platform. The frustum is then lifted, allowing the concrete to spread before the flow table is raised and dropped. The spread of the concrete is subsequently measured, as shown in Figure 2.3b, providing an indication of workability.

2.1.3 Concrete unit weight

Concrete can also be specified on the basis of unit weight or density (ρ); for example,

- Lightweight concrete typically has a mass in the order of 300–2200 kg/m^3. This class of concrete is typically of low compressive strength and is, therefore, applied in scenarios where minimising weight is more important than strength, such as in the slabs of high-rise structures or in bridge decks. Lightweight concrete is also commonly applied in the manufacture of masonry units for use in a range of structural and non-structural applications.
- Normal-weight concrete is the most commonly specified weight class and has a unit weight of between ~2200 and ~2600 kg/m^3.
- Finally, heavyweight concrete has a mass of more than 2600 kg/m^3 and unit masses of up to 8900 kg/m^3 can be achieved by specifying steel or lead shot as aggregate. Heavyweight concrete has limited application and is typically specified as ballast or in specialist applications such as radiation shielding.

2.1.4 Fibre-reinforced concrete

A final class of special concretes is fibre-reinforced concrete (*FRC*), achieved by adding between 0.5% and 6% by volume of either synthetic or steel fibres to the concrete. Originally, fibres were added in relatively small volumes to NSC or HSC for crack control. However, high volumes (typically up to 3%) of fibres can be incorporated into self-compacting *UHPC* mortars that do not contain coarse aggregates. This yields ultra-high-performance FRC (*UHPFRC*) that display strain hardening behaviour in tension and significant post-cracking tensile strength and ductility.

2.2 COMPONENTS OF CONCRETE

The batching of the components of concrete (aggregates, binders, water, admixtures and fibres) to form a concrete mix that achieves a set of performance criteria is a complex task which itself constitutes a dedicated profession – the concrete technologist. However, as a structural engineer, to understand the performance of concrete at member or structural level, it is important to have a basic understanding of the components of concrete and how they can be combined to manipulate required design properties.

2.2.1 Aggregates

In Figure 2.1, the skeletal structure of concrete consists of aggregate particles which occupy approximately 60%–75% of the volume of concrete and make up between 70% and 85% of its mass (Neville and Brooks 2010). Aggregates are produced in a range of different sizes; their distribution is termed the grading, commonly expressed using a grading curve (Figure 2.4) that describes the fraction of material passing through sieves with a fixed aperture size. Aggregates are most broadly classified as either coarse or fine

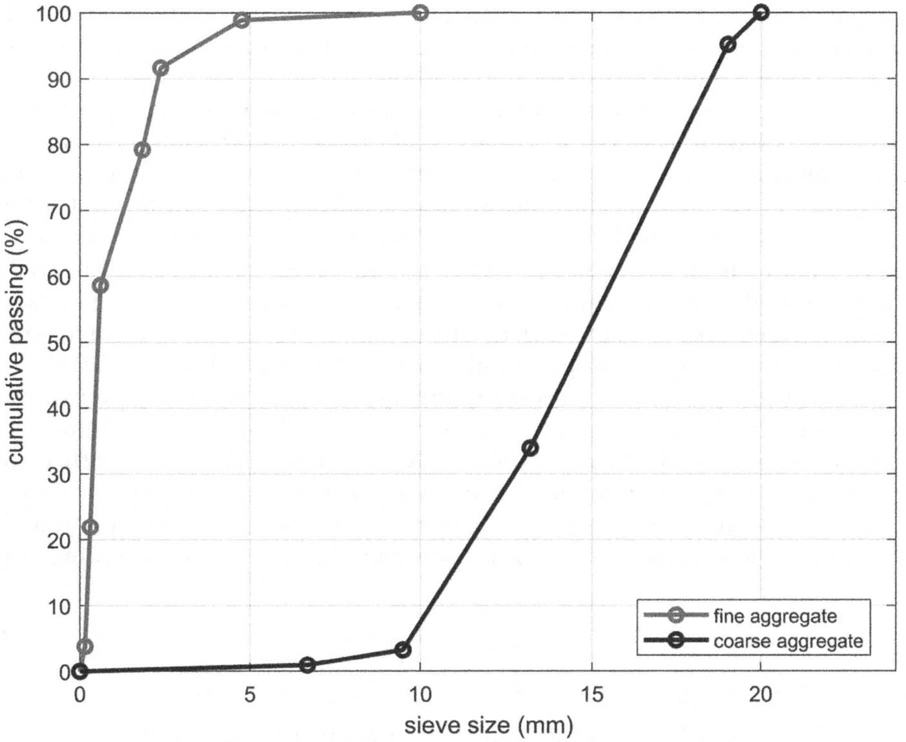

Figure 2.4 Typical aggregate grading curve.

depending on their size, with coarse aggregate defined as material greater than 4.75 mm and fine aggregates as those less than 4.75 mm.

The specific aggregate grading, as shown in Figure 2.4, is an important factor in determining concrete material properties as it heavily influences the degree of packing of the aggregate particles, with smaller particles filling the space between larger particles. This degree of packing is important as the space between aggregate particles must be filled by cement paste, which is typically weaker and more porous than aggregate. Consequently, by improving the packing of aggregate particles, it is possible both to improve concrete compressive strength and to minimise behaviours such as concrete shrinkage which occur as a result of moisture exchange between the concrete and the surrounding environment. From the perspective of the concrete producer, minimising the volume of cement paste within the concrete is also important from a financial perspective as cement is typically significantly more expensive than aggregates.

In a similar way to concrete, aggregates are also classified based on their bulk density, into either lightweight (less than 1200 kg/m^3), normal weight (2200–2450 kg/m^3), or heavyweight (up to 4000 kg/m^3). The vast majority of concrete is batched using normal weight aggregates, which are sourced either by mining and crushing sedimentary or igneous rocks, or by mining rocks that are already broken up by environmental actions such as erosion. Lightweight aggregates are typically highly porous and can be derived either from natural sources, such as scoria or pumice, or from manufactured sources, such as pulverised fuel ash. The high porosity of lightweight aggregates typically means they have low strength and are therefore applied only in the production of lightweight concrete. Heavyweight aggregates, such as steel or lead shot, are typically only used in scenarios where heavyweight concrete is sought, as otherwise they simply add to the cost of material transport and handling.

Aggregates are further classified based on their shape, which ranges from rounded to elongated, as shown in Figure 2.5. The classification of aggregate shape is generally based on appearance and is important in mix design because particle shape impacts the properties of concrete in both the wet and hardened states. In the wet state, the shape of aggregates impacts their ease of movement relative to each other, meaning that concrete produced using rounded aggregates has improved workability compared with concrete produced using angular aggregates. The shape of aggregates also impacts the maximum packing density of the skeletal structure and, hence, like aggregate grading, it can impact the mechanical properties of concrete (Neville and Brooks 2010).

The strength of aggregates is generally not explicitly considered because the parent material from which aggregates are derived is typically significantly stronger than the mortar that surrounds them. Consequently, failure typically occurs either around aggregates or through the interfacial transition zone (*ITZ*), which is the softer and more porous boundary between the aggregate particles and the mortar that holds them together. An example of the *ITZ* is shown in Figure 2.6 using scanning electron

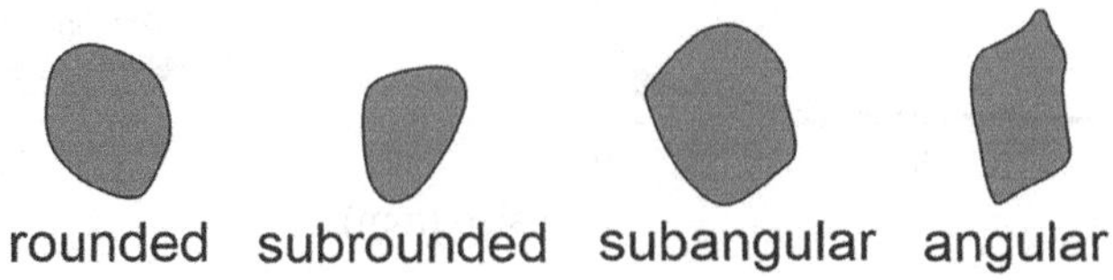

Figure 2.5 Aggregate shape.

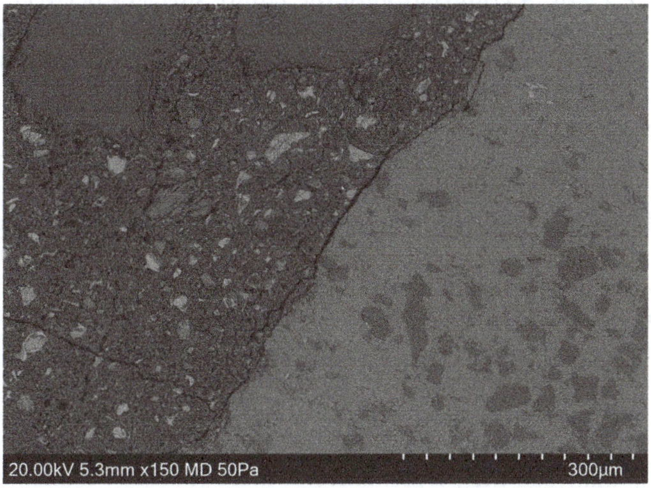

Figure 2.6 Interfacial transition zone (*ITZ*) between aggregate and the mortar phase. (Photo courtesy of Lloyd Dadd.)

(a)

(b)

Figure 2.7 Direct shear failure: (a) concrete around aggregates; (b) concrete through aggregates. (Photo courtesy of Lloyd Dadd.)

microscopy. The exception to this is for very HSC or concrete with weak aggregates, in which case cracks may form through the aggregates. This is demonstrated in Figure 2.7, which compares the failure surface of NSC made with basalt aggregates tested under direct shear (Figure 2.7a) with that of concrete produced with softer recycled concrete aggregates (Figure 2.7b). A comparison of the failure surfaces between these two tests shows that when failure occurs through the aggregates, there is generally a smoother failure plane, and this may impact aggregate interlock, as explained in Figure 1.3.

Sourcing natural aggregates has a significant environmental impact, both through the utilisation of natural resources and the destruction of natural ecosystems. Given the demand for concrete and the broader uses for aggregate in both civil works and road construction, easily accessible, low-cost sources of aggregates have been largely utilised. Globally, there is either already a shortage of natural aggregates or one will soon emerge (Přikryl 2021). The low availability of natural aggregates coupled with an abundance of waste materials has recently led to extensive research into the potential to use alternatives to natural aggregates, such as construction and demolition waste (crushed concrete and masonry); crumb rubber sourced from tyres; and shredded plastic sourced from municipal waste.

Significant work has been devoted to developing concrete mix designs using alternative fillers that achieve a specific compressive strength, as demonstrated by Shobeiri et al. (2024), who analysed the performance of more than 5300 concrete mix designs. However, further work is required to transition these alternative aggregates from laboratory materials into practice. Being derived from wastes, there is often greater variability in mechanical and durability properties, and their impact on mechanical performance has not yet been quantified to the degree to which performance can be reliably predicted. There is also a significant need to consider how the resulting change in concrete material properties impacts the mechanisms that drive member-level performance described in Chapter 1. The cascade of implications from aggregate performance to concrete performance to member and structure performance is critical to ensuring that these alternative aggregates do in fact yield an environmental benefit, and this will be discussed in more detail in Part 6.

2.2.2 Binders

It is not the purpose of this book to provide a detailed description of cement chemistry, as such detail is arguably better left to a chemical engineer or materials scientist. Instead, here, a basic overview of the role of concrete binders is provided, to the extent required for understanding the remainder of the topics presented in this book; the reader is referred to dedicated texts such as Hewlett and Liska (2019) and Taylor (1997) for an in-depth treatment.

The primary binder in concrete has traditionally been Ordinary Portland Cement (*OPC*), but to modify the performance of the concrete, reduce costs and increasingly lessen its environmental impact, alternative binders are increasingly being specified. Alternative binders can broadly be split into two categories: those that are utilised in conjunction with *OPC* as supplementary cementitious materials (*SCMs*), and those that are used as primary binders in the absence of *OPC*, typically as part of alkali-activated or geopolymer concretes. Although the materials used as *SCMs* and binders in geopolymers are the same (typically ground granulated blast furnace slag (*GGBFS*) and fly ash (FA)), the underlying reaction process is different.

Significantly, when considering concretes containing SCMs, although these materials may be classified as being the same (and share the same name), their underlying chemical composition can vary significantly. Figure 2.8 illustrates this, plotting a ternary diagram of the relative proportions of reactive compounds typically identified as

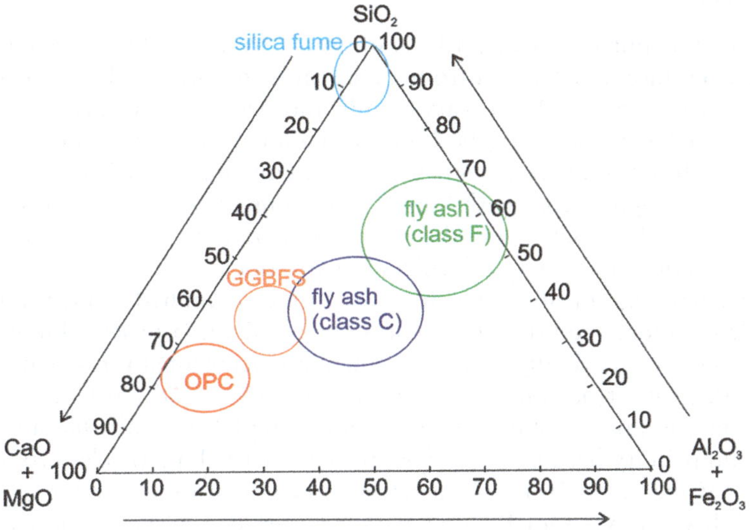

Figure 2.8 Variation in major reactive oxides in *OPC* and *SCMs*. (Modified from Xie and Visintin 2018.)

important in a concrete binder. These compounds, shown on the axes, include calcium oxide (CaO), magnesium oxide (MgO), aluminium oxide (Al_2O_3), iron oxide (Fe_2O_3) and silicon dioxide (SiO_2), for a range of different *SCMs* (FA and *GGBFS*) as well as for *OPC* (Xie and Visintin 2018).

Figure 2.8 highlights that manufactured binders such as *OPC* have a very narrow range of chemical composition; therefore, their specific impact on concrete performance can be reliably predicted. In contrast, binders produced from naturally occurring materials, such as FA, have a wide range of chemical composition despite being classified as the same material. Let us now consider how *OPC* and the most common types of *SCM* can be used in concrete production.

2.2.2.1 *Ordinary Portland Cement*

OPC is the most widely accessible and widely used binder. It is produced using limestone, clay and sand to provide sources of calcium oxide (CaO), silica (SiO_2), alumina (Al_2O_3) and iron oxide (Fe_2O_3). These materials are heated to between 1400°C and 1600°C, at which point a series of chemical reactions occur through a process known as calcination, resulting in the production of the mineral known as clinker. This clinker is then cooled and ground to achieve an appropriate particle size distribution, becoming known as *OPC*. The *OPC* can be blended with other materials, such as limestone and slags, to modify the properties of the cement or to reduce the quantity of clinker required to produce a given mass of cement. Although the sustainability of concrete as a material will be explored in more detail in Chapter 13, it is worth noting here that the significant environmental impact of concrete stems from the firing of the kiln and the chemical reactions that occur during clinker formation. Methods for minimising the carbon footprint of concrete are therefore largely split between increasing the efficiency of cement kilns and minimising the use of *OPC*.

2.2.2.2 Fly ash

FA, also known as pulverised fuel ash, is obtained from the flue gases emitted when burning coal. Because it is sourced from a natural material, the chemical composition and therefore reactivity of FA can vary significantly as shown in Figure 2.8 (Xie and Visintin 2018). FA can, however, be broadly classified into two major classes: Class F, derived from bituminous coal with high silica content, and Class C, derived from sub-bituminous or lignite coal with higher CaO content. As a result of its high CaO content, Class C FA has some hydraulic reactivity (reactivity with water) in addition to pozzolanic reactivity (reactivity with calcium hydroxide).

FA is a pozzolan high in silica, which reacts with calcium hydroxide produced when cement hardens to form additional cementing materials, in a process known as a pozzolanic reaction. FA is generally used as a cement replacement at rates of approximately 30% and at this rate it has an impact on both the fresh and hardened properties of concrete. In the fresh or wet state, concrete containing FA generally has a higher workability and pumpability because the spherical shape of the FA particles provides a lubricating effect (Wang et al. 2003). As a result, mixes require less water to achieve the same workability, allowing for higher concrete strengths (because concrete strength is inversely proportional to the water-to-cement ratio ($w{:}c$)). Replacing hydraulic cement with pozzolanic FA changes the rate of reaction, but this does not result in an overall reduction in strength, and the change depends on the class of FA used. As a result, mixes with Class F FA will generally have lower early-age strengths but higher late-age strengths than mixes with only cement, and mixes with Class C FA will have increased early and late age strengths.

The addition of FA also has a significant effect on both time-dependent material properties and concrete durability, which is crucial for quantifying service-life (Razmi et al. 2025). The combined impact of reducing water content and the production of additional reaction products reduces concrete permeability and, therefore, reduces time-dependent behaviours such as shrinkage and creep which are dependent on moisture exchange with the environment (Neville and Brooks 2010).

2.2.2.3 Ground granulated blast furnace slag

GGBFS is a by-product of the iron making process. Slag is removed from molten iron and is cooled by jets of water to form a granular material with a particle size similar to that of fine aggregate. The granulated material is then ground to a fineness similar to *OPC*. *GGBFS* can then be either blended with *OPC* prior to packaging or added at the batching plant to act as an *SCM*.

The chemical composition of *GGBFS*, as shown in Figure 2.8 (Xie and Visintin 2018), is high in calcium and silica. Consequently, it possesses latent hydraulic properties, allowing it to react upon contact with water, although generally at a rate not practical for use as a primary binder in concrete. In an alkaline environment, such as that produced in *OPC*, the hydration of *GGBFS* is accelerated, and hence it can be used in high volumes as a cement replacement where early-age strength is not required.

GGBFS can be used at cement replacement ratios of up to 90% but it is most commonly used at levels between 20% and 40%. The addition of *GGBFS* to concrete mixes impacts both the fresh and hardened properties, similarly to FA. In the fresh state, the most significant impact is on the rate of reaction and therefore the setting time;

high volumes of *GGBFS* are typically used to delay hydration. In the hardened state, the development of concrete compressive strength is generally delayed because of the reduced cement content, but final compressive strengths remain unaffected (Topçu 2013). Similarly to *FA*, the addition of *GGBFS* significantly reduces permeability, leading to an improvement in concrete durability. For this reason, it is often blended into bagged cement products such as sulphate-resisting cements.

2.2.2.4 Silica fume

Silica fume (*SF*), also known as micro silica, is derived from silicon and ferrosilicon smelting. The material is very high in silica (see Figure 2.8) and has a particle size distribution approximately 100 times smaller than that of cement. As a result of its chemical composition and very small particle size, SF is a highly reactive pozzolan, following a general reaction mechanism similar to that of FA. The very fine particle size also allows SF to fill the voids between cement particles. This effect is particularly apparent in the ITZ between aggregate and cement particles, which is often more porous than the surrounding concrete.

As a result of its high reactivity, SF is generally added to concrete mixes at 10% or less by mass (Xie and Visintin 2018). In the fresh state, the addition of SF leads to a reduction in concrete workability. Consequently, mixes containing SF generally include admixtures to improve workability (as will be discussed in Section 2.2.3). In the hardened state, the pozzolanic nature of SF and its ability to increase the density of concrete both contribute to increased concrete compressive strength (Siddique 2011). SF is, therefore, an essential component in the design of UHPC*s* with compressive strengths well above 100 MPa.

SF is also highly beneficial to concrete durability, as its reduced pore size and decreased permeability lead to a reduction in concrete shrinkage and creep by minimising moisture exchange with the environment (Siddique 2011). This reduction in permeability also reduces the potential for the concrete to react with deleterious chemicals.

2.2.2.5 Geopolymer binders

When considering concrete materials, it is no longer possible to consider only *OPC* based concretes. As a result of the push to minimise the environmental impact of concrete, binders that are not based on *OPC* are increasingly being considered. In the past two decades, the vast majority of research into alternative binders has focused on alkali-activated concretes, with geopolymer concretes being a specific class. The distinction between geopolymer concretes and alkali-activated concretes, although important in terms of chemistry, is not important in terms of civil engineering and, hence, the term geopolymer is generally taken as a catch-all term (Xie et al. 2020).

Geopolymers are produced by reacting an aluminosilicate source, typically *FA* or *GGBFS*, with a very concentrated alkali solution known as an activator. The activator is typically a mix of sodium hydroxide and sodium silicates. As a result of this reaction, an alkali aluminosilicate gel is produced, which acts as the binder in concrete, much like calcium silicate hydrate gel acts as the binder in *OPC* based concretes.

Significant research has been conducted to develop geopolymer concrete mix designs, and there is growing research into material and member performance, but with the exception of a small number of structures, geopolymer concrete usage to date has largely been confined to the lab. Importantly for this book, research has shown that the same strain based and partial-interaction mechanisms that control the behaviour of *OPC* based concrete can be applied to geopolymer concretes (e.g. Visintin et al. 2017).

2.2.3 Admixtures

When preparing high-performance concretes or in situations that are non-standard, the properties of concrete can be manipulated by adding chemical admixtures. These admixtures are proprietary and there are often many forms of the same class of admixture available. It is generally recommended that when using admixtures, tests are undertaken to fully quantify the impact of their use on material properties, particularly where multiple admixtures may be used in combination. Here, a basic introduction to the specific classes of admixtures and their use is provided. For a detailed description of their chemical composition and the mechanism of action, the reader is referred to concrete technology texts such as Neville and Brooks (2010) or directly to the admixture manufacturer product documentation.

Plasticisers also known as **water-reducing agents** reduce the mixing water required to achieve a given slump by approximately 10%. Water-reducing agents are important because concrete strength is inversely proportional to the ratio of *w:c* in the concrete mix design. That is, lower *w:c* produce higher concrete strengths, but have lower workability. Plasticisers can, therefore, be used either to achieve higher concrete strengths by maintaining workable mix designs with a lower *w:c* or to ensure the concrete has adequate workability for placement in regions of congested reinforcement. The importance of maintaining workability and therefore compaction is shown in Figure 2.9, where the cross-section of concrete with poor workability (and therefore compaction) in Figure 2.9a can be compared to that with good workability and compaction in

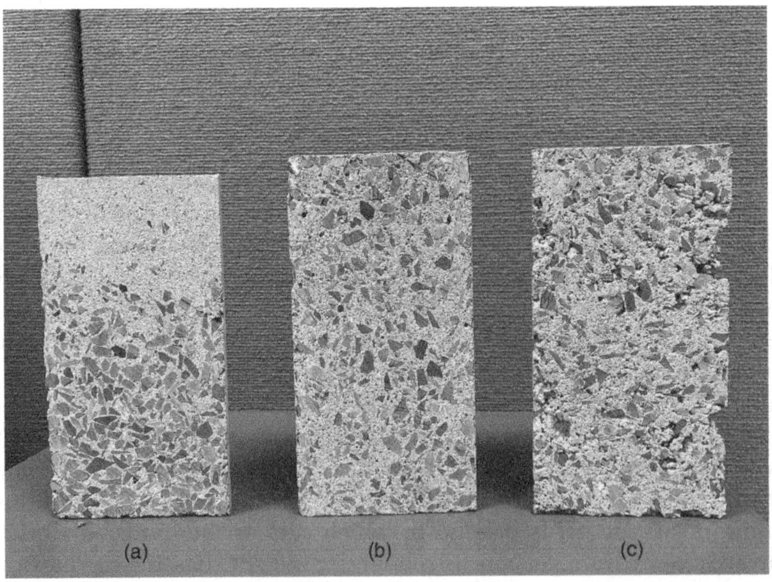

Figure 2.9 Impact of compaction on hardened concrete: (a) under-compacted (honeycombed); (b) well-compacted; (c) over-compacted (segregated). (Photo courtesy of Lloyd Dadd.)

Figure 2.9b. The concrete with poor compaction can be observed to have a high porosity, termed 'honeycombed', and will display poor strength and durability compared to well-compacted concrete.

Superplasticisers also known as *high-range water-reducing agents* are similar to plasticisers but allow a significantly higher reduction in mixing water (of up to 35%). They are, therefore, important in the production of HSC and UHPCs, where *w:c* may be approximately 0.2, at which point the concrete has insufficient workability to be compacted. Superplasticisers are also essential in the production of self-compacting concrete and ensure flowability without segregation (separation of the mortar from the aggregates). The impact of segregation is shown in Figure 2.9c, where the dense aggregates are observed to have separated from the mortar and have sunk to the bottom of the prism resulting in non-homogeneous material properties.

Retarders delay the initial set of concrete, providing a longer period of workability. This can be important in scenarios such as hot weather concreting, where setting time is reduced or when long travel times are required between the concrete batching plant and the location of placement. Although retarders can delay the early strength gain of concrete, later strength gain occurs at a faster rate and so the final strength is not significantly different. Retarders can often be found blended with superplasticisers such that the admixture has both a plasticising and retarding effect.

Accelerating admixtures increase the early strength gain of concrete and are, therefore, useful when setting is delayed by cold weather, which would otherwise postpone concrete finishing. Accelerating admixtures are not intended to reduce setting time and are distinct from *set-accelerating admixtures*, which are intended to significantly reduce setting time by inducing a phenomenon known as flash setting. Such admixtures are essential for activities such as extruding mortars in 3D printing.

Shrinkage-reducing agents are added to concrete to reduce shrinkage. These can be further categorised by their mechanism of action; for example, they can be expansive agents that compensate for shrinkage by causing the concrete to initially expand. Alternatively, shrinkage-reducing agents can be surfactants that lower the surface tension between pores and the water that fills them. As a result, the capillary tension that forms within the pores as water evaporates is minimised, and shrinkage is reduced. Further information on the mechanisms of concrete shrinkage and its relationship to capillary tension is provided in Section 2.5.1.

2.2.4 Fibre reinforcement

Fibre reinforcement, when added to concrete, acts as discrete, unanchored reinforcement, with each fibre having a random embedment length and orientation relative to a cracked plane. A fibre transfers stress across the crack as a result of partial-interaction load-slip behaviour until it is either pulled out or ruptured. Depending on the volume of fibres (V_f) and their dispersion throughout the concrete matrix, fibres can either have a relatively minor impact on concrete material properties or significantly increase them by transferring substantial stress across a crack. It is also worth noting that fibres can substantially increase the compressive strength and ductility of the concrete.

As a result of the random orientation, embedment length and local density of fibres within the concrete, material properties are generally derived at the concrete material scale. It is, however, possible to simulate the behaviour at the single-fibre level and then apply statistical approaches to randomly distribute fibres, allowing material models to be built up from the sub-concrete scale (e.g. Li et al. 1991). Although this chapter and book will focus on the quantification of material properties at the concrete scale, it is important to understand the behaviour of a single fibre, since manipulating the properties of an individual fibre directly affects those at the concrete scale.

Fibre reinforcement can broadly be classified by material type. In practice, fibres are generally either synthetic (e.g. polypropylene) or metallic, but within the research community, the use of natural fibres such as coir, flax or wood pulp is being considered as a more sustainable alternative that may utilise a waste material.

Fibres may be further classified based on shape and aspect ratio. It is common practice to relate the performance of a fibre to a reinforcing index (RI), where

$$RI = V_f\left(\frac{l_f}{d_f}\right) \tag{2.1}$$

and in which V_f is the fibre volume fraction within the concrete, l_f is the fibre length and d_f is the fibre diameter, where l_f/d_f is the fibre aspect ratio (Neves and Fernandes de Almerida 2005). The RI has been extensively used to benchmark the performance of different FRCs, as it captures the total amount of fibres in the mix (and therefore crossing a cracked plane) by considering both the fibre volume fraction and the aspect ratio. This is important as the product of these two variables is often assumed to approximate the number of fibres that would cross the crack. Further, research has shown that the ability to distribute fibres uniformly within a concrete mix is highly dependent on both the fibre volume and the fibre aspect ratio, as larger fibre volumes or higher aspect ratios are more likely to cause fibre balling and poor concrete compaction (Le Hoang and Fehling 2017).

Importantly, the behaviour of a single fibre within a mass of concrete can be simulated using the same tension stiffening partial-interaction mechanics used to describe reinforcement pull-out, as described in Section 1.4.1, because the force developed within the fibre is a function of the local bond between the fibre and the surrounding concrete. Typical fibre shapes are shown in Figure 2.10, which shows that hooks, enlargements or changes in surface texture have been added to increase the bond between the fibre and the concrete, thereby resulting in a larger force transferred across a given crack width. While increasing the bond through the addition of anchorage is beneficial, fibre rupture is a less ductile failure mechanism than fibre pull-out; hence, there is an upper limit to the benefit of anchorage.

While it is theoretically possible to add fibres in very high volume fractions, in practice fibre volumes are typically between 0.5% and 1.5% for concretes containing coarse aggregate and up to ~3.5% for concretes such as *UHPFRC* that do not contain coarse aggregate. At higher volume fractions, the benefits of fibres are not necessarily realised

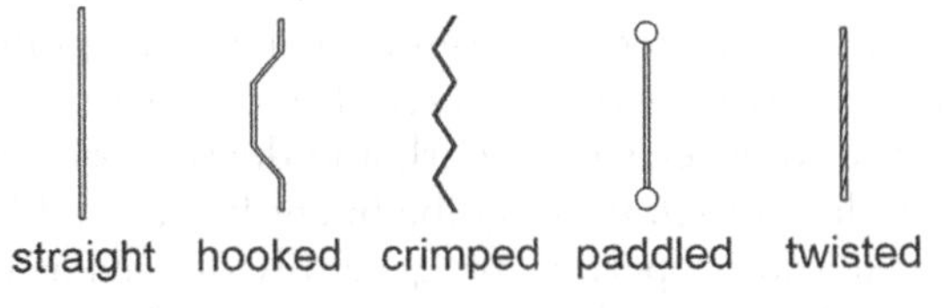

Figure 2.10 Fibre reinforcement.

because they tend to ball within the mix, leading to both poor fibre distribution and poor concrete compaction.

2.3 CONCRETE IN COMPRESSION

In this section, we will consider the properties of concrete in compression when subjected to instantaneous loading. First, we will consider the most general way of describing the behaviour under compressive load by examining the unconfined axial stress–strain ($\sigma_{ax}/\varepsilon_{ax}$) relationship. Having examined the complete stress–strain relationship, the key parameters used in design, including the elastic modulus (E_c), concrete compressive strength (f_c), the strain corresponding to f_c (ε_0) and Poisson's ratio (υ), are then examined in more detail, followed by a general discussion of common approaches used to estimate them. Finally, the impact of confining concrete on improving material properties is discussed from the perspective of strain-based methods.

2.3.1 Complete stress–strain relationship

Let us first consider the complete axial stress–strain ($\sigma_{ax}/\varepsilon_{ax}$) relationship in Figure 2.11a, which is typical of NSC and HSC. The complete $\sigma_{ax}/\varepsilon_{ax}$ relationship is normally obtained by testing cylindrical specimens with an aspect ratio of 2:1, as shown in Figure 1.19b and Figure 2.2, to minimise restraint from the loading platens, thus allowing the obtained material properties to approximate those for unconfined concrete.

When considering Figure 2.11, it is important to remember and emphasise that concrete compressive failure – specifically, the softening or descending branch – is a mechanism as illustrated in Figures 1.7, 1.9b, 1.17, 1.20 and 1.27b. Therefore, although the falling branch in Figure 2.11a, beyond the strain ε_o, can be expressed as a strain-based property, this is not the case. Rather, concrete softening is caused by a *PI* mechanism governed by the *PI* material properties in Chapter 3, and the falling branch of the $\sigma_{ax}/\varepsilon_{ax}$ is more correctly defined as an effective stress–strain relationship. This is very important because, as already explained in Section 1.4.3, the falling branch, such as those shown in Figure 2.11, is a size-dependent property, depending on the length of the specimen tested, unlike strain-based properties, such as the rising branch, which are not size dependent. These effective strain-based properties of the falling branch

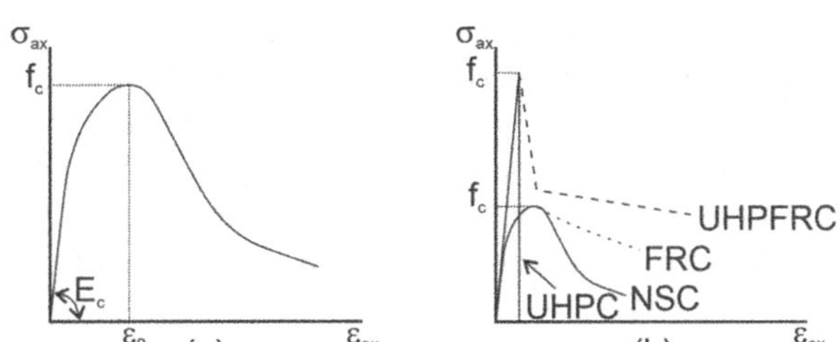

Figure 2.11 Generic shape of axial stress–strain relationship: (a) *NSC* and *HSC* concrete without fibres; (b) stress–strain relationship of *NSC FRC* and *UHPFRC* compared to *NSC* and *UHPC* without fibres.

are included in this chapter on strain-based properties as they are used in Chapter 4 to construct the true *PI* size-dependent properties.

For NSC and HSC, as shown in Figure 2.11a, the complete concrete $\sigma_{ax}/\varepsilon_{ax}$ relationship has commonly been approximated using the continuous relationship first proposed by Popovics (1973):

$$\sigma_{ax} = f_c \frac{n\left(\frac{\varepsilon_{ax}}{\varepsilon_0}\right)}{(n-1)+\left(\frac{\varepsilon_{ax}}{\varepsilon_0}\right)^n} \tag{2.2}$$

in which ε_0 is the strain corresponding to the concrete compressive strength f_c in Figure 2.11, and which, in the absence of specific test results, can be approximated by the following general expression derived by Chen et al. (2014) based on fitting experimental data:

$$\varepsilon_0 = 4.76\times10^{-6} f_c + 2.13\times10^{-3} \tag{2.3}$$

where the concrete compressive strength is in units of MPa, and where n is the ductility factor that heavily influences the shape of the descending branch and is given by

$$n = \frac{E_c}{E_c - \frac{f_c}{\varepsilon_0}} \tag{2.4}$$

and is seen to be related to the difference between the concrete elastic modulus E_c and the concrete's secant modulus at the peak strength (f_c/ε_0).

In Figure 2.11b, the $\sigma_{ax}/\varepsilon_{ax}$ relationships of concrete without fibres (*NSC* and *UHPC*) are compared with those of concrete with fibres (*FRC* and *UHPFRC*). In this comparison, it can be seen that for the *NSC* and *UHPC*), the addition of fibres has little impact on the concrete elastic modulus and compressive strength but does increase concrete ductility. Similar behaviour occurs when comparing *UHPC* to *UHPFRC*, where for concrete without fibres (*UHPC*), the behaviour is essentially elastic–brittle, but for concrete with fibres (*UHPFRC*), there is significant residual strength and ductility once concrete softening occurs.

Generalised relationships to describe the $\sigma_{ax}/\varepsilon_{ax}$ relationship for *FRC* have not yet been developed to the same extent as for concrete without fibres. That is, while several models have been proposed by incorporating the *RI* from Eq. 2.1 into Popovics' equation (Eq. 2.2), these models have generally been calibrated to very small datasets and generally only to a single fibre type (Ezeldin and Balaguru 1992; Someh and Saeki 1996; Mansur et al. 1999; Nataraja et al. 1999; Bhargava et al. 2006; Ou et al. 2012; Lee et al. 2015). It is, therefore, suggested in design standards and model codes such as (Standards Australia 2018) that models used to predict the $\sigma_{ax}/\varepsilon_{ax}$ relationship of concretes without fibres should be applied to concretes with fibres.

As an example of the impact of fibres on the $\sigma_{ax}/\varepsilon_{ax}$ relationship, let us consider the model of Ou et al. (2012) which has been calibrated for concrete with compressive strengths between 32 and 48 MPa, fibre contents between 0% and 3.4%, and hooked fibre aspect ratios between 50 and 100. Within this calibration range Ou et al. (2012) were able to modify Popovics' expression in Eq.2.2 by replacing the strain at peak stress ε_0 with

$$\varepsilon_{0-FRC} = (\varepsilon_0) + 0.0007RI \quad (2.5)$$

and the ductility factor n with

$$n_{FRC} = 0.71RI^2 - 2RI + 3.05 \quad (2.6)$$

In Eqs. 2.5 and 2.6, the RI is derived from Eq. 2.1.

Figure 2.12 presents the results of a parametric study investigating the impact of f_c and RI on the $\sigma_{ax}/\varepsilon_{ax}$ relationship. In Figure 2.12a, f_c is set to 32 MPa and the RI is 0, 0.75, 1.5 or 3. In Figure 2.12b, the analysis is repeated for 48 MPa to cover the range of the calibration. The results of this parametric study show that increases in f_c and RI have opposing effects on ductility when RI is less than 3. That is, as f_c increases, ductility decreases, and as RI increases, ductility increases. This is to be expected as the fibres act as reinforcement crossing the sliding planes that form once concrete softening commences (Figure 1.17), and hence, increasing RI is equivalent to increasing the reinforcement crossing these planes. When the fibre content is very high such that RI is 3, the strain at peak stress increases considerably and the ductility of the concrete decreases. This behaviour occurs because very high quantities of fibre reinforcement can lead to fibre balling, which results in poorly compacted, honeycombed concrete, such as that shown in Figure 2.9.

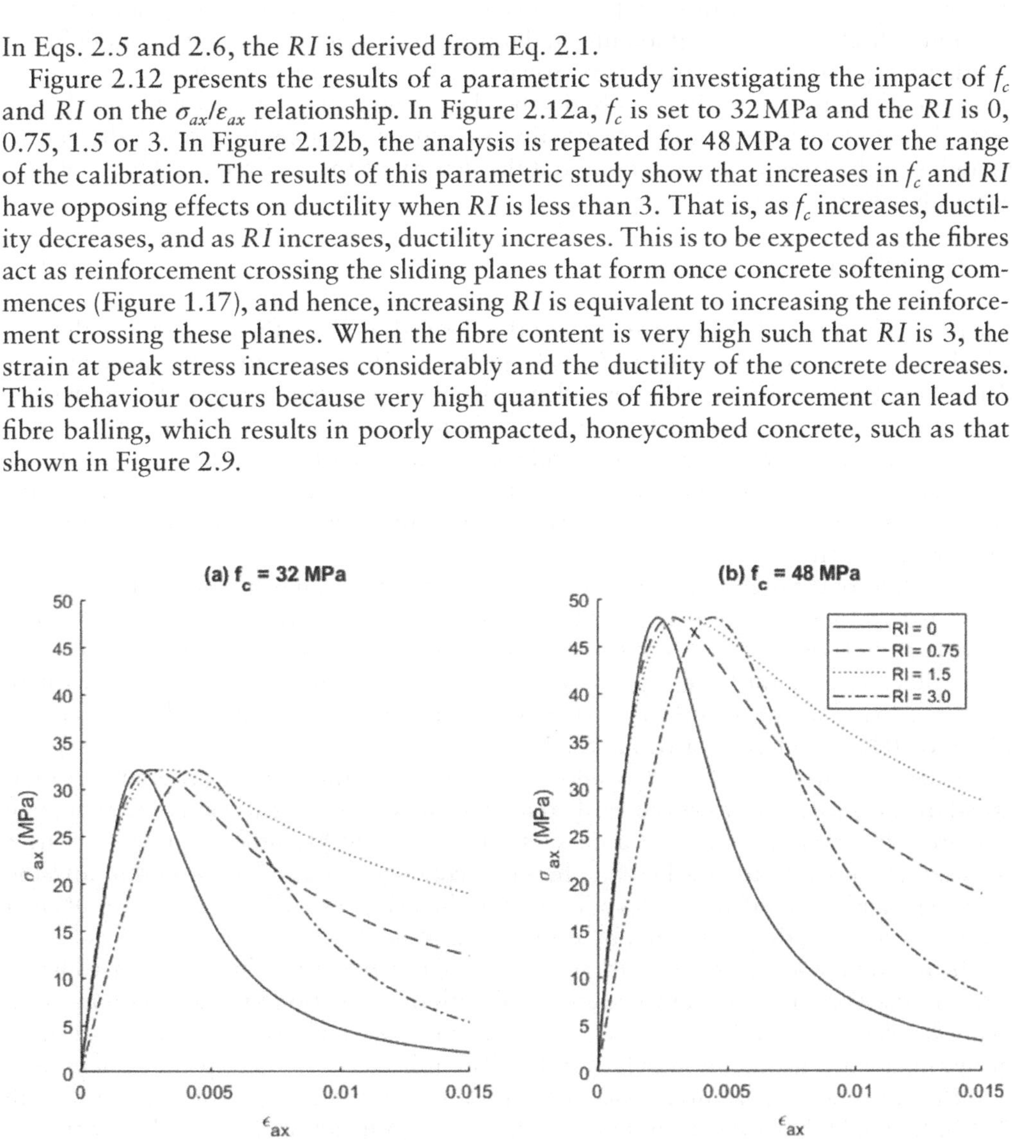

Figure 2.12 Impact of fibre reinforcement on $\sigma_{ax}/\varepsilon_{ax}$ relationship: (a) of f_c and RI; (b) of coarse aggregate and; (c) of hooked-end fibres.

It should be noted that all these results were from a limited dataset and have only been used to illustrate the effect of fibres and not as a design tool.

Let us now consider each of the key aspects of the concrete stress–strain relationships in Figure 2.11 in more detail.

2.3.2 Elastic modulus

Concrete, being a brittle material, behaves non-linearly, and although not significant under low loads, some permanent set develops upon unloading. As a result of this hysteresis, it is common to calculate the elastic modulus of concrete by subjecting a concrete cylinder to a series of loading and unloading cycles, and taking the modulus of elasticity, E_c, as the secant stiffness between the cycle peak and trough.

As an alternative to testing to obtain the elastic modulus, national reinforced concrete design standards provide empirically derived expressions that relate the elastic modulus to the compressive strength and the density of the concrete, ρ. These models generally take the form:

$$E_c = a\rho^b \left(f_c'\right)^c \tag{2.7}$$

where coefficients a, b and c are empirically derived, and coefficient c is less than 1, which limits the rate of increase in E_c as f_c' increases. The reason for the dependency on density (which is equivalent to a dependency on aggregate type) relates to the composite nature of the concrete such that, when determining E_c, concrete can be broadly considered a composite of stiffer aggregates and softer cement paste. Hence, as the density of the concrete increases through the addition of greater quantities of aggregate, so too does the modulus of elasticity; similarly, as the stiffness of the aggregate increases, so too does the stiffness of the concrete.

Regardless of the national design code used to estimate E_c, it is important to note that these material models have significant scatter, and as a result, codes such as AS3600 provide bounds of ±20%. It is important to note that the elastic modulus also needs to be modified to account for creep occurring under long-term or sustained loads, where sustained loads lasting as little as a day may result in a reduction in the elastic modulus of up to 20% (Standards Australia 2022).

To demonstrate the range of results that can be obtained from different national standards, Figure 2.13 shows the predicted elastic modulus as a function of the concrete compressive strength, assuming a concrete density of 2400 kg/m^3. Significant scatter is observed between approaches for the same concrete compressive strength, and the change in aggregate stiffness incorporated into the FIB approach (Fib 2013) results in a scatter larger than that suggested by AS3600 (±20%).

The instantaneous elastic modulus increases slightly over time due to increased hardening of the cement paste (not to be confused with the reduction in elastic modulus that occurs under sustained load due to creep). This increase in elastic modulus over time is generally defined using a power law and is also influenced by the type of cement used and the ambient temperature, which affects the rate at which the cement hardens (Fib 2013). It should, however, be noted that the most significant changes in E_c occur over the first few years and that the impact of changes in temperature within a reasonable working range is small. For detailed modelling approaches of the changes in E_c with time, temperature and cement type, the reader is referred to Fib (2013).

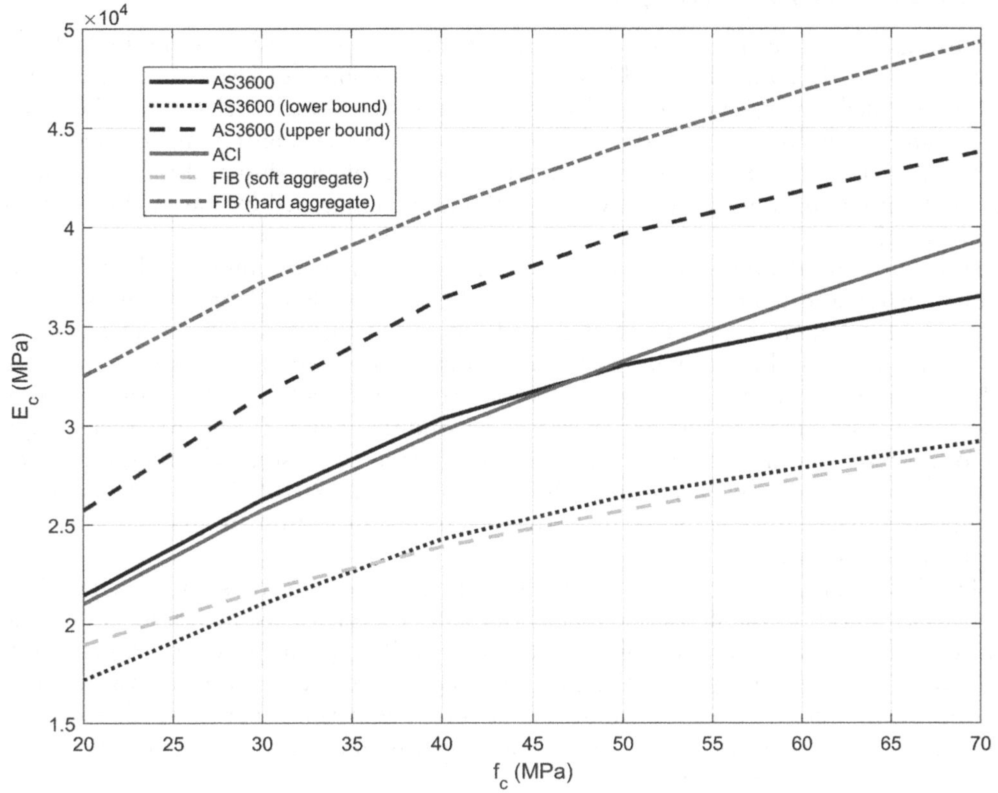

Figure 2.13 Comparison of predicted elastic modulus.

As was the case for $\sigma_{ax}/\varepsilon_{ax}$ models for *FRC*, few models are available to predict the elastic modulus of FRC ($E_{c\text{-}FRC}$), and those that have been developed generally have limited calibration ranges (Ezeldin and Balaguru 1992; Barros and Figueiras 1999; Neves and Almeida 2005; Aslani and Natoori 2013). As an example, the model of Aslani and Natoori (2013) given by Eq. 2.8 is considered, as this has the widest calibration range (f_c between 21.8 and 138 MPa, V_f between 0.13 and 6.3 and l_f/d_f between 37 and 85.7)

$$E_{c-FRC} = 1.25E_c + 0.0242RI \tag{2.8}$$

In Eq. 2.8, the elastic modulus of *FRC* is taken as 1.25 times the elastic modulus of concrete without fibres of the same compressive strength and is further increased as a proportion of the *RI* given by Eq. 2.1. It is notable that the 1.25 multiplier applied to E_c is at least an order of magnitude larger than the second term dependent on *RI*. This limited dependency on *RI* is likely because, in the elastic region of the $\sigma_{ax}/\varepsilon_{ax}$ relationship, only micro-cracks may have formed, and therefore, there is very little engagement of the fibres. It should be reiterated that these results were derived from a limited dataset and have only been used to illustrate the effect of fibres.

2.3.3 Compressive strength

Let us again consider the axial stress–strain relationship for *NSC* in Figure 2.11a. It can be observed that as the axial load is increased beyond the elastic region (which is commonly defined as between 0% and 40% f_c), the stress–strain relationship becomes increasingly non-linear until the concrete compressive strength f_c is reached. The increasing non-linearity of the concrete stress–strain relationship and the attainment of the limiting concrete compressive strength can be explained by the propagation of micro-cracks.

During the hardening of concrete, micro-cracks form at the interface between the cement paste and aggregates as a result of differential changes in volume brought about by the hydration process and the differences in the stress–strain relationships of the aggregate and the cement paste. Importantly, these micro-cracks remain stable when subjected to instantaneous loads of approximately 30%–40% of the concrete compressive strength. As the load increases beyond this point, the micro-cracks begin to join, creating a distinct crack pattern; the opening of these cracks then limits the compressive strength of the concrete (Neville and Brooks 2010). That is, the concrete compressive strength can be related to the partial-interaction mechanism of shear friction in Section 1.3.1.

The hydraulic and pozzolanic reactions that give concrete its strength take place over time, and this time-dependent behaviour can be expressed similarly to the change in concrete's elastic modulus over time. Specifically, the increase in strength follows a power relationship such that the most significant increases occur over a short period, with minor increases continuing over very long periods. For detailed modelling approaches that relate the change in f_c with time, temperature and cement type, the reader is referred to Fib (2013).

2.3.4 Strain-based behaviour under combined loading

When subjected to lateral compressive or tensile stresses, the axial $\sigma_{ax}/\varepsilon_{ax}$ behaviour of concrete can be significantly modified. This is shown in Figure 2.14, where an unconfined concrete $\sigma_{ax}/\varepsilon_{ax}$ relationship is presented, modified to account for various degrees of lateral compressive and tensile stresses (σ_{lat}).

The two lower graphs in Figure 2.14 combine lateral tension with axial compression. Comparing these graphs with the laterally unconfined curve, it can be seen that lateral tension: increases the non-linearity of the rising branch (particularly at the early stages of loading); reduces the compressive or peak strength; and reduces the ductility, meaning the flattening part of the curve or the extent of strain (ε_{ax}) over which stress (σ_{ax})

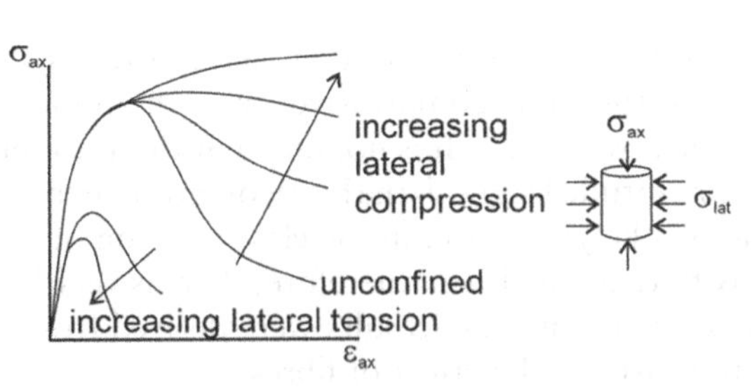

Figure 2.14 Change in $\sigma_{ax}/\varepsilon_{ax}$ with changes in lateral σ_{lat}.

is sustained. Conversely, the upper graphs illustrate that lateral compression has little impact on elastic behaviour, but substantially increases both the compressive strength and ductility. The role of confinement is, therefore, essential in the design of *RC* elements, particularly where ductility is concerned. The cause of this change lies in the shear friction mechanism acting on the sliding plane shown in Figures 1.16 and 1.17.

The mirror image of the left-hand free body in Figure 1.16 is shown in Figure 2.15. The free body forces consist of the applied axial compressive force N which can be converted to an axial stress σ_{ax}; the applied lateral tensile force T or the applied lateral compressive force C which can be converted to a lateral stress σ_{lat}; the induced sliding forces along the sliding plane $N \cos \alpha$, $C \cos \alpha$ and $T \cos \alpha$ that can be converted to a shear stress along the sliding plane τ_{SF}; and the induced confinement forces normal to the sliding plane $N \sin \alpha$, $C \sin \alpha$ and $T \sin \alpha$ which can be converted to a confining stress across the sliding plane σ_N. Figure 2.15 also shows that the interface slip Δ_{SF} causes the prism to contract, resulting in an effective axial compressive strain that, when combined with the material strain, contributes to ε_{ax}. These forces and displacements control the behaviour at the onset of sliding (i.e. the strength) and during sliding (i.e. the ductility) through the shear friction properties along the sliding plane, as follows.

The shear friction properties from Section 1.3.1 and Figure 1.4 are summarised in Figure 2.16 where τ_{SF} is the shear stress along a sliding plane and Δ_{SF} is the interface slip along the sliding plane for different amounts of confinement σ_N normal to the

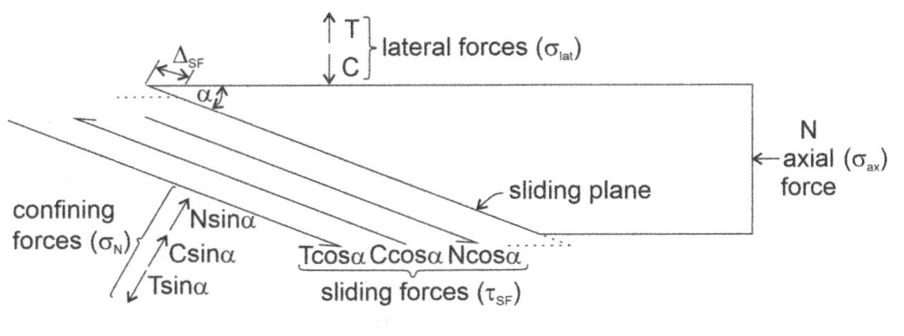

Figure 2.15 Forces acting on a confined free body.

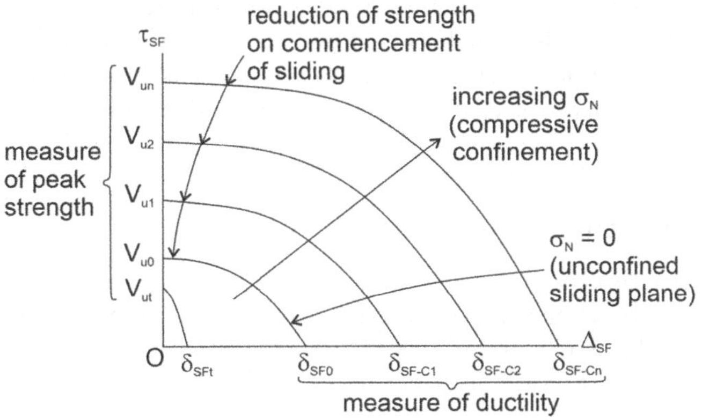

Figure 2.16 Shear friction strength and ductility.

sliding plane. It is these properties that control the behaviour of the sliding plane in Figure 2.15. In Figure 2.16, the intercepts on the ordinate for each variation, denoted V_u, represent the shear stresses at the onset of sliding and consequently are a measure of the peak strength for a particular confinement across the sliding plane σ_N. Once sliding commences, there may be a short plateau but invariably the strength reduces to zero at δ_{SF}; δ_{SF} is the extent of slip within which shear stresses are transferred across the interface; beyond this point, no shear stresses are present. Hence δ_{SF} is a measure of the ductility of the sliding plane. For example, when Δ_{SF} in Figure 2.15 is equal to or greater than δ_{SF} in Figure 2.16, uncontrollable failure occurs. When δ_{SF} is small, the effective axial strain due to sliding is small, and vice versa. It should be borne in mind that this effective axial strain due to sliding is in addition to the material strain. It can be seen in Figure 2.16 that compressive confinement normal to the sliding plane increases both the peak strength V_u and the ductility δ_{SF}.

It can be seen in Figure 2.15 that when only the axial force N is applied, that is, when the concrete specimen is unconfined by the lateral forces C or T, then the induced shear force is $N \cos \alpha$ and the induced confinement across the sliding plane is a compressive force $N \sin \alpha$. When the lateral compressive force C is applied in conjunction with N, then the shear force induced by N, that is, $N \cos \alpha$, reduces by $C \cos \alpha$, which is beneficial, and the confinement force across the sliding plane $N \sin \alpha$ increases by $C \sin \alpha$, which is also beneficial. Based on the shear friction properties, this means that both the peak strength and ductility increase, as can be seen in Figure 2.14. When a lateral tensile force T is applied instead of C, then this increases the shear force by $T \cos \alpha$, which reduces the peak strength, and also reduces the confinement by $T \sin \alpha$, thereby reducing the ductility, as can also be seen in Figure 2.14.

Micro-cracking causes the non-linearity in the ascending branch in Figure 2.14. Once sliding starts at V_u in Figure 2.16, the axial stress reduces, as in Figure 2.14. This is because the *SF* properties shown in Figure 2.16 reduce after the onset of sliding. Hence, it is generally assumed that the commencement of sliding occurs at the peak axial stress. If τ_{SF} in Figure 2.16 were to increase with Δ, the strength could increase after sliding, which is unlikely to be a material property but can occur if there is sufficient passive confinement, such as from stirrups. This is also discussed in Section 1.4.4 and further in Chapter 5, which deals with the confining mechanism.

While a shear friction approach has been presented here to explain the observed changes in the $\sigma_{ax}/\varepsilon_{ax}$ relationship for various degrees of lateral stresses in Figure 2.14, significant research has led to the development of strain-based models that also account for confinement. These models, such as that presented in (Fib 2013), are generally complex and are, therefore, not presented here. However, it will be shown in Chapter 3 that they can be very useful for extracting the more fundamental shear friction material properties.

2.3.5 Poisson's ratio

Like other construction materials, concrete has a positive Poisson's ratio and therefore expands laterally when it is subjected to contracting axial strains and vice versa. Poisson's ratio in the elastic region for all forms of concrete is typically taken as 0.2 regardless of the direction of loading. Outside of the elastic region, the lateral dilation that occurs with axial shortening is no longer a strain-based property but an outcome of the sliding mechanism illustrated in Figures 1.7, 1.17 and 1.20 and the accompanying text.

2.4 CONCRETE IN TENSION

In Figure 2.17a, the stress–strain relationship of plain concrete obtained under direct tension is elastic–brittle, with an ascending branch of the same modulus as that obtained under compression until the cracking strength is reached. The stress–strain behaviour of concrete in tension is driven by very similar processes to those described in compression. When subjected to an external load, micro-cracks present within the composite structure of the concrete propagate. Then, depending on the magnitude of the load, they are either arrested as the crack extends into a local region that requires additional energy for further propagation or if the applied load is sufficient, micro-cracks distributed throughout the concrete link up and fracture occurs. If tested under direct tension under highly controlled conditions, it is possible for a small softening branch to be recorded, a result of stress transferred across the crack by aggregate bridging, but this is generally neglected in design.

For concrete containing fibres, as in Figures 2.17b and c, the tensile behaviour of concrete is markedly different from that of plain concrete in Figure 2.17a, as it is highly dependent on the volume and type of fibres present within the concrete. In Figure 2.17b, for *FRC* and in Figure 2.17c for *UHPFRC*, the behaviour is split into a pre-cracking strain-based stress–strain relationship and a post-cracking slip-based stress–crack width (w) relationship; the latter will be discussed in more detail in Chapter 3 as it is a slip-based material property. In the pre-cracking region, behaviour can further be split into materials exhibiting linear elastic behaviour with a sudden failure, as shown by *O–A* in Figure 2.17b (which occurs for *FRC*s with small volumes of fibres); or materials where strain hardening occurs, as shown by the linear behaviour *C–D* at modulus E_{sh} in Figure 2.17c (which occurs for *UHPFRC* with large volumes of fibres), with micro-cracks occurring at Point *C* and crack localisation (a single expanding micro-crack) at Point *D* (Visintin and Oehlers 2018).

Before examining the tensile stress–strain and stress–crack width relationships in Figures 2.17b and c in more detail, it is worth first considering the relationship between

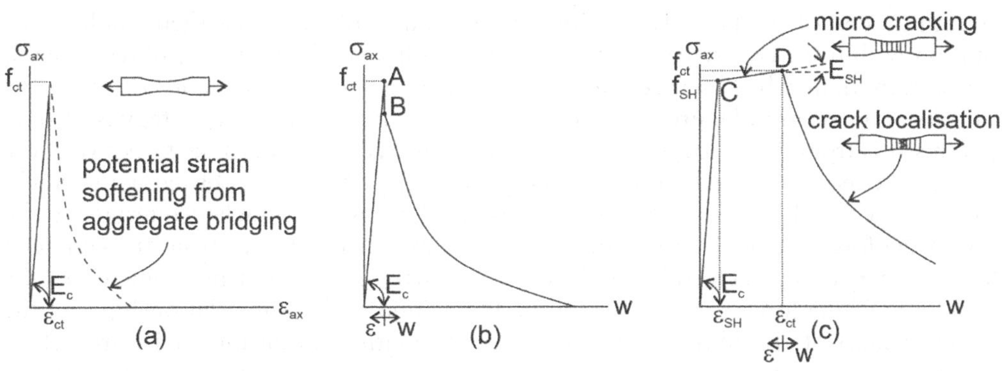

Figure 2.17 Uniaxial tensile stress-strain relationship of: (a) plain concrete; (b) *FRC*; and (c) *UHPFRC*.

an individual fibre within a concrete matrix and a traditional reinforcing bar. That is, the behaviour of an individual fibre is the same as that of an individual reinforcing bar as explained in Section 1.4.1 with the help of Figure 1.12. Let us replace the bar in Figure 1.12 with a single fibre, in which case the behaviour of *FRC* can be deduced as follows. At a micro-crack the bond stresses build up as in Figure 1.12c and the concrete stress builds up as in Figure 1.12e. If the fibre stress at the micro-crack is sufficient, then another micro-crack will form at a distance S_{pr} where the tensile capacity of the concrete is achieved as shown in Figure 1.12e. It is worth noting that S_{pr} is less than L_{crit}, which is the length required to achieve *IC* debonding and thus the maximum spacing of the micro-cracks. A succession of micro-cracks will form along *C–D* in Figure 2.17c until one micro-crack widens at the point of crack localisation (Point *D*). If the fibre stress at the micro-crack is insufficient to cause further micro-cracks, then that micro-crack will be the start of crack localisation, as at Point *A* in Figure 2.17b, and there will be no strain hardening.

Of course, many individual fibres cross a micro-crack, and these fibres are randomly orientated and have random anchorage lengths. Hence S_{pr} will only be an indication of the micro-crack spacing. One micro-crack will eventually substantially widen relative to the others; this will be the 'initial crack' in Figure 1.12a that initiates the tension stiffening analysis of the reinforcing bar and controls the primary crack spacings for that individual bar. It should be noted that debonding is not an undesirable phenomenon as it allows the concrete to fail in tension in a ductile fashion. In fact, in most cases, it is the preferred mode of failure compared with fibre fracture. It can be seen that the tensile behaviour of concrete post-cracking is not a *FI* strain-based material property but a *PI* material property that could theoretically be derived from *PI* mechanics. This approach has been taken (Li et al. 1991), but because of the random nature of the mechanics, the *PI* properties are determined from experimental testing and are thus treated as pseudo strain-based properties. It is also worth noting that it is reasonable to distinguish between micro-cracking due to fibres and macro-cracking due to reinforcement as the latter is at least an order of magnitude larger than the former; hence micro-cracking occurs at much smaller slips than macro-cracking.

Returning now to the behaviour of concrete with a small amount of fibres where no strain hardening occurs, as shown in Figure 2.17b. The rising branch of the tensile stress–strain relationship displays essentially linear elastic material properties until a crack localises at the tensile strength f_{ct}. Once a crack forms, fibres that cross the crack are engaged, and stress is transferred across the crack via the partial-interaction mechanism described above. Depending on the volume of fibres and their random orientation within the concrete, it is possible that with a small volume of fibres, a small tensile strength may be observed due to fibres bridging micro-cracks. For *FRC* with low volumes of fibres, as in Figure 2.17b, once the crack forms at point A, a stepwise reduction in stress from *A* to *B* occurs; this represents the reduction in capacity between the uncracked concrete at crack initiation and the capacity of the fibres bridging the crack. Upon further loading, the total stress transferred across the cracked interface is the sum of the force carried by each individual fibre as it is pulled out from the concrete; the force carried by each individual fibre is a partial-interaction problem where the magnitude of the force is a function of its local bond stress–slip relationship and the bonded length of the fibre. In the stress–crack width relationship in Figure 2.17b, the force continues to reduce until the fibres pull out and can no longer transfer stress across the crack.

Let us now consider concretes that exhibit strain hardening behaviour, as shown in Figure 2.17c. Similarly to *FRC*s that do not strain harden, the concrete behaves linearly elastically until a crack initiates. However, in strain hardening materials, the volume of fibres crossing the crack means that the cracked section is of equal strength to, or stronger than, the uncracked section, and thus, no stepwise reduction in stress occurs. Upon further displacement of the specimen, stress builds up in the fibres at the crack face and is transferred to the concrete via partial-interaction mechanics. In strain hardening materials, the transfer of stress from the fibres to the concrete via bond is sufficient for a second crack to form, and given that the transfer of stress occurs over the fibre length, it is expected that micro-cracks will form along the specimen at a spacing less than the fibre length. This micro-cracking phase is observed as the strain hardening branch extending from points C (f_{SH}, ε_{SH}) to point $D(f_{ct}, \varepsilon_{ct})$ with a modulus of E_{SH}, as shown in Figure 2.17c. It continues until a single crack localises, at which point the total force that can be transferred across the crack begins to reduce, leading to a reduction in the total force applied to the specimen. Following the localisation of a crack, the behaviour of a strain hardening *UHPFRC* is driven by the same fibre pull-out mechanisms as observed in *FRC*, but the stress transferred across the crack is generally higher because of the higher fibre volumes.

Importantly, when considering the strain hardening and stress–crack width behaviour of FRC, a significant scatter in test results arises because of the random orientation and density of fibres. It is also essential to ensure that the testing approach does not significantly change the expected performance relative to that expected within a larger element; this will be discussed in more detail in the following sections.

2.4.1 Methods for quantifying tensile strength

2.4.1.1 Forms of tensile testing

Three distinct test methods are commonly applied to quantify the tensile behaviour of concrete as shown in Figure 2.18: the direct tensile test; the splitting or Brazilian indirect tensile test; and the flexural indirect tensile test.

The direct tensile test, as the name implies, usually subjects a dog bone-shaped specimen to an axial tensile load as shown in Figure 2.18a. The major benefit of this

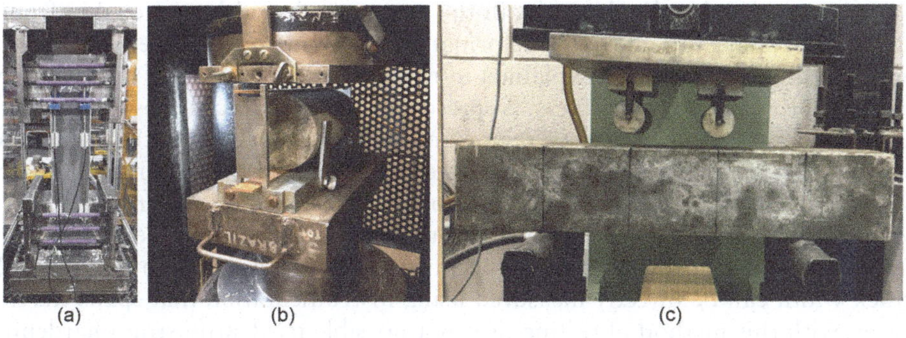

Figure 2.18 Tensile tests of concrete: (a) direct tensile test; (b) split tensile test; (c) flexural tensile test. (Photos courtesy of Jon Ayoub and Lloyd Dadd.)

approach is that it directly yields measurements of tensile strength, and if instrumented to do so, can also directly yield the stress–strain relationship. Specimens are typically dog bone-shaped with an expanded grip area and a reduced-size testing shank. Care must be taken to design grips such that stress concentrations occurring during loading do not cause failure outside of the test region and do not influence the test results. Care must also be taken to ensure that loads are applied concentrically such that bending stresses are not introduced.

When designing a direct tensile test for either conventional or FRC, it is important to consider specimen size. When considering concrete without fibres, the number of initial defects contained within the specimen is proportional to its size. Consequently, testing larger specimens will minimise the scatter of results and ensure that the test results tend towards the actual material property.

2.4.1.2 Fibre-reinforced concrete

When considering FRC, specimen size is further complicated by the need to ensure that the size and shape of the specimen does not unduly influence the density or orientation of the fibres. It is generally recommended that the specimen cross-section dimensions should be at least 2.5 times the maximum length of the fibre in order to achieve this. At present there is not yet broad agreement on specimen size for FRC; unfortunately, a significant quantity of research has been published in which the direct tensile properties of the concrete have been obtained from specimens with the smallest cross-sectional dimension either less than or only slightly larger than the length of the fibres. This is unfortunate because not only is there now a strong precedent for this form of testing, but the total pool of test data has been influenced by upper bound test results where the degree of strain hardening and the post-cracking tensile strength are influenced by a preferential alignment of fibres that simply cannot be achieved at the member level. FRC direct tension specimens can be divided into unnotched specimens for the study of the strain hardening behaviour *C–D* in Figure 2.17c and notched specimens for the specific study of the stress–crack width behaviour from *D* onwards.

For unnotched specimens such as that shown in Figure 2.18a, displacements can be measured over an extended region in which multiple micro-cracks may form. Given the small nature of these cracks, the strain hardening region as described above is generally treated as a strain-based material property with an average stress–strain relationship. Once a crack is localised, to obtain the stress–crack width relationship from test results, it is necessary to remove the deformation occurring outside the localised crack; methods for doing this will be explained in more detail in Chapter 3 but they rely on knowledge of the unloading stiffness. At present there is insufficient testing to quantify damage that may occur in the elastic region and so E_c is a best estimate of the unloading stiffness. Further research is required in this regard, not only for quantifying the stress–crack width behaviour but also for analysis of tension stiffening under monotonic and cyclic loading and for member-level analysis under high- and low-cycle fatigue.

Importantly, when testing notched specimens such as that shown in Figure 2.19, the crack width can be directly measured using instrumentation that spans the notch. However, with this method of testing, it is not possible to identify strain hardening as multiple cracks cannot form within the test region.

Figure 2.19 Notched tension test specimen directly measuring stress–crack width. (Photo courtesy of Lloyd Dadd.)

2.4.1.3 Tensile strength from indirect tensile tests

Due to challenges in setting up direct tensile tests, primarily concerning alignment, indirect tests are commonly employed to induce tensile failure, from which the tensile strength can be inferred. The simplest form of indirect tensile test is the split or Brazilian tensile test, in which a standard cylindrical specimen is laid horizontally and a line load is applied until failure occurs, as shown in Figure 2.18b. The splitting behaviour for a single concentrated load, described in Section 1.4.5 and illustrated in Figure 1.25, can be used to explain the behaviour of the Brazilian tensile test shown in Figure 2.18b. The forces induced by the upper bearing pressure in Figure 2.18b are shown in Figure 1.25a. The dispersal of the downward concentrated force (F_{dow}) into the concrete prism induces a lateral tensile force (P_{tr}), which is balanced by a concentrated lateral compressive force of equal magnitude. In the Brazilian tensile test, an upward concentrated compressive force, also of F_{dow}, produces lateral forces of magnitude P_{tr}. Hence, under the Brazilian test setup, a high compressive stress is present at the supports, but towards the centre of the specimen lateral tensile stresses are essentially uniform. Based on the ultimate load at which failure occurs (P_{ult}), the splitting tensile stress is expressed as

$$f_{ct.sp} = \frac{2P_{ult}}{\pi Dl} \tag{2.9}$$

in which D is the diameter of the cylinder and l is the length of the cylinder.

The flexural tensile strength, or modulus of rupture ($f_{ct.f}$), is found by testing an unreinforced prismatic beam under four-point bending, as shown in Figure 2.18c. Given that the specimen is unreinforced, once a crack is formed there is no mechanism to restrain its growth, and the fracture thus extends to full depth. The tensile stress at the outer fibre corresponding to the peak load can, therefore, be taken as the flexural tensile strength, and from elementary elastic theory $\sigma = My/I$.

For FRC, the flexural tensile test has proved to be a popular method for extracting the stress–strain and stress–crack width behaviour, requiring the use of an inverse analysis to infer the stress–crack width behaviour because the crack width and, therefore, the stress carried by the fibres vary along the crack height. For both conventional and FRC, neither of the indirect test methods provides a uniform distribution of tensile stresses. Therefore, these approaches yield an estimate of the tensile strength rather than the intrinsic material property, which can only be obtained from direct tensile testing ($f'_{ct.d}$).

2.4.2 Code approximations for tensile strength

In the absence of testing, national codes of practice provide relationships between the compressive and tensile strength. These relationships are generally in the form of

$$f'_{ct.f} = af_c'^{b} \tag{2.10}$$

$$f'_{ct.sp} = cf_c'^{d} \tag{2.11}$$

$$f'_{ct.d} = ef_c'^{f} \tag{2.12}$$

where the coefficients a, b, c, d, e and f are empirically derived and the exponents b, d and f are less than 1, reflecting that the tensile strengths of HSC are not significantly larger than those of NSC.

Figure 2.20 compares the different types of tensile strength obtained from codes of practice (Standards Australia 2018; ACI 2008; Fib 2013). It can be seen that there is, in general, agreement that the splitting and flexural tensile strengths represent upper-bound estimates of the direct tensile strength, owing to the non-linear strain profile that exists in the indirect test methods. It is also observed that, in general, the code approaches yield very similar results.

When considering FRCs, the tensile behaviour must be divided into strain-based behaviour (which occurs prior to the localisation of a single crack) and displacement-based behaviour (which occurs after the localisation of a single crack). This is because, following the localisation of a single crack, behaviour is controlled by the partial-interaction bond-slip behaviour of fibres crossing the crack. This behaviour will be discussed in detail in Chapter 3, which presents partial-interaction material properties, and hence, only the strain-based rising branch of the tensile stress–strain relationship shown in Figure 2.21 is considered here.

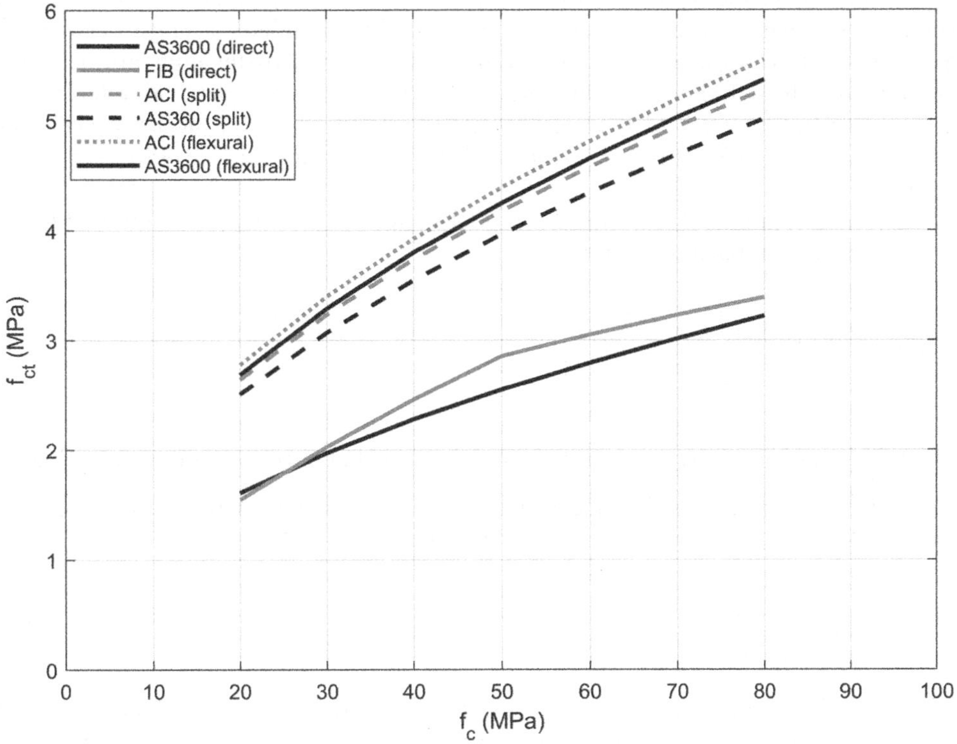

Figure 2.20 Variation in tensile strength with compressive strength and test method.

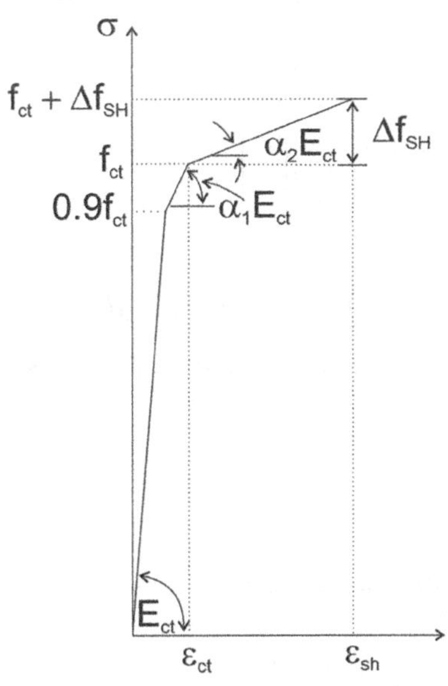

Figure 2.21 Strain-based tensile properties of *FRC*. (According to Kim et al. 2023.)

Given that the tensile behaviour of any *FRC* is uniquely dependent on the properties of the concrete matrix, the fibre reinforcement and the bonded interface between the two materials, developing generic material models that are broadly applicable to a wide range of materials is challenging, and code-based models are not yet available. Instead, here we consider the work of Kim et al. (2023), who proposed a modified version of the fib model code 2010 (fib 2013) approach through calibration of 48 experimental datasets.

As shown in Figure 2.21, a bilinear model is applied for strain softening *FRC*, in which the tensile strength is given by

$$f_{ct} = 0.535 f_r \left[\left(\frac{f_c}{f_0} \right)^{0.8} \left(\frac{\rho_c}{\rho_0} \right)^{0.4} \beta_f^{0.2} \right]^{0.2} \tag{2.13}$$

in which $\frac{\rho_c}{\rho_0}$ is the density of the concrete, normalised by the control value (ρ_0=2300 kg/m^3), and β_f and f_r are parameters that account for different fibre performance. The β_f parameter controls the maximum force that can be transferred and is given by

$$\beta_f = g V_f \left(\frac{l_f}{d_f} \right)^{0.1} \sqrt{\frac{\tau_f}{f_c}} \tag{2.14}$$

in which g is the snubbing factor that accounts for the highly localised friction that can occur at the crack face when a fibre is pulled at an inclined angle and τ_f is the interfacial bond strength. Both parameters g and τ_f must be determined from experimental tests; in the absence of further information, g can be taken as 2.5 for steel fibres and τ_f generally varies between 1 and 2 MPa. Within the calibration set used by Kim et al. (2023), β_f was found to vary between 0.17 (for 24 MPa concrete with V_f = 0.5%) and 4.59 (for 47 MPa concrete with V_f = 2%).

In Eq. 2.13, the parameter f_r is given by

$$f_r = 1.02 \left(1 + \beta_f^{0.75} \right) \left[f_c^{0.6} \left(\frac{d_0}{d_a} \right)^{0.1} \left(\frac{\rho_c}{\rho_0} \right)^{1.5} \right]^{0.7} \tag{2.15}$$

where $\frac{d_0}{d_a}$ is the ratio of the control aggregate size (d_0=25 mm) to the maximum aggregate size, and all other parameters are as defined previously.

In Figure 2.21, ε_{ct} is defined as

$$\varepsilon_{ct} = \frac{0.9 f_{ct}}{E_{cf}} + \frac{0.9 f_{ct}}{\alpha_l E_{cf}} \tag{2.16}$$

where E_{cf} is the elastic modulus of the *FRC* in tension and is given by

$$E_{cf} = \left(1 + 0.098 \beta_f \right) \left[8470 f_c^{\frac{1}{3}} \left(\frac{\rho_c}{\rho_0} \right)^{1.17} \right] \tag{2.17}$$

$$\alpha_1 = 0.38\left[\left(\frac{f_c}{f_0}\right)^{1.4}\left(\frac{\rho_c}{\rho_0}\right)^{1.1}\beta_f^{-0.1}\right]^{-0.76} \tag{2.18}$$

where all parameters are as defined previously, and f_c is in units of MPa.

For strain hardening concretes in Figure 2.21, the increase in tensile strength due to strain hardening (Δf_{SH}) is given by

$$\Delta f_{SH} = \alpha_2 E_{cf}\left(\varepsilon_{SH} - \varepsilon_{ct}\right) \tag{2.19}$$

where

$$\alpha_2 = 0.029 ln\left[\left(\frac{f_c}{f_0}\right)^{-0.1}\left(\frac{\rho_c}{\rho_0}\right)^{1.1}\beta_f^{1.4}\right] - 0.037 \tag{2.20}$$

and

$$\varepsilon_{SH} = \varepsilon_{ct} + \frac{0.5}{3d_a} \tag{2.21}$$

and all variables are as defined previously.

2.5 TIME-DEPENDENT PROPERTIES OF CONCRETE

Until this point, we have mainly considered the behaviour of concrete at an instant in time and ignored the interactions with the environment that may change the mechanical responses. However, the impact of time and changes in environmental conditions can lead to changes in concrete behaviour that are equal to or larger than those resulting from the application of an external load. To examine this interaction, consider the relationship between concrete strain and time shown in Figure 2.22a, which is read in conjunction with the relationships for sustained compressive stress (σ_0) and time in (Figure 2.22b) and temperature (T) and time in Figure 2.22(c). Consider a concrete specimen where the temperature at time t_{t0} is T_1 and changes between T_1 and T_2 during times t_{t1} and t_{t2}, as depicted in Figure 2.22c. A constant compressive stress is applied at time t_0 and released at t_{f1}, as shown in Figure 2.22b, and shrinkage is first measured at time τ_d in Figure 2.22a.

Under the temperature, load and shrinkage conditions in Figure 2.22, the variation in strain with time is shown in Figure 2.22a. It can be seen that the shrinkage strain, which is first measured at time t_d, increases at a decreasing rate throughout the entire time period. At the arbitrary time t_t, the shrinkage strain has been labelled $\varepsilon_{sh}(t_{t1})$. As the specimen is first loaded at time t_0, a step change in the total strain occurs, equal to the applied instantaneous strain $\varepsilon_{ci}(t_0)$, and this instantaneous strain remains constant for the entire duration of its application between t_0 and t_{f1}. Immediately after the load is applied to the specimen at t_0, it begins to creep and the creep strain increases at a decreasing rate until the applied load is removed at time t_{f1}. At the arbitrary time t_t, the strain due to creep caused by the load applied at t_0 is $\varepsilon_{cc}(t_t, t_0)$. The temperature drops between t_{t1} and t_{t2} and this reduction in temperature causes a change in strain $\varepsilon_T(t_{t1})$.

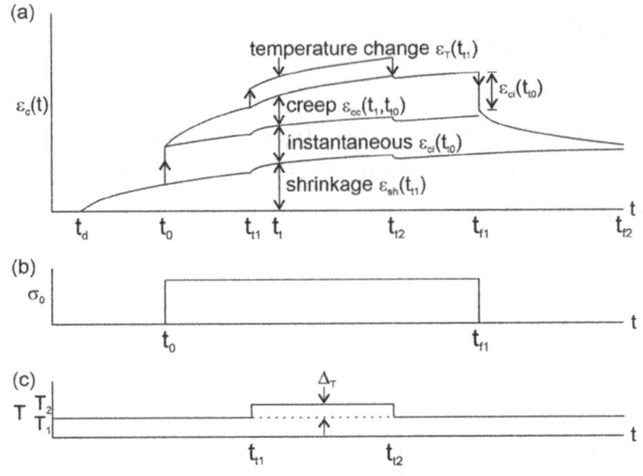

Figure 2.22 Time-dependent variation of concrete strain: (a) variation of total concrete strain $\varepsilon_c(t)$; (b) variation in applied stress (σ_0); and (c) variation in temperature (T).

It is also worth noting that the shrinkage and creep strains depend on the temperature. Varying the temperature will not only cause a variation in strain due to thermal expansion or contraction ε_T but also change the rate at which the creep and shrinkage strains develop.

On unloading at time t_{f1} in Figure 2.22a, there is an instantaneous reduction in strain equal to the initial instantaneous strain, $\varepsilon_{ci}(t_0)$ followed by a gradual recovery of the creep strain. As shown at time t_{f2}, the creep strain after the removal of the applied load never fully recovers, meaning the total strain remains above the shrinkage strain; this additional strain above the shrinkage strain results from underlying damage due to the applied load.

From Figure 2.22a, the total strain at any time t is given by

$$\varepsilon_c(t) = \varepsilon_{ci}(t_0) + \varepsilon_{cc}(t, t_0) + \varepsilon_{sh}(t) + \varepsilon_T(t) \tag{2.22}$$

Following the discussion of the elastic response of concrete to an instantaneous load through the definition of an elastic modulus in Section 2.3.2, we now consider the time-dependent deformations of concrete due to shrinkage, thermal effects and creep.

2.5.1 Concrete shrinkage

Concrete shrinkage is a time-dependent change in deformation or strain that occurs in the absence of load. At the material level, unrestrained concrete will shorten (shrink) over time without generating stresses; however, if this deformation is restrained either externally by supports or internally by reinforcement, stresses develop, and cracking may occur depending on the magnitude of the deformation (Gilbert and Ranzi 2010).

It should be noted that the above concept of shortening without stress, although simple, is very important and forms the fundamental principle of the base-line approach in Section 1.5.2, as illustrated in Figure 1.29; it is not an approximation but an exact mechanical solution. The base-line concept is illustrated for the half specimen in

Figure 1.29f when subjected to a Euler–Bernoulli deformation in Figure 1.29g. The specimen prior to shrinkage is of length $L_{seg}/2$. Longitudinal shrinkage causes the unrestrained segment face *A–A* to contract to *B–B*. Thus, strain and consequently stress is induced by the distance between the linear Euler–Bernoulli deformation and the curved concrete face *B–B*. In contrast, the reinforcement is assumed not to shrink, so the deformation between the Euler–Bernoulli deformation and the face *A–A* induces stresses in the reinforcement. It is also worth noting that thermal expansion and contraction and concrete expansion (e.g. through the use of expansive concrete) can also be analysed using base-line theory, as illustrated in Figure 1.29k.

As mentioned in Section 1.5.2, shrinkage is not uniform longitudinally, as denoted by the curve *B–B* in Figure 1.29g, nor laterally (i.e. it varies between the segments in Figure 1.29i. However, it is often idealised as uniform to facilitate a simple solution. The mechanisms of shrinkage are complex and varied, but in general, the total shrinkage strain can be divided into the following categories.

Plastic shrinkage occurs as water evaporates from the surface of the concrete during hardening. If the rate of water evaporating from the surface is faster than the rate at which water can rise to the surface (a process known as bleeding), the pores within the concrete desiccate. As the pores desiccate, their volume reduces, resulting in the formation of tensile stresses in the mortar surrounding the pore (Radocea 1994). Plastic shrinkage is especially important in hot weather concreting or where high wind speeds are present, as this will increase the rate of evaporation, which is why concreting is preferred during cooler and calmer periods of the day. Plastic shrinkage is also common in concrete such as *UHPCs* which have very low *w:c* (meaning there is little water in the pore space) and no coarse aggregate (thereby removing the skeleton of concrete which does not shrink). Hence, without care, 'elephant-skin' cracking can result, as shown in Figure 2.23, where, as the name implies, the concrete surface appears to be wrinkled

Figure 2.23 Cracking of *UHPC* due to plastic shrinkage. (Photo courtesy of Phillip Visintin.)

and cracked. In both scenarios, plastic shrinkage can be minimised by appropriate curing that either limits evaporation or provides moisture to the concrete surface. Because plastic shrinkage is associated with concrete in its plastic form, it typically only occurs during the first hours after water addition.

Chemical shrinkage occurs during the cement hydration reaction, predominantly during the first days and weeks after water addition. This form of shrinkage occurs because the volume of hydration products is smaller than the volume of the unhydrated materials (Gilbert and Ranzi 2010). In the initial stages of chemical shrinkage, while the concrete is still plastic, it causes a measurable change in its dimensions. After the concrete has hardened, chemical shrinkage causes an increase in pore volume within the concrete rather than a measurable change in the overall external dimensions (Kovler and Zhutovsky 2006).

Autogenous shrinkage predominantly occurs during the first few months after hardening and in the absence of any interaction with the external environment. This form of shrinkage is a result of the concrete self-desiccating; that is, ongoing hydration consumes water within the pores, generating a negative pore pressure that puts the surrounding concrete into tension (Li and Li 2014). For NSCs, autogenous shrinkage is generally at least an order of magnitude smaller than subsequent drying shrinkage and is therefore of minor importance. This is not the case for *UHPC*, where, because of the high binder content, autogenous shrinkage can be of the same order of magnitude and therefore needs to be explicitly considered (Sun et al. 2022).

Drying shrinkage occurs over many months and years and is caused by the loss of moisture from the concrete to the surrounding environment due to diffusion. This loss of water results in a change in capillary pressure and subsequent shrinkage, which can be modelled using the same fundamental physics as described above for autogenous shrinkage.

Importantly, given that the moisture exchange with the environment is a diffusion-driven process, drying shrinkage strains can be expressed as a function of pore humidity at a given temperature, according to Fick's second law (Sakata 1983):

$$\frac{\partial RH}{\partial t} = \frac{\partial}{\partial x}\left[D(RH)\frac{\partial RH}{\partial x}\right] \tag{2.23}$$

In lay terms, Fick's second law shows that drying shrinkage is a function of the change in humidity within the concrete. This is driven by the diffusion of water from areas of high humidity to areas of low humidity within the concrete, bearing in mind that humidity is highest in the middle and lowest at the surface, where it can evaporate.

In Eq. 2.23, the movement of moisture within the concrete over time is modelled by defining the change in relative humidity (∂RH) that occurs over small increments in time (∂t) and space (∂x). The rate of change of RH with respect to space and time is controlled by the diffusion coefficient, which must be determined experimentally and for which codes of practice, such as fib (2013), define general models.

At the boundary between the concrete and the atmosphere, the rate of diffusion changes because of a step change in the humidity difference between the environment and the concrete surface. To account for this change, a surface factor can be applied to simulate the moisture transfer from the concrete surface to the environment, according to

$$D(RH)\frac{\partial RH}{\partial x} = f(RH_{en} - RH_s) \tag{2.24}$$

in which RH_{en} is the environmental humidity surrounding the concrete surface, which is at a humidity of RH_s. The surface factor can be taken as 0.55 cm/day for NSC, according to Sakata (1983). The above equation represents the rate of change of evaporation as a function of the difference in humidity.

Any analytical or numerical method can be applied to solve the differential equation given by Eq. 2.23. Here, a simple finite difference scheme is proposed because it can be solved in a spreadsheet. To do this, a Taylor expansion is applied to simplify Eq. 2.23 to express the distribution of relative humidity as a function of space and time. By applying the Taylor expansion, the derivatives in Eq. 2.23 are approximated by discretising them from continuous functions into discrete differences over space and time. That is, Eq. 2.32 becomes

$$RH_i^{t+1} = RH_i^{t+1} + \frac{D_1 \Delta t}{\Delta x}\left(RH_{i+1}^t - 2RH_1^t + RH_{i-1}^t\right) \tag{2.25}$$

where the formulation in Eq. 2.25 gives the relative humidity at a specific point i at the new time $t+1$, given that the relative humidity at the surrounding points $RH_{i+1}^t RH_{i-1}^t$is known at the current time t. As described in (Walker et al. 2009), the technique applied to implement Eq. 2.25 is relatively straightforward, in that Eq. 2.25 is applied incrementally for all times and all locations, with defined starting conditions such that the solution at time $t+1$ then serves as the starting point for the next time step.

Once the distribution of pore humidity within the concrete is determined, it can be linked to the shrinkage strain using an empirical conversion (Bažant and Yunping 1994)

$$\varepsilon_{sh} = \varepsilon_{sh\infty} g_s(t) \cdot (1 - RH) \tag{2.26}$$

in which $\varepsilon_{sh\infty}$ is the ultimate shrinkage strain after the concrete has equilibrated with the external moisture and $g_s(t)$ is the ratio of the initial elastic modulus to the elastic modulus at time t.

An example of typical distributions of shrinkage when solving Eqs. 2.23–2.26 is shown in Figure 2.24 following the work of Hasan et al. (2019), where, in part a), a

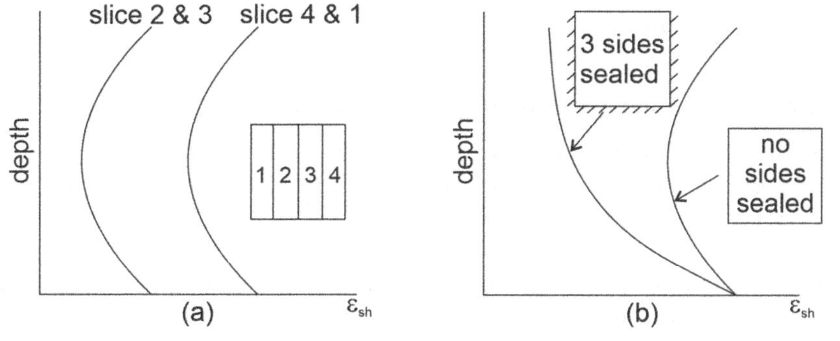

Figure 2.24 Shrinkage: (a) non-linear distribution of shrinkage within a specimen; (b) impact of boundary conditions on distribution of shrinkage within a specimen.

beam has been divided into four slices across its breadth and the distribution of shrinkage strains throughout the depth is shown assuming all boundaries are exposed to the atmosphere. In this scenario, it is observed that shrinkage is highly non-linear throughout the member depth and along its breadth because the edges of the beam dry at a faster rate than the core, resulting in higher shrinkage strain in these locations. These variations can be incorporated into the flexural analysis in Figure 1.29.

The same beam as in Figure 2.24a is analysed in Figure 2.24b, but in this case, the surface exposure is limited on three sides, as shown. The results are for a slice through the centre of the beam. The results for 'no sides sealed' are similar to those in Figure 2.24a, as expected. With limited exposure, as in 'three sides sealed', the moisture is forced to move in only one direction, causing the distribution of shrinkage to be more linear throughout the depth, with lower shrinkage at the top surface because all moisture must evaporate from the unsealed surface, thereby increasing the humidity at this surface relative to the rest of the section.

In addition to being dependent on environmental humidity, the ability of surfaces to exchange moisture depends on the environment and member size. Drying shrinkage strains are also strongly related to concrete mix design, which significantly impacts permeability and, consequently, moisture diffusion. Further, the quantity and quality of coarse aggregates within the mix also play an important role in concrete shrinkage, as these particles are inert and, therefore, act as a shrinkage-resistant skeleton.

Naturally, if concrete is exposed to the environment, there will be scenarios in which the environmental humidity is higher than the concrete's humidity. In this case, moisture exchange occurs in the opposite direction and the concrete swells (extends) rather than shrinks (contracts). The process of swelling cannot, however, fully compensate for shrinkage.

Carbonation shrinkage occurs as the atmospheric CO_2 reacts with moisture in the concrete, leading to degradation of the cement matrix, resulting in shrinkage. The rate at which concrete carbonation occurs is typically very slow and so carbonation shrinkage occurs over many years (Kovler and Zhutovsky 2006).

Thermal shrinkage and expansion occur as the temperature of concrete changes, with the resulting change in strain given by

$$\varepsilon_T = \alpha_T \Delta T \tag{2.27}$$

in which α_T is the coefficient of thermal expansion in units of K^{-1} and ΔT is the change in temperature in Kelvin.

2.5.2 Approaches for predicting concrete shrinkage strains

Given the complexity of the mechanisms that drive concrete shrinkage and the challenges in experimentally measuring it, unsurprisingly, there are significant difficulties in developing generic material models that allow the prediction of concrete shrinkage strains. While there are numerous models of varying complexity based on research (e.g. Bažant and Jirásek 2018), code approaches generally divide the total shrinkage strain ε_{cs} into an autogenous component ε_{cse} and a drying component ε_{csd}.

The autogenous component generally follows an exponential form, meaning that the final autogenous shrinkage strain occurs over very short time periods. The maximum autogenous strain is a function of the reactivity of the binder, which is generally expressed as a function of the concrete compressive strength and which may be modified when the cement type is changed or SCMs are introduced, as both can impact the rate of reaction and therefore autogenous shrinkage.

As described above, drying shrinkage is a diffusion process; therefore, codes of practice generally use a simplified approach to determine the drying shrinkage strain, employing an exponential function to express the change in shrinkage strain with time. This function is modified to account for the amount of concrete exposed to the environment and the environmental conditions (temperature and humidity).

2.5.3 Concrete creep

Concrete creep is a time-dependent change in the stiffness of the concrete that occurs in the presence of load. That is, over time and in the presence of a sustained stress, the concrete strain will increase as shown in Figure 2.22. Importantly, creep occurs even when loads are sustained for very short periods, with durations significantly less than one day leading to concrete creep. This is significant, because as shown in Figure 2.22, concrete creep causes irreversible damage that results in residual strains even after loads have been removed. For analyses, such as the flexural analyses illustrated in Figure 1.29, creep is commonly incorporated by simply reducing the strain-based modulus, in contrast to shrinkage, which requires a change in dimensions. The impact of concrete creep on the behaviour of reinforced concrete sections, therefore, is an increase in the rotation of a segment or the curvature of a cross-section, leading to increased member deflection and crack width over time.

The mechanism of concrete creep, much like concrete shrinkage, can be divided into two major components:

- **Basic creep** occurs in the absence of moisture exchange with the environment and is caused by two major actions: (i) the stress-induced movement of water towards larger diameter pores, which in turn induces stress and damage to the microstructure; and (ii) the slip of the cement paste microstructure, facilitated by absorbed water (Bažant and Jirásek 2018).
- **Drying creep** occurs in the presence of moisture exchange with the environment and its mechanics are harder to fully quantify, but it is likely the result of the impact of moisture on the local distribution of stress, which leads to damage and re-formation of chemical bonds within the cement. Given that creep is dependent on moisture movement, as described above for concrete shrinkage, changing member dimensions and surrounding environmental conditions also have an impact on the rate of moisture exchange and, therefore, drying creep (Neville and Brooks 2010; Gilbert and Ranzi 2010).

Importantly, given that the mechanisms of creep are generally related to damage to cement and that aggregates generally exhibit very little creep under typical serviceability loads, the use of stiff aggregates tends to act as a stiff skeleton, resisting creep. Further, the magnitude of concrete creep is significantly impacted by the magnitude of the applied load and the age at which the load is applied. Hence, the larger the applied load, the higher the magnitude of creep; and the earlier the load is applied, the greater the creep. Again, these relationships are relatively intuitive when considering that the mechanisms of creep are related to stress-induced moisture movements and damage generated by stress.

2.5.4 Approaches to predicting concrete creep

As with concrete shrinkage, complex models have been developed to predict concrete creep strains, and substantial information on research-based models can be found in sources such as Bažant and Jirásek (2018). Alternatively, code-based models are generally based on the definition of a concrete creep coefficient φ_{cc}, which is defined as the ratio of the creep strain at time t ($\varepsilon_{cc}(t, t_0)$) to the strain induced by the instantaneous stress at time t_0 ($\varepsilon_i(t_0)$):

$$\varphi_{cc}(t,t_0) = \frac{\varepsilon_{cc}(t,t_0)}{\varepsilon_i(t_0)} \tag{2.28}$$

Hence, using the creep coefficient φ_{cc}, an effective elastic modulus that accounts for creep at any time t resulting from a load applied at t_0 can be defined as

$$E_c(t,t_0) = \frac{E_c(t_0)}{\left[1+\varphi_{cc}(t,t_0)\right]} \tag{2.29}$$

Models for the design creep coefficient φ_{cc} are empirically developed and generally contain modifiers to account for the degree of exposure to the environment, the environmental conditions (temperature and humidity), the age at initial loading and the magnitude of the initial loading. Of significant importance regarding creep is the impact of the time of initial loading and the loading magnitude, as large loads applied at early ages can significantly increase the magnitude of creep throughout the lifespan of the concrete element.

2.6 MECHANICAL PROPERTIES OF REINFORCEMENT

Given the very low tensile strength of concrete, it is rarely used in an unreinforced state; rather, it is reinforced with materials that have relatively high tensile strengths, the reinforcement acting to transfer forces across concrete cracks.

Concrete reinforcement can generally be divided according to its placement, that is

i. Internal reinforcement such as steel or *FRP* bars or prestressing tendons.
ii. External reinforcement typically used for the strengthening of existing structures (covered in Chapter 11). External reinforcement typically takes the form of steel or *FRP* plates or fabrics that may be adhesively or mechanically anchored to the concrete surface.

In this chapter, the instantaneous and time-dependent material properties of steel and *FRP* are discussed; the impact of durability (corrosion of steel) is covered in Chapter 12.

2.6.1 Steel reinforcement stress–strain relationship

Reinforcing bars are provided on the basis of minimum tensile strengths and can be further classified according to ductility. Low-ductility bars are most commonly used in the production of welded wire fabrics that can serve as primary reinforcement for strength or secondary reinforcement to limit crack widths. When specified as primary reinforcement, it is common for codes of practice to apply larger capacity reduction factors for low-ductility bars (Standards Australia 2018), as it is commonly believed that the ductility of the reinforcement material is directly responsible for member-level ductility, leading to moment redistribution. However, as shown in Chapter 1 and further in Chapter 8, material ductility is not required to achieve ductile members. This is because the partial-interaction load-slip mechanisms of tension stiffening, reinforcement debonding and concrete softening provide significant ductility regardless of the ductility of the reinforcing bar.

At present, the yield stress of bars commonly used in practice varies between 250 and 500 MPa, though the vast majority are either 400 or 500 MPa. Ribbed reinforcing bars with yield stresses above 750 MPa are also available, the higher yield strength allowing for a smaller total area of reinforcement. This allows for a reduction in the congestion of reinforcement in joints, and the minimisation of the environmental impact of reinforcement by reducing the mass that must be used to achieve a given level of performance. Prestressed reinforcement typically consists of strands produced from cold-drawn wires spun together; these strands typically have a yield strength significantly higher than conventional reinforcing bars (between 1300 and 1600 MPa).

The stress–strain relationship for steel reinforcement is shown in its generic form in Figure 2.25, where the extent of the yield plateau and the material's ductility are generally defined by the way the steel is produced (hot rolled or cold worked). Cold-worked steels are generally stronger but have reduced ductility. This occurs because cold working increases the number of dislocations within the crystalline structure of the metal, making it harder for the crystals to displace relative to one another during deformation.

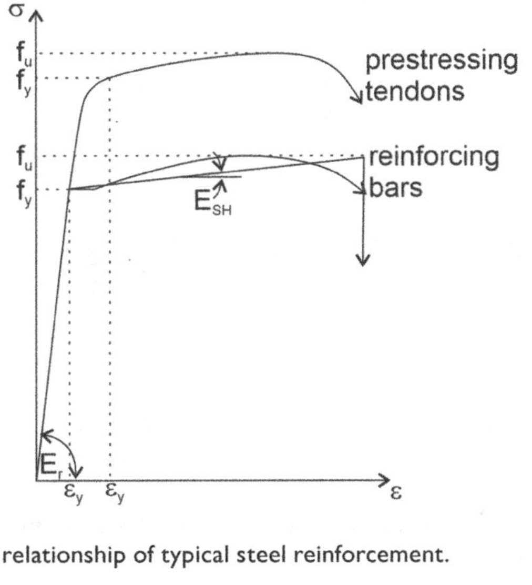

Figure 2.25 Stress–strain relationship of typical steel reinforcement.

Although the stress–strain relationship of steel is shown as non-linear in Figure 2.25, it is uncommon to use a non-linear stress–strain relationship for design. In practice, a bilinear relationship is used, as shown in Figure 2.25 and given by Eq. 2.32, which consists of a linear path with a modulus of E_r up to the yield stress f_y and strain ε_y, followed by a linear path with the strain hardening modulus E_{sh} up to fracture f_u.

$$\sigma = \begin{cases} E_r\varepsilon; \ \varepsilon \le \varepsilon_y \\ f_y + E_{SH}\left(\varepsilon - \varepsilon_y\right); \ \varepsilon > \varepsilon_y \end{cases} \tag{2.30}$$

where, in Eq. 2.32, E_r is the elastic modulus, which is generally taken as 200 GPa; E_{sh} is the strain hardening modulus, which can generally be taken as approximately 10% of E_r. For routine design, it is common to further simplify the relationship to an elastic–perfectly plastic model by setting the strain hardening modulus to zero.

Although it is common practice to assume that the stress–strain relationship of steel reinforcing bars is identical in tension and compression, in reality, it may not be possible to reach their full strength or ductility when loaded in compression. This is because reinforcing bars are slender, and as such, may buckle if not restrained along their length. For example, consider the concentrically loaded reinforced concrete column in Figure 2.26a subjected to an axial load less than that required to cause the concrete to crush. In this scenario, the concrete cover is intact, and hence, the reinforcing bars are continuously supported along their entire length by the cover concrete. Once the load P_{ax} reaches that required to cause the concrete to crush in compression at a strain ε_0, the cover concrete begins to spall. Hence, in Figure 2.26b, the reinforcing bars in compression are unsupported along their length and are free to buckle along the effective length L_e. Importantly, given that spalling of the cover concrete generally occurs at a strain similar to or greater than the yield strain, buckling of reinforcing bars is generally considered to occur after the reinforcement has yielded.

This behaviour is important because even though the yield strength may not be compromised by buckling, after buckling has occurred, the reinforcement effectively strain softens as shown in Figure 2.27. Hence, the assumption that reinforcement is

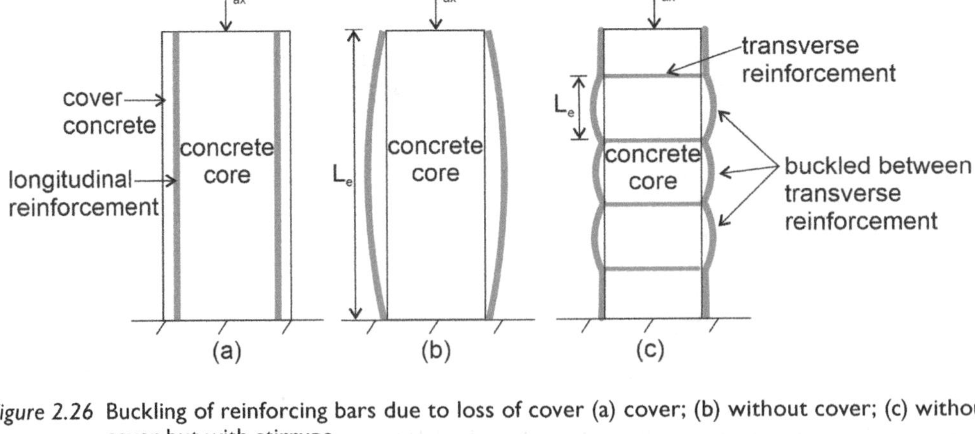

Figure 2.26 Buckling of reinforcing bars due to loss of cover (a) cover; (b) without cover; (c) without cover but with stirrups.

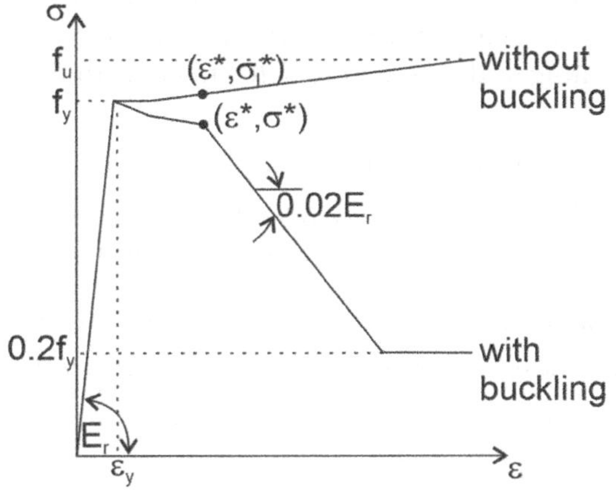

Figure 2.27 Stress–strain relationship of steel with and without buckling.

elastic–perfectly plastic, commonly applied in analysis techniques, does not apply unless reinforcement buckling is prevented.

Adding transverse reinforcement in Fig. 2.26c reduces the effective length L_e over which buckling occurs, and as a result, the residual strength and ductility of the bar are improved. Hence, the presence of transverse reinforcement can lead to a significant improvement in the performance of longitudinal reinforcement by preventing strain softening due to buckling.

Using the results of experimental testing, Dhakal and Maekawa (2002) developed a piecewise compressive stress–strain relationship for steel reinforcing bars that accounts for the reduction in strength caused by buckling over a length L_e, as shown in Figure 2.26. Prior to yield, linear elastic conditions apply, and Eq. 2.30 is used. After the yield point, a linear descending relationship, as shown in Figure 2.27, is followed until an intermediate strain ε^* and stress σ^* point is reached, after which a linear descending modulus of $0.02E_r$ applies, with a lower limit on the stress of $0.2f_y$. This can be expressed mathematically as

$$\frac{\sigma}{\sigma_l^*} = 1 - \left(1 - \frac{\sigma^*}{\sigma_l^*}\right)\left(\frac{\varepsilon - \varepsilon_y}{\varepsilon^* - \varepsilon_y}\right); \quad \varepsilon_y < \varepsilon \le \varepsilon^* \tag{2.31}$$

$$\sigma = \max\left(\sigma^* - 0.02E_r\left(\varepsilon - \varepsilon^*\right), 0.2f_y\right); \quad \varepsilon > \varepsilon^* \tag{2.32}$$

To apply Eqs. 2.31 and 2.32, the intermediate strain ε^* and the ratio of $\frac{\sigma^*}{\sigma_l^*}$ have been determined empirically by Dhakal and Maekawa (2002) to be

$$\frac{\varepsilon^*}{\varepsilon_y} = 55 - 2.3\sqrt{\frac{f_y}{100}}\frac{L_e}{D}; \quad \frac{\varepsilon}{\varepsilon^*} \ge 7 \tag{2.33}$$

$$\frac{\sigma^*}{\sigma_l^*} = \alpha\left(1.1 - 0.016\sqrt{\frac{f_y}{100}}\frac{L_e}{D}\right); \quad \sigma^* \ge 0.2f_y \tag{2.34}$$

where σ^* is the stress at strain ε^* if buckling were not to occur; α is a calibratable factor which was found to be between 0.75 and 1 and is taken as closer to 1 for materials with significant strain hardening and closer to 0.75 for those with limited strain hardening.

Worked Example WE-2.1: Reinforcement buckling

Let us apply the reinforcement buckling model of Dhakal and Maekawa (2002) using Eqs. 2.31–2.34 by considering a 16 mm bar supported by transverse reinforcement at 250 mm spacing, where L_e is taken as 250 mm. The yield stress of the bar, f_y, is taken as 500 MPa, and the strain hardening modulus, E_{sh}, as 1000 MPa$\alpha = 0.75$.

The ratio of the critical strain to the yield strain $\frac{\varepsilon^*}{\varepsilon_y}$ in Figure 2.27 is

$$\frac{\varepsilon^*}{\varepsilon_y} = 55 - 2.3\sqrt{\frac{500}{100}}\frac{250}{16}; \quad \frac{\varepsilon^*}{\varepsilon_y} \geq 7 = -25.3 \tag{E2.1.1}$$

$$\therefore \frac{\varepsilon^*}{\varepsilon_y} \geq 7 \tag{E2.1.2}$$

and

$$\varepsilon^* = 0.0175 \tag{E2.1.3}$$

Having identified the critical strain ε^*, the corresponding stress for the unbuckled bar can be determined using Eq. 2.30:

$$\sigma_l^* = 500 + 1000(0.0175 - 0.0025) = 515 \text{ MPa} \tag{E2.1.4}$$

The ratio of the stress of the buckled bar to that of the unbuckled bar at the critical strain is $\frac{\sigma^*}{\sigma_l^*}$ is then determined using Eq. 2.34:

$$\frac{\sigma^*}{\sigma_l^*} = 0.75\left(1.1 - 0.016\sqrt{\frac{500}{100}}\frac{250}{16}\right); \quad \sigma^* \geq 0.2 \cdot 500 \tag{E2.1.5}$$

$$\frac{\sigma^*}{\sigma_l^*} = 0.406 \tag{E2.1.6}$$

Checking the lower limit of Eq. 2.34,

$$\sigma^* = 0.406 \cdot 515 = 205 \text{ MPa} > 0.2 \cdot 500 = 100 \text{ MPa} \tag{E2.1.7}$$

Now, by applying Eqs. 2.31 and 2.32 and using the results from Eqs. E2.1.3 and E2.1.6 for a range of different strains ε, the complete stress–strain relationship is obtained as shown in Figure 2.28, where it can be seen that the bar strain softens immediately after reaching the yield stress.

Repeating the analysis for a range of different effective lengths, it can be seen that the buckling resistance is highly dependent on L_e, but that the relative reduction in residual strength decreases as L_e increases.

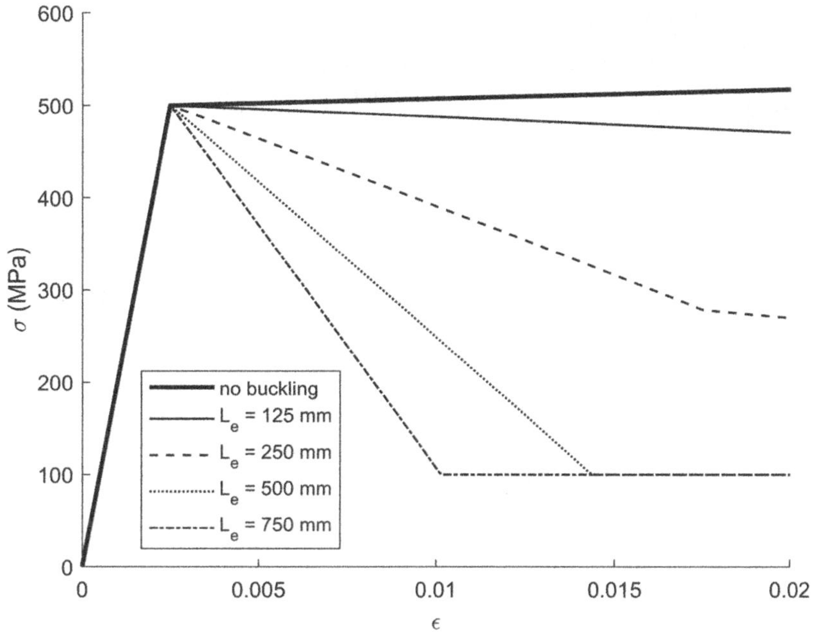

Figure 2.28 Stress–strain of steel reinforcing bar with various effective lengths.

2.6.2 Steel reinforcement relaxation and thermal effects

In the above discussion on reinforcement, it has been assumed that the stress–strain response occurs instantaneously. However, steel, much like concrete, experiences time- and temperature-dependent changes in behaviour. The mechanisms that cause these time- and temperature-dependent changes in behaviour are related to the continued movement of the grain structure of the material at the micro-scale, and for a detailed understanding, the reader is referred to more detailed texts on the science of metallurgy.

Of principal importance for the structural engineer is that, while the time- and temperature-dependent deformations of steel are generally not important for the design of reinforced concrete elements, they are critical to the design of prestressed concrete elements. This is because the time-dependent deformations of steel are highly dependent on the magnitude of the sustained stress; at the serviceability limit state, this is small for all non-prestressed reinforcement.

To explain the impact of relaxation on prestressed reinforcement, let us consider the construction and behaviour of a pre-tensioned concrete beam using Figure 2.29. In Figure 2.29a, a prestressed concrete beam is constructed in a pre-tensioning bed that consists of a rigid frame between which the prestressing tendon is first loaded to the force P_{pr} (or stress σ_{pr}) by jacking against the frame.

After the tendon is stressed, a concrete beam is cast around the tendon and left to cure. Once the concrete has sufficiently cured, the jack is released and the pre-tensioning force is transferred from the prestressing bed to the concrete beam with a loss of force from P_{pr} to P_{pr}^{*} due to the contraction of the concrete beam. That is, the prestressing force is

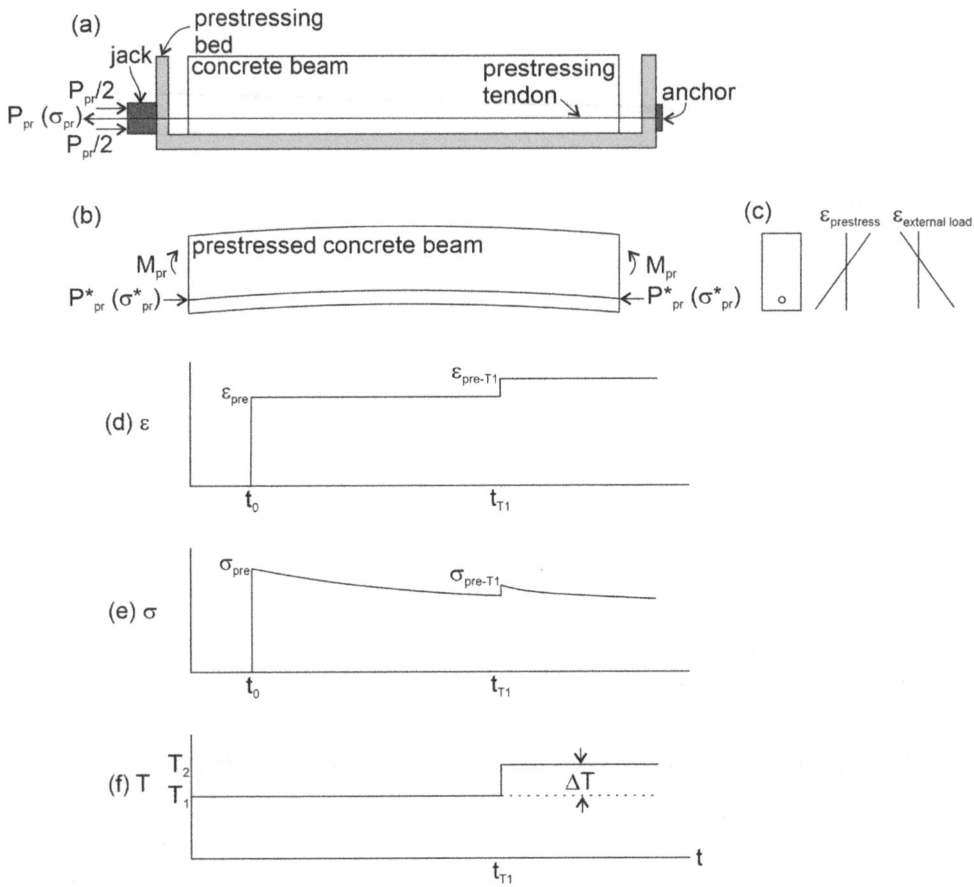

Figure 2.29 Relaxation of prestressing tendon (a) prestressing bed; (b) prestressed beam external actions; (c) strain distribution from prestress and external load; (d) variation in strain with time; (e) variation in stress with time; (f) variation in temperature with time.

no longer resisted by the prestressing bed, but rather by the reinforced concrete beam through the bond developed between the prestressing tendon and the surrounding concrete. As a result of this transfer of forces, the prestressed beam can be thought of as a column subjected to an eccentric load as shown in Figure 2.29b. As shown in Figure 2.29c, the strains developed within the section at the transfer of prestress have the opposite sense to those developed upon application of the external load, and therefore, the action of prestressing significantly increases the capacity of a prestressed concrete beam relative to the same cross-section without prestressing. It should be noted that prestressing, both pre-tensioning and post-tensioning, is dealt with in great detail in Chapter 9.

Now consider the impact of time in Figure 2.29. The prestressing force P^*_{pr} is applied at time t_0, causing the prestressing tendon to strain to ε_{pr} (Figure 2.29d) and inducing a stress σ_{pr} (Figure 2.29e); this occurs when the temperature Figure 2.29f is T_1 and changes at time t_{T1}. Between t_0 and t_{T1}, there is a reduction in strain (Figure 2.29d) and in tendon stress (Figure 2.29e) due to steel relaxation. As the relaxation of the tendon reduces the level of prestress, the initial contracting strain induced at the bottom of the beam in Figure 2.29c is reduced, and consequently, the capacity of the member is also reduced.

If we now consider the impact of changing the temperature to T_2 (Figure 2.29f), the increase in temperature by ΔT causes a change in the tendon's extension due to thermal expansion (Figure 2.29d). This increase in strain due to thermal expansion can be determined using Eq. 2.29, with the coefficient of thermal expansion α_T taken as that of steel, typically in the range of $11 \cdot 10^{-6} - 13 \cdot 10^{-6}$/K. In addition to increasing the tendon strain, the rate of relaxation in Figure 2.29e is also increased as higher temperatures allow for easier movement of the grain microstructure of the steel.

It is important to note that in this conceptual example, it has been assumed that the concrete section is very stiff, such that the tendon strain does not change over time. In reality, this is not the case, as the mechanisms of concrete shrinkage and creep cause the concrete to deform over time and in response to changes in environmental conditions. There is, therefore, a complex interaction between concrete shrinkage and creep, tendon relaxation and the performance of prestressed concrete members, which must be considered in design. Failure to adequately consider time effects including concrete shrinkage and creep when designing prestressed concrete can lead to significant failures; for example, the collapse of the 240.8 m central span of the Koror–Babeldoab Bridge in Palau, which led to multiple fatalities.

For a given initial stress, the relaxation of prestressing tendons is determined empirically by extending a tendon between a rigid frame, such as the prestressing bed in Figure 2.29a, and recording the change in stress over time and in response to temperature changes. The degree of relaxation is often provided directly by manufacturers of prestressing tendons or in codes of practice, where code-based models modify a basic relaxation parameter with factors to account for the magnitude of sustained load, the duration of prestressing and the environmental temperature.

2.6.3 FRP bars, sections and fabrics

The application of FRPs in civil engineering has experienced significant growth in the last 30 years. Applications typically fall into two general areas: (i) as internal reinforcing bars for new construction; and (ii) as external reinforcement for retrofitting (strengthening) applications. FRPs are composite materials consisting of strong and stiff fibres embedded in a relatively weak and flexible resin matrix. As composites, the properties of *FRPs* are a function of those of the fibres and the resin matrix, and because fibres are generally aligned in a uniaxial direction, this imparts strong anisotropic behaviour to the composites.

In civil engineering applications, bars and sections are typically available in a pultruded form, whereby the final structural shapes are produced in a factory setting through a process that combines pulling and weaving of the fibres after impregnation with the matrix. The alternative to pultrusion is wet layup of fibres, in which dry fibre fabric is impregnated with the matrix on site to produce the final desired form. Various types of fibres have been developed and the typical ranges of material properties are presented in Table 2.1.

Table 2.1 Typical ranges of strength and elastic modulus for different fibre types

Fibre type	*Tensile strength (MPa)*	*Elastic modulus (GPa)*
Carbon	2500–4000	215–650
Glass	1800–3500	69–85
Basalt	2800–3100	85–87
Aramid	2700–3500	60–175
Large rupture strain, e.g. polyethylene terephthalate (PET) or polyethylene naphtholate (PEN)	140–220	5–25
Natural fibres, e.g. jute or flax	200–800	8–75

To produce *FRP*, the fibres are bound by a matrix intended to be durable and ideally possessing an elongation capacity greater than that of the fibres, so that the full capacity of the fibres can be utilised. Most commonly, the matrix is either an epoxy resin, vinyl ester or polyester. Resins generally have tensile strengths of 30–100 MPa and elastic moduli of approximately 2.0–3.5 GPa.

While a significant number of fibre types are available, in practice, glass *FRP* (*GFRP*) is currently used almost universally as internal reinforcement due to its low cost. Carbon *FRP* (*CFRP*), however, is generally restricted to use as prestressing tendons because of its high strength and associated high cost. For wet layup applications common in strengthening projects, *CFRP* is most commonly used because of its very high specific strength, but other fibre types are increasingly being considered and are the focus of research.

2.6.4 FRP reinforcement stress–strain relationship

Typical stress–strain relationships for commonly used *FRP*s are provided in Figure 2.30, where approximate upper and lower bounds are given for each material type. A comparison of these bounds shows that material properties can vary widely. Significantly, unlike steel, which exhibits significant plasticity prior to failure, *FRP*s (with the exception of large rupture strain *FRP*s) are elastic–brittle, and this lack of material ductility needs to be considered in design. It should be remembered that the rotational ductility of an *RC* member does not rely solely on the yield of the tension reinforcement. In fact, it will be shown in Chapter 6 that concrete crushing is the primary component controlling rotation and that an *FRP* reinforcement member can be designed for significant amounts of rotation and, consequently, moment redistribution, as described in Chapter 8. Hence, the brittle nature of *FRP* reinforcement need not and should not restrict its use, but it does require a new approach to design.

It is also important to note that, as a composite material, both strength and stiffness are functions of the properties of the fibres and the matrix. While the strength of the *FRP* composite is strongly controlled by the fibres, the full fibre strength is only

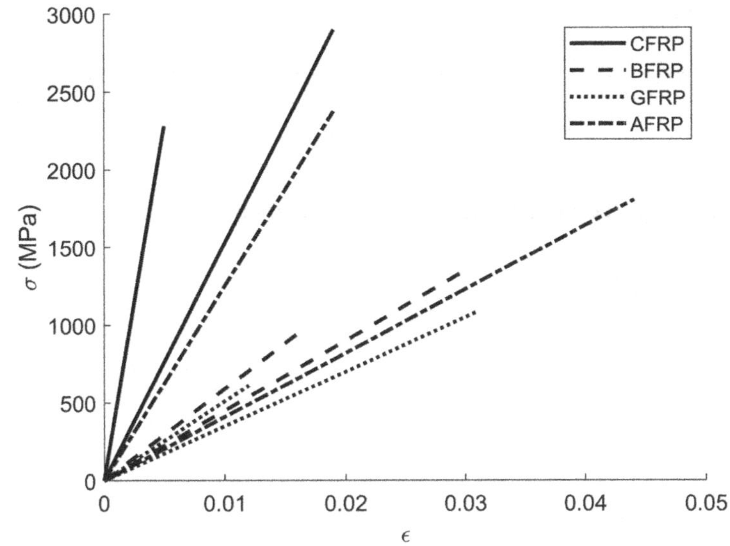

Figure 2.30 Typical upper and lower bound stress-strain for different types of *FRP*.

achieved if the matrix is sufficiently strong to ensure the load is shared between the individual fibre strands. The following rule of mixtures (Campbell, 2010) can be used to provide an upper bound to the ultimate tensile strength of the composite, $f_{rup \cdot frp}$,

$$f_{rup.frp} \leq f \cdot f_{rup.fib} + (1-f) f_{rup.mat} \tag{2.35}$$

in which f is the fibre volume fraction, given by the ratio of the fibre volume (V_f) to the total *FRP* volume (including the matrix volume, V_{mat}), as follows:

$$f = \frac{V_{fib}}{V_{fib} + V_{mat}} = \frac{A_{fib}}{A_{fib} + A_{mat}} \tag{2.36}$$

Let us examine this rule of mixtures more closely. The components of the composite *FRP* material are shown in Figure 2.31a, where, in any cross-section, two components are observed a matrix with cross-sectional area A_{mat} and modulus E_{mat}; and fibres with cross-sectional area A_{fib} and modulus E_{fib}; such that the modular ratio is

$$n = \frac{E_{fib}}{E_{mat}} \tag{2.37}$$

An element of the *FRP* is stretched by εL, as shown in Figure 2.31b, inducing a force in the fibres F_{fib} and an associated stress σ_{fib}, and a force in the matrix F_{mat} and a stress σ_{mat}. The total force is F_{frp} and the associated stress is σ_{frp}.

The behaviour of each of the components of the *FRP* is shown in Figure 2.32, where the behaviour of the fibre component *O–A* is shown as linear elastic with a rupture

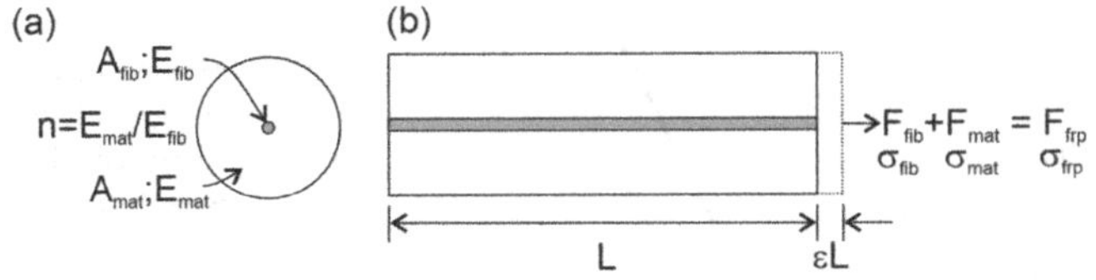

Figure 2.31 Composite properties (a) cross section; (b) elevation.

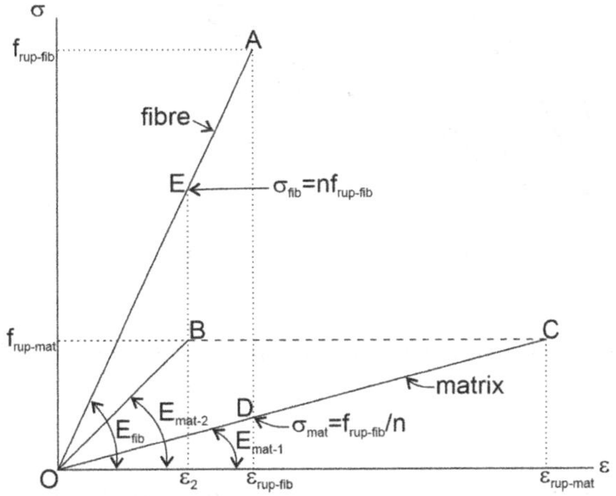

Figure 2.32 Rule of mixtures.

strain $\varepsilon_{rup\text{-}fib}$ and a rupture stress $f_{rup\text{-}fib}$, and the behaviour of the matrix is O–C with a rupture strain $\varepsilon_{rup\text{-}mat}$ and a rupture stress $f_{rup\text{-}mat}$. It should also be noted that, from simple geometry, at the rupture strain $\varepsilon_{rup\text{-}fib}$, the stress in the matrix σ_{mat} at Point *D* is $f_{rup\text{-}fib}/n$ (as shown); and at strain ε_2, the stress in the fibre σ_{fib} at Point *E* is $nf_{rup\text{-}mat}$ (also shown).

From Figure 2.31b, at a strain ε, the total force F_{frp} is equal to the force in the matrix F_{mat} plus the force in the fibres F_{fib}, which can be converted to stresses to give

$$\sigma_{frp} = \frac{\sigma_{mat}A_{mat} + \sigma_{fib}A_{fib}}{A_{mat} + A_{fib}} \tag{2.38}$$

The fibre volume fraction *f* in Eq. 2.38 is also the ratio of the cross-sections, as shown in the equation. Substituting this definition of *f* into Eq. 2.38 gives

$$\sigma_{frp} = \sigma_{mat}\left(1-f\right) + \sigma_{fib}\cdot f \tag{2.39}$$

When the rupture strain of the fibre $\varepsilon_{rup\text{-}fib}$ is less than the rupture strain of the matrix $\varepsilon_{rup\text{-}mat}$, as shown in Figure 2.32 for variations O–A and O–C, then $\varepsilon_{rup\text{-}fib}$ controls failure. At this strain, and from Figure 2.32, the stress in the fibres is $f_{rup\text{-}fib}$ and the stress in the matrix is $f_{rup\text{-}fib}/n$. That is, when $\varepsilon_{rup\text{-}fib} < \varepsilon_{rup\text{-}mat}$, then

$$f_{rup-frp} = f_{rup-fib}\cdot f + f_{rup-fib}\left(1-f\right)/n \tag{2.40}$$

Conversely, when $\varepsilon_{rup\text{-}mat} < \varepsilon_{rup\text{-}fib}$, such as O–B in Figure 2.32, then

$$f_{rup-frp} = f_{rup-mat}\left(1-f\right) + nf_{rup-mat} \tag{2.41}$$

Comparing Eqs. 2.40 and 2.41 with Eq. 2.35, it can be seen that the rule of mixtures equation is simply an upper bound when both components reach their maximum strength at the same strain, an unlikely scenario when both materials are linear elastic and brittle. If, for example, one of these components were elastic–plastic, such as O–B–C for the matrix in Figure 2.32, then both components could achieve their maximum strength at the fibre rupture strain $\varepsilon_{rup\text{-}fib}$.

Equations 2.40 and 2.41 may appear trivial but give insight into long-term behaviour. If the matrix material creeps with time, this reduces $E_{mat\text{-}1}$ in Figure 2.32, which in turn increases n. Hence, from Eq. 2.40, when $\varepsilon_{rup\text{-}fib} < \varepsilon_{rup\text{-}mat}$, the rupture strength reduces, but from Eq. 2.41, when $\varepsilon_{rup\text{-}mat} < \varepsilon_{rup\text{-}fib}$, the rupture strength increases. If the bond/slip between the fibre and the matrix softens as a form of creep or deterioration, but does not debond, then this will in effect reduce E_{mat}; that is, it acts as a form of creep and, therefore, as discussed for creep above, it can be either detrimental or beneficial.

Substituting the stresses σ in Eq. 2.39 with their appropriate $E\varepsilon$ values and cancelling out ε gives

$$E_{frp} = E_{mat}\left(1-f\right) + f \cdot E_{fib} \tag{2.42}$$

which gives the rule of mixtures (Campbell, 2010) for estimating the elastic modulus as

$$E_{frp} \leq f \cdot E_{fib} + \left(1-f\right)E_{mat} \tag{2.43}$$

Prior to rupture, this upper bound represents the true modulus. If an element ruptures, that component goes to zero, yielding the true modulus of the *FRP*.

Also of importance for the use of *FRP*s in structural engineering is that the material properties are generally highly anisotropic because the fibres are typically orientated in the direction of loading. Further, tests have shown that the compressive strength of pultruded bars is significantly lower than their tensile strength. Extensive testing has not yielded a single approach to quantifying the compressive strength of bars. However, it has identified that failure is typically due to transverse fracture of the matrix caused by the bar's expansion from Poisson's effect, and the buckling of individual fibres within the matrix, which can also induce transverse shear stresses sufficient to cause shear failure.

Worked Example WE-2.2: Rule of mixtures

If an *FRP* bar is to be produced with 60% fibres (tensile strength: 2000 MPa, elastic modulus: 300 GPa) and a matrix (tensile strength: 100 MPa, elastic modulus: 2 GPa), let us determine the bar's tensile strength and elastic modulus.

Using Hooke's law, the rupture strain of the fibre is

$$\varepsilon_{rup-fib} = \frac{2000}{300{,}000} = 0.00667 \tag{E2.2.1}$$

and similarly, the rupture strain of the matrix is

$$\varepsilon_{rup-mat} = \frac{100}{2000} = 0.05 \tag{E2.2.2}$$

Given $\varepsilon_{rup\text{-}fib} < \varepsilon_{rup\text{-}mat}$, Eq. 2.42 can be used to determine the rupture strength:

$$f_{rup-frp} = (2000)(0.6) + \frac{2000(1-0.6)}{\frac{300,000}{2000}} = 1205.3 \text{ MPa} \tag{E2.2.3}$$

The elastic modulus can be determined using Eq. 2.45

$$E_{frp} \leq 0.6 \cdot 300 + (1-0.6)2 \leq 180.8 \text{ MPa} \tag{E2.2.4}$$

2.6.5 *FRP* reinforcement time and thermal effects

As with steel, *FRP* is not immune to time-dependent behaviour. When subjected to an applied load, *FRP* materials creep. This occurs because, although the fibres themselves do not creep significantly, the matrix generally has a very low modulus of elasticity and undergoes significant creep. As a result, over time, the stress initially resisted by the composite (fibres and matrix) must be entirely resisted by the fibres.

This is shown in Figure 2.33. At a temperature of T_1 (Figure 2.33c), a bar is loaded to a sustained stress of σ_0 (Figure 2.33b), which causes an initial strain of

$$\varepsilon_0 = \frac{\sigma}{E_{mat}(1-f) + f \cdot E_{fib}} \tag{2.44}$$

Over time, the matrix undergoes creep, resulting in a loss of stiffness. Hence, to maintain the same overall stress, the load is transferred from the matrix into the fibres.

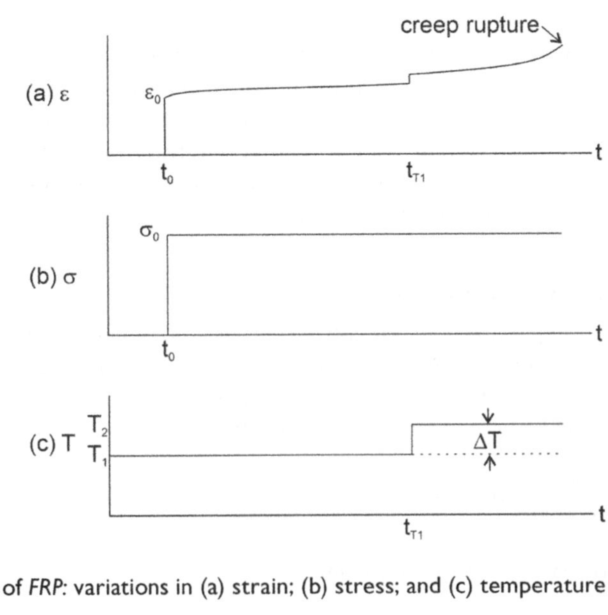

Figure 2.33 Creep of *FRP*: variations in (a) strain; (b) stress; and (c) temperature with time.

At time t_{t1}, the temperature is increased from T_1 to T_2 (Figure 2.33c), causing an instantaneous change in strain in the bar (Figure 2.33a), the magnitude of which is dependent on the coefficient of thermal expansion according to Eq. 2.29. The rate of creep increases with temperature, and as the duration of loading increases, the stiffness of the matrix tends to zero such that the stress is resisted entirely by the fibres.

Importantly, in this case, the strain in the fibres can rise to a level where the fibres can no longer resist the applied stress, and the *FRP* ruptures in a process known as creep rupture. To minimise the likelihood of creep rupture, it is common practice to ensure that the sustained stress in Figure 2.33b is kept between 30% and 80% of the rupture stress for *GFRP* and CFRP bars and tendons respectively.

Similar to steel bars, when FRP bars are used in prestressed concrete such that they are subjected to a constant strain, the material relaxes, causing a reduction in stress. Given the scarcity of test data on *FRP* creep and relaxation, generic material models are not currently available and manufacturer guidance is generally provided on the basis of product testing.

When considering strain changes due to temperature, it is also important to note that because *FRP* is a composite, the coefficient of thermal expansion is different in the longitudinal and transverse directions. Typical ranges for longitudinal and transverse thermal expansion coefficients are between −9 and 10 in the longitudinal direction and 21 and 104 in the transverse direction; these also vary with fibre type (FIB 2007). It is therefore best to refer to specialist texts such as FIB (2007) and to the manufacturer for specific details. It is important to note, however, that the coefficient of thermal expansion can be negative in the longitudinal direction for materials such as *CFRP* and *AFRP*. This means that as the temperature increases, although the bar is expanding significantly in the transverse direction, it is contracting in the longitudinal direction. In this scenario, a reduction in strain would occur in Figure 2.33 at time t_{T1} even though the temperature is increasing. The significant transverse expansion of *CFRP* and *AFRP* may be important for bars embedded in concrete, as the transverse expansion may generate pressures sufficient to crack the concrete cover.

2.7 NOTATION AND ABBREVIATIONS

a	mass of aggregate in a concrete mix in kg/m^3
AFRP	aramid fibre-reinforced polymer
Ag	gross area of a cross-section
C	compressive force
CFRP	carbon fibre-reinforced polymer
c	mass of cement in a concrete mix in kg/m^3
D	diameter
$D(RH)$	moisture diffusion coefficient at relative humidity RH
$D1$	maximum diffusion coefficient (when RH is 100%)
df	diameter of fibres
E	modulus
Ec	elastic modulus of concrete
$Ec(t)$	elastic modulus at time t
$E_c(t,t_0)$	elastic modulus at time t allowing for creep due to sustained load applied at t_0

Ecf	E for FRC in tension
$Ecm28$	28-day mean elastic modulus
E_{fib}	elastic modulus of fibres in FRP
E_{mat}	elastic modulus of matrix in FRP
Er	E for steel reinforcement
E_{SH}	strain hardening modulus of reinforcement
FA	fly ash
FI	full interaction
FRC	fibre-reinforced concrete
FRP	fibre-reinforced polymer
f	fibre volume fraction of FRP
fc	compressive strength of unrestrained concrete
f'_c	characteristic concrete compressive strength
f_{cm}	mean concrete compressive strength
f_{cm28}	mean 28 day compressive strength
f_{cmi}	mean in-situ concrete compressive strength
f_{ct}	concrete tensile strength
f'_{ct}	characteristic direct tensile strength
$f_{ct.f}$	flexural tensile strength
$f'_{ct.f}$	characteristic flexural tensile strength
$f_{ct.sp}$	split or Brazilian tensile strength
$f'_{ct.sp}$	characteristic split tensile strength
f_{ctm}	mean tensile strength
$f_{rup.fib}$	rupture stress of fibres in FRP
$f_{rup.frp}$	rupture stress of FRP
$f_{rup.mat}$	rupture stress of matrix in FRP
f_{SH}	stress at the initiation of concrete strain hardening
f_u	ultimate stress of reinforcement
f_y	yield stress of steel
$GFRP$	glass fibre-reinforced polymer
$GGBFS$	ground granulated blast furnace slag
$g_s(t)$	ratio of the initial elastic modulus to the elastic modulus at time t
H	type of fibre category
ITZ	interfacial transition zone
$J(t,t_0)$	compliance function at time t due to sustained load applied at time t_0
HSC	high-strength concrete
l	length
L_e	effective length
L_{seg}	length of rectangular segment
l_f	length of fibres
M	bending moment
N	axial force
NSC	normal-strength concrete
n	modular ratio
OPC	Ordinary Portland Cement
P_{ax}	axial load on column

P_{pr}	prestressing force
PI	partial interaction
RC	reinforced concrete
PET	polyethylene terephthalate
PEN	polyethylene naphtholate
P_{ult}	force at failure
P_{tr}	transverse forces induced by splitting action
q_1	compliance function due to instantaneous applied load
R	degree of relaxation
RH	relative humidity
RH_{en}	relative humidity of the environment
RH_s	relative humidity of the concrete surface
RI	reinforcing index of fibre reinforcement
S_{pr}	crack spacing
SCM	supplementary cementitious material
SF	silica fume; shear friction
t	time in days
T	temperature in °C; tensile force
t_d	time at which shrinkage strain readings are initiated
t_f	time at which sustained stress is released
t_h	hypothetical thickness used when determining concrete shrinkage and creep
t_s	time at which drying of concrete commenced
t_0	time at which sustained stress σ_0 is applied
$t_{0,adj}$	adjusted t_0 to allow for the change in rate of strength development with temperature and cement class
$UHPC$	ultra-high-performance concrete
$UHPFRC$	ultra-high-performance fibre-reinforced concrete
u_e	perimeter of cross-section exposed to the environment
V_f	volume of fibres in concrete; volume of fibres in *FRP*
V_{mat}	total volume
V_u	shear at the onset of sliding
V/S	volume to surface area ratio
w	mass of water in a concrete mix in kg/m^3; crack width
$w{:}c$	water-to-cement ratio
α	angle of sliding plane
$\propto T$	coefficient of thermal expansion
$\propto T_L$	coefficient of thermal expansion in the longitudinal direction
$\propto T_T$	coefficient of thermal expansion in the transverse direction
Δf_{sh}	change in strength over strain hardening region
Δt_i	length of time period i
ΔT	change in temperature
Δ_{SF}	slip along sliding plane
δ	slip
ε	strain
ε_{ax}	axial strain
ε_{ci}	instantaneous strain applied to concrete

$\varepsilon_{cc}(t, t_0)$	concrete creep strain at time t due to sustained stress applied at time t_0
$\varepsilon_{csd.b}$	basic drying shrinkage strain
ε_{ct}	concrete tensile (macro) cracking stress
$\varepsilon_i(t_0)$	instantaneous strain applied at time t_0
ε_{sh}	concrete shrinkage strain
$\varepsilon_{sh\infty}$	ultimate shrinkage strain
ε_{SH}	strain at the initiation of concrete tensile strain hardening
ε_T	strain induced by change in temperature
ε_y	strain at which yield stress is first reached
ε_0	strain corresponding to stress f_c
$\varepsilon_{0\text{-}FRC}$	ε_0 for concrete with fibres
ε^*	reinforcement stress after buckling
ε^*_{cse}	final autogenous concrete shrinkage strain
$\varepsilon^*_{csd.b}$	final basic drying shrinkage strain
ε^*_l	reinforcement stress of unbuckled bar
υ	Poisson's ratio
ρ	concrete density
$\varphi(t, t_0)$	creep coefficient at time t as a result of a sustained load applied at time t_0
σ	stress
σ_{ax}	axial stress
σ_{conf}	lateral confinement
σ_{lat}	lateral confining stress
σ_N	stress normal to sliding plane or potential sliding plane; confining stress
σ_0	sustained stress
σ^*	reinforcement stress after buckling
σ^*_l	reinforcement stress of unbuckled bar
τ	shear stress
τ_c	shear stress due to σ_c
τ_{SF}	shear stress in *SF* material properties
τ_{sh}	shrinkage strain half-time

REFERENCES AND DEVELOPMENT

American Concrete Institute (ACI), 2002. ACI PRC-211-01: Standard practice for selecting proportions for normal, heavyweight and mass concrete. ACI Committee 211. American Concrete Institute Farmington Hills.

American Concrete Institute (ACI), 2008. ACI 209.2R-08: Guide for modelling and calculating shrinkage and creep in hardened concrete. ACI Committee 209. American Concrete Institute Farmington Hills.

Aslani, F. and Natoori, M., 2013. Stress-strain relationships for steel fiber reinforced self compacting concrete. *Structural Engineering and Mechanics, An International Journal*, *46*(2), 295–322.

Barros, J.A. and Figueiras, J.A., 1999. Flexural behavior of SFRC: testing and modeling. *Journal of Materials in Civil Engineering*, *11*(4), 331–339.

Bažant, Z.P. and Jirásek, M., 2018. *Creep and Hygrothermal Effects in Concrete Structures* (Vol. 225). Dordrecht, The Netherlands: Springer.

Bažant, Z.P. and Yunping, X.I., 1994. Drying creep of concrete: constitutive model and new experiments separating its mechanisms. *Materials and Structures*, *27*, 3–14.

Bhargava, P., Sharma, U.K. and Kaushik, S.K., 2006. Compressive stress-strain behavior of small scale steel fibre reinforced high strength concrete cylinders. *Journal of Advanced Concrete Technology*, *4*(1), 109–121.

Campbell, F.C., 2010. *Structural Composite Materials*. Materials Park, OH: ASM International.

Chen, Y., Visintin, P., Oehlers, D.J. and Alengaram, U.J., 2014. Size-dependent stress-strain model for unconfined concrete. *Journal of Structural Engineering*, *140*(4), 04013088.

Dhakal, R.P. and Maekawa, K., 2002. Modeling for postyield buckling of reinforcement. *Journal of Structural Engineering*, *128*(9), 1139–1147.

Ezeldin, A.S. and Balaguru, P.N., 1992. Normal-and high-strength fiber-reinforced concrete under compression. *Journal of Materials in Civil Engineering*, *4*(4), 415–429.

Federation for Structural Concrete (*fib*), 2013. *Fib Model Code for Concrete Structures 2010*. Lausanne, Switzerland: Wiley.

Gilbert, R.I. and Ranzi, G., 2010. *Time-Dependent Behaviour of Concrete Structures*. Boca Raton, FL: CRC Press.

International Federation for Structural Concrete (*fib*), 2007. Bulletin 40 FRP reinforcement in RC structures Technical report prepared by a working party of Task Group 9.3 FRP (Fibre Reinforced Polymer) reinforcement for concrete structures.

Kim, H.Y., Yang, K.H. and Lee, H.J., 2023. Tensile stress–strain relationship of fiber-reinforced concrete with different densities. *Construction and Building Materials*, 385, 131526.

Kovler, K. and Zhutovsky, S., 2006. Overview and future trends of shrinkage research. *Materials and Structures*, *39*(9), 827–847.

Le Hoang, A. and Fehling, E., 2017. Influence of steel fiber content and aspect ratio on the uniaxial tensile and compressive behavior of ultra high performance concrete. *Construction and Building Materials*, *153*, 790–806.

Lee, S.C., Oh, J.H. and Cho, J.Y., 2015. Compressive behavior of fiber-reinforced concrete with end-hooked steel fibers. *Materials*, *8*(4), 1442–1458.

Li, V.C., Wang, Y. and Backer, S., 1991. A micromechanical model of tension-softening and bridging toughening of short random fiber reinforced brittle matrix composites. *Journal of the Mechanics and Physics of Solids*, *39*(5), 607–625.

Li, Y. and Li, J., 2014. Capillary tension theory for prediction of early autogenous shrinkage of self-consolidating concrete. *Construction and Building Materials*, *53*, 511–516.

Hasan, N.M.S., Visintin, P., Oehlers, D.J., Bennett, T. and Sobuz, M.H.R., 2019. Time dependent deflection of RC beams allowing for partial interaction and non-linear shrinkage. *Materials and Structures*, *52*, 1–15.

Hewlett, P. and Liska, M. eds., 2019. *Lea's Chemistry of Cement and Concrete*. Oxford: Butterworth-Heinemann.

Mansur, M.A., Chin, M.S. and Wee, T.H., 1999. Stress-strain relationship of high-strength fiber concrete in compression. *Journal of Materials in Civil Engineering*, *11*(1), 21–29.

Nataraja, M.C., Dhang, N. and Gupta, A.P., 1999. Stress–strain curves for steel-fiber reinforced concrete under compression. *Cement and Concrete Composites*, *21*(5–6), 383–390.

Neves, R.D. and Fernandes de Almeida, J.C.O., 2005. Compressive behaviour of steel fibre reinforced concrete. *Structural Concrete*, *6*(1), 1–8.

Neville, A.M. and Brooks, J.J., 2010. *Concrete Technology*, Second Edition. Essex, England: Pearsons.

Ou, Y.C., Tsai, M.S., Liu, K.Y. and Chang, K.C., 2012. Compressive behavior of steel-fiber-reinforced concrete with a high reinforcing index. *Journal of Materials in Civil Engineering*, *24*(2), pp.207–215.

Popovics, S., 1973. A numerical approach to the complete stress-strain curve of concrete. *Cement and Concrete Research*, *3*(5), 583–599.

Přikryl, R., 2021. Geomaterials as construction aggregates: a state-of-the-art. *Bulletin of Engineering Geology and the Environment*, *80*(12), 8831–8845.

Radocea, A., 1994. A model of plastic shrinkage. *Magazine of Concrete Research*, *46*(167), 125–132.

Razmi, A., Bennett, T., Xie, T. and Visintin, P., 2025. A phenomenological model for chloride diffusion coefficient in concretes with traditional and blended binders and alternative fillers. *Construction and Building Materials*, *461*, 139783.

Sakata, K., 1983. A study on moisture diffusion in drying and drying shrinkage of concrete. *Cement and Concrete Research*, *13*(2), 216–224.

Shobeiri, V., Bennett, B., Xie, T. and Visintin, P., 2024. Mix design optimization of waste-based aggregate concrete for natural resource utilization and global warming potential. *Journal of Cleaner Production*, *449*, 141756.

Siddique, R., 2011. Utilization of silica fume in concrete: review of hardened properties. *Resources, Conservation and Recycling*, *55*(11), 923–932.

Someh, A.K. and Saeki, N., 1996. Prediction for the stress-strain curve of steel fiber reinforced concrete. *Proceedings of the Japan Concrete Institute*, *18*, 1149–1154.

Standards Australia, 2018, AS3600:2018 Concrete Structures. Sydney, Australia.

Standards Australia, 2022, AS3600:2018 Sup 1:2022 Concrete Structures – Commentary, Sydney, Australia.

Sun, M., Visintin, P. and Bennett, T., 2022. The effect of specimen size on autogenous and total shrinkage of ultra-high performance concrete (UHPC). *Construction and Building Materials*, *327*, 126952.

Taylor, H.F., 1997. *Cement Chemistry* (Vol. 2, p. 459). London: Thomas Telford.

Topçu, İ.B., 2013. High-volume ground granulated blast furnace slag (GGBFS) concrete. In *Eco-Efficient Concrete*, Pacheco-Torgal, F., Jalali, S., Labrincha, J. and John, V.M. (eds.) (pp. 218–240). Philadelphia, PA: Woodhead Publishing.

Visintin, P., Ali, M.M., Albitar, M. and Lucas, W., 2017. Shear behaviour of geopolymer concrete beams without stirrups. *Construction and Building Materials*, *148*, 10–21.

Visintin, P. and Oehlers, D.J., 2018. Fundamental mechanics that govern the flexural behaviour of reinforced concrete beams with fibre-reinforced concrete. *Advances in Structural Engineering*, *21*(7), 1088–1102.

Walker, D., Leonard, M., Metcalfe, A., Lambert, M., 2009. *Engineering Modelling and Analysis*. New York: CRC Press.

Wang, A., Zhang, C. and Sun, W., 2003. Fly ash effects: I. The morphological effect of fly ash. *Cement and Concrete Research*, *33*(12), 2023–2029.

Xie, T. and Visintin, P., 2018. A unified approach for mix design of concrete containing supplementary cementitious materials based on reactivity moduli. *Journal of Cleaner Production*, *203*, 68–82.

Xie, T., Visintin, P., Zhao, X. and Gravina, R., 2020. Mix design and mechanical properties of geopolymer and alkali activated concrete: Review of the state-of-the-art and the development of a new unified approach. *Construction and Building Materials*, *256*, 119380.

Chapter 3

Partial interaction slip-based material properties

This chapter addresses the shear friction (SF) mechanism that governs the behaviour of all cracks in RC members. It also examines bond slip (BS) between the reinforcement and the adjacent concrete, which is a special case of the SF mechanism.

It has been shown in Section 1.5 that the fundamental free body diagram that governs virtually all behaviour of reinforced concrete members is that shown in Figure 1.28, which is partially replicated in Figure 3.1. This free body is subjected to a moment $M_{seg}+V_{seg}L_{seg}$, axial force N_{seg} and shear force V_{seg} as shown on the *LHS* of the free body. These stress resultants induce corresponding stress resultants on the *RHS* of the free body along any arbitrary plane at an angle β, as shown. The question is how to determine the effects of these stress resultants: what deformations do they produce, how are they resisted and can they be resisted?

Assuming the moment on the *RHS* is resisted through Euler–Bernoulli deformation, whilst the free body or segment remains uncracked, the *FI* strain-based material properties in Chapter 2 can be used to determine these effects. It has been shown in Chapter 2 that these strain-based properties are quite versatile, allowing for phenomena such as creep, shrinkage, temperature changes, material buckling and corrosion.

Strain-based properties cannot directly allow for the cracking shown in Figure 3.1. Where concrete softening starts, via the formation of a wedge and subsequent sliding as shown, the behaviour is governed by the concrete's partial-interaction (*PI*) *SF* properties, which are covered in the first part of this chapter. Sliding can also occur along the inclined plane and is therefore also governed by the SF properties. Once the concrete cracks in tension, the bars bridging the crack slip relative to the adjacent concrete, and the forces in the bar are then controlled by the bar–concrete interface *PI BS* properties, which are covered in the second part of this chapter. Should the concrete contain fibres,

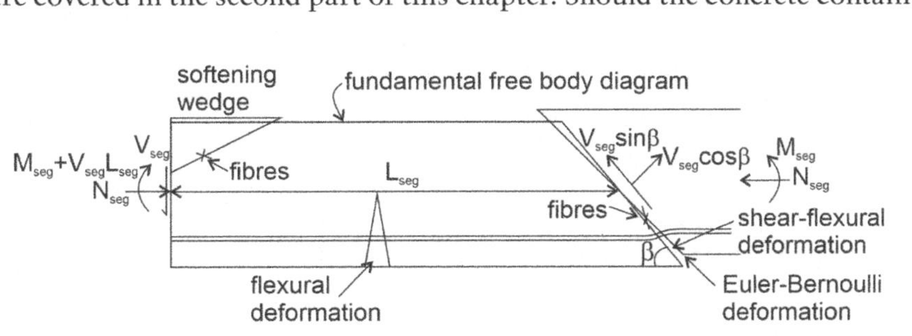

Figure 3.1 Fundamental free body diagram.

DOI: 10.1201/9781003711407-5

the fibres bridging the crack also resist crack widening. This resistance depends on the *PI* fibre BS properties, which define the fibre-reinforced concrete (FRC) post-cracking mechanism, covered in the third part of this chapter.

As explained in Section 1.5, it should be borne in mind that Figure 3.1 is the generic fundamental free body diagram. When β is 90°, the segment corresponds to that in Figure 1.29, which is often associated with a pure flexural analysis; as such, it is also governed by both the *FI* strain-based material properties and the *PI* slip-based material properties described below. It is also worth noting that the following three *PI* material properties, namely *SF*, *BS* and *FRC*, can be determined directly from tests, which is why this chapter on slip-based properties has been placed after Chapter 2 on strain-based properties. However, these three slip-based properties also depend on the tension stiffening mechanism, which is fully explained in Chapter 4 and can be used to extract the slip-based properties indirectly. Hence, it is suggested that the reader may wish to familiarise themselves with Chapter 4 before reading this chapter.

3.1 SHEAR FRICTION (*SF*)

3.1.1 SF mechanism

The SF mechanism has already been explained in Section 1.3.1 with the help of Figure 1.3. This figure is shown in Figure 3.2 with the addition of transverse reinforcing bars crossing the sliding plane.

The free body shown in Figure 3.2 is subjected to an external active force normal to the sliding plane, labelled *A*, and an external transverse shear force *V*, which induces a transverse slip Δ_{SF}. Due to the surface roughness, this slip Δ_{SF} causes crack widening w_{SF}, as shown, and this crack widening causes the transverse reinforcement to stretch, inducing a passive internal tensile force *P*. Consequently, through local equilibrium, as explained in Section 1.4.4 with the help of Figures 1.21 and 1.22, an internal passive compressive force *P* is induced across the sliding plane. The force in the reinforcing bars depends on the crack width w_{SF} and is controlled by the tension stiffening mechanism, as explained in Section 1.4.1 with the help of Figure 1.12. It is also worth bearing in mind that should there be any fibres crossing the sliding crack, they too will go into internal tension, inducing an additional internal compressive force *F* across the crack, in the same way as the lateral reinforcement. Finally, the shear displacement Δ_{SF} is resisted locally by the dowel force V_{dow}, as explained in Section 1.4.5 with the help of Eq. 1.2.

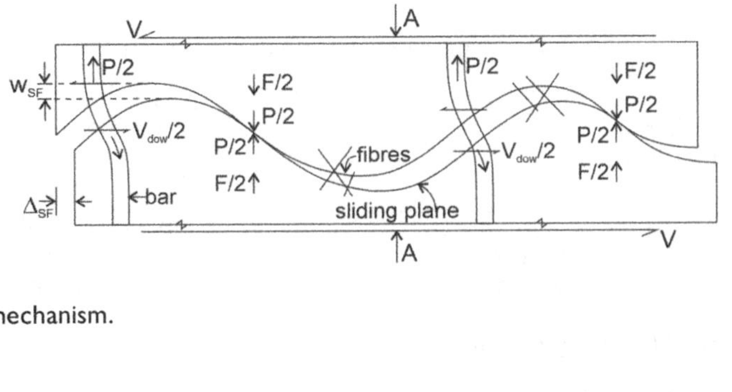

Figure 3.2 PI SF mechanism.

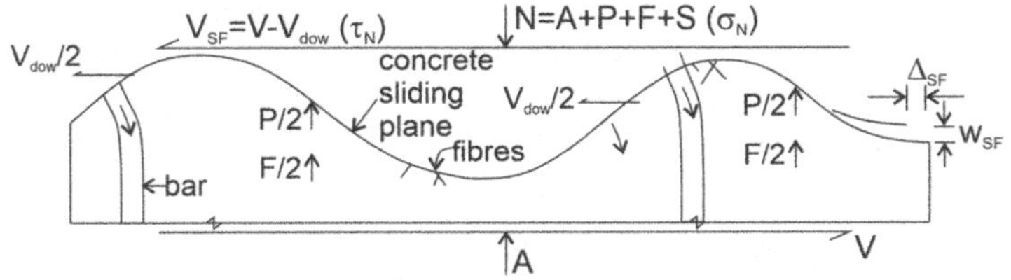

Figure 3.3 Forces on concrete sliding plane.

The ability of the sliding plane to resist the externally applied shear force V shown in Figure 3.2 depends on the forces in the sliding plane and the properties of the sliding plane. Consider the free body of the bottom half of the sliding plane shown in Figure 3.3. Along the sliding plane, the externally applied shear force V is resisted by the dowel action of the bars V_{dow} and the shear resistance of the concrete surface V_{SF}.

The question is what part of the shear is resisted by the dowel action V_{dow} and what part by SF V_{SF}. The solution is not easy, but it is realised that if the bar resists shear, its axial capacity must reduce, and vice versa. The dowel strength is not easy to quantify, as its maximum strength, while potentially given by Eq. 1.2, can be severely reduced by concrete failure through splitting, as given by Eq. 1.4. For a safe analysis, this book ignores any benefits from dowel action and assumes that the bar fails only in tension. Hence, the concrete along the sliding plane resists all the shear force; that is, V_{SF} equals V.

The next question concerns whether the sliding plane in Figure 3.3 can resist V_{SF}, and this depends to a great extent on the normal force N on the sliding plane. The normal force N acting on the concrete surface of the sliding plane is equal to the sum of the applied axial force A, and the passive forces induced by the tension stiffening action of the bar P and the fibres F. A fourth passive confining force due to shrinkage S also occurs when the sliding plane is encircled by concrete, for example, when a reinforcing bar is embedded in concrete. Hence the normal stress σ_N can be derived from the following forces:

$$N = A + P + F + S \tag{3.1}$$

The SF properties are often presented as in Figure 3.4. The variations are generally given for constant values of σ_N, such as those drawn with an unbroken line (e.g. V_{u1} $-V_{f1}$), where the normal forces are a combination of the active and passive forces as in Eq. 3.1. The variation for unconfined concrete, where the term 'unconfined' indicates that the normal force N in Eq. 3.1 is zero, has been labelled σ_{N0}. The upper ordinate is the interface shear stress τ_{SF}, which is derived from V_{SF}. The maximum value of the ordinate, which occurs at zero slip, is V_u, the shear stress required to first form the crack along the sliding plane. After which, the strength reduces to the frictional component V_f. The lower ordinate is the crack width w_{SF} and the abscissa is the interface slip Δ_{SF}.

SF theory is commonly used to simulate the behaviour of concrete-to-concrete interfaces when subjected to loads that generate both normal (σ_N) and shear (τ_{SF}) stresses. SF theory can, therefore, be applied to a wide range of scenarios and scales. For example, *SF* theory can be applied directly at the member level to quantify the shear strength of elements or the strength of joints constructed from different types of concrete or concrete cast at different times. While at the material level, SF theory can be used to

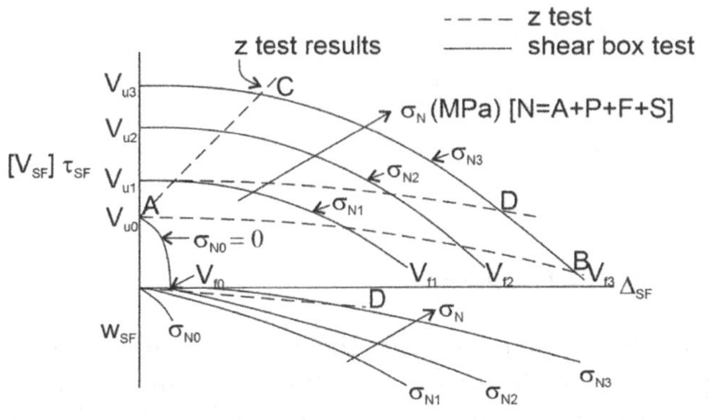

Figure 3.4 Schematic representation of SF properties.

simulate the ductility of concrete or the impact of concrete confinement on the axial and lateral stress–strain behaviour of concrete. Section 3.1.2 will first establish the general factors that impact the SF material properties, then examine approaches to either directly or indirectly quantify them from test results in Sections 3.1.3 and 3.1.4, and finally present general material models available in codes of practice.

3.1.2 SF material properties

3.1.2.1 Parameters that affect SF properties

The SF mechanism that controls sliding can, in its most simple form, be reduced to an elementary friction problem where the lateral force (V_{SF}) required to resist sliding is proportional to the coefficient of friction (μ) and the normal force (N) (as defined in Eq. 3.1):

$$V_{SF} = \mu N \tag{3.2}$$

However, when applying a lateral force V_{SF} to the cracked plane in Figure 3.3, the actions of sliding (Δ_{SF}) and crack opening (w_{SF}) cause damage to the concrete surface such that μ cannot be taken as a constant. Similarly, the normal force N in Eqs. 3.1 and 3.2 is affected not only by interface sliding but also by crack opening due to flexure. Hence, although it is convenient to think of SF in terms of elementary friction mechanics (and it will be shown in Section 3.1.4 that this is how some national codes of practice deal with the problem), in reality, the non-linear SF properties shown in Figure 3.4, which connect the shear stress (τ_{SF}), shear slip (Δ_{SF}), crack opening (w_{SF}) and normal stress (σ_N), are required.

Before examining how to determine the SF material properties in Figure 3.4, let us first use the simplified form of behaviour, that is Eq. 3.2, to examine the parameters that most significantly affect their magnitude.

Surface roughness: Increasing the surface roughness is akin to increasing the coefficient of friction in Eq. 3.2. Hence, when designing joints between wet and hardened concrete, it is common practice to specify surface treatments to increase the roughness of the concrete, such as exposing the aggregate.

Concrete and aggregate strength: The impact of concrete and aggregate strength is closely related to that of surface roughness. That is, the shear crack or sliding plane in Figure 3.2 is more likely to pass through the aggregates if the mortar strength is higher than the aggregate strength, which will result in a smoother crack surface and, therefore, a reduction in μ in Eq. 3.2. If the surface is smoother, there is also a reduced crack opening w_{SF} because the maximum height the aggregates need to ride over is reduced, thereby reducing the benefits of tension stiffening.

Concrete confinement: In Eqs. 3.1 and 3.2, increasing the quantity of reinforcement across a cracked plane is equivalent to increasing the confining force N, which acts to clamp the surfaces together and increases the force V_{SF} required to achieve a given displacement Δ_{SF} in Figure 3.2. Given that the surfaces are clamped together with a higher force, increasing confinement also reduces the crack width (w_{SF}) for all given displacements Δ_{SF}.

Fibre reinforcement: Although there have been few test campaigns considering fibre-reinforced and ultra-high-performance FRC, those that have been undertaken (e.g. Sturm et al., 2019) show that fibre reinforcement is highly effective at restraining crack opening (w_{SF}) and shear sliding (Δ_{SF}). This is because the fibres act to transfer significant stresses across the crack and hence act in a similar way to increased concrete confinement, as indicated in Eq. 3.1.

3.1.2.2 Typical SF material properties

Typical SF material properties extracted from test results are shown in Figure 3.5 for concrete with a strength of 44 MPa, produced with aggregate having a maximum size of 20 mm and subjected to normal confining stresses (σ_N) ranging between 5% and 20% of f_c, as tested by Dadd et al. (2025).

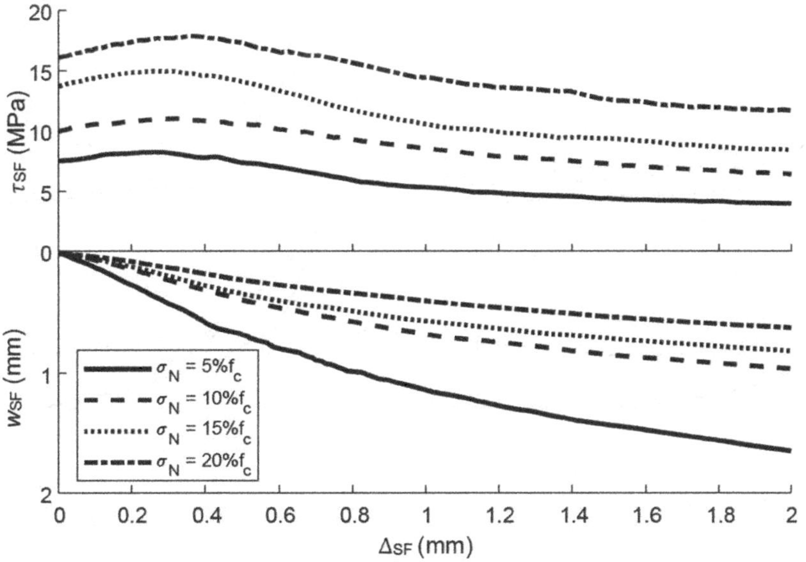

Figure 3.5 SF properties of plain concrete.

Figure 3.5 shows an initial small increase in shear stress after the initiation of shear sliding, followed by a gradual decline in shear stress before reaching a stable plateau. This initial increase in shear stress is not seen in all test results. The rate and magnitude of this initial increase in shear stress are functions of the mortar's strength relative to the aggregate, as well as the aggregate grading, both of which impact the roughness of the cracked plane (see Figure 3.2). The magnitude of the initial rise after cracking, as shown in Figure 3.5, is also impacted by the degree of confinement, which increases friction by clamping the crack faces together. Given the uncertainty regarding the magnitude of this shear stress increase, a lower bound solution is typically found when fitting models by ignoring stresses above the shear stress at the initiation of sliding (v_u). Figure 3.5 also reveals that the majority of the reduction in shear stress occurs over approximately 10% of the maximum aggregate diameter, after which the shear stresses remain relatively stable. This is because significant smoothing of the crack plane occurs with relatively small slips, due to damage caused by the hard aggregates as they bear against softer mortar.

Considering the relationship between Δ_{SF} and w_{SF} in Figure 3.5, it can be seen that the rate of crack opening is highly non-linear; it is large for small confining stresses but rapidly reduces as the confining stress increases. It is also apparent that for small confinements, the crack width continues to increase at a much higher rate than for larger confinements. This is likely due to greater damage, that is a reduction in roughness, to the crack surface under higher levels of confinement.

Figure 3.6a shows the SF properties for an *UHPC* without fibres and with a compressive strength of 172 MPa. In contrast, Figure 3.6b presents these properties for an ultra-high-performance FRC (*UHPFRC*) with 2% microfibres by volume and a compressive strength of 164 MPa. Full details of the results of these tests can be found in

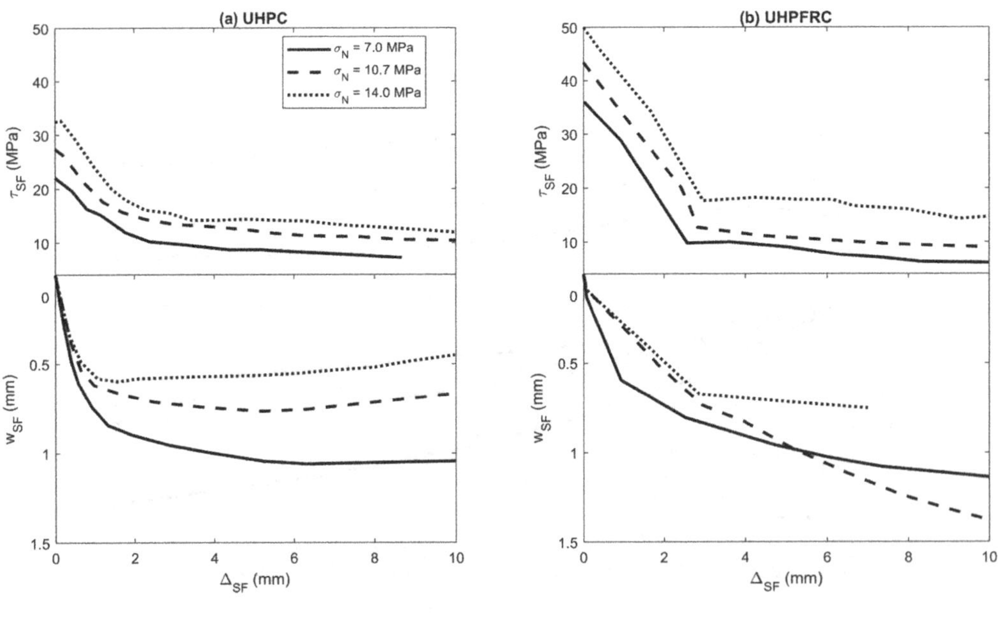

Figure 3.6 SF properties of fibre concrete according to Sturm et al. (2018) for (a) UHPC; (b) UHPFRC. (Modified from Sturm et al. 2018.)

Sturm et al. (2018). In both cases, confining stresses ranging between 4.2% and 8.4% of f_c are applied.

It is noteworthy that, compared to the results shown in Figure 3.5, there is no increase in shear strength immediately after cracking. This can be explained by the much smoother surface of the cracked plane created in *UHPC* due to the absence of coarse aggregate, as the maximum aggregate size in these concretes is that of the fine aggregate. It is also notable that v_u does not increase linearly with compressive strength; despite having a compressive strength nearly four times higher, v_u is approximately three times larger than that of the 44 MPa concrete in Figure 3.5.

When examining the relationship between Δ_{SF} and w_{SF} in Figure 3.6a, it is observed that after a peak crack width is reached, w_{SF} reduces as Δ_{SF} increases. This is particularly apparent when confinement is high. This behaviour is attributed to the loss of the surface layer of the concrete as it is ground away and ejected from the crack during shear sliding.

Comparing the results in Figure 3.6b for concrete with 2% microfibres to those in Figure 3.6a for the same concrete mix design but without fibres, it is clear that the presence of fibres significantly increases the initial shear strength by providing additional confinement across the crack, as shown in Figure 3.2 and Eq. 3.1. As the shear displacement increases, this confinement reduces due to fibre pull-out; hence, the shear stress tends towards that of the concrete without fibres.

Comparing the crack opening behaviour of the concrete with fibres to that without fibres in Figure 3.6, it is observed that in the former there is no longer a reduction in the crack opening due to sliding. This can be explained simply by the toughness of the fibres compared to the mortar. That is, the fibres are significantly tougher than the mortar and hence, even if they were to fully pull out, they would remain in the crack opening and would have to ride over each other as the shear displacement is increased.

3.1.3 Quantification of SF material properties

SF material properties can be directly measured through direct shear tests (e.g. Anderson 1960). Alternatively, they can be deduced indirectly by applying the SF mechanism to simulate the strength and deformation of other types of tests where behaviour is controlled by SF, such as confined cylinder tests (Chen et al. 2015a). Here, both approaches will be considered by briefly describing the approach taken to testing and extracting test results; for specific test protocols, the reader is referred to papers dedicated to the topic. Importantly, both direct and inverse analysis techniques are presented. Although inverse analysis introduces additional statistical uncertainty, it is nevertheless a valuable tool for the development of material models, particularly given the complexity of undertaking direct shear tests and the limited test data available.

3.1.3.1 Direct shear tests

Direct shear tests aim to apply a direct shear across a potential cracked plane while measuring the applied stresses and deformations. A number of direct shear test apparatus have been developed since testing began in the early 1960s, with the most common being the Z-test developed by Anderson (1960) which is named after the shape of the test specimen shown in Figure 3.7a. The Z-test specimen is shaped to allow load

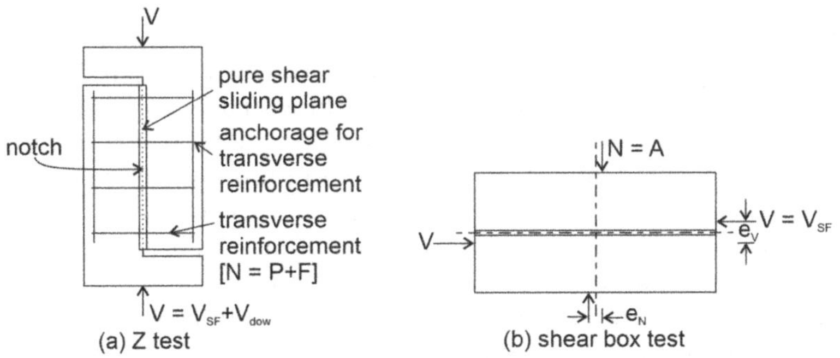

Figure 3.7 Typical SF test set-ups using (a) Z-test and (b) a shear box.

to be applied to the centre of the specimen, which is therefore a region or line of pure shear, initiating a crack along the specimen height as indicated. To better facilitate a vertical crack in line with the applied force and therefore in pure shear, it is also common to form a V-shaped notch along the specimen height to reduce the thickness of the specimen.

During testing, the load V in Figure 3.7a is increased and it is assumed that the interface shear stress τ_{SF} is uniform across the cross-sectional area of the crack plane. Similarly, the relative vertical displacement of the two halves of the test specimen Δ_{SF} and the crack opening w_{SF} can be directly measured using displacement transducers. In this test set-up, the stress normal to the crack is applied passively via the lateral reinforcement crossing the cracked plane; that is, N in Figure 3.3 is equal to the passive force P in Eq. 3.1.

It is commonly assumed in deriving the confining stress σ_N that the reinforcement is fully yielded at all stages of loading such that

$$\sigma_N = \rho_N f_y \tag{3.3}$$

in which ρ_N is the reinforcement ratio and f_y is the reinforcement yield stress. Consequently, it is assumed that a single relationship between τ_{SH}–Δ_{SH}–w_{SH} at a specific σ_N, such as the V_u-V_f variations shown in Figure 3.4, can be obtained from a Z-test. Hence, it is assumed that repeating the Z-test with various reinforcement ratios and consequently various confinements σ_N will give a full series of τ_{SH}–Δ_{SH}–w_{SH} as in Figure 3.4. Unfortunately, this is not the case.

As explained using Figure 3.2, stretching the reinforcement to eventual yield requires a transverse slip Δ_{SF} which, due to surface roughness, causes a separation w_{SF}. Let us assume that the Z specimen is tested under displacement control. Hence, when the sliding plane crack first forms in a Z section, the load at which this occurs corresponds to V_{u0} in Figure 3.4 when σ_N is zero. Increasing the applied displacement after cracking will then cause a slip Δ_{SF} and a separation w_{SF} which stretches the reinforcement in Figure 3.3 to induce the confining force P. It can be seen that this type of confinement is passive; that is, it gradually increases with displacement. It is certainly not active, meaning it is certainly not applied all the time. This means that the very important *SF* data from Z-tests can only be extracted using *PI* theory, as explained in Section 5.4 on the passive confinement mechanism, and which incorporates the use of tension stiffening as discussed in Chapter 4. For the transverse reinforcement, it would simply mean using the P/Δ relationship for the reinforcement bar from Section 4.2.2.2, which can be derived from numerical simulations or from closed-form solutions, where P is the axial force in the bar and Δ is the half crack width, that is, $w_{SF}/2$. Hence, knowing the crack width w_{SF} from the tests, it would be a straightforward procedure to derive the confining force N and consequently σ_N. The results will resemble the dashed line A–B in Figure 3.4, which emanates from V_{u0} for unconfined concrete and then follows the path with increasing confinement. For large amounts of transverse reinforcement, the path A–B will rise to A–C, and as the reinforcement is further increased, the path will rise further. All this information can be used to plot the *SF* family of curves.

When considering the results of models developed using Z-specimen test results, because the full yield stress is not developed at the initiation of cracking nor immediately after cracking, but builds up as a result of PI, neither the assumption that σ_N is proportional to the transverse reinforcement yield force nor that σ_N is constant throughout testing holds true in reality. As a result of these assumptions, σ_N is overestimated until the reinforcement yields and then underestimated if the reinforcement exhibits significant strain hardening.

To overcome the issues of a continuously changing $\sigma_{N,}$ researchers have increasingly transitioned to using a shear box test (Figure 3.7b) to determine the SF properties (e.g. Dadd et al. 2025). In this approach, an unreinforced prism is encased in steel platens that allow for the simultaneous application of axial and lateral force, enabling a direct shear force V_{SF} to be applied while maintaining a constant normal force N. Hence, the shear box test approach directly obtains the τ_{SH}–Δ_{SH}–w_{SH} relationship for a given σ_N (as shown in Figure 3.4). By repeating testing for a range of normal stresses, the full τ_{SH}–Δ_{SH}–w_{SH} relationship can be obtained. There is one difficulty: in most shear box tests, it is impossible to align V as shown in Figure 3.7b to obtain a pure shear region. The challenge lies in reducing the eccentricity e_v as much as possible to minimise e_N and achieve the most uniform confinement possible. Of course, one simple solution is to rotate the Z specimen (shown in Figure 3.7a) by 90° and test it as in the shear box test (Figure 3.7b) to ensure a pure shear region. The shear box set-up has one further advantage: it does not require any transverse reinforcement across the sliding plane, so the problems with dowel action and passive confinement do not occur.

3.1.3.2 Inverse analysis of confined concrete properties

Given the complexity and cost of the testing required to directly obtain the SF material properties, here we consider an inverse analysis approach developed by Chen et al. (2015a) based on the application of the SF mechanism to simulate the confinement of concrete. That is, the inverse analysis approach allows for the utilisation of the significant research devoted to calibrating strain-based models to quantify the confined stress–strain behaviour of concrete.

There are two types of *SF* sliding mechanisms to consider in the softening process: those that form cones (as shown in Figure 3.8a),referred to as circumferential wedges, which tend to occur in stubby specimens such as cubes (e.g. Figure 1.19c); and those that form a single sliding plane (as shown in Figure 3.8b and also in Figures 1.17 and 1.20), which tend to occur in more slender specimens where platen restraints are less.

Figure 3.8 Concrete failure: (a) circumferential wedge; (b) single sliding plane. (Reproduced from Visintin et al. 2015.)

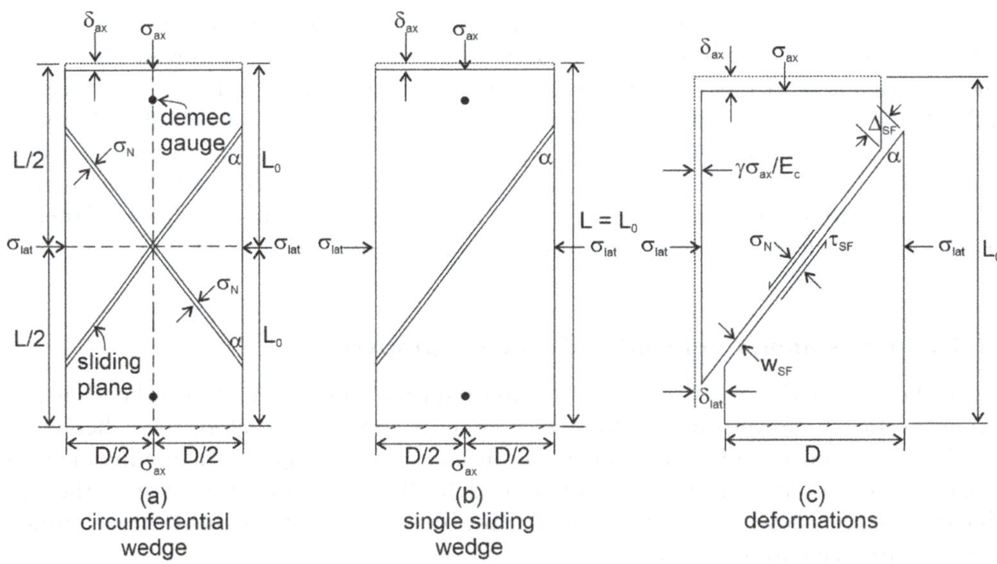

Figure 3.9 Wedges: (a) circumferential wedge, (b) single sliding plane, and (c) deformations.

The circumferential wedge in Figure 3.8a is shown schematically in Figure 3.9a within a cylinder of diameter D and length L, with the conical sliding plane occurring at an angle α. The cylinder is subjected to an axial stress σ_{ax} and a lateral stress σ_{lat}.

Let us assume that the axial stress σ_{ax} causes the whole cylinder to contract by δ_{ax}, meaning that the effective strain ε_{ax} is δ_{ax}/L. As explained in Section 1.4.3, concrete softening is a size-dependent phenomenon; that is, the stress–effective strain relationship depends on the length of the element being considered. Consequently, the effective strain derived from the total deformation δ_{ax}/L would be different from the effective strain measured using the demec gauges shown, which are spaced at a distance less than L. The single sliding wedge is shown schematically in Figure 3.9b, and once again, the effective strain depends on the length over which contraction is measured.

The deformation of an element whose length L_0 encompasses one sliding plane is shown in Figure 3.9c. For the single sliding plane in Figure 3.9b, L_0 would be L if δ_{ax} is the deformation of the whole cylinder, or it could be the distance between the demec gauges if these were used to measure the deformation. However, in the circumferential wedge in Figure 3.9a, let L_0 be the length of one of the quadrants shown, as by symmetry, each quadrant must have the same stress–strain relationship.

Let us assume that the deformations in Figure 3.9c are those of the *RHS* top quadrant shown in Figure 3.9a. This quadrant is first subjected to a lateral confining stress σ_{lat} and then an increasing axial stress σ_{ax} as is common when conducting tests on triaxially confined concrete. Prior to the formation of concrete sliding planes, which occurs at the peak stress f_c, the axial stress–strain relationship ($\sigma_{ax}/\varepsilon_{ax}$) and the axial stress–lateral strain relationship ($\sigma_{ax}/\varepsilon_{lat}$) of the confined concrete are strain-based material properties and can be determined using strain-based material models such as the elastic modulus E_c and Poisson's ratio γ, such that the axial strain is given by

$$\varepsilon_{ax-mat} = \frac{\sigma_{ax}}{E_c} \tag{3.4}$$

and the lateral strain by

$$\varepsilon_{lat-mat} = \gamma \frac{\sigma_{ax}}{E_c} \tag{3.5}$$

Once concrete sliding planes form in Figure 3.9a, strain-based material properties no longer directly apply, and the SF mechanism defined in Section 1.3.1 must be utilised. This is done by considering compatibility and equilibrium within the quarter cylinder, as shown in Figure 3.9c.

Let us first consider compatibility in Figure 3.9c. From Figures 3.5 and 3.6, it can be seen that w_{SF} is an order of magnitude smaller than Δ_{SF}; therefore, as a good approximation, let us ignore the vertical component of w_{SF} (this is a further example of making simplifications to obtain a tractable solution). Consequently, the axial contraction of the cylinder (δ_{ax}), as shown in Figure 3.9, is given by the following summation:

$$\delta_{ax} = \frac{\sigma_{ax}}{E_c} L_0 + \Delta_{SF} \cos\alpha \tag{3.6}$$

where L_0 is the length of the prism that contains one sliding plane and over which the contraction is measured. The first term is the material contraction over the length L_0. The second term is the vertical component of the slip across the sliding plane, specifically, the vertical component of the wedge sliding component Δ_{SF}. Furthermore, α is the angle formed by the sliding plane, which, in the absence of experimental observation, can be approximated from Chen et al. (2015a) as

$$\alpha = 26° + \frac{\sigma_{lat}}{f_c} \times 20° \tag{3.7}$$

where f_c is the unconfined cylinder strength.

Dividing Eq. 3.6 by the length of the segment L_0, the total axial strain after the formation of sliding planes is given by

$$\varepsilon_{ax} = \frac{\sigma_{ax}}{E_c} + \frac{\Delta_{SF} \cos\alpha}{L_0} \tag{3.8}$$

In the lateral direction in Figure 3.9, as the material lateral expansion is an order of magnitude smaller than the axial deformation, w_{SF} cannot be ignored. Hence, the total lateral deformation δ_{lat} over the half diameter $D/2$ is given by the sum of the material dilation; the horizontal component of the crack opening along the sliding plane w_{SH}; and the horizontal component of the sliding component Δ_{SF}. Hence,

$$\delta_{lat} = \gamma \frac{\sigma_{ax}}{E_c}\frac{D}{2} + \frac{w_{SF}}{\cos\alpha} + \Delta_{SF} \sin\alpha \tag{3.9}$$

Consequently, dividing by the half diameter $D/2$, over which the dilation is measured, the lateral strain is

$$\varepsilon_{lat} = \gamma \frac{\sigma_{ax}}{E_c} + 2\left(\frac{\frac{w_{SF}}{\cos\alpha} + \Delta_{SF} \sin\alpha}{D}\right) \tag{3.10}$$

Now consider equilibrium of the free body in Figure 3.9c. Resolving in the shear and normal directions, the shear and normal stress along the sliding plane can be related to the axial and lateral stresses by

$$\tau_{SF} = \sigma_{ax} \sin\alpha \cos\alpha - 2\sigma_{lat} \sin\alpha \cos\alpha \tag{3.11}$$

$$\sigma_N = \sigma_{ax} \sin^2\alpha + 2\sigma_{lat} \cos^2\alpha \tag{3.12}$$

It should be noted that the analysis of the single sliding plane in Figure 3.9b is the same as for the circumferential wedge, except that the lateral strain occurs over a width D and not $D/2$ (as for the quadrant). Hence for the single sliding plane Eqs. 3.4–3.8 apply, and Eq. 3.10 should be replaced by

$$\varepsilon_{lat} = \gamma \frac{\sigma_{ax}}{E_c} + \left(\frac{\frac{w_{SF}}{\cos\alpha} + \Delta_{SF} \sin\alpha}{D}\right) \tag{3.13}$$

Having established these governing equations of compatibility and equilibrium for the circumferential wedge, the following steps are needed to perform an inverse analysis to extract the SF material properties.

1. The first step of the analysis is to gather the results of triaxially confined cylinder tests or to generate $\sigma_{ax}/\varepsilon_{ax}$ relationships using existing empirically derived strain-based material models for confined concrete. For this analysis, in addition to these $\sigma_{ax}/\varepsilon_{ax}$ relationships, the elastic modulus E_c and Poisson's ratio γ must also be known, as well as the dimensions of the test specimen used to generate the test data or to calibrate the material model. To extract the SF properties, $\sigma_{ax}/\varepsilon_{ax}$ relationships need to be constructed for a range of different confinement levels σ_{lat}; the precision of the extracted SF properties will be increased by considering a larger number of confinement levels.
2. Now, for a known $\sigma_{ax}/\varepsilon_{ax}$ relationship for a given σ_{lat} from Step 1, τ_{SF} and σ_N can be determined from Eqs. 3.11 and 3.12, where α, in the absence of experimental measurement of the sliding plane angle, can be approximated according to Eq. 3.7.
3. For any given point on the $\sigma_{ax}/\varepsilon_{ax}$ after the commencement of concrete softening, the shear displacement Δ_{SF} can be obtained by solving Eq. 3.8.
4. For any given point on the $\sigma_{ax}/\varepsilon_{ax}$ where the corresponding lateral strain ε_{lat} is known and Δ_{SF} has been determined from Step 3, w_{SF} can be determined by solving Eq. 3.10.
5. Steps 2–4 are repeated for all axial stresses to generate the relationship between τ_{SF}–Δ_{SF}–w_{SF}–σ_N for a given σ_{lat}. The process is then repeated for all given confinement levels, yielding the SF properties in a similar form to that shown in Figures 3.5 and 3.6.

Worked Example WE-3.1: Extracting SF properties for plain concrete

Let us consider the extraction of the concrete SF material properties using the results of strain-based material models for concrete subjected to active triaxial confinement. Research has shown (Xiao et al. 2010) that the unconfined concrete $\sigma_{ax}/\varepsilon_{ax}$ relationship defined by Popovics (1973) in Section 2.3.1 and Eqs. 2.2 and 2.4 can be reliably used to predict the relationship of actively confined concrete. In Eqs. 2.2 and 2.4, the unconfined strength f_c is replaced with the confined strength f_{cc}, and the peak strain at which this occurs that is the corresponding peak strain ε_0 is replaced by ε_{cc}, the peak strain at which f_{cc} occurs. This gives the stress/strain relationship for confined concrete $\sigma_{ax}/\varepsilon_{ax}$ as

$$\sigma_{ax} = f_{cc}\frac{n\left(\dfrac{\varepsilon_{ax}}{\varepsilon_{cc}}\right)}{(n-1)+\left(\dfrac{\varepsilon_{ax}}{\varepsilon_{cc}}\right)^{n}} \tag{E3.1.1}$$

in which

$$n = \frac{E_c}{E_c - \dfrac{f_{cc}}{\varepsilon_{cc}}} \tag{E3.1.2}$$

In this model, based on the work of Richart et al. (1928) and Richart et al. (1929), the confined compressive strength f_{cc} and corresponding strain (ε_{cc}) are taken to vary with the lateral stress σ_{lat} according to

$$f_{cc} = f_c\left[1+3.65\left(\frac{\sigma_{lat}}{f_c}\right)^{0.82}\right] \tag{E3.1.3}$$

$$\varepsilon_{cc} = \varepsilon_0\left[1+14.74\left(\frac{\sigma_{lat}}{f_c}\right)^{1.03}\right] \tag{E3.1.4}$$

where, from Section 2.4.1 and Eq. 2.3,

$$\varepsilon_0 = 4.76\times10^{-6} f_c + 2.13\times10^{-3} \tag{E3.1.5}$$

and Poisson's ratio can be taken as 0.2.

Furthermore, taking the empirical model of Jiang and Teng (2007) to relate the axial and lateral strain for the confined concrete as a function of the unconfined strain at peak stress ε_0 and the confinement ratio $\frac{\sigma_{lat}}{f_c}$

$$\frac{\varepsilon_{ax}}{\varepsilon_0} = 0.85\left[1+8\left(\frac{\sigma_{lat}}{f_c}\right)\right]\left\{\left[1+0.75\left(\frac{-\varepsilon_{lat}}{\varepsilon_0}\right)\right]^{0.7} - e^{\left[-7\left(\frac{-\varepsilon_{lat}}{\varepsilon_0}\right)\right]}\right\} \tag{E3.1.6}$$

For the purpose of this example, considering a single test result for concrete with an unconfined compressive strength of 40 MPa and a lateral confinement of 4 MPa, from Eq. E3.1.3, the confined concrete compressive strength is

$$f_{cc} = f_c\left[1+3.65\left(\frac{\sigma_{lat}}{f_c}\right)^{0.82}\right] = 40\left[1+3.65\left(\frac{4}{40}\right)^{0.82}\right] = 62.1\text{ MPa} \tag{E3.1.7}$$

From Eq. E3.1.5, the unconfined strain at peak stress is

$$\varepsilon_0 = 4.76\times10^{-6}\times40 + 2.13\times10^{-3} = 0.000232 \tag{E3.1.8}$$

which can be substituted into Eq. E3.1.4 to give

$$\varepsilon_{cc} = \varepsilon_0\left[1+14.74\left(\frac{\sigma_{lat}}{f_c}\right)^{1.03}\right] = 0.000232\left[1+14.74\left(\frac{4}{40}\right)^{1.03}\right] = 0.0055 \tag{E3.1.9}$$

Taking the elastic modulus of unconfined concrete to be 31,975 MPa from Example E2.1 in Section 2.4.2.1, Eqs. E3.1.1 and E3.1.2 can be solved using a spreadsheet to obtain the confined stress–strain relationship shown on the *RHS* of Figure 3.10. Similarly, using a spreadsheet to solve Eq. E3.1.6, the corresponding axial stress–lateral-strain relationship is also shown on the *LHS* of Figure 3.10. Also plotted are the curves for unconfined concrete where it can be seen that a confinement of 10% of the compressive strength can increase the compressive strength by more than

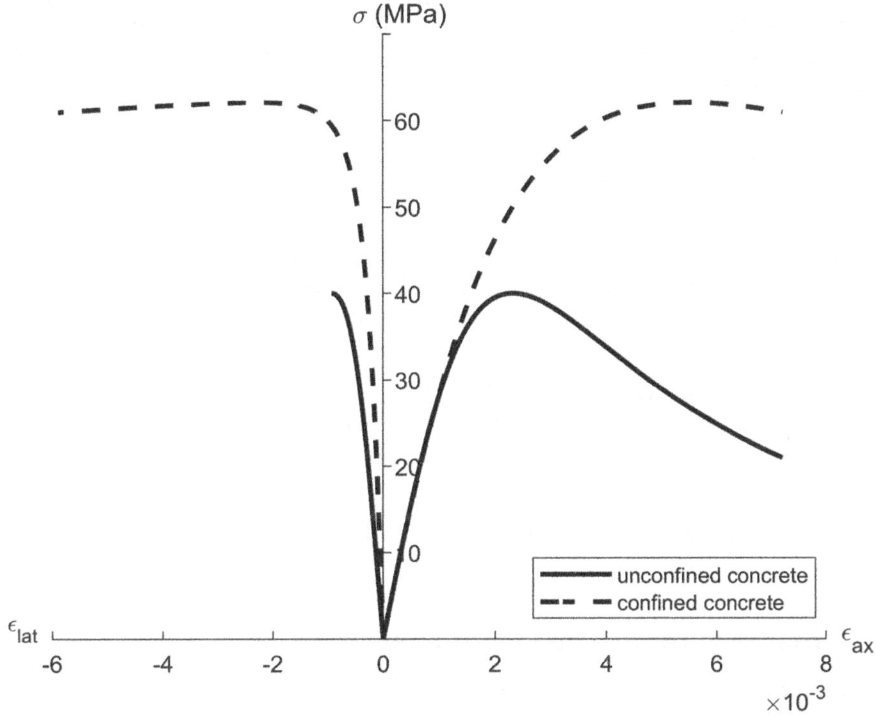

Figure 3.10 Axial and lateral stress strain relationship of actively confined concrete used in Example WE-3.1.

50%. Also of interest in Figure 3.10 is that the $\sigma_{ax}/\varepsilon_{lat}$ relationship for unconfined concrete is only shown up until the concrete begins to crush; beyond this point reliable models for the lateral strain are not available as the concrete cover quickly spalls and falls off.

For the purpose of this example, let us derive the *SF* properties at a specific axial stress σ_{ax}=61 MPa. It is possible to solve Eqs. 3.11 and 3.12 to obtain the stresses normal and parallel to the sliding plane if the angle of the sliding plane (α) is known. From Eq. 3.7

$$\alpha = 26° + \frac{4}{40} \times 20° = 30° \quad \text{(E3.1.10)}$$

Substituting α, σ_{lat} and the chosen σ_{ax}=61 MPa into Eq. 3.11 gives

$$\tau_{SF} = 61 \times \sin 30 \cos 30 - 2 \times 4 \sin 30 \cos 30 = 22.95 \text{ MPa} \quad \text{(E3.1.11)}$$

and into Eq. 3.12 gives

$$\sigma_N = 61 \times \sin^2 30 + 2 \times 4 \cos^2 30 = 21.25 \text{ MPa} \quad \text{(E3.1.12)}$$

The total axial strain is given by Eq. 3.8.

The axial strain corresponding to an axial stress of 61 MPa from Figure 3.10 is 0.00716 on the falling branch. Let us assume that the model was calibrated using cylinders with a total height of 200 mm such that L_0 is 100 mm. Then rearranging Eq. 3.8 and inserting these values gives

$$\Delta_{SF} = \frac{100\left(0.00716 - \frac{61}{31,975}\right)}{\cos 30} = 0.606 \text{ mm} \tag{E3.1.13}$$

Reading the lateral strain ε_{lat} corresponding to an axial stress of 61 MPa from Figure 3.10 as 0.00575 on the falling branch, and assuming that the model was calibrated using cylinders with a diameter of 100 mm, a rearranged Eq. 3.13 yields

$$w_{SF} = \cos 30\left[\frac{100\left(0.00575 - 0.2\frac{61}{31,975}\right)}{2} - 0.274 \sin 30\right] = 0.351 \tag{E3.1.14}$$

and so it can be seen how the *SF* properties can be derived for the entire falling branch.

3.1.4 SF material models for plain concrete

SF material models can generally be split into two categories:

i. Models that seek to define the shear strength at the initiation of sliding, that is V_u in Figure 3.4, which for example may be used to predict the shear capacity of a reinforced concrete member prior to sliding.
ii. Models that seek to define the full-spectrum τ_{SH}–Δ_{SH}–w_{SH}–σ_N relationship in Figure 3.4, which may be used, for example, to predict the degree of confinement in a column or the shear capacity at the commencement of sliding.

3.1.4.1 *Initiation of sliding models $V_{uz\delta}$*

Models to predict the shear strength at the initiation of sliding have predominantly been developed based on calibration to test data obtained from Z-test specimens. While a large number of models have been developed, they generally agree that the most influential parameters controlling shear strength are the concrete compressive strength and the degree of confinement (Santos and Júlio 2012). Hence, design-oriented models are generally in the form of

$$V_{u\text{-}z\delta} = af_c + b\sigma_N \tag{3.14}$$

where $V_{u\text{-}z\delta}$ is the shear strength at zero interface slip, that is, $\Delta_{SF\_}$ is zero; the coefficients a and b are calibratable factors; depending on the specific model, the concrete strength term af_c may be replaced with the concrete tensile strength; and the confinement term σ_N may be replaced with the reinforcement yield force term $\rho_N f_y$ where ρ_N is the area of reinforcement as a proportion of the area of the potential sliding plane.

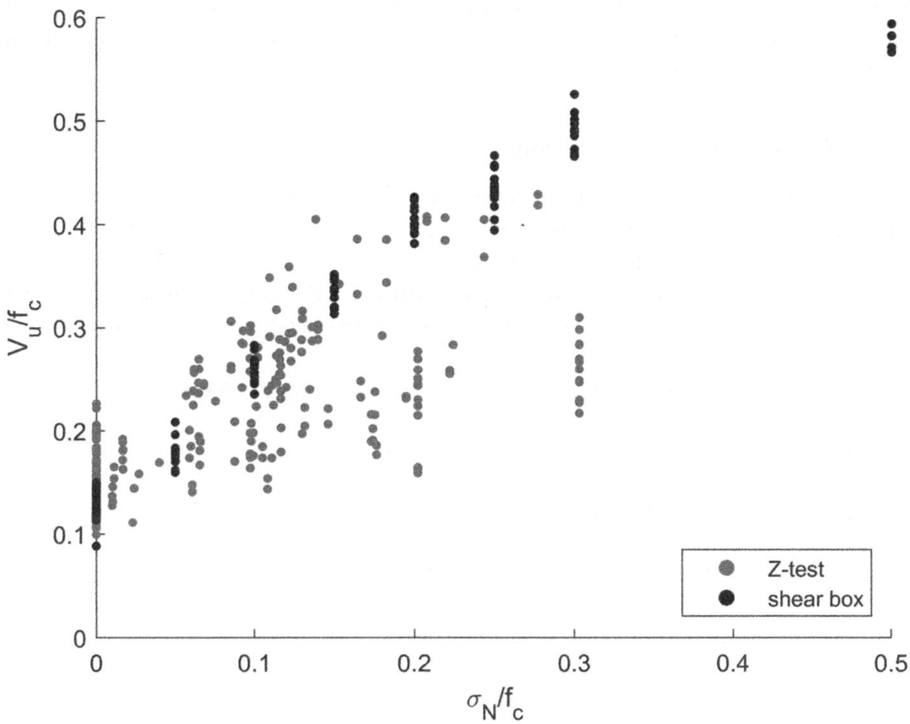

Figure 3.11 Database of shear strength tests. (Adapted from Dadd et al. 2025.)

When compared to global databases or raw test data, such as that shown in Figure 3.11, it can be observed that at higher levels of confinement there is some levelling off of the shear strength if the internal reinforcement yields. Because of this, models such as the following by ACI-318 (2019) include an upper limit to the shear strength $V_{u.max}$.

$$V_{u-z\delta} = \mu\rho_N f_y \leq V_{u.max} \tag{3.15}$$

where the friction coefficient μ varies between 0.6 (for fresh concrete placed against unroughened hardened concrete) and 1.0 (for monolithic concrete). An upper limit V_{u-max} for Eq. 3.15 also exists and varies with concrete surface treatment. For example, for monolithic concrete, $V_{u.max}$ is taken as $(3.3+0.08f_c)$ MPa, which must be within the range of $0.2f_c$ and 11 MPa.

It should be noted, however, that when considering only tests conducted using the shear box test method in Figure 3.7, this levelling-off limit to strength is less apparent. Furthermore, following the reprocessing of test data using *PI* theory to account for the variation in σ_N, Chen et al. (2015b) found the behaviour to be approximately linear regardless of the degree of confinement, that is,

$$V_{u-\delta} = 0.175f_c + 0.918\sigma_N \tag{3.16}$$

It should be noted, however, that only a small amount of test data was available for the analysis by Chen et al. (2015b) and hence further work is required to identify the upper limit to the shear strength.

3.1.4.2 Full-spectrum sliding models

Now considering the full-spectrum models that describe the τ_{SF}–Δ_{SF}–w_{SF}–σ_N relationship, which are generally only found in research literature. Hao et al. (2020) calibrated the following model to describe the relationship between τ_{SF}–Δ_{SF}–σ_N based on an inverse analysis of confined test results. The analysis ignored the impact of w_{SF} as its impact was found to be small. From this analysis, the following piecewise model was proposed.

$$\tau_{SF} = \begin{cases} \left[(0.00112f_c - 0.337)\sigma_N - 0.000784f_c^2 + 0.0152f_c + 0.556\right]\Delta_{SF}\ldots \\ \quad + 1.50\sigma_N + 0.105f_c;\ \textit{when}\ \sigma_N < \sigma_{N0} \\ \left[(0.00112f_c - 0.0636)\sigma_N - 0.000784f_c^2 + 0.0620f_c + 0.556\right]\Delta_{SF}\ldots \\ \quad + 0.498\sigma_N + 0.290f_c;\ \textit{when}\ \sigma_{N0} \le \sigma_N < 3\sigma_{N0} \end{cases} \tag{3.17}$$

where

$$\sigma_{N0} = f_c \cos^2 26° \tag{3.18}$$

Alternatively, Chen et al. (2015a) proposed a complete τ_{SF}–Δ_{SF}–w_{SF}–σ_N relationship where the relationship between τ_{SF}, Δ_{SF} and σ_N is given by

$$\frac{\tau_{SF}}{V_u} = \frac{2a(\Delta_{SF} + b)}{(\Delta_{SF} + b)^2 + a^2} \tag{3.19}$$

in which V_u is given by

$$\frac{V_u}{f_c} = -0.6502\left(\frac{\sigma_N}{f_c}\right)^2 + 1.104\left(\frac{\sigma_N}{f_c}\right) + 0.1659 \tag{3.20}$$

and the calibratable parameters a and b are

$$a = 4\left(\frac{\sigma_N}{f_c}\right) - 0.2 \tag{3.21}$$

$$b = 0.1\left(\frac{\sigma_N}{f_c}\right) + 0.945 \tag{3.22}$$

The relationship between Δ_{SH}, w_{SH} and σ_N is given by

$$w_{SF} = \begin{cases} 0; & 0 \le \Delta_{SF} < \Delta_{sep} \\ \left(\Delta_{SF} - \Delta_{sep}\right); & \Delta_{SF} > \Delta_{sep} \end{cases} \tag{3.23}$$

where Δ_{sep} is defined as the slip at which the crack begins to open and is given by

$$\Delta_{sep} = \Delta_0 \left[1 + 10\left(\frac{\sigma_N}{f_c}\right)^2\right] \tag{3.24}$$

in which

$$\Delta_0 = 0.429\left(\frac{\sigma_N}{f_c}\right)^2 + 0.529\left(\frac{\sigma_N}{f_c}\right) + 0.013 \tag{3.25}$$

Worked Example WE-3.2: Plotting *SF* properties for plain concrete

To give an appreciation of the variation in different SF material property models, let us consider a 40 MPa concrete that is subjected to normal stresses σ_N ranging from 5 to 15 MPa. Using the model of Hao et al. (2020) the relationship between Δ_{SF}–τ_{SF}–σ_N can be constructed by simple substitution into Eq. 3.17 to yield the family of curves shown in Figure 3.12.

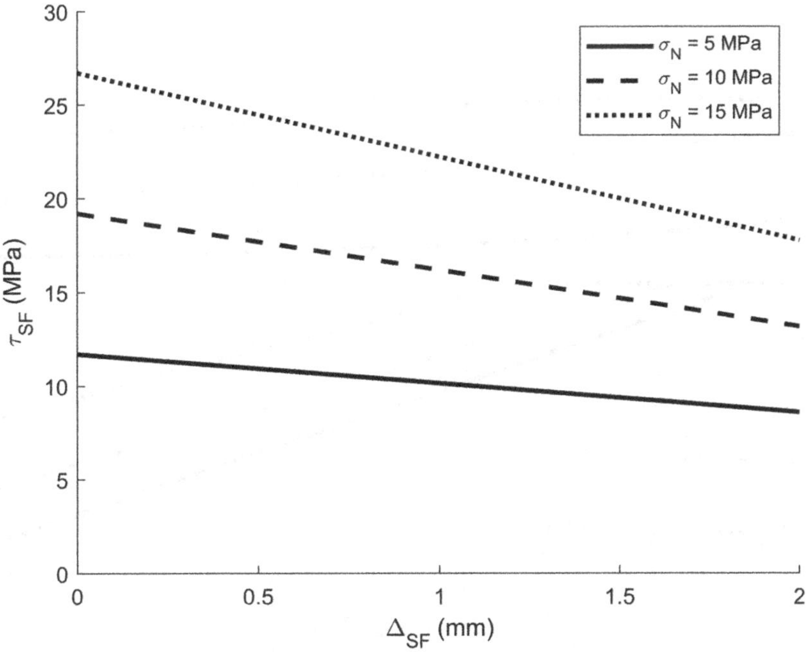

Figure 3.12 Δ_{SH}–τ_{SF}–σ_N relationship according to Hao et al. (2020).

Now applying the model of Chen et al. (2015a) for the same concrete with 5 MPa normal stress, the shear stress V_u can be determined from Eq. 3.20 as

$$V_u = 40\left[-0.6502\left(\frac{5}{40}\right)^2 + 1.104\left(\frac{5}{40}\right) + 0.1659\right] = 11.75 \text{ MPa} \tag{E3.2.1}$$

Solving for coefficients a and b

$$a = 4\left(\frac{5}{40}\right) - 0.2 = 0.30 \tag{E3.2.2}$$

$$b = 0.1\left(\frac{5}{40}\right) + 0.945 = 0.96 \tag{E3.2.3}$$

Substituting V_u, a and b into Eq. 3.19 for any value of Δ_{SF} then allows the construction of the relationship between τ_{SF}–Δ_{SF}–w_{SF}–σ_N in Figure 3.13.

Interestingly, when comparing the results in Figure 3.13 with those in Figure 3.12, it can be seen that the models of Hao et al. (2020) and Chen et al. (2015a) vary significantly in magnitude even at zero slip; this is because the numerator of Eq. 3.19 is significantly less than 1 due to the a coefficient in Eq. 3.19. It is also interesting to

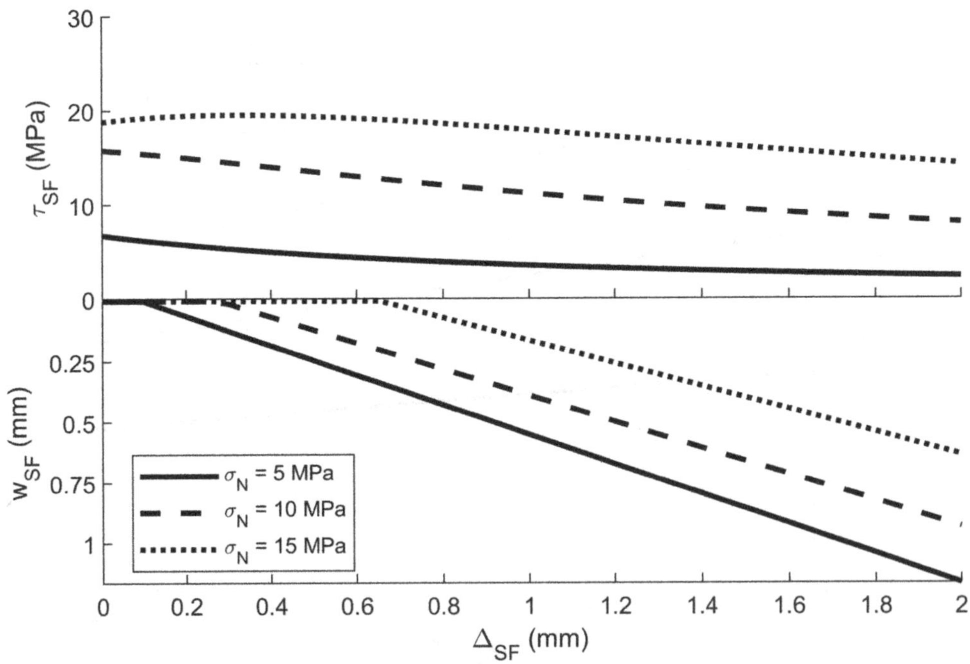

Figure 3.13 Δ_{SF}–τ_{SF}–σ_N relationship according to Chen et al. (2015a).

note that the shape of the Δ_{SF}–τ_{SF}–σ_N developed by Chen et al. (2015a) captures the initial increase in τ_{SH} at low slips seen in the test results in Figure 3.5 while Hao et al. (2020) uses a descending model with the starting point fixed at V_u.

The relationship between Δ_{SF}–w_{SF}–σ_N is now developed using Chen's approach. The slip corresponding to a zero crack width, Δ_{sep} can be determined using Eqs. 3.24 and 3.25. That is,

$$\Delta_0 = 0.429\left(\frac{5}{40}\right)^2 + 0.529\left(\frac{5}{40}\right) + 0.013 = 0.0858 \tag{E3.2.4}$$

$$\therefore \Delta_{sep} = 0.0858\left[1 + 10\left(\frac{5}{40}\right)^2\right] = 0.1 \tag{E3.2.5}$$

Substituting the result of Eq. E3.2.5 into Eq. 3.23, the crack opening can be determined for any slip Δ_{SF} to produce the relationships shown in Figure 3.13. Comparing the general shape of Chen's model in Figure 3.13 to the test results in Figure 3.5, it is seen that Chen's approach suggests crack openings commence at significantly larger slips than observed. Given that Chen's approach was developed from curve fitting test data, this outcome implies more work may be required to calibrate generic SF material models.

3.1.5 *SF* material models for FRC

Research to quantify the SF properties of FRC is scarce compared to that for plain concrete; as a result, little test data is available for the construction of material models. Therefore, we consider the model of Sturm et al. (2018) that was calibrated using a very small quantity of data. Specifically, using the shear box test approach, Sturm et al. (2018) tested 16 *UHPFRC* specimens. These specimens either contained no fibres, or 2% fibres of straight microfibres, hooked fibres, or a blend of the two fibre types. At the time of testing, the concrete compressive strength varied between 150 and 172 MPa.

Based on the results of this testing, a Δ_{SF}–τ_{SF}–σ_N model was developed for *UHPC* without fibres as follows:

$$\frac{\tau_{SF}}{V_{u\text{-}UHPC}} = \frac{a\Delta_{SF} + b}{\Delta_{SF} + b} \tag{3.26}$$

where the shear strength $V_{u\text{-}UHPC}$ is given as a function of the concrete compressive strength:

$$\frac{V_{u\text{-}UHPC}}{f_c} = 1.48\frac{\sigma_N}{f_c} + 0.0705 \tag{3.27}$$

and the calibrable coefficients a and b are

$$a = \min\left(6.11\frac{\sigma_N}{f_c} - 0.065, 0.304\right) \tag{3.28}$$

$$b = 0.183\left(\frac{\sigma_N}{f_c}\right)^{-0.654} \tag{3.29}$$

The test results were then used to calibrate the following model for Δ_{SF}–w_{SF}–σ_N:

$$w_{SF} = w_{SF-m-UHPC}\frac{1.15\frac{\Delta_{SF}}{\Delta_{SF-cw}}}{\frac{\Delta_{SF}}{\Delta_{SF-cw}}+0.15} \tag{3.30}$$

where the maximum crack width for *UHPC* without fibres $w_{SF-m-UHPC}$ is

$$w_{SF-m-UHPC} = 1.48 - 10.9\left(\frac{\sigma_N}{f_c}\right) \geq 0 \tag{3.31}$$

and which occurs at the corresponding critical slip Δ_{SF-cw}

$$\Delta_{SF-cw} = 13.7 - 146\left(\frac{\sigma_N}{f_c}\right) \geq 0 \tag{3.32}$$

Similarly, a Δ_{SF}–τ_{SF}–σ_N model (Eqs. 3.33–3.36) was developed for *UHPFRC* with fibres, noting that no significant difference was observed based on fibre type; consequently, this parameter is not included in the model. Further, because only a single volume of fibres was considered, the model is only applicable to *UHPFRC* with 2% fibres.

$$\frac{\tau_{SF}}{V_{u-UHPFRC}} = 1 - 0.233\Delta_{SF} \tag{3.33}$$

where the shear strength is given by $V_{u-UHPFRC}$:

$$\frac{V_{u-UHPFRC}}{f_c} = 1.83\left(\frac{\sigma_N}{f_c}\right) + 0.146 \tag{3.34}$$

The relationship between Δ_{SF}–w_{SF}–σ_N for *UHPFRC* is given by

$$w_{SF} = \left[1.16 - 6.15\left(\frac{\sigma_N}{f_c}\right)\right]\left(\frac{\Delta_{SF}}{\Delta_{SF-c}}\right) \leq 1.72 - 12.4\left(\frac{\sigma_N}{f_c}\right) \tag{3.35}$$

where the maximum crack width occurs at a critical slipΔ_{SF-c}, which is obtained by solving the following quadratic equation:

$$0.233\Delta_{SF-c}^2 + \left(\frac{V_{u-UHPC}}{V_{u-UHPFRC}}a + 0.233b - 1\right)\Delta_{SF-c} - b\left(V_{u-UHPFRC} - V_{u-UHPC}\right) = 0 \tag{3.36}$$

which is itself a function of the SF properties of the *UHPC* without fibres given by Eqs. 3.27–3.29.

To illustrate the magnitude of the impact of adding 2% fibres, the SF properties for *UHPC* and *UHPFRC* are shown in Figure 3.14: for 5 MPa confinement (solid line) and

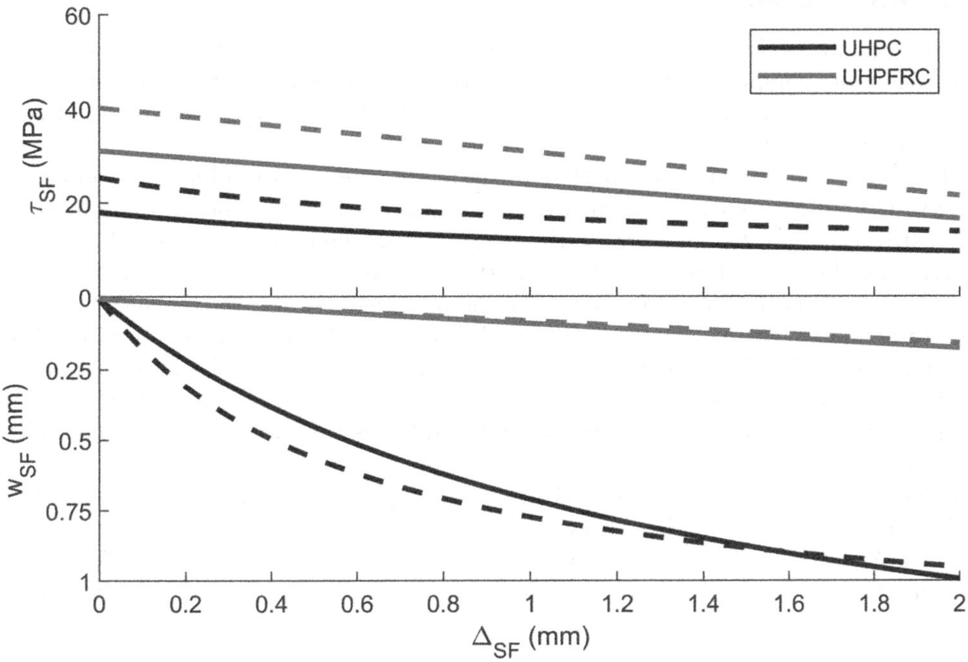

Figure 3.14 Comparison of shear friction properties with and without fibres.

10 MPa confinement (broken line). Of note is the large increase in V_u that is achieved through the addition of fibres (approximately 50%) and that the increase in shear stress due to fibres is maintained over large shear slips. Also of interest in Figure 3.14 is that the contributions of fibres to crack opening significantly outweigh changes to normal stress. The significant impact that fibres have on the SF properties provides the foundation for explaining the benefits that fibres offer in improving the post-peak ductility of concrete at a material scale, as well as the improvements in concrete ductility and shear strength at the member scale, as introduced in Chapter 1.

3.2 BOND SLIP (*BS*)

In Section 1.3.2, the local BS (τ/δ) material properties were shown to directly control the global load slip (P/Δ) relationship of any form of bonded internal or external reinforcement through the mechanics of PI. It was further shown that this PI load slip behaviour was critical to explaining the performance of reinforced concrete at virtually all load levels, including the definition of tension stiffening and crack opening at the serviceability limit state, and the debonding of reinforcement at the ultimate limit state. Hence, in this chapter, we will examine the factors that impact the τ/δ material properties and approaches to quantify them, either directly or indirectly, through experimental work or the use of established material models.

3.2.1 BS mechanism

The BS mechanism is simply an SF mechanism, as described in Section 3.1.1. However, it is much more complex as it has an additional confinement force due to shrinkage S (in Eq. 3.1), and all confinement forces are passive and difficult to quantify. Consequently, BS properties need to be quantified empirically.

The simplest BS mechanism is that of a steel or FRP plate adhesively bonded to an external concrete surface, as illustrated in Figure 3.15a. In general, the adhesive is much stronger than the concrete, so the shear sliding plane occurs within the adjacent concrete as shown. Referring to Eq. 3.1, there are no active or applied confinement forces (A), no passive confinement forces due to transverse reinforcement (P), and no constraints due to shrinkage (S) across the sliding plane. If the concrete contains fibres, they may cross the sliding plane, and hence there may be a small amount of passive confinement (once slip occurs) due to these fibres (F). However, the available anchorage length of fibres adjacent to the plate is small, and thus this confinement is most likely negligible.

When the axial force P (in Figure 3.15a) is applied, a crack will form at the end (i.e. at the crack face) when the shear capacity for unconfined concrete, V_{u0} (in Figure 3.4), is reached. As the load P is increased, the crack tip will extend into the uncracked region at a shear stress V_{u0} at the crack tip. However in the cracked region, the shear stress

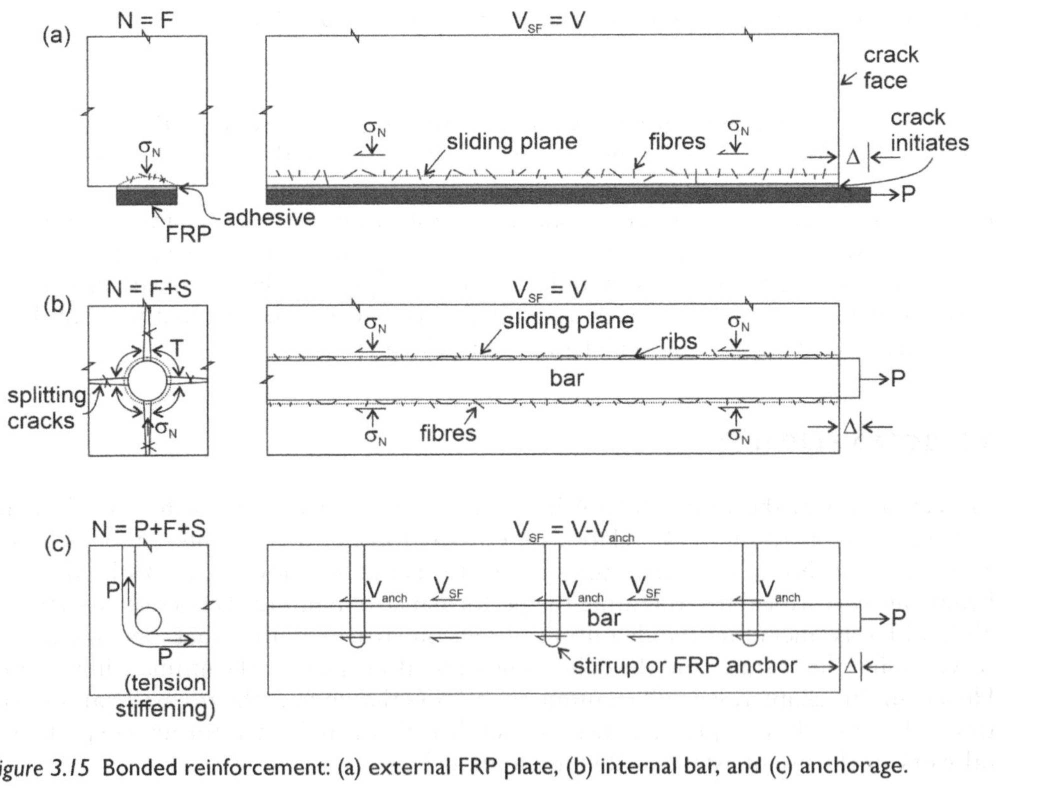

Figure 3.15 Bonded reinforcement: (a) external FRP plate, (b) internal bar, and (c) anchorage.

will reduce along the line V_{u0}–V_{f0} in Figure 3.4 or, when there is a small amount of fibre confinement, along a line such as $V_{u0} - B$.

The behaviour of an internal reinforcing bar, as shown in Figure 3.15b, is much more complex. There is now an additional confinement force due to shrinkage which causes circumferential tension *T*, leading to an additional normal force *S* that enhances the *SF* properties. The reinforcing ribs enhance the 'aggregate interlock' mechanism, thereby further improving the *SF* properties. Hence, when the axial force *P* is initially applied, some confinement will be present, and the sliding plane will first form under this confinement, such as at V_{u1} in Figure 3.4. The crack tip of the sliding plane will then penetrate into the uncracked region at V_{u1}. However, in the cracked region, crack opening w_{SF} will be resisted by an increase in *T*, which increases the confinement. Consequently, the shear capacity will follow a path such as $V_{u1} - D$, where confinement is increasing. Eventually, *T* may cause splitting cracks, as shown in Figure 3.15b, which will reduce the confinement, though this detrimental effect may be mitigated if fibres are present.

Stirrups, as shown in Figure 3.15c, will cause further confinement *P* through tension stiffening. Consequently, the confinement force *N* is now equal to $P + F + S$, which further enhances the *SF* properties. In analysis, one could assume that within the region close to the stirrups, the *SF* properties are enhanced by the stirrups. Alternatively, if there is a strong connection between the longitudinal reinforcement and the stirrups, these could be treated as mechanical anchors whose properties could be determined experimentally, as with stud shear connectors (Oehlers 1980) or *FRP* anchors (Zhang and Smith 2012).

The *SF* properties, as illustrated in Figure 3.4, vary with interface slip, and consequently, bond strengths also vary with slip. Examples of experimentally determined BS properties, under varying levels of confinement, are shown in Figure 3.16. These examples

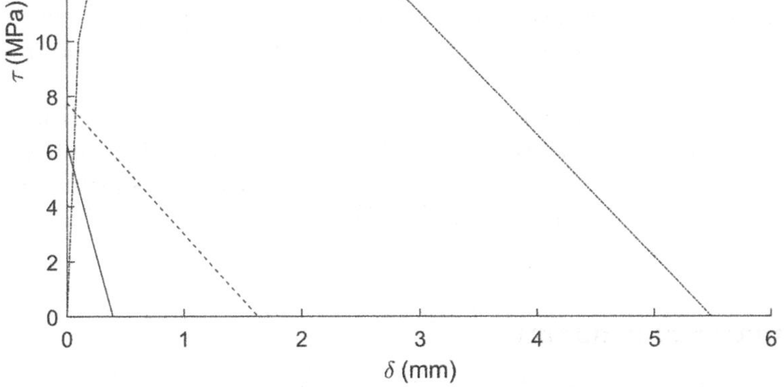

Figure 3.16 Typical bond-slip properties.

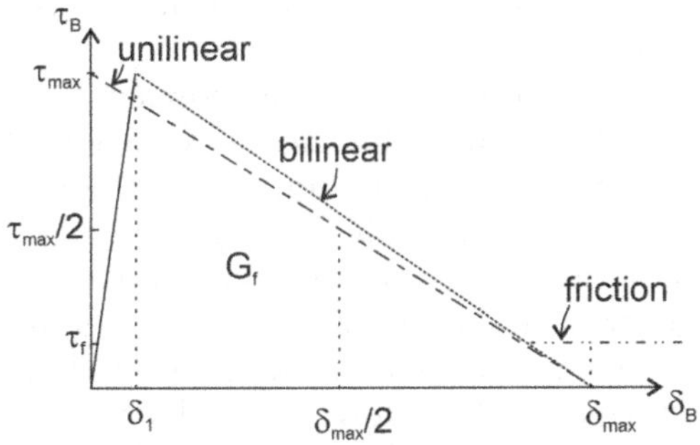

Figure 3.17 Idealised bond-slip properties.

are derived by considering an externally bonded plate, a near-surface mounted (NSM) plate and an internal reinforcing bar, all with the same cross-sectional area; the bond properties have been determined using the material models described in Sections 3.2.4.3 and 3.2.4.4. Most significantly, when comparing the results in Figure 3.16, it is observed that increasing the normal stress applied to the failure plane, as illustrated in Figure 3.15, leads to an increase in strength, but more importantly, to an increase in ductility.

To further understand the BS behaviour, let us idealise the BS properties as shown in Figure 3.17, where the ascending branch is represented by an unbroken line and the descending branch by a broken line. As slip increases, the shear stress increases rapidly up to a maximum value of τ_{max} at a slip δ_1, after which the strength gradually reduces, approaching zero at δ_{max}. It is often assumed that a frictional component τ_f exists at large slips, as shown. The area under the bilinear ascending and descending region represents the energy G_f.

Let us apply the BS relationship shown in Figure 3.17 to the pull tests in Figure 3.15b, where the circumference of the sliding plane is C. The results are shown in Figure 3.18, where the abscissa L is the distance from the crack face. For ease of understanding, let us assume that the slip varies linearly from the crack face as shown in Figure 3.18a. When the slip at the crack face is δ_1, the slip distribution is the line δ_1–L_1 in Figure 3.18a. From the BS relationship shown in Figure 3.17 and for this slip distribution, the bond stress distribution is given by the line τ_{max}–L_1 in Figure 3.18b, where the shear stress lies within the ascending branch of the BS relationship shown in Figure 3.17. Integrating this shear stress distribution over the surface of the sliding plane of length L_1, that is, the surface area is CL_1, gives the force in the bar $P_{\delta 1}$ when the slip at the crack face is δ_1. That is $P_{\delta 1} = \tau_{max} L_1 C/2$.

Increasing the slip at the crack face to δ_{max} in Figure 3.18a, the full BS relationship shown in Figure 3.17 is developed as O–A–L_{crit} in Figure 3.18b. Integration of the shear stress over L_{crit} is the IC debonding resistance P_{IC}, which is proportional to G_f in Figure 3.17. This proportionality highlights the importance of G_f, which represents the area of the BS relationship.

3.2.2 BS material properties

The τ/δ relationship that describes the shear stress (τ) transferred from the reinforcement to the concrete for any local slip (δ) is shown in Figure 3.19a, where the generic shape for a ribbed steel reinforcing bar embedded in concrete is represented by the

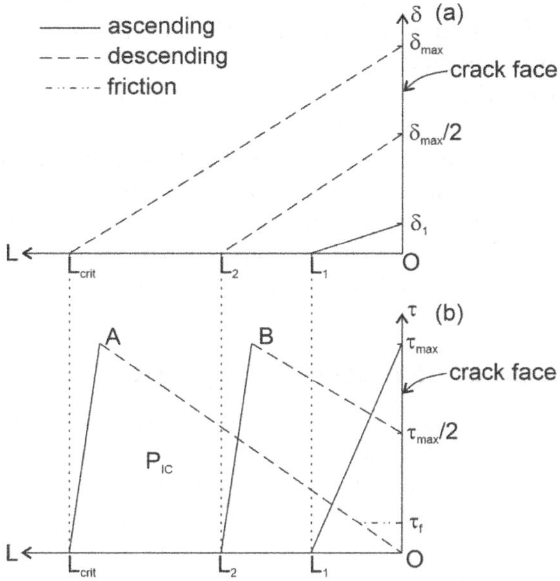

Figure 3.18 Distribution of (a) bond slip and (b) bond stress in a pull test.

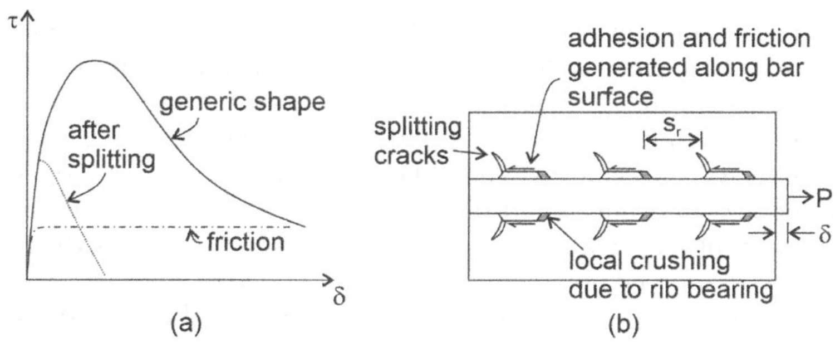

Figure 3.19 Bond slip: (a) generic shape of τ/δ material properties; (b) mechanism of shear transfer between reinforcement and concrete.

unbroken line. The mechanics of the transfer of stresses between the reinforcement and concrete can best be explained by considering Figure 3.19b, which shows a small portion of reinforcement with ribs at a spacing s_r encased in concrete after being pulled a distance δ. The transfer of stresses is a result of the combined action of (FIB 2000)

i. adhesion between the reinforcement and concrete,
ii. bearing forces that are generated as the ribs begin to slide against the concrete and
iii. friction that is generated as the reinforcement slides and the concrete between the ribs on the bar is damaged.

The non-linear nature of the τ/σ relationship in Figure 3.19a is, therefore, the result of the varying magnitudes of each of the above actions. Hence, the shape of the τ/δ

relationship can vary significantly depending on the surface treatment of the reinforcement; the strength of the concrete; and the detailing conditions that provide confinement to the reinforcement. For example, if the reinforcement is placed in a region of little cover, the tensile stresses generated as a result of bearing forces in Figure 3.19b can cause the concrete to split or rip, and consequently, due to the reduction in confinement, the bearing and friction forces rapidly reduce, as shown in Figure 3.19a using a dotted line. Similarly, if a smooth bar is used instead of a ribbed bar, the mechanism of bearing is removed and only adhesion and friction are present. This significantly reduces the maximum bond stress but leads to a very stable frictional component, shown as a dashed line in Figure 3.19a.

3.2.3 Quantification of *BS* material properties

The τ/δ material properties can be determined experimentally through bond tests, which are separated into those conducted on short bond lengths and those on long bond lengths. The term 'long' or 'short' is relative to L_{crit} in Figure 3.18, which is the length required to develop the full bond stress distribution. Hence, 'long' refers to bond lengths greater than L_{crit} where the full bond stress distribution (as shown in Figure 3.17) can be achieved (e.g. O–A–L_{crit} in Figure 3.18b). Conversely, 'short' refers to bond lengths smaller than L_{crit} where the full bond-stress distribution is not achieved (e.g. $(\tau_{max}/2)$–B–L_2 and τ_{max}–L_1).

The bond properties can be extracted using either of these test methods (long or short). However, in common practice, the short bond length test is applied in scenarios where the τ/δ relationship is developed over large slips or involves very high bond stresses, potentially causing the reinforcement to yield or rupture before debonding, as is often the case with internal reinforcing bars. Conversely, the long bond length test is generally conducted in scenarios where the full bond stress is developed over small ranges of slip or the maximum bond stress is low, such that debonding occurs before the reinforcement ruptures. This latter scenario is most common for external reinforcement such as fibre-reinforced polymer (*FRP*) sheets bonded to concrete. Another important distinction is that τ/δ properties can be directly extracted from tests conducted on short bond lengths, whereas inverse analysis is required for tests on long bond lengths.

3.2.3.1 *Short bond length tests*

An example of a pull test conducted on a short bonded length is shown in Figure 3.20a, where a reinforcing bar embedded in concrete is pulled from the concrete by reacting against its top surface. The bonded length, denoted as L_{bd}, should be sized so that both the slip δ and shear stress τ along L_{bd} can be assumed to be uniform, as shown in Figure 3.20b and c. This requires that L_{bd} is at least an order of magnitude smaller than L_{crit} in Figure 3.18. Furthermore, to ensure an accurate average strength, L_{bd} should be an order of magnitude larger than the rib spacing s_r, though this is not always possible. The reinforcement is placed in sleeves at either end of the bonded region to ensure no bond occurs in these regions.

For short bond tests, L_{bd} is chosen to be as short as possible while still ensuring that the impact of local bar variations is averaged. As a rule of thumb, L_{bd} is generally less than or equal to six times the bar diameter (d_b) for internal reinforcement. A lower

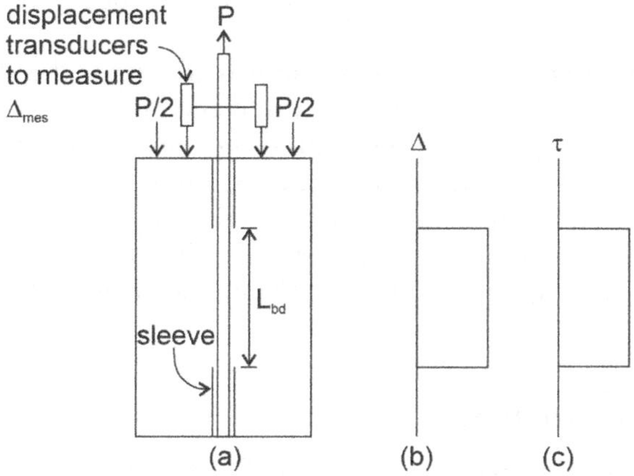

Figure 3.20 Short bonded length pull test: (a) specimen; (b) distribution of local slip; (c) distribution of shear stress (b).

limit of two times the rib spacing (s_r) in Figure 3.19 is also generally imposed; for very short bonded lengths, additional scatter is introduced as individual ribs are included or excluded from the test region. It should, however, be noted that bond lengths in the order of $2s_r$ are only required for concretes such as ultra-high-performance concrete where, for longer lengths, the reinforcement yields prior to pull-out; this is undesirable because plastic deformation of the reinforcement impacts the τ/σ relationship (Sturm and Visintin 2019).

When conducting the pull-out test in Figure 3.20, a load P is applied to pull the reinforcement from the concrete by reacting against its surface. The slip of the reinforcement relative to the top surface of the concrete Δ_{mes} is measured using displacement transducers which can be located either at the top or bottom of the specimen. For tests where the slip of the reinforcement is measured at the top of the specimen, the elastic deformation of the reinforcement between the slip reference point and the start of the bonded length (L_{ref}) must be subtracted from the measured slip (Δ_{mes}). That is, in Figure 3.20a, Δ_{mes} is converted to the actual slip Δ using

$$\Delta = \Delta_{mes} - L_{ref} \frac{P}{E_r A_r} \tag{3.37}$$

where P is the load corresponding to Δ_{mes}, E_r is the elastic modulus of the reinforcement and A_r is the cross-sectional area of the reinforcement.

In Figure 3.20b, the slip of the reinforcement along the bonded length is shown to be constant, which is a consequence of the bonded length being short. As a result, the bond stress in Figure 3.20c is also constant over the bonded length, such that τ can be determined directly from the applied load P as

$$\tau = \frac{P}{L_p L_{bd}} \tag{3.38}$$

in which P is the applied load, L_p is the width of the bonded perimeter and L_{bd} is the bonded length.

It is critical to note that if the bonded length is too long, the slip in Figure 3.20b would not be constant and, hence, neither would the bond stress in Figure 3.20c. In this case, application of Eq. 3.38 would lead to the quantification of an average bond stress (τ_{avg}) in Figure 3.20c, meaning that the maximum or minimum stress would not be captured. It is therefore important to remember that the guideline recommending a bonded length less than $6L_{db}$ is merely a rule of thumb. Modelling as described in Section 3.2.3.2 should be applied to confirm the bonded length is indeed short prior to applying Eq. 3.38.

3.2.3.2 Long bond length tests

Now consider the pull test specimen in Figure 3.21a, where the bonded length L_{bd} is long. In this case, the corresponding distribution of slip δ in Figure 3.21b and bond stress τ continuously varies along the bonded length. In this scenario, the test is carried out in an identical manner as for the short bonded length; that is, the reinforcement is pulled at the loaded end, and the slip of the loaded or free end is measured during loading until the reinforcement is extracted from the concrete that is it is fully debonded.

Importantly, in the case of a long bonded length, the average bond stress τ_{avg} is significantly lower than the maximum bond stress τ in Figure 3.21c and the end slip bears no simple relationship to the slip distribution. Hence, to determine the τ/δ relationship, an inverse analysis must be performed using the numerical model of the PI tension stiffening mechanism, as discussed in detail in Section 4.2.

To perform the inverse analysis, the following steps are required:

i. The measured slip Δ_{mes} must first be processed using Eq. 3.37 to obtain the actual slip Δ by removing the elastic deformation of the unbonded reinforcing bar.
ii. A τ/δ relationship is guessed, such as that shown in Figure 3.17 with appropriate values, and the numerical *PI* tension stiffening model (Section 4.2.2) is run using the boundary conditions specified in Section 4.2.2.2 to obtain a predicted P/Δ relationship.

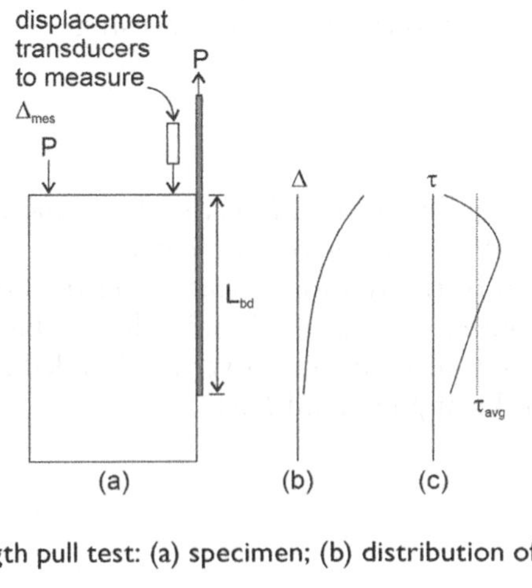

Figure 3.21 Long bonded length pull test: (a) specimen; (b) distribution of local slip; (c) distribution of shear stress.

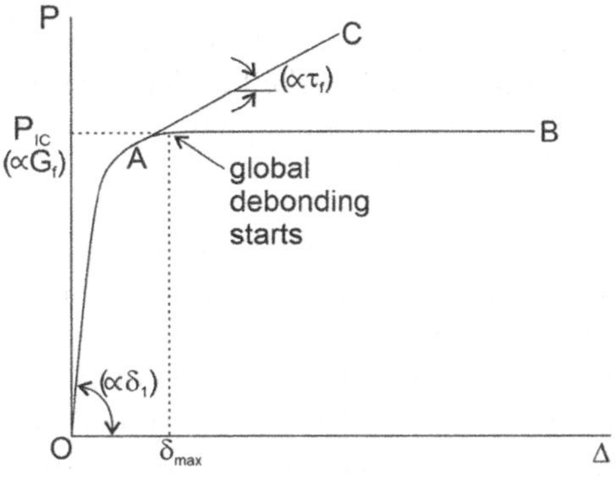

Figure 3.22 Deriving bond slip from test results.

iii. The predicted P/Δ relationship is compared to the experimental P/Δ relationship, and statistics of fit, such as the root mean square error, are applied to measure the goodness of fit between the predicted and experimental P/Δ relationships.
iv. The τ/δ relationship is updated, and the goodness of fit is recalculated until an optimum fit is obtained.

When updating the guess for τ/δ in step (iv), key indicators can be applied; these will be demonstrated by considering a pull test conducted on an externally bonded *FRP* sheet, the results of which are shown as *O–A–B* in Figure 3.22. For *FRP* sheets externally bonded to concrete, it is common practice to assume the τ/δ relationship is bilinear (as in Figure 3.17), with key points being the maximum bond stress τ_{max} (occurring at a slip of δ_1) and the debonding slip δ_{max} (at which τ is zero). When guessing the τ/δ relationship, the following indications from Figure 3.22 can be applied:

i. The turning point along *O–A–B* is expected to occur at a slip of δ_{max}; hence, δ_{max} can be directly determined from the experimental results.
ii. The maximum force is the *IC* debonding resistance, P_{IC}. For the bilinear τ/δ, this is derived in Section 4.3.3.4 and is given by the following equation, where E_p, A_p and L_p are the modulus, cross-sectional area and bond perimeter, respectively, of the bonded plate.

$$P_{IC} = \sqrt{2G_f}\sqrt{E_p A_p L_p} \tag{3.39}$$

Hence, this should give a good indication of G_f. As we now have a good indication of both τ_{max} and G_f from Figure 3.17, this provides another indicator for δ_{max}.
iii. The slope of the rising branch prior to P_{IC} can be changed by shifting δ_1 in Figure 3.17; a reduction in δ_1 leads to an increase in stiffness.
iv. If a secondary rising branch *A–C* is found following the turning point (shown as a broken line in Figure 3.22), then the τ/δ relationship in Figure 3.17 must include a frictional component (τ_f), and its magnitude is proportional to the slope of the P/δ relationship after P_{IC}.

Hence, all the BS parameters in Figure 3.22 can theoretically be detected when the bonded length is greater than or equal to L_{crit}. Unlike the case of short bonded lengths, there is no simple rule of thumb to predict the bonded length required to achieve long bonded length conditions, and there is no immediate transition between short and long bonded lengths. To restate this important point,

- When conducting tests with short bonded lengths, the experimentally measured P/Δ relationship will be the same shape as the τ/δ relationship because the slip is constant along the bonded length.
- Conversely, for the same τ/δ relationship when considering long bonded lengths, there will be a clear turning point in the experimentally measured P/Δ relationship corresponding to the initiation of debonding, after which the force either remains constant (no friction) or increases at a constant but significantly slower rate (with friction).
- Between these two scenarios, there is, however, a significant region in which the P/Δ relationship cannot be determined using the short bond length conditions. This is because the bond length is insufficient, making it more difficult to pick key features such as δ_{max} and P_{IC} since the bond stress is not fully developed when slip commences along the full plate length.

Hence, when designing long bonded length tests, it is best to ensure that the bonded length L_{bd} is significantly larger than the length L_{crit} required to achieve the debonding force P_{IC}, which Yuan et al. (2004) approximated by

$$L_{crit} = \frac{\pi}{2\sqrt{\dfrac{\tau_{max} L_p}{\delta_{max} E_r A_r}}} \tag{3.40}$$

in which L_p is the perimeter length of the failure plane in Figure 3.15 (being about the width of the plate in Figure 3.15a and about the diameter of the bar in Figure 3.15b), and all the other variables have been previously defined. Given that L_{crit} is dependent on the BS properties being determined (i.e. τ_{max} and δ_{max}), these properties must be estimated from the results of similar tests or experience. In the absence of prior knowledge, it is necessary to conduct tests with increasing bonded lengths until the location of the turning point on the P/Δ relationship stabilizes, at which point L_{crit} is known.

When performing this inverse analysis, it is critical to remember that it is impossible to obtain a perfect fit between the experimental and predicted P/Δ relationships. This is partly because of the natural variation that occurs between test results and partly because of the simplified shape of the assumed τ/δ relationship compared to the naturally occurring relationship. While it would appear that this limitation can be addressed by simply increasing the non-linearity of the assumed τ/δ relationship, this is not the case, as excessively increasing the non-linearity of the τ/δ relationship without increasing the number of observations to which the model is calibrated leads to an unavoidable statistical uncertainty in the fit.

3.2.4 BS material models

In Section 3.2.1, the mechanisms of transferring stresses between the reinforcement and concrete were shown to be adhesion between the two materials; bearing of the bar ribs against the surrounding concrete; and friction between the reinforcement and surrounding concrete, generated as the concrete is damaged by the bar. The magnitudes of the forces generated by these three mechanisms are highly dependent on

i. the type of reinforcement (e.g. reinforcing bars encased in concrete versus plates adhesively bonded to the concrete surface), which can impact the degree of confinement provided to the reinforcement;
ii. the bar surface treatment (e.g. ribbed bars generating significant bearing forces versus smooth bars generating no bearing forces); and
iii. the strength of the reinforcement, concrete and adhesion, which controls the location of failure and, consequently, the degree of friction. For example, an externally bonded plate where failure occurs in the concrete has a significantly rougher failure plane than if it occurred in the adhesive between the plate and the concrete.

Owing to the dependence on reinforcement type, bar surface treatment and material strength, τ/δ relationships are generally presented for subsets of materials as follows.

3.2.4.1 BS for ribbed steel bars encased in plain concrete

The FIB Model Code 2010 (FIB 2013) provides a detailed model for the τ/δ relationship between ribbed steel bars and plain concrete of various strengths. The general shape is shown in Figure 3.23 and follows the work of Eligehausen et al. (1982), as given by Eq. 3.41, where the units are in *N* and *mm*.

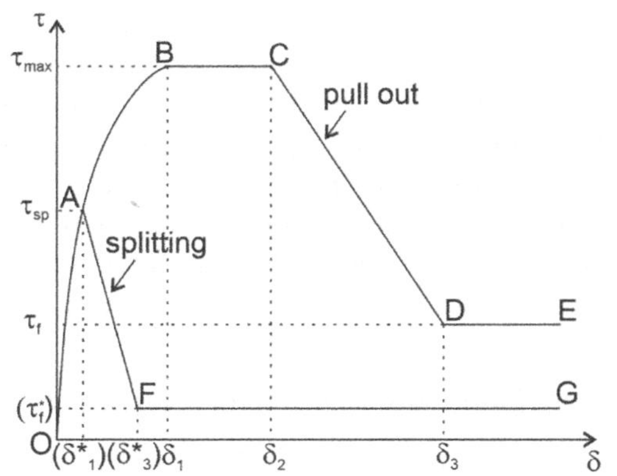

Figure 3.23 Generic τ/δ relationship for ribbed steel bars encased in concrete.

$$\tau = \begin{cases} \tau_{max}\left(\dfrac{\delta}{\delta_1}\right)^{\alpha}; & \text{when } 0 \le \delta < \delta_1 \\ \tau_{max}; & \text{when } \delta_1 \le \delta < \delta_2 \\ \tau_{max} - \left(\tau_{max} - \tau_f\right)\left(\dfrac{\delta - \delta_2}{\delta_3 - \delta_2}\right); & \delta_2 \le \delta < \delta_3 \\ \tau_f & \text{when } \delta > \delta_3 \end{cases} \tag{3.41}$$

where the first term represents the non-linear variation *O–A–B* in Figure 3.23 up to a maximum shear stress τ_{max} at a slip δ_1; the second term is the plateau *B–C* up to a slip δ_2; the third term is the degradation *C–D* up to a slip δ_3; and the fourth is the frictional term *D–E* at τ_f from δ_3 onwards. This is an upper bound for the strength and occurs when the surrounding concrete does not fail in tension by splitting before the bar is pulled out. These are the 'pull-out' properties. It is also useful to note that the form of Eq. 3.41 is highly adaptable and can easily be transformed into the trilinear or bilinear relationships used for *FRP* plates by setting the coefficient α, which controls the non-linearity of the rising branch *O–A–B*, to 1 and by removing the plateau at τ_{max} by setting δ_2 equal to $\delta_{1.}$

Importantly, while Figure 3.23 provides the generic shape of the τ/δ bond properties, in the case of pull-out failure, it is modified depending on whether failure occurs via pull-out of the reinforcement or whether the concrete cover splits along the length of the bar. Splitting occurs when the reinforcement has limited cover or is either unconfined or poorly confined, and if splitting does occur, both the strength and ductility of the τ/δ relationship in Figure 3.23 are significantly reduced. The mechanics of splitting can be modelled using thick wall theory (Tepfers 1979), where splitting occurs if the circumferential tensile stresses that are generated as the reinforcement pulls from the concrete are sufficient to cause the stress at the concrete surface to reach the tensile cracking stress.

Extensive research has been conducted to develop generic values for the key points identified in Figure 3.23 (τ_{max}, $\tau_f\,\delta_1\;\delta_2\;\delta_3$). The most well-known and widely accepted are those provided by the FIB Model Code 2010 (FIB 2012), which the reader can consult for specific values applicable to various bond conditions. Of most significance in the specific calibration of bond models for steel bars encased in concrete is that splitting causes an approximate halving of the maximum bond stress $\tau_{max,}$ and a significant reduction in bond ductility (Harajli 2004; FIB 2012). This is represented by the change from OABCDE to OAFG in Figure 3.23.

3.2.4.2 BS for ribbed steel bars encased in fibre concrete

The inclusion of fibre reinforcement in concrete can significantly improve the bond between the reinforcement and the concrete, particularly when the cover to the reinforcement is small. This is because the fibres act to restrain splitting cracks and provide confinement in a similar way to transverse reinforcement. Given the large number of

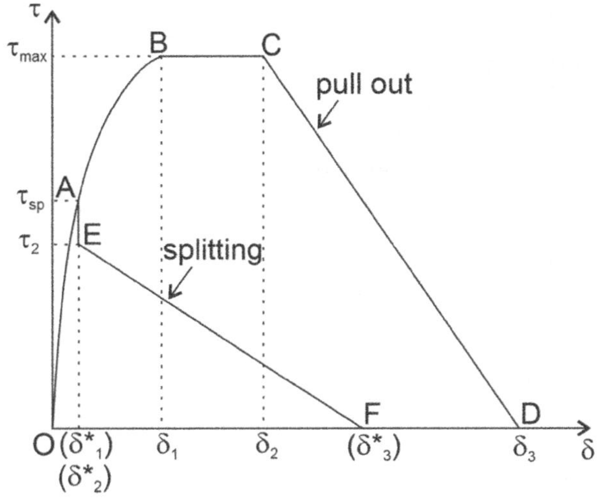

Figure 3.24 Generic τ/δ relationship for ribbed steel bars encased in *UHPFRC*. (Sturm and Visintin 2019.)

potential combinations of fibre type and dosage, it is again recommended that tests be conducted to obtain bond properties specific to the concrete; however, in the absence of testing, the FIB approach defined above for concrete without fibres provides a good lower bound solution.

When considering *UHPFRC*, the high concrete compressive strength and high volume of fibres leads to a very strong bond between the reinforcement and concrete; as described previously, embedment lengths of as little as two times the bar diameter have been shown to cause yielding of the reinforcement. As a result, it is necessary to recalibrate a new bond model for *UHPFRC* with micro, macro or blended micro and macro fibres. Sturm and Visintin (2019) have developed such models, shown in Figure 3.24, where the 'pull-out' model applies when cover to the reinforcement for central bars exceeds 50 mm or for corner reinforcing bars exceeds 75 mm, and the 'splitting' model applies to bars with less cover.

The following general expressions for the 'pull out' in Figure 3.24 (i.e. *O–A–B–C–D*) are the same as those for the 'pull out' for plain concrete in Figure 3.23, except that there is no frictional component. The units are also *N* and *mm*.

$$\tau = \begin{cases} \tau_{max}\left(\dfrac{\delta}{\delta_1}\right)^{\alpha}; & 0 \le \delta < \delta_1 \\ \tau_{max}; & \delta_1 \le \delta < \delta_2 \\ \tau_{max}\left(1 - \dfrac{\delta - \delta_2}{\delta_3 - \delta_2}\right); & \delta_2 \le \delta < \delta_3 \\ 0; & \delta > \delta_3 \end{cases} \tag{3.42}$$

where

$$\tau_{max} = (0.0018c + 0.186)f_c \tag{3.43}$$

$$\delta_1 = (-0.0002c + 0.001)f_c + 0.0386c + 0.0715 \tag{3.44}$$

$$\delta_2 = 2.42 \tag{3.45}$$

$$\delta_3 = 0.179\tau_{max} + 0.74 \tag{3.46}$$

and

$$\alpha = 0.0183d_b - 0.0055\tau_{max} + 0.304 \tag{3.47}$$

in which c is the minimum cover and d_b the bar diameter. It is interesting to note that Eqs. 3.42–3.47 are independent of fibre type. This may be because the range of crack widths experienced while the bars were pulled out was too small for the variation in fibre type to have an impact. It should, however, also be noted that the data available to calibrate a model is limited, and so further testing over a wide range of fibre types and volumes is required.

The splitting strength τ_{sp} is given by the same expression as for τ_{max} (Eq. 3.43), where τ_{max} is replaced by τ_{sp} when the cover is less than 50 mm for corner bars or less than 65 mm for bars located away from the corner. The slip at which splitting occurs at Point A in Figure 3.24 δ_1^* is determined by solving Eq. 3.44 using the known concrete cover c and concrete compressive strength f_c. At the onset of splitting, the bond stress drops from Point A to Point E (τ_{sp} to τ_2) where τ_2 is given by

$$\tau_2 = (0.0008c + 0.941)\tau_{max} \le \tau_{max} \tag{3.48}$$

after which the strength reduces linearly from Point E to Point F, at δ_3^* of which the strength is given by

$$\delta_3 = 0.179\tau_{sp} + 0.74 \tag{3.49}$$

Worked Example WE-3.3: Bond properties for embedded steel bars

To demonstrate the significant improvement that the combination of high concrete strength and fibre reinforcement can have on the bond between concrete and internal reinforcement, let us compare the bond properties obtained from the FIB (2013) model for conventional concrete and the Sturm et al. (2018) and Sturm and

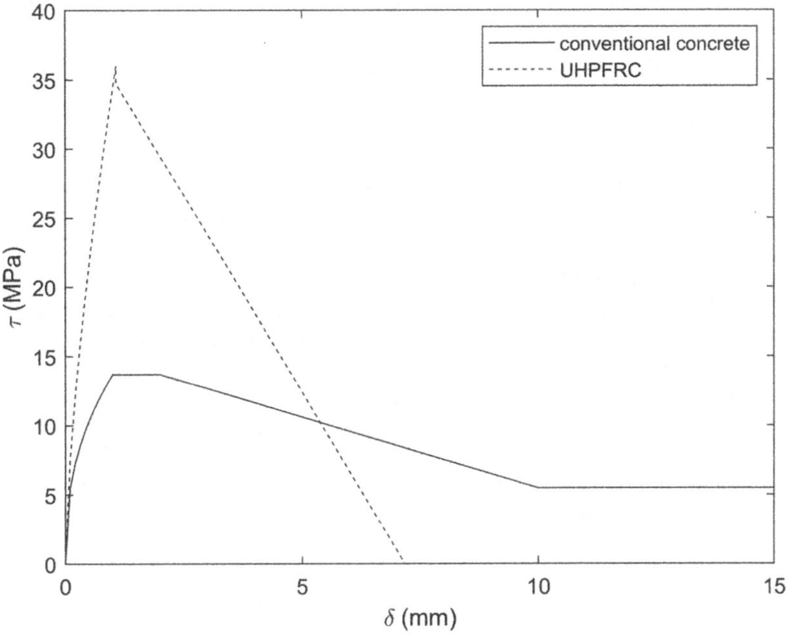

Figure 3.25 Example τ/δ relationships for conventional concrete and *UHPFRC*.

Visintin (2019) model for *UHPFRC*. To do this, let us consider the scenario of a 16mm steel bar with a rib spacing s_r of 10mm, which is located at the corner of a beam such that it has a clear cover of 30mm to both the side and the bottom of the beam. Let us compare the bond properties obtained from a concrete with a mean concrete strength of 30MPa and a *UHPFRC* with a mean concrete strength of 150MPa and a fibre volume fraction of between 2% and 3% (the range over which the model was calibrated).

Assuming good bond conditions and following the approach outlined in FIB (2012) yields the bond properties shown as a solid line in Figure 3.25 where $\tau_{max} = 2.5\sqrt{f_{cm}} = 13.69$ MPa and $\tau_f = 0.4\tau_{max} = 5.48$MPa

For the *UHPFRC*, because the bar is located in the corner of the beam and has a cover of less than 50mm, splitting will occur. In this scenario, the result of Eq. 3.43 is taken as the splitting strength such that

$$\tau_{sp} = (0.0018 \times 30 + 0.186)150 = 36 \text{ MPa} \tag{E3.3.1}$$

Similarly, the result of Eq. 3.44 is determined as δ_1^* such that

$$\delta_1^* = (-0.0002 \times 30 + 0.001) \times 30 + 0.0386 \times 30 + 0.0715 = 1.08 \text{ (E3.3.2)}$$

and the non-linearity of the rising branch controlled by the α parameter in Eq. 3.42 is given by solving Eq. 3.47 to yield

$$\alpha = 0.0183 \times 30 - 0.0055 \times 36 + 0.304 = 0.655 \tag{E3.3.3}$$

Now, the reduction in bond stress post splitting is given by Eq. 3.48

$$\tau_2 = (0.0008 \times 30 + 0.941)36 = 34.74 \text{ MPa} \tag{E3.3.4}$$

and the slip at which the bond stress reduces to zero is

$$\delta_3 = 0.179 \times 36 + 0.74 = 7.18 \text{ mm} \tag{E3.3.5}$$

Substituting the results of Eqs. E3.3.1–E3.3.5 into Eq. 3.42, the full τ/δ relationship as shown in Figure 3.25 is obtained. Comparing the bond properties of conventional concrete and *UHPFRC* in Figure 3.25, it can be observed that, despite splitting occurring in the *UHPFRC*, the confining action of the fibres results in a bond strength still more than 2.5 times larger than that of the same bar in a conventional 30 MPa concrete. Also of interest are the very stiff bond properties of *UHPFRC*, which lead to a rapid transfer of stress from the reinforcement to the surrounding concrete. This results in both a larger number of smaller cracks forming and a reduction in the bonded length required to develop a given stress in the reinforcement.

It can also be seen in Figure 3.25 that splitting has not significantly reduced the maximum slip of the bar, which can be taken as equal to the rib spacing (10 mm for conventional concrete). When interpreting the results of δ_3=7.18 mm in Eq. E3.3.5, it should also be remembered that the crack width is taken as twice the slip of the reinforcing bar from the crack face. Hence, a slip equal to δ_3 is equivalent to a crack width of between 14 and 20 mm, making it highly likely the bar will rupture before debonding.

3.2.4.3 BS for FRP bars encased in plain concrete

When considering *FRP* reinforcement encased in concrete, there is not yet an agreed-upon calibrated model that defines the τ/δ under a range of detailing conditions. Codes of practice such as FIB (2013), instead, suggest undertaking experimental work to yield results specific to the materials and detailing conditions.

To help address this limitation, Li et al. (2020) compiled a database of existing test results quantifying the bond between *FRP* bars and concrete that had been tested before and after exposure to environmental conditions that cause bond degradation. Considering the control specimens that had not degraded, it was possible to calibrate a model in the generic form given by Eq. 3.50, which, as shown in Figure 3.26, is similar to the FIB model for ribbed steel bars after splitting (Figure 3.23), assuming friction is ignored.

$$\tau = \begin{cases} \tau_{max}\left(\dfrac{\delta}{\delta_1}\right)^{\alpha} ; 0 \le \delta < \delta_1 \\ \tau_{max}\left(1 - \dfrac{\delta - \delta_1}{\delta_3 - \delta_1}\right); \delta_1 \le \delta < \delta_3 \\ 0; \delta > \delta_3 \end{cases} \tag{3.50}$$

Examples of the τ/δ equations derived from the model by Li et al. (2020) are shown in Figure 3.26.

Due to the limited test data, this model made no attempt to account for the influence of bar surface treatment; however, work by Cosenza et al. (1997) has indicated that bar surface treatment is a significant factor influencing both the strength and ductility of the FRP-to-concrete bond. This is evident when comparing the different bar surface

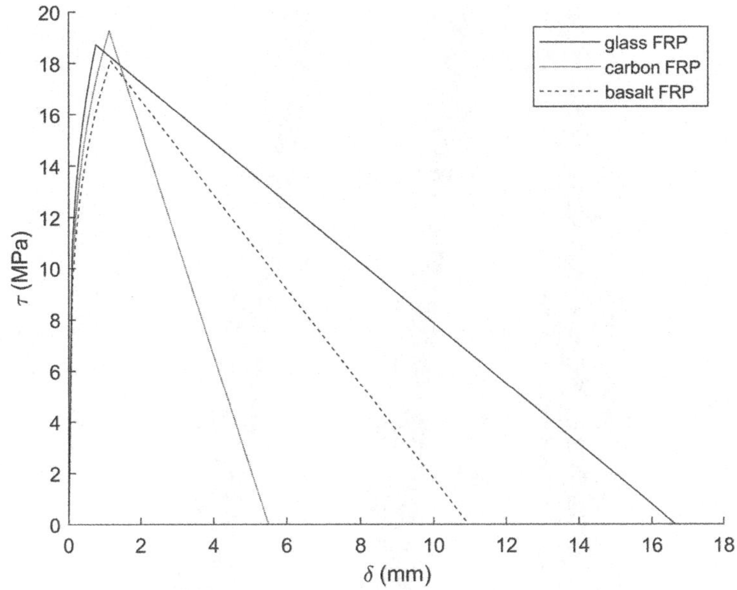

Figure 3.26 Local τ/δ bond properties for *FRP* bars. (According to Li et al. 2020.)

treatments shown in Figure 3.27. Specifically, for sand-coated reinforcement, the predominant mechanism of stress transfer is friction, while for bars with ribs or indents, significant bearing forces can be generated, similar to those in steel reinforcement.

3.2.4.4 BS of external and NSM FRP reinforcement to concrete

Numerous very good models of the BS properties of *EB* plates (e.g. Lu et al. 2005) are available in codes of practice and journal publications. The following model (Seracino et al. 2007; Oehlers et al. 2008) is described below for its ability to integrate most, if not all, of the mechanics governing BS behaviour. It covers both *EB* plates and *NSM* plates; the interaction between adjacent plates; is based on the tension stiffening mechanics in Chapter 4, particularly the bilinear model; and explains the variation in the strength of the debonding failure plane in terms of the mechanics of SF and confinement. Externally bonded (*EB*) *FRP* plates are bonded to the concrete surface (Figure 3.28a), while *NSM FRP* plates are bonded adjacent to the concrete (Figure 3.28b).

The mechanism of bond transfer is largely defined by the failure mode. For well-prepared surfaces, where the epoxy used to bond the plate is sufficiently strong, bond failure occurs within the concrete, with a surface layer shearing off along with the *FRP* plate (Figure 3.29). Alternatively, when the surface preparation of the concrete is poor or the epoxy is insufficiently strong, failure can occur in the layer between the epoxy and the plate or between the plate and the concrete.

For the externally bonded plate shown in Figures 3.28a and 3.29b, there is virtually no confinement force, N in Eq. 3.1, across the failure plane, except for occasional fibres crossing it. Hence, in the absence of any additional mechanical anchorage, it is not possible to reliably provide a normal confinement force to the plate. Therefore, in

Figure 3.27 FRP bar surface treatment from left to right: smooth, sand coated and ribbed. (Photo courtesy of Phillip Visintin.)

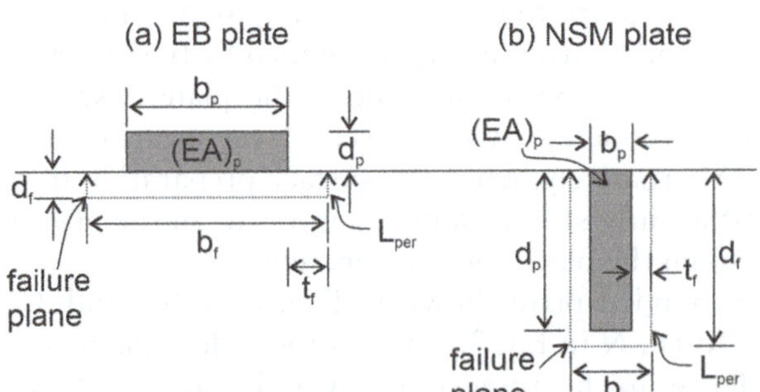

Figure 3.28 Bonded FRP plates in (a) externally bonded and (b) near surface mounted configurations.

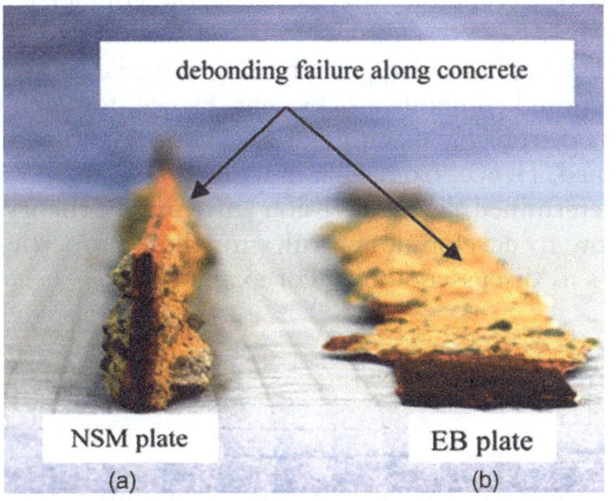

Figure 3.29 Failure planes. (Reproduced from M.S. Mohamed Ali et al 2008.)

general, no residual frictional component, as shown in Figure 3.17, is provided, despite the potential for some friction to occur as the rough concrete surfaces bear against each other, similar to the SF mechanism. In contrast, *for NSM* plates, as shown in Figures 3.28b and 3.29a, substantial passive confinement can occur due to aggregate interlock (*P* in Eq. 3.1) and possibly shrinkage (*S*).

As shown in Figure 3.16 for externally bonded plates, δ_{max} is generally too small to allow the BS properties to be determined directly through tests. Hence, they are generally determined indirectly through the mechanical analysis of tension stiffening tests. For externally bonded reinforcement such as the *FRP* plates in Figure 3.28a, the unilinear or bilinear τ/δ relationship shown in Figure 3.17 is generally applied. For both the unilinear and bilinear BS relationship, the intermediate crack debonding resistance P_{IC} is given by Eq. 3.39 (Yuan et al. 2004), which can be written as

$$P_{IC} = \sqrt{\tau_{max}\delta_{max}}\sqrt{E_pA_pL_{per}} = \sqrt{2G_f}\sqrt{E_pA_pL_{per}} \tag{3.51}$$

where E_pA_p is the axial rigidity of the plate and G_f is the fracture energy, as shown in Figure 3.17. Although we are dealing specifically with *FRP* plates in this section, E_pA_p could be the flexural rigidity of any type of plate such as a steel plate. The minimum length required to develop this maximum bond force, L_{crit} (also given by Yuan et al. 2004), is provided by Eq. 3.40:

$$L_{crit} = \frac{\pi}{2\sqrt{\frac{\tau_{max}}{\delta_{max}}}\sqrt{\frac{L_{per}}{E_pA_p}}} \tag{3.52}$$

where L_{per} is the perimeter length of the failure plane over which debonding occurs, as shown in Figure 3.28. It can be seen that L_{per} is slightly longer than the contact length between the plate and the concrete to allow for failure within the concrete as shown in Figure 3.29. It is suggested (Seracino et al. 2007) that d_f and t_f be taken as 1 mm, although zero would be acceptable; that is, using the contact length would be only slightly conservative.

As shown in Figure 3.22, P_{IC} and δ_{max} can be determined directly from tests, irrespective of whether a frictional component is present. Hence, for each individual test, τ_{max} δ_{max} can be determined from P_{IC} using Eq. 3.51. As δ_{max} is known from the test, τ_{max} can then be determined. Hence, from each individual test, the BS properties shown in Figure 3.17 can be determined through tension stiffening *PI* theory.

The question is how to develop design rules for these properties that allow for the varying confinements as discussed above. For example, it has been assumed (Seracino et al. 2007) that

$$\tau_{max}\delta_{max} = C\left(\frac{d_f}{b_f}\right)^m f_c^n \tag{3.53}$$

such that the first term – that is, the aspect ratio of the failure plane – allows for increasing confinement with depth, while the second term accounts for the concrete strength. Based on this approach, it was found that the mean value is

$$\tau_{max}\delta_{max} = 0.976\left(\frac{d_f}{b_f}\right)^{0.526} f_c^{0.6} \tag{3.54}$$

and the 95% characteristic value is given by reducing the coefficient from 0.976 to 0.830, where the units are in N and mm. From Eq. 3.54, and using δ_{max} from the tests reported in the journal,

$$\tau_{max} = \left(0.802 + 0.078\frac{d_f}{b_f}\right) f_c^{0.6} \tag{3.55}$$

and consequently

$$\delta_{max} = \frac{0.976\left(\frac{d_f}{b_f}\right)^{0.526}}{0.802 + 0.078\frac{d_f}{b_f}} \tag{3.56}$$

This approach (Seracino et al. 2007) was taken a step further (Oehlers et al. 2008) to allow for the interaction between adjacent plates, as depicted in Figure 3.30 and illustrated by a pull test in Figure 3.31. In theory, the entrapped concrete between the plates should increase the axial rigidity of the E_pA_p component in Eq. 3.51, such that the *IC* debonding resistance of a group of plates is

$$P_{IC-grp} = \sqrt{\tau_{max}\delta_{max}}\sqrt{L_{per}}\sqrt{nE_cA_c + (n-1)k_{EAC}E_cA_c} \tag{3.57}$$

where n is the number of plates, E_cA_c is the axial rigidity of the concrete between the adjacent plates (as shown in Figure 3.31) and k_{EAC} is a reduction factor, ranging from

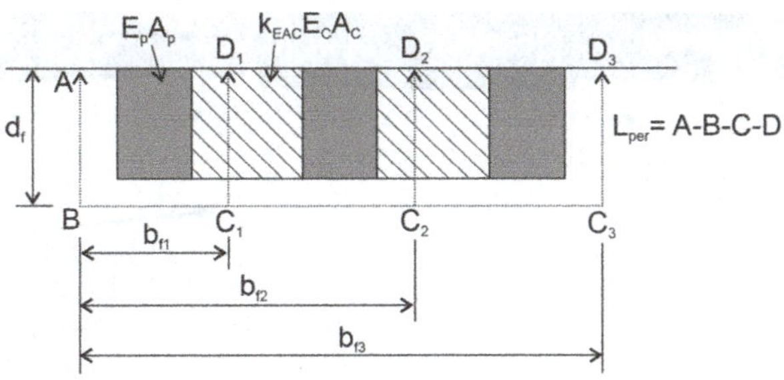

Figure 3.30 Interaction between adjacent plates.

Figure 3.31 Axial rigidity between adjacent plates. (Reproduced from Oehlers et al. 2008.)

zero to one, to allow for the reduction in E_cA_c due to the transverse cracking which can be seen in Figure 3.32.

As expected, it was empirically found that k_{EAC} decreased as the moment in the beam increased, as follows:

$$\left(k_{EAC}\right)_{mean} = 0.45 - 0.030\frac{M_u}{M_{cr}} \tag{3.58}$$

where M_u is the moment at debonding and M_{cr} is the moment causing a flexural crack. The 95% characteristic value can be obtained by replacing the 0.45 coefficient with 0.31. Hence, k_{EAC} tends to zero at the ultimate limit and is approximately 0.42 at

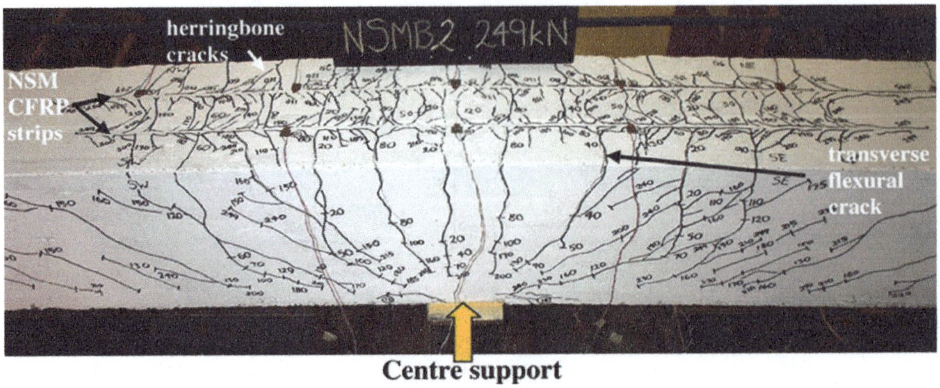

Figure 3.32 Cracking of concrete between adjacent *NSM* plates. (Reproduced from Oehlers et al. 2008.)

serviceability. When there are groups of plates (as in Figure 3.30), each possible failure plane must be checked for debonding, such as A–B–C_1–D_1, A–B–C_2–D_2, and so on.

Worked Example WE-3.4: Bond of external and NSM strips

In this example, we will examine the difference in bond properties obtained when externally bonding an *FRP* strip versus near-surface mounting. For this example, consider the scenarios where a carbon *FRP* strip with dimensions of 1 mm×7mm is either externally bonded in a horizontal orientation (as in Figure 3.28a) or NSM in a vertical orientation (as in Figure 3.28b). In this scenario, the plates are bonded to a concrete substrate with a compressive strength $f_{c}=$ 30MPa and an elastic modulus E_c=30,000MPa.

Firstly, consider the externally bonded plate. The maximum bond stress τ_{max} and maximum slip at the initiation of debonding δ_{max} are given by the empirical Eqs. 3.55 and 3.56, both are functions of the aspect ratio of the failure plane d_f/b_f (see Figure 3.28).

For the externally bonded scenario, if the failure plane is assumed to occur at a distance of 1mm from the plate surface, b_f = 9mm and a conservative value of $d_{f}=$ 1mm. Hence, solving Eqs. 3.55 and 3.56 yields

$$\tau_{max} = \left(0.802 + 0.078\frac{1}{9}\right)30^{0.6} = 6.24 \text{ MPa} \tag{E3.4.1}$$

and

$$\delta_{max} = \frac{0.976\left(\frac{1}{9}\right)^{0.526}}{0.802 + 0.078\left(\frac{1}{9}\right)} = 0.396 \text{ mm} \tag{E3.4.2}$$

Taking the perimeter of the failure plane to be 1mm beyond the plate dimensions, the failure perimeter L_{per}=11mm. The bond properties from Eqs. E3.4.1 and E3.4.2

can then be substituted, along with the axial rigidity of the plate, into Eqs. 3.51 and 3.52 to determine the *IC* debonding force and the critical bond length required to reach this force.

$$P_{IC} = \sqrt{6.24}\sqrt{0.396}\sqrt{200{,}000 \times 7 \times 11} = 6.17\text{ kN} \tag{E3.4.3}$$

$$L_{crit} = \frac{\pi}{2\sqrt{\dfrac{6.24}{0.396}}\sqrt{\dfrac{11}{200{,}000 \times 7}}} = 141.2\text{ mm} \tag{E3.4.4}$$

Repeating the process for the NSM strip, the same process is followed, but with the aspect ratio d_f / b_f now taken as 8/3 and L_{per} as 19 mm.

$$\tau_{max} = \left(0.802 + 0.078\frac{8}{3}\right)30^{0.6} = 7.77\text{ MPa} \tag{E3.4.5}$$

$$\delta_{max} = \frac{0.976\left(\dfrac{8}{3}\right)^{0.526}}{0.802 + 0.078\left(\dfrac{8}{3}\right)} = 1.63\text{ mm} \tag{E3.4.6}$$

$$P_{IC} = \sqrt{6.24 \times 0.396}\sqrt{200{,}000 \times 7 \times 11} = 18.4\text{ kN} \tag{E3.4.7}$$

$$L_{crit} = \frac{\pi}{2\sqrt{\dfrac{6.24}{0.396}}\sqrt{\dfrac{11}{200{,}000 \times 7}}} = 195.3\text{ mm} \tag{E3.4.8}$$

Of significant interest here is that changing the orientation of the plate has increased τ_{max} by 24.5% and δ_{max} by 411%. This change in bond properties has increased P_{IC} by 298% and L_{crit} by 38%. That is, by confining the strip within the concrete, the slip required to cause debonding has significantly increased, which in turn has substantially increased the force that can be developed in the strip. Importantly, this large increase in force has come with a relatively small increase in the critical bond length. That is, although the force has increased by nearly a factor of 3, for practical detailing of the reinforcement placement, the bonded length is virtually the same. It is also worth noting that the *CFRP* stress in this scenario has increased from 881to 2629 MPa; thus, by near-surface mounting the strip, the high tensile strength of the carbon fibre is better utilised. Good utilisation of the full capacity of materials is particularly important when considering environmental sustainability, as will be discussed in Chapter 13. This is because the energy consumed and carbon dioxide equivalent emissions generated during the production of the *CFRP* are not insignificant, and a proportion of these are 'wasted' if the material is poorly utilised.

If near-surface mounting a plate of a given cross-sectional area is always beneficial compared to externally bonding, one might ask why an engineer would ever choose to externally bond the plate. The answer to this question is related to durability. That is, as discussed in Chapter 11, the resistance of a reinforcing bar to corrosion is directly proportional to the thickness of the concrete cover.

In scenarios where reinforcement corrosion is of particular concern, cutting a groove for an *NSM* strip reduces the cover concrete, and this may cause an unacceptable reduction in the resistance to corrosion. Alternatively, in scenarios where corrosion is not of significant concern, the concrete cover may be small, and hence, there may not be sufficient room within the cover concrete to place the *NSM* strip. That is, although the mechanics of quantifying the difference in bond strength are straightforward, the engineer is also required to apply some judgement in detailing the reinforcement.

3.3 FIBRE CONCRETE POST-CRACKING TENSILE MECHANISM

3.3.1 Fibre concrete material properties

The pre-cracking tensile properties, specifically the strain-based properties, of fibre concrete are given and discussed in Section 2.4.1.2 and are similar to those of plain concrete. For concrete containing fibres, the post-cracking tensile behaviour is markedly different from that of plain concrete as it is highly dependent on the volume and type of fibres present in the concrete. Typical behaviours are shown in Figure 3.33a for *FRC* and in Figure 3.33b for *UHPFRC*. It can be seen that the behaviour is divided into a pre-cracking stress–strain relationship and a post-cracking stress–crack width (w) relationship. Concrete with small volumes of fibres, such as the *FRC* in Figure 3.33a, only has a strain softening branch, whereas concrete with large volumes of fibres, such as *UHPFRC* in Figure 3.33b, can undergo a period of strain hardening prior to strain softening.

Consider *FRC* with low volumes of fibres as shown in Figure 3.33a. Upon crack initiation at point A, a stepwise reduction in stress from A to B occurs, as the stress transferred by the fibres across the crack is less than the tensile strength f_{ct} of the uncracked plane. Upon further displacement, the total stress transferred across the cracked interface now becomes a PI problem where the magnitude of force available is a function of the force transferred by each fibre through bond as it pulls from the concrete, and which, in this case, reduces with slip from B onwards.

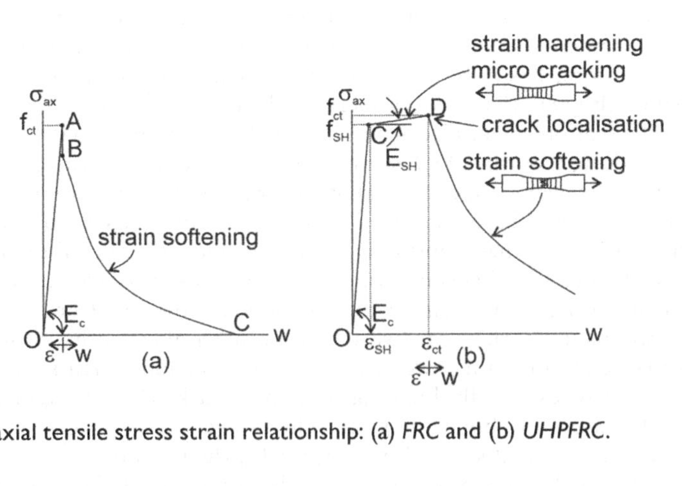

Figure 3.33 Uniaxial tensile stress strain relationship: (a) *FRC* and (b) *UHPFRC*.

Now consider concretes that exhibit strain hardening behaviour, such as *UHPFRC* shown in Figure 3.33b. At crack initiation, the volume of fibres crossing the crack means that the cracked section is of equal strength to, or stronger than, the uncracked section, and so the stress remains equal to or greater than that of the uncracked concrete. In strain hardening materials, the transfer of stress from the fibres to the concrete via bond is sufficient to cause many very closely spaced cracks of very small width, a phenomenon typically called the micro-cracking phase, which is modelled as a material behaviour using a strain hardening modulus defined by the branch extending from points $C(f_{SH}, \varepsilon_{SH})$ to point $D(f_{ct}, \varepsilon_{ct})$ and having a modulus of E_{SH} in Figure 3.33b. As the load continues to increase, the fibres pull out and eventually a tipping point is reached at one of the micro-cracks where fibre pull-out reduces the force transferred via bond, and this crack is then said to localise and continues to widen in a similar way to strain softening *FRC*s.

3.3.2 Fibre concrete tension stiffening mechanism

The post-cracking fibre concrete mechanism is a tension stiffening behaviour where each fibre has a different orientation relative to the crack face and a different embedment length. The question is what mechanisms can explain (i) why the fibres tend to straighten across the crack as shown in Figure 1.13; (ii) why a uniform distribution of micro-cracks can form; (iii) why these micro-cracks form with increasing applied load; (iv) and why the strength eventually reduces.

Figure 3.34a–c show the fibres to the *RHS* of an initial crack and, on the *LHS*, the forces acting on the fibres at the *RHS* crack face. Let P_{cr} be the axial force (i.e. a force perpendicular to the crack face) required to cause the initial crack. The fibres crossing the crack face have different orientations α (which includes orientation into the page), varying embedment lengths L and span varying crack widths w.

Figure 3.34a shows a fibre at right angles to the crack face with an embedment length L that is smaller than the length L_{crit} required to achieve the maximum possible bond force P_{IC}. If the maximum force in the fibre P is less than P_{cr} (the force required to cause another primary fibre crack), then as the crack widens, the embedded length

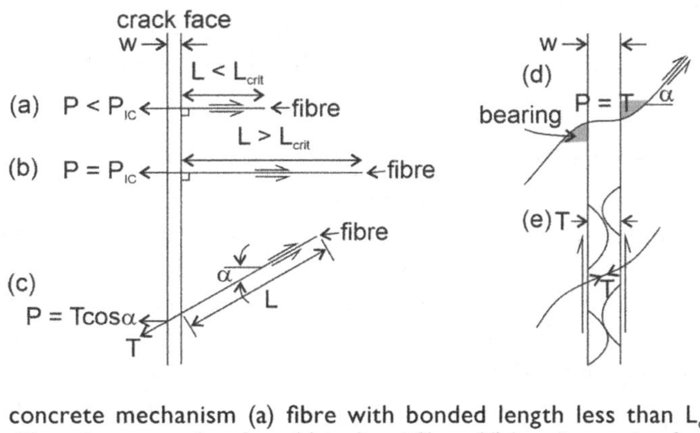

Figure 3.34 Fibre concrete mechanism (a) fibre with bonded length less than L_{crit}; (b) fibre with bonded length greater than L_{crit}; (c) inclined fibre; (d) bearing region for an inclined fibre; (e) fibre subject to shear deformation.

reduces to $L-w/2$ (where w is the crack width) and, consequently the force across the crack P reduces. Now consider a fibre with an embedded length $L> L_{crit}$ as shown in Figure 3.34b. In this case, the full bond strength P_{IC} can be achieved. If this is not sufficient to cause a primary micro-crack to form at P_{cr} (which depends on the tensile strength of the concrete), then as the crack widens, P remains at P_{IC} but the fibre gradually debonds until the debonded region is within L_{crit} of the fibre end, after which the fibre force reduces. Both this mechanism and that for the short length in Figure 3.34a account for the fourth phenomenon, namely, a reduction in strength. In contrast, if P_{cr} can be achieved (i.e. $P_{cr}<P_{IC}$), then a primary micro-crack will form at a spacing S_p. This sequence of events will lead to the formation of equally spaced primary micro-cracks at a constant load P_{cr} and a spacing S_p, which is analogous to the primary crack spacing in Figures 1.7 and 1.10. This mechanism accounts for the second phenomenon, a uniform distribution of micro-cracks, but contradicts the third phenomenon, a constant load.

Now consider an inclined fibre as in Figure 3.34c. If the force in the fibre is T, then the contribution to the tensile strength of the specimen will be $T \cos \alpha$. As the crack widens to w in Figure 3.34d, the fibre will straighten across the crack width as shown, increasing its contribution to the tensile capacity of the concrete from $T \cos \alpha$ to T. This mechanism explains the first and third phenomena, namely, the straightening of the fibres along the axis of the applied load and the increase in load with the development of micro-cracks. It is worth noting that the bearing of the fibre against the concrete in Figure 3.34d follows the same mechanism as that shown in Figure 3.15c. It is also worth noting that if the crack is subjected to shear deformations as in Figure 3.34e, then the opening of the crack due to aggregate interlock as in Figure 3.2 would tend to orientate the fibres more perpendicularly to the crack face.

3.3.3 Extracting tensile crack width properties

Given the very large combination of potential concrete types, fibre types, fibre lengths and fibre volumes, a generic stress–strain and stress–crack width model for *FRC* and *UHPFRC* has yet to be developed. Instead, national codes of practice suggest that modelling should be based on tensile stress–crack width (σ/w) relationships obtained from testing. Importantly, tensile-stress–crack width (σ/w) relationships can be determined either directly from tension tests or indirectly from flexural tests using an inverse analysis; both approaches will be discussed in the following sections.

3.3.3.1 Direct tension tests

When using the direct tension test, two test methods are possible: (i) the dog bone test in Figure 3.35a and (ii) the notched prism test in Figure 3.35b. Both the dog bone and notched prism direct tension tests (Figure 3.35) are conducted by applying a uniaxial tensile load to the specimen until failure. It is very important in these tests to ensure that the applied load P is aligned such that a uniform axial stress is imposed. If this is not done, the bending induced on the section produces a non-uniform crack width, and the measured average stress will correspond to this variation, rather than reflecting the true σ/w relationship.

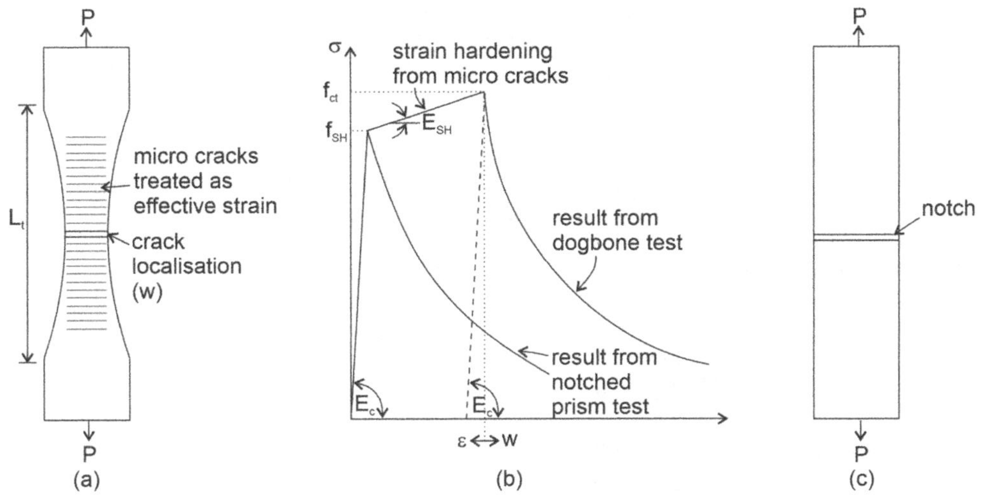

Figure 3.35 Extracting tensile crack width properties: (a) dog bone test; (b) notched direct tension prism; (c) tensile stress–strain and stress–crack width relationship.

In the case of the dog bone specimen in Figure 3.35a, micro-cracks can form along the length of the dog bone shank prior to the localisation of a single crack. Hence, with the dog bone-shaped specimen, it is possible to extract the strain hardening modulus E_{SH}, as shown in Figure 3.35c, directly from test results. That is, from elementary mechanics, the strain hardening modulus can be determined using displacement readings measured over the gauge length L_t in Figure 3.35a. This strain hardening region is bounded by a stress f_{SH} at the commencement of strain hardening and by f_{ct} at the initiation of crack localisation, as shown in Figure 3.35c.

Following the localisation of a single crack, the stress–crack width relationship can be obtained from the experimental load and deformation readings by subtracting the elastic deformation (which accounts for elastic and micro-cracking behaviour), assuming that the unloading branch occurs with a stiffness of E_c in Figure 3.35c.

In the notched prism shown in Figure 3.35b, the presence of the notch causes a single crack to localise at a stress f_{ct}, which is equal to f_{SH} from the dog bone test. That is, because of the change in cross-section at the location of the notch, only a single crack forms within the specimen. As a result, the increase in strength that arises due to the micro-cracking tension stiffening mechanism cannot occur and, hence, no strain hardening is observed. Using the notched prism approach, it is therefore not possible to obtain the strain hardening modulus E_{SH}, but it is possible to place displacement transducers across the notch to directly measure w.

The complexity of setting up a direct tension test has led to the suggestion that inverse analysis be applied to infer the tensile stress-crack width behaviour from the results of bending tests conducted on unreinforced prisms, and this approach will be discussed in Section 3.3.3.2.

3.3.3.2 Inverse analysis using flexural tension tests

A criticism of the direct tension test approach is that it is complicated to set up and perform, particularly regarding the alignment of axial forces to nullify any moments

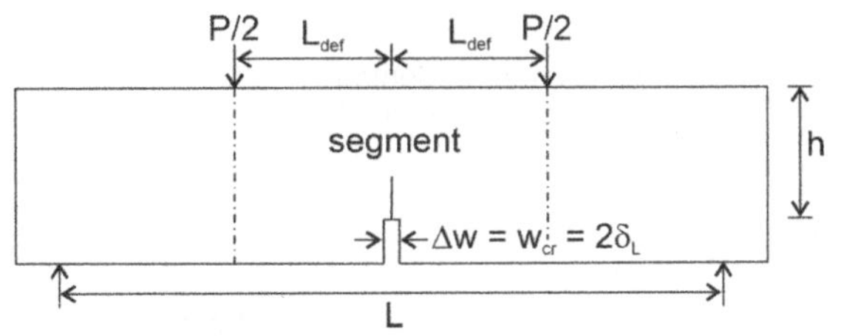

Figure 3.36 Notched beam.

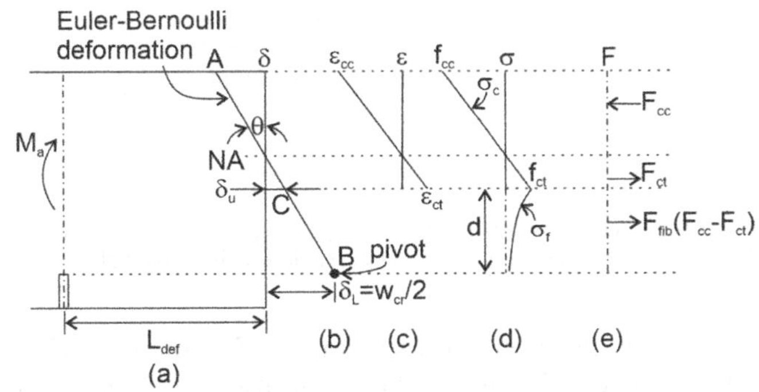

Figure 3.37 Segmental analysis (a) segment; (b) Euler Bernoulli displacement δ;(c) strain profile ε; (d) stress profile σ; (e) force profile F.

caused by eccentric alignment. An alternative approach is the beam test shown in Figure 3.36, where a notch is formed midspan to induce a flexural crack. The depth of the section at the notch is h, which is less than the depth of the beam. Because of this reduced size, micro-cracks will not occur, so the test set-up only measures the fibre stress–crack width relationship. The opening of the notch $\Delta w\_$ is measured directly to give the crack width at the base of the crack (w_{cr}), which is twice the slip of the fibres from a crack face ($2\delta_L$). Let us assume we know the concrete's modulus (E_c), its tensile strength (f_{ct}) at a strain (ε_{ct}) and, of course, all the dimensions of the beam, although the segmental approach can easily cope with non-linearities. The flexural segmental analysis, first introduced in Section 1.5 and fully explained in Chapter 6, will be used to extract the fibre stress–crack width relationship.

It is suggested that the beam in Figure 3.36 be loaded under displacement control to allow a gradual crack extension and opening. Consider a small segment of the beam, of length L_{def} on either side of the notch, located sufficiently far from the applied loads to avoid affecting the strain distributions within the segment. As the segment is symmetrical about the notch, only half needs to be considered, as shown in the free body diagram in Figure 3.37a, where M_a is the applied moment. In this test, the relationship between the applied moment M_a and the half crack width at the soffit of the segment δ_L is determined experimentally.

The test in Figure 3.36 will first be analysed using the standard approach developed in Chapter 6, which assumes all material properties are known, to determine the M/δ_L relationship, as δ_L is the measured property in the test. A Euler–Bernoulli displacement A–B is applied to the RHS with a guessed rotation θ_{gu}, as shown in Figure 3.37b. As we are dealing with half the segment, the deformation at the base δ_{L1} is half the measured crack width w_{cr1} for a specific applied moment M_1. Consequently, the Euler–Bernoulli deformation is fixed at δ_{L1}; that is, we need to pivot the Euler–Bernoulli deformation about Point B at δ_{L1} to find a solution. We need to vary the guessed rotation θ_{gu} until the resultant forces in Figure 3.37e sum to zero, thereby achieving translational equilibrium. Thus, for a specific θ_{gu}, dividing the displacement distribution δ in Figure 3.37b by L_{def} gives the strain profile in Figure 3.37c, where ε_{ct} is the strain at which the concrete fails in tension. Below this strain level, the concrete has cracked, so no strain profile is shown; we are assuming that the concrete on either side of the crack has a negligible strain, and consequently negligible deformation relative to the crack width, which is reasonable. From these strains and E_c, the stress profile in the concrete in Figure 3.37d can be determined, ranging from compression f_{cc} to the tensile capacity f_{ct}. Knowing the width of the beam, these concrete stresses can be used to determine the resultant forces in each stress block and their positions; specifically, the centroids at which the resultants act, which for this case is at one-third the height of the block. These resultants are shown in Figure 3.37e, where the resultant compressive force in the concrete is F_{cc} and the resultant tensile force in the concrete is F_{ct}. From the variation in the crack width C–B in Figure 3.37b and from the fibre stress–crack width relationship (which we assume is known for this specific analysis), the distribution of the fibre stress σ_f in Figure 3.37d can be determined, and consequently the fibre force F_{fib} in Figure 3.37e and its position. If all the forces in Figure 3.37e sum to zero (i.e. there is translational equilibrium), then our guessed rotation θ_{gu} is correct; otherwise, θ_{gu} is varied until equilibrium is achieved. This will give one point of the M/δ_L relationship; then δ_L is varied to obtain more points.

The question is how to adapt the standard procedure from the previous paragraph to determine the fibre stress–crack width property. From the test in Figure 3.36, we already know the M_a/δ_L relationship, but the fibre properties are unknown. So instead of ensuring translational equilibrium by summing the resultant forces in Figure 3.37e, equilibrium is achieved by ensuring the theoretical moment matches the measured moment; that is, by achieving rotational equilibrium. It can be seen in Figure 3.37e that the fibre force:

$$F_{fib} = F_{cc} - F_{ct} \tag{3.59}$$

is known once the concrete forces F_{cc} and F_{ct} are known. Hence, we do not need the fibre material properties to determine F_{fib}. Therefore, the procedure for extracting the fibre properties initially follows the standard approach described above.

For a specific applied moment M_1 and a corresponding half crack width δ_{L1}, an initial value for $\_\theta_1$ (shown in Figure 3.37b) is guessed. From this displacement and the consequent strain in Figure 3.37c, the resultant forces F_{cc} and F_{ct} (as depicted in Figure 3.37e) and their positions are determined. Then Eq. 3.59 gives F_{fib} in Figure 3.37e. We now know the resultant fibre force F_{fib}, but its precise position or level is unknown. To overcome this problem, we have to assume that this fibre stress is uniform such that the resultant acts at half the depth of the stress block. As an example, this is illustrated in Figure 3.38b for deformation labelled δ_{L1} in Figure 3.38a. Being uniform, the resultant fibre force in the block, F_{f1}, acts at half the depth of the stress block, d_1, as shown. Based on a uniform stress block, F_f in Figure 3.37e acts at half the depth from level

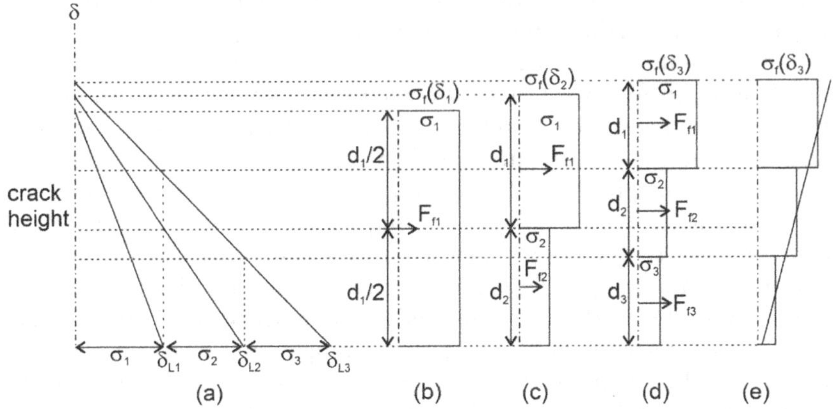

Figure 3.38 Rectangular fibre stress blocks (a) displacement profiles; (b) single stress block for displacement δ_{L1}; (c) stress block for δ_{L2}; (d) stress block for δ_{L3}; (e) linearisation of stress block for δ_{L3}.

C to B in Figure 3.37b hence the level of F_{fib} in Figure 3.37e is now known. As all the resultant forces and their respective levels of action in Figure 3.37e are now known, moments can be taken about any level. If the calculated moment does not equal the applied moment M_1(associated with δ_{L1}), θ_1 can be varied until rotational equilibrium is achieved. Hence, a value for the fibre stress σ_1 corresponding to δ_1 (as shown in Figure 3.38b) is obtained, which means one point on the fibre stress–crack width property has now been determined.

The above analysis appears at first glance to be rather crude, as we are substituting a uniform stress block for a variable stress block; however, the key is to gradually increase δ_L in very small increments. For example, repeating the analysis in Figure 3.37 but for the profile δ_{L2} in Figure 3.38a will give a fibre force F_{fib2} in Figure 3.37. From the distribution of half crack widths for δ_{L2} in Figure 3.38a, the distribution of stresses is shown in Figure 3.38c. For half crack widths less than δ_1, the stress block at a stress σ_1 is known, so F_{f1} and its position are known. Hence, F_{f2}, which equals $F_{fib2}-F_{f2}$, is now known. When rotational equilibrium is achieved, this gives the next point on the material properties, which is σ_2 for δ_2. The distribution of stress for δ_{L3} in Figure 3.38a is shown in Figure 3.38d. It can be seen that the smaller the increase in δ_L the greater the accuracy. It should be noted that there is no need to assume rectangular stress blocks as shown in Figure 3.38; any simple shape with an easily definable centroid could be assumed. Another way of visualising the results of the analysis is shown in Figure 3.38a, where the stress is assumed to be σ_1 when the half crack width is less than δ_{L1}, σ_2 when it lies between δ_{l1} and δ_{L2}, and so on.

It is suggested that an improvement to the test in Figure 3.36 is to measure the contraction of the segment near the top fibre, as well as the crack width and the applied moment, as illustrated in the free body diagram in Figure 3.39a. Hence, for an applied moment M_a and the associated half crack width $\delta_{L,}$ this would fix the Euler–Bernoulli deformation as shown, and consequently, the rotation θ would be fixed. For the fixed Euler–Bernoulli deformation in Figure 3.39a, we can determine the strain and stress distributions as in the standard approach described above. Consequently, we can

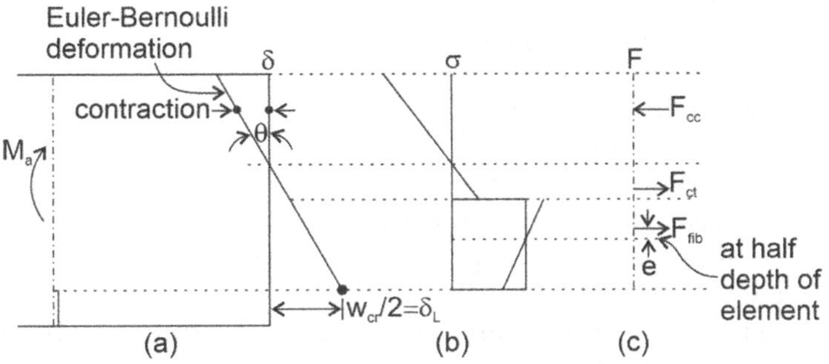

Figure 3.39 Fixed Euler–Bernoulli deformation (a) elevation and displacement profile; (b) stress distribution; (c) force distribution.

determine the forces in Figure 3.39c for that fixed Euler–Bernoulli deformation. The levels at which F_{cc} and F_{ct} act are known from the stress distribution, and hence, the level of F_{fib} can be adjusted until the internal moment in Figure 3.39c is equal to the applied moment for that specific half crack width. Hence, the eccentricity e to the centre of the fibre stress block in Figure 3.39c is known, so the linear variation in the stress block shown in Figure 3.39b can be calculated. Increasing the half crack widths as shown in Figure 3.38a will give a more uniform variation in stress as shown in Figure 3.38e, and allow larger increments in δ_L to achieve the same accuracy as with the uniform block approach. Furthermore, there is no need to iterate to find a solution.

3.4 NOTATION AND ABBREVIATIONS

A	axial force; applied axial force
a	calibration coefficient for SF properties
A_c	cross-sectional area of concrete
A_p	area of plate
A_r	area of reinforcement
BS	bond slip
b	calibration coefficient for SF properties
b_c	width of concrete to which plate is bonded
b_f	width of failure plane
b_p	width of plate
C	circumference of sliding plane
c	minimum reinforcement cover; reinforcement cover
D	diameter; diameter of prism
d	half depth of block under uniform fibre stress
d_b	diameter of reinforcing bar; depth of plate
d_p	depth of failure plane
e	eccentricity
F_f	resultant force in block with constant fibre stress
$(EA)_p$	axial rigidity of plate
ESH	strain hardening modulus during micro-crack development

E_c	elastic modulus of concrete
E_p	elastic modulus of plate
E_r	elastic modulus of reinforcement
F	force induced by fibres
FI	full interaction
FRC	fibre-reinforced concrete
f_c	concrete compressive strength
f_{cc}	confined concrete compressive strength
f_{ct}	concrete tensile strength
f_{SH}	tensile stress at the initiation of micro-cracking in UHPFRC
f_y	reinforcement yield stress
G_f	fracture energy
k_{EAC}	reduction factor for P_{IC} strength
L	length; prism length
L_{bd}	bonded length of reinforcement
L_{crit}	bonded length required to achieve P_{IC}
L_{def}	half constant moment length
L_p	length of perimeter of reinforcing bar
L_{per}	perimeter of failure plane
L_{ref}	reference length
L_{seg}	length of segment
L_t	transducer gauge length
L_0	segment length encompassing sliding plane
l_f	length of fibre
M_a	applied moment
M_{cr}	moment at first cracking
M_{seg}	moment applied to segment
M_u	applied moment at debonding
N	normal force
N_{seg}	normal force applied to segment
n	factor controlling the ductility of the concrete stress–strain relationship
P	applied load; passive force induced by tension stiffening
P_{cr}	force to cause a primary micro-crack
PI	partial interaction
P_{IC}	IC debonding force
S	force arising due to shrinkage
SF	shear friction
s_r	spacing of ribs on reinforcement
t_f	distance of failure plane to the plate
T	circumferential tension; axial force in single fibre
t	tension
V	shear force
V_{seg}	shear force applied to segment
V_{SF}	lateral force to resist sliding; SF property; shear to cause slip
$V_{u\text{-}z\delta}$	shear strength at the initiation of shear sliding
$V_{u\text{-}UHPC}$	V_u for $UHPC$
$V_{u\text{-}UHPFRC}$	V_u for $UHPFRC$
V_{dow}	force due to dowel action

$V_{u\text{-}max}$	upper bound to V_u when using predictive model
V_{u0}	V_u when σ_N is zero
w	crack width
w_{cr}	width of crack at top of notch
w_{SF}	crack opening due to shear friction
α	angle of sliding plane; parameter influencing the non-linearity of the bond stress slip relationship; fibre orientation
β	angle of inclined shear crack
γ	Poisson's ratio
δ	slip
δ_{ax}	axial contraction
δ_B	interface slip across bonded interface
δ_L	half crack width at top of notch; slip of fibre from crack face
δ_{lat}	lateral deformation; lateral movement
δ_{max}	slip at which shear stress is zero
δ	local slip of reinforcement
δ_1	slip at τ_{max}
δ_1, δ_2, δ_3	slips defining shape of bond stress slip relationship
Δ	reinforcement slip at the loaded end or crack face
Δ_{mes}	measured slip
Δ_{SF}	lateral slip across a cracked sliding plane
Δ_{sep}	slip at which crack begins to open
Δ_w	opening on notch
Δ_0	coefficient in Δ_{sep} equation
ε	strain
ε_{ax}	axial strain
$\varepsilon_{ax\text{-}mat}$	material component of the axial strain
ε_{cc}	strain corresponding to f_{cc} for confined concrete; maximum compressive strain when under flexure
ε_{ct}	strain at peak strength f_{ct} after strain hardening
ε_{lat}	lateral strain
$\varepsilon_{ax\text{-}mat}$	material component of the lateral strain
ε_0	strain corresponding to f_c for unconfined concrete
μ	coefficient of friction
ρ_N	reinforcement ratio for reinforcement providing normal stress to cracked plane; cross-sectional area of reinforcement as proportion of the cross-sectional area of the potential sliding plane
ρ_r	reinforcement ratio
σ_{ax}	axial stress
σ_f	fibre stress
σ_{lat}	lateral confining stress
σ_N	normal force across a cracked sliding plane; stress due to N
τ	bond stress between reinforcement and concrete
τ_{avg}	average bond stress
τ_f	frictional component of bond stress
τ_{max}	maximum bond stress
τ_N	shear stress for a normal force N
τ_{SF}	shear stress transferred across a cracked sliding plane due to shear friction

REFERENCES AND DEVELOPMENT

ACI Committee 318, 2019. Building Code Requirements for Structural Concrete (ACI318-19) and Commentary (ACI318R-19). American Concrete Institute, Farmington Hills, MI.

Anderson, A.R., 1960. Composite designs in precast and cast-in-place concrete. *Progressive Architecture*, 41(9), 172–179.

Chen, Y., Visintin, P. and Oehlers, D.J., 2015a. Concrete shear-friction material properties: derivation from actively confined compression cylinder tests. *Advances in Structural Engineering*, 18(8), 1173–1185.

Chen, Y., Zhang, T., Visintin, P. and Oehlers, D.J., 2015b. Concrete shear-friction material properties: application to shear capacity of RC beams of all sizes. *Advances in Structural Engineering*, 18(8), 1187–1198.

Cosenza, E., Manfredi, G. and Realfonzo, R., 1997. Behaviour and modelling of bond of FRP rebars to concrete. *Journal of composites for Construction*, 1(2), 40–51.

Dadd, L., Xie, T., Bennett, B. and Visintin, P., 2025. Shear friction in single-generation and multigeneration recycled aggregate concrete. *Journal of Materials in Civil Engineering*, 37(3), 04024551.

Eligehausen, R., Popov, E.P. and Bertero, V.V., 1982. *Local bond stress-slip relationships of deformed bars under generalized excitations* (Report No. UCB/EERC-83/23). Earthquake Engineering Research Center, University of California, Berkeley. NIST NEHRP.

Hao, X., Visintin, P. and Oehlers, D.J., 2020. Closed-form solutions for quantifying the ductility of concrete beams with passively restrained concrete. *Journal of Structural Engineering*, 146(8), 04020154.

International Federation for Structural Concrete (fib), 2000. Bond of reinforcement in concrete state-of-the-art report bulletin 10. Lausanne, Switzerland.

International Federation for Structural Concrete (fib), 2013. Fib model code for concrete structures 2010. Lausanne, Switzerland.

Jiang, T. and Teng, J.G. 2007. Analysis-oriented stress-strain models for FRP confined concrete. *Engineering Structures*, 29(11), 2968–2986.

Li, J., Gravina, R.J., Smith, S.T. and Visintin, P., 2020. Bond strength and bond stress-slip analysis of FRP bar to concrete incorporating environmental durability. *Construction and Building Materials*, 261, 119860.

Lu, X.Z., Teng, J.G., Ye, L.P. and Jiang, J.J., 2005. Bond–slip models for FRP sheets/plates bonded to concrete. *Engineering Structures*, 27(6), 920–937.

Mohamed Ali, M.S., Oehlers, D.J., Griffith, M.C. and Seracino R., 2008. Interfacial stress transfer of near-surface mounted FRP to concrete. *Engineering Structures*, 30, 1861–1868.

Oehlers, D.J., 1980. *Stud shear connectors for composite beams* (Doctoral dissertation, University of Warwick).

Oehlers, D.J., Rashid, R. and Seracino, R., 2008. IC debonding resistance of groups of FRP NSM strips in reinforced concrete beams. *Construction and Building Materials*, 22, 1574–1582.

Popovics, S., 1973. A numerical approach to the complete stress-strain curve of concrete. *Cement and Concrete Research*, 3(5), 583–599.

Richart, F.E., Brandtzæg, A. and Brown, R.L., 1928. *A Study of the Failure of Concrete under Combined Compressive Stresses*. Urbana, IL: University of Illinois. Engineering Experiment Station. Bulletin; no. 185.

Richart, F.E., Brantzaeg, A. and Brown, R.L., 1929. *The Failure of Plain and Spirally Reinforced Concrete in Compression*. Urbana, IL: University of Illinois, Engineering Experiment Station, Bulletin No. 185.

Santos, P.M. and Júlio, E.N., 2012. A state-of-the-art review on shear-friction. *Engineering Structures*, 45, 435–448.

Seracino, R., Raizal Saifulnaz, M.R. and Oehlers, D.J., 2007. Generic debonding resistance of EB and NSM plate-to-concrete joints. *Journal of Composites for Construction*, 11(1), 62–70.

Sturm, A.B. and Visintin, P., 2019. Local bond slip behaviour of steel reinforcing bars embedded in ultra-high performance fibre reinforced concrete. *Structural Concrete*, 20(1), 108–122.

Sturm, A.B., Visintin, P., Farries, K. and Oehlers, D.J., 2018. New testing approach for extracting the shear friction material properties of ultra-high-performance fibre-reinforced concrete. *Journal of Materials in Civil Engineering*, 30(10), 04018235.

Tepfers, R., 1979. Cracking of concrete cover along anchored deformed reinforcing bars. *Magazine of Concrete Research*, 31(106), 3–12.

Visintin, P., Chen, Y. and Oehlers, D.J., 2015. Simulating the behaviour of FRP-confined cylinders using the shear-friction mechanism. *Journal of Composites for Construction*, 19(6), 04015014.

Xiao, Q.G., Teng, J.G. and Yu, T., 2010. Behaviour and modelling of confined high-strength concrete. *Journal of Composites for Construction*, 14(3), 249–259.

Yuan, H., Teng, J.G., Seracino, R., Wu, Z.S. and Yao, J., 2004. Full-range behaviour of FRP-to-concrete bonded joints. *Engineering Structures*, 26(5), 553–565.

Zhang, H.W. and Smith, S.T., 2012. FRP-to-concrete joint assemblages anchored with multiple FRP anchors. *Composite Structures*, 94(2), 403–414.

Part 3

Partial-interaction elemental mechanics

Having covered in Part 2 the material and pseudo-material properties of a particle of the material, we can now look at a slightly larger region of the reinforced concrete member that represents an element of the member where a physical mechanism occurs. The results of these localised mechanisms can then be used in Part 4, where a segment of the member is analysed, or they can provide pseudo-material properties for use in strain-based analyses, such as the use of smeared cracks and concrete hinge regions in finite element analyses.

Chapter 4 deals with the tension stiffening region, which is the region in the vicinity of the tension reinforcement in a member where the concrete is under flexural tension. The results can be used in a strain-based analysis to either account for the increase in the effective axial stiffness of the tensile reinforcement due to the concrete bonded to it or, alternatively, to assume that the concrete can resist tension after cracking, which is a smeared crack approach.

Chapter 5 covers the concrete softening mechanism of concrete in compression. The analysis quantifies the size of the softening region and the stress/strain properties for use within that specific area. As such, the results could be used as pseudo-properties in strain-based analyses, such as finite element analyses, to simulate the hinge region. Finally, this chapter addresses the beneficial effects of confinement on the concrete in compression, which is of particular importance in a hinge region.

DOI: 10.1201/9781003711407-6

Chapter 4

Reinforced concrete tension-stiffening mechanism

It cannot be overstated how important the tension stiffening mechanism is to the behaviour of reinforced concrete members. It occurs wherever reinforcement is used to resist tension for both flexure and shear and is associated with the cracks in Figures 1.1, 1.7, 1.8, 1.10, 1.27 and 1.30. As illustrated in Figure 1.12 and described in Section 1.4.1, tension stiffening controls the force that the reinforcement can resist and defines crack spacing and crack widths. By understanding the tension stiffening mechanism, the engineer has a tool to control crack widths and consequently design for durability at the serviceability limit state. Tension stiffening is also very important at the ultimate limit state as it can be used to design for concrete confinement as illustrated in Figures 1.20 and 1.21 and explained in Section 1.4.4. Furthermore, it is a major parameter in controlling the rotation of hinges as illustrated in Figure 1.27b and explained in Section 1.5. Hence, tension stiffening provides engineers with the tools to design for ductile failures such as that required for earthquake resistance.

In this chapter, the tension stiffening mechanism is first explained in full using a numerical model (Section 4.2) that simulates a tension stiffening test as illustrated in Figure 1.10. This numerical model is useful as it illustrates diagrammatically all the fundamental mechanics that control tension stiffening. Furthermore, it can also be used with any material property, whether linear or non-linear, to obtain accurate solutions. The mechanics controlling crack development are first explained visually using the basic numerical model. Having introduced the basic mechanics, they are then expanded to include time effects, encompassing not only creep and shrinkage but also concrete expansion, thermal gradients and tendon relaxation. Finally, the model is further expanded to include the beneficial effect of fibres in the concrete.

Having covered all the mechanics of the numerical model in Section 4.2, a more mathematical approach is used to develop an example of a closed-form solution in Section 4.3. Great care has been taken to link the explanations of the numerical and mathematical approaches closely, so a deep understanding of the numerical approach will help in understanding the more mathematical approach. As has already been mentioned, the numerical approach is very useful as it can cope with any variation or non-linearity of the material properties. It can also be used in parametric studies to determine the important variables that control tension stiffening. However, it does not easily identify the major parameters that control the behaviour directly. In contrast, closed-form solutions do provide the parameters that control the behaviour directly and consequently are very helpful in understanding what controls the behaviour. However,

DOI: 10.1201/9781003711407-7

the more complex the material properties, the more complex the solutions, leading to a simplification of the material properties and an accompanying loss of accuracy.

Let us start with the properties of the tension stiffening prism in Figures 1.10 and 1.12, as it is used in both the numerical and mathematical models.

4.1 TENSION STIFFENING PRISM PROPERTIES

To quantify the deformation of reinforced concrete, it is necessary to simulate the tension region in *RC* beams such as those in Figures 1.7, 1.8, 127 and 1.30. It can be seen in these photographs that the cracks adjacent to the tension faces of the beams are near vertical (i.e. almost perpendicular to the tension face) before they propagate towards the applied load. Hence it is common practice in tension stiffening research to test concentrically loaded prisms as in Figure 1.10 and then to simulate their mechanics both numerically and theoretically (Gupta and Maestrini 1990; Wu et al. 1991; Choi and Cheung 1996; Marti et al. 1998; Yuan et al. 2004).

Tension stiffening theory is based on the transfer of pure axial forces between the reinforcement and surrounding concrete. Hence, for analysis, a concentrically loaded prism must be considered, ensuring no flexure or transverse shear (Gupta and Maestrini 1990; Wu et al. 1991; Choi and Cheung 1996; Marti et al. 1998; Yuan et al. 2004; Muhamad et al. 2011; Visintin et al. 2013; Visintin and Oehlers 2017; Knight et al. 2013, 2014; Sturm et al. 2018a). Consider the cross-section of the tension region of a beam of width b in Figure 4.1a where the reinforcing bars are located a distance d_r from

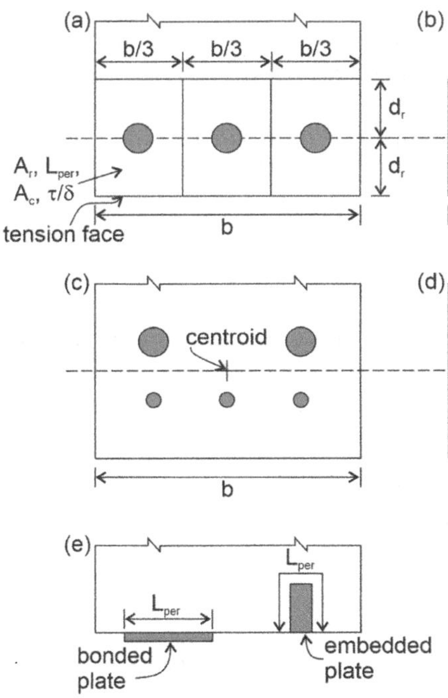

Figure 4.1 Tension stiffening prism cross-sections (a) bars of equal diameter; (b) equivalent bar for scenario shown in (a); (c) bars of different diameter; (d) equivalent bar for scenario shown in (c); (e) externally bonded and near surface.

the tension face. Each reinforcing bar has a cross-sectional area A_r, a circumference L_{per} and a bond-slip property τ/δ that can be determined using the *PI* material properties presented in Section 3.2.4 or from experimental results as described in Section 3.2.3. Because of the need for symmetry, the cross-sectional area of the concrete A_c should be taken such that each reinforcing bar acts at the centroid of a concrete area of $(b/3)\times(2d_r)$. Hence, the resultant longitudinal or axial forces acting on the concrete and reinforcement are concentric, thus ensuring that each of the three tension stiffening prisms is concentrically loaded as required. An alternative solution, yielding identical results, is shown in Figure 4.1b. Here, the equivalent single bar has the same cross-sectional area as the three single bars; a perimeter which is the sum of the perimeters of the three bars; a concrete cross-sectional area that is the sum of those of the three prisms in Figure 4.1a; and the bond slip (τ/δ) remains unchanged, as the total perimeter length in Figure 4.1a and b is conserved.

Problems arise when symmetry is lost, as shown in Figure 4.1c (Muhamad et al. 2011). In this case, the bars can be replaced with a single bar at their centroid, as shown in Figure 4.1d, where d_r is now the distance to the centroid; A_c remains $2d_rb$; A_r is the sum of all the bar cross-sections; L_{per} is the sum of all the perimeter lengths and τ/δ represents the bond-slip properties suitable for all the different types of bars. Further difficulties may arise when bonding plates (which could be steel or *FRP*) to the surface, as shown in Figure 4.1e. What, for example, should A_c be when a plate is bonded to the surface as shown? Thankfully, for externally bonded or near-surface mounted plates, it will be shown that quantifying A_c is of little importance, or indeed, no importance at all. This is because, in typical plating scenarios, A_c is much greater than A_r; hence, the concrete strains are much less than the bar strains and can, therefore, be ignored with no significant loss of accuracy.

The interface bond-slip properties have been introduced in Figure 1.6 in Section 1.3.2 and in detail in Section 3.2.4.1. These τ/δ properties are explained, quantified and discussed in detail in Section 3.2 of Chapter 3. These experimentally derived properties can be used directly in the ensuing numerical model and often have the generic shape *O–A–B–C* (Muhamad et al. 2012), as shown in Figure 4.2. In this shape, τ_{max} represents the maximum shear stress at slip δ_1 and δ_{max} is the interface slip beyond which the shear stress is considered zero. This is often simplified to the bilinear variation *O–A–C* (Muhamad et al. 2012) or a bilinear variation with a friction component *O–A–H–I* (Vaculik et al. 2018). Sometimes, a plateau *A–G* from δ_1 to δ_2 is incorporated, or a frictional component *H–I* from δ_3 onwards. When studying the stiffness of

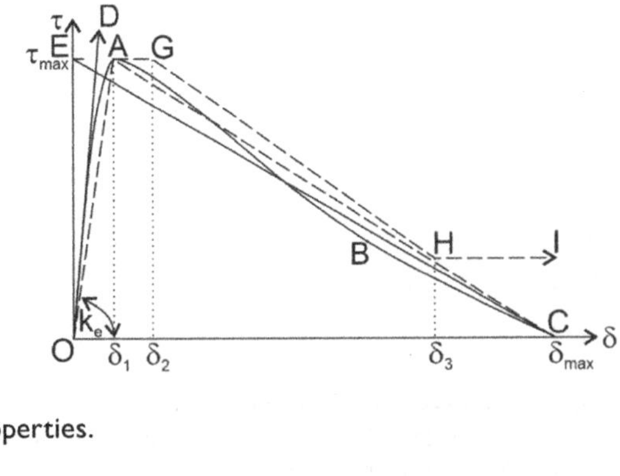

Figure 4.2 Bond-slip properties.

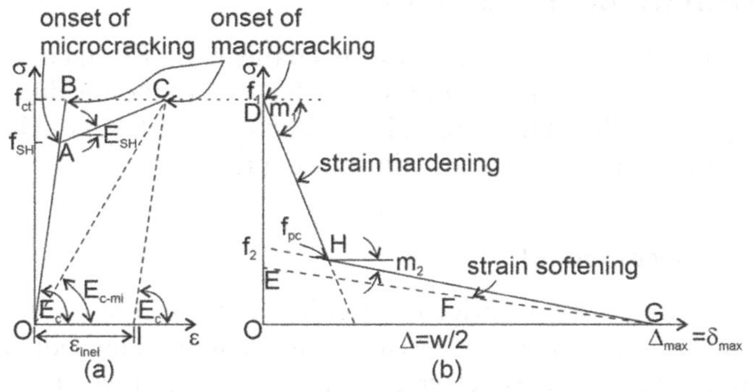

Figure 4.3 Fibre concrete properties (a) stress strain; (b) stress crack width.

a member at serviceability, a linear ascending branch *O–D* with stiffness k_e may be considered appropriate. While studying crack widths, the increased complexity of the non-linear branch *O–A–C* may be necessary. In contrast, if considering the debonding of reinforcement at the ultimate limit state, it may be more appropriate to use a simple linear descending branch *E–C*.

The effect of fibres in concrete, as shown in Figure 1.13, has been introduced in Section 1.4.2. It has been shown, using Figure 1.14, that the material properties need to be described using both strain-based and slip-based properties. On initial loading and whilst micro-cracks develop, a strain-based stress/strain relationship can be used, such as *O–A* in Figure 4.3a, with a concrete modulus of E_c. Micro-cracking can be accommodated with a reduced modulus $E_{c\text{-}mi}$, or by using a strain hardening modulus E_{SH} in conjunction with E_c, as shown.

Once a macro-crack forms at f_{ct} (Figure 4.3), the reduction in strength can be accommodated using linear descending branches, as shown in Figure 4.3b. When there are few fibres, the behaviour may be idealised as following *O–A–B* in Figure 4.3a and subsequently *D–E–F–G* in Figure 4.3b. In this scenario, a macro-crack occurs at f_{ct}, after which there is a step change. The equivalent concrete stress due to the fibres then follows *E–F–G* in Figure 4.3b, reducing to zero at the crack-face displacement $\Delta_{max}=\delta_{max}$. This behaviour is referred to as strain softening. When there are substantial fibres, strain hardening may occur. A micro-crack may form at Point *A* (corresponding to f_{SH} in Figure 4.3a), after which the strength may rise to f_{ct} at Point *C*. The behaviour may then be idealised as *O–A–C* followed by *D–H–G* in Figure 4.3b. Alternatively, the reduced modulus $E_{c\text{-}mi}$ can be used, with the strength following the path *D–H–G* at the onset of a macro-crack.

4.2 NUMERICAL MODEL FOR TENSION STIFFENING

Through the development of a numerical model (Muhamad et al. 2011; Visintin et al. 2013; Visintin and Oehlers 2017; Knight et al. 2013, 2014), the full mechanics of tension stiffening are first described visually. These mechanics are then used in Section 4.3 to derive closed-form solutions.

4.2.1 Numerical model mechanics

Let us first consider the tension stiffening mechanism in *RC* members constructed from normal concrete without fibres, and in which there are no residual strains such as shrinkage or thermal strains. Once the full mechanics have been explained using this model, the effects of residual strains and fibres are introduced.

4.2.1.1 *Basic numerical model*

Consider the prism in Figure 1.10 prior to cracking. Let the ends be fully restrained so that slip cannot occur at the ends. The prism is pulled apart longitudinally, so that one crack forms at mid-length; this is referred to as the initial crack. Let us consider the prism on the right-hand side (RHS)of the initial crack, bearing in mind the behaviour of the prism on the left-hand side (LHS) is a mirror image. The *LHS* of the prism is shown in Figure 4.4a, where the initial crack has a width *w*. For the purpose of analysis, it is assumed that the *RHS* extends as much as required to a full-interaction (*FI)* region, as shown in Figure 4.4b; this is referred to as an infinite prism. The cross-section of the prism is shown in Figure 4.4c.

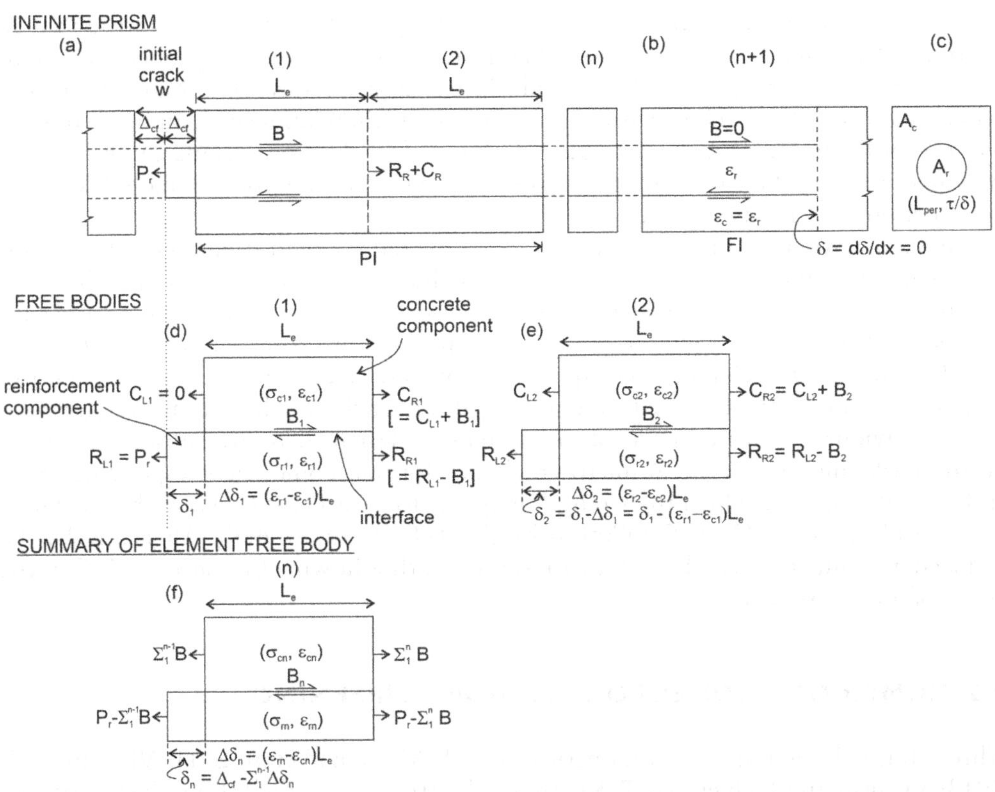

Figure 4.4 Numerical analysis of tension stiffening (a) infinite prism to the left of the initial crack and to the right of the initial crack; (b) prism in a FI region; (c) cross-section of prism; (d) free body element 1 of the prism adjacent to initial crack; (e) free body element 2 adjacent to element 1; (f) generic forces acting on a free body.

The prism in Figure 4.4a is pulled apart such that the force in the reinforcement at the initial crack is P_r and the half-crack width is Δ_{cf} as shown (Muhamad et al. 2011). Hence, the crack width w, which is the distance between the crack faces, is $2\Delta_{cf}$. At the crack face, the half-crack width Δ_{cf} is also the slip of the reinforcement. It is a question of determining the force P_r that causes the half-crack width Δ_{cr} or vice versa. In this particular numerical model, this is achieved iteratively by varying one of the parameters until a solution is found by considering the boundaries of the prism on the *LHS* (Figure 4.4a) and at the *RHS* (Figure 4.4b), as will be explained. If P_r is fixed and Δ_{cf} is varied, then the analysis is performed under load control. In contrast, if Δ_{cf} is fixed and P_r is varied to find a solution, then the analysis is performed under displacement control; this is often more convenient as it can cope with reductions in P_r that might occur as the reinforcing bar unloads under cyclic loading or if debonding occurs on short bonded lengths. In this instance, let us describe the analysis under displacement control, which involves fixing Δ_{cf} at the *LHS* and varying P_r to find a solution. To perform the numerical analysis, the prism is sliced into n very narrow elements of length L_e, labelled (1) to (n). As mentioned previously, the prism must be loaded concentrically such that resultant forces in the reinforcement R_R and the concrete C_R are in line with P_r, as shown. Bearing in mind that the tension stiffening prism must be symmetric (as explained with the help of Figure 4.1) and to aid in understanding the theory, the concrete and reinforcement components are shown adjacent to each other in Figure 4.4 (d–f) to clearly show the individual forces and displacements.

The free-body diagram of the first element in Figure 4.4a is shown in Figure 4.4d. The forces on the *LHS* of the element have been labelled with a subscript L and those on the *RHS* with R. The force on the *LHS* in the reinforcement R_{L1} is equal to P_r and the force in the concrete on the *LHS*, C_{L1}, is zero, as it is assumed that the crack width is wide enough for the stress due to aggregate interlocking across the crack face to be zero. The interface slip at the crack face Δ_{cf}, is equal to the slip in the first element, δ_1. As we are in a displacement-mode analysis, the force at the crack face, P_r, will be varied to find a solution for the displacement δ_1 (It is important to note that the notation δ_1 refers to the local slip in the first element, not the slip corresponding to τ_{max} in the local bond-slip properties as shown in Figure 3.17, Section 3.2). As the element slice L_e is very small, the slip δ_1 can be assumed to be uniform across the interface between the reinforcement and concrete components. From the bond-slip properties τ/δ in Figure 4.2, the interface shear τ_1 for the slip δ_1 can be determined. The interface bond force B_1 in Figure 4.4d is simply the interface surface area $L_{per}L_e$ multiplied by the bond stress τ_1, which is

$$B = L_{per}L_e\tau \tag{4.1}$$

The interface bond force in the first element B_1 due to τ_1 and δ_1 in Figure 4.4d is now known. The direction of the shear force B_1 that the bond induces is important. In this figure, the shear force above the interface has been drawn directed to the right, as this is the shear force that resists the reinforcing bar from being pulled out; the opposite applies to the concrete component. From free-body equilibrium, the force on the *RHS* of the concrete component C_{R1} has increased by B_1 to $C_{L1}+B_1$. Thus, the increase in stress in the concrete component $\Delta\sigma_c$ is equal to B/A_c. Substituting B from Eq. 4.1 and dividing by L_e gives the increase in concrete stress per unit length, $d\sigma_c/dx$, as

$$\frac{\Delta\sigma_c}{L_e} = \frac{d\sigma_c}{dx} = \frac{L_{per}\tau}{A_c} \tag{4.2}$$

Similarly, for the reinforcement component, the reduction in stress $\Delta\sigma_r$ per unit length is given by

$$\frac{\Delta\sigma_r}{L_e} = \frac{d\sigma_r}{dx} = \frac{L_{per}\tau}{A_r} \tag{4.3}$$

Dividing the force $C_{R1}=C_{L1}+B_1$ on the *RHS* of Element 1 (Figure 4.4d) by the cross-sectional area of the concrete component A_c gives the concrete stress on the *RHS*. As the concrete stress on the *LHS* is zero, the mean concrete stress in the first element, σ_{c1}, can be determined. From its modulus at this stress, the mean concrete strain, ε_{c1}, can then be determined; both are shown bracketed. Furthermore, the force on the *RHS* of the reinforcement component, R_{R1}, is reduced by B_1 to $R_{L1}-B_1$, as shown. Hence, with the forces on both sides of the reinforcement now known, the mean reinforcement stress, σ_{r1}, and strain, ε_{r1}, can also be determined; these are shown in brackets. It should be noted that the axial stresses σ_c and σ_r are assumed to be uniform across the components' cross-sections, that is there is no shear lag.

All the forces and deformations in the free bodies of Element 1 in Figure 4.4d are now known. So let us move on to the adjacent Element 2. At the boundary between Elements 1 and 2, the concrete forces C_{R1} and C_{L2} are equal, as are the reinforcement forces R_{R1} and R_{L2}. In every element, the sum of the concrete force C and reinforcement force R sums to the applied force P_r, which can be written in terms of stress as

$$\sigma_c A_c + \sigma_r A_r = P_r \tag{4.4}$$

To further analyse Element 2 (Figure 4.4e), the slip δ_2 is required, which can be determined from the slip–strain as follows.

The deformations in an element in Figure 4.4e are shown in Figure 4.5. These do not depict the actual deformations in Figure 4.4d and e, but rather their relative deformations,

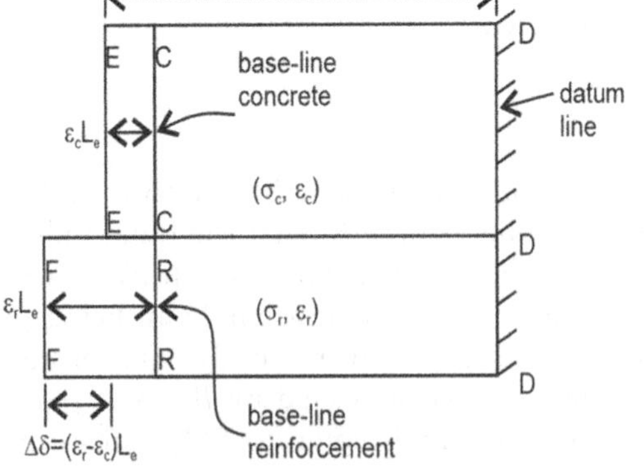

Figure 4.5 PI slip–strains in an element no residual strains.

by fixing the *RHS* in Figure 4.5, which is shown and labelled as the datum line *D–D*. Hence all deformations on the *LHS* are relative to the *RHS* datum line *D–D*. Before the concrete is stressed, the *LHS* of the concrete component lies at *C–C*. This is referred to as the concrete base-line – the position of the unstressed and unstrained concrete. Hence, to accommodate the concrete tensile stress σ_c (due to slip in Figure 4.4d), an expansion $\varepsilon_c L_e$ from the base-line *C–C* to *E–E* is required, as shown in Figure 4.5. Similarly, the reinforcement base-line is *R–R*. To accommodate the reinforcement tensile stress σ_r, an expansion $\varepsilon_r L_e$ from the reinforcement base-line *R–R* to *F–F* is required, as shown.

Hence, the change in slip between the two components of the free-body (i.e. $\Delta\delta$ in Figure 4.5) is equal to the strain difference $(\varepsilon_r - \varepsilon_c)$ multiplied by the length L_e:

$$\Delta\delta = (\varepsilon_r - \varepsilon_c) L_e \tag{4.5}$$

Dividing Eq. 4.5 by L_e gives the change in slip per unit length, referred to as the slip–strain $d\delta/dx$:

$$\frac{\Delta\delta}{L_e} = \frac{d\delta}{dx} = \varepsilon_r - \varepsilon_c \tag{4.6}$$

Having now defined slip–strain, we can return to the deformations in Figure 4.4d and e. The slip in the first element is δ_1; the change in slip, $\Delta\delta_1$ is given by Eq. 4.5 with the addition of subscript *1 to denote the first element.* Therefore, the slip in the second element, δ_2, is equal to $\delta_1 - \Delta\delta_1$, as shown. As we know the slip in the second element this allows us to determine the bond force B_2, which in turn provides C_{R2} and R_{R2}, and consequently, σ_{c2} and σ_{r2} and so on and so forth. In summary, the nth element is shown in Figure 4.4f, where its slip is given by

$$\delta_n = \Delta_{cf} - \sum_{1}^{n-1} \Delta\delta_n L_e \tag{4.7}$$

4.2.1.2 Inclusion of time-dependent properties

The numerical procedure described using Figure 4.4 can be easily adapted to account for material variations with time (Knight et al. 2013; Visintin et al. 2013), such as concrete creep and tendon relaxation, and for the residual strains resulting from concrete shrinkage, expansion and thermal gradients. Creep and relaxation can be accounted for by simply changing the material moduli, as described in Section 2.5.3. Concrete shrinkage or expansion and thermal gradients can be accounted for by determining the residual strains they induce and subsequently adjusting the slip–strain as follows.

Figure 4.5 has already been used to show the stresses and deformations that occur in an element (as depicted in Figure 4.4d) that is free from residual stresses due to time-dependent variations in material properties. It should be noted that we are not dealing with deformations due to creep or relaxation resulting from applied stresses, but rather with changes in material properties over time in the absence of applied stresses. These will be referred to as residual strains. It was shown in Figure 4.5 that *C–C* and *R–R* are the base-lines for both the concrete and reinforcement components; that is, they represent the position of the *LHS* of the free bodies when neither component is stressed. To produce a stress in either component requires a displacement from these base-lines. In this example without residual strains, these base-lines are at the same position.

Now consider the effect of concrete shrinkage and start with an element where there is FI between both components, that is, there is no slip across the interface. Analysis of the *FI* behaviour is shown in Figure 4.6a. To analyse the *FI* behaviour, we need to start by considering the no-interaction (*NI*) behaviour. When there is *NI* between the concrete and reinforcement, the base-line for the reinforcement is *R–R* in Figure 4.6a, which is equal to the unstressed length of the reinforcement. Applying a shrinkage strain ε_{sh}, which can be quantified using the models in Section 2.5.2, causes the concrete element, initially of length L_e, to contract by $\varepsilon_{c\text{-}sh}L_e$ as shown. Recalling that shrinkage is a strain in the concrete that does not induce a stress, the base-line for the concrete is the position where there is no internal stress in the concrete, that is, *C–C* in Figure 4.6a. Again, with the conditions of *FI* imposed, ensuring no interface slip, the *LHS* must move to some position *L–L* between *C–C* and *R–R* as shown. As a result of this movement away from their respective base-lines *R–R* and *C–C*, the reinforcement goes into compression with a contracting strain $\varepsilon_{r\text{-}sh}$ and the concrete goes into tension with an extending strain $\varepsilon_{c\text{-}sh}$. The position *L–L* in both components can be found by simply moving *L–L* horizontally, that is to the left or right, until the tensile force in the concrete component of area A_c equals the compressive force in the reinforcement component of area A_r, that is, there is internal force equilibrium. Hence, whilst the materials remain linear elastic, the concrete force $A_cE_c\varepsilon_{c\text{-}sh}$ has the same magnitude as the reinforcement force $A_rE_r\varepsilon_{r\text{-}sh}$. Furthermore, bearing in mind that the difference between $\varepsilon_{c\text{-}sh}$ and $\varepsilon_{r\text{-}sh}$ is ε_{sh}, this gives the strain in the concrete whilst there is *FI* (Knight et al. 2013; Visintin et al. 2013) as

$$\left(\varepsilon_{c-sh}\right) = \frac{\varepsilon_{sh}}{1+\dfrac{E_cA_c}{E_rA_r}} \tag{4.8}$$

and the strain in the reinforcement whilst there is *FI* as

$$\left(\varepsilon_{r-sh}\right) = \frac{\varepsilon_{sh}}{1+\dfrac{E_rA_r}{E_cA_c}} \tag{4.9}$$

It should be noted that we are dealing here with local internal equilibrium, that is, the stresses or forces involved do not affect global equilibrium. This local equilibrium is the same type of local mechanism as that occurring in confinement across a sliding plane due to transverse reinforcement as illustrated in Figure 1.21 and described in Section 1.4.4.

As explained in the previous paragraph, shrinkage will cause the *LHS* of the element in Figure 4.6a to move to *L–L* whilst the element is under *FI*. Now consider the effect of *PI*, that is interface slip, as shown in Figure 4.6b. Prior to interface slip, the prism is bound by *L–L* on the *LHS* and *D–D* on the *RHS*. From the *PI* behaviour in Figure 4.4d, interface slip requires, for equilibrium, a concrete stress σ_c; from the material modulus E_c, the strain at this stress is ε_c. Therefore, the concrete component must expand by ε_cL_e from its base-line *C–C* in Figure 4.6b to achieve the required force in

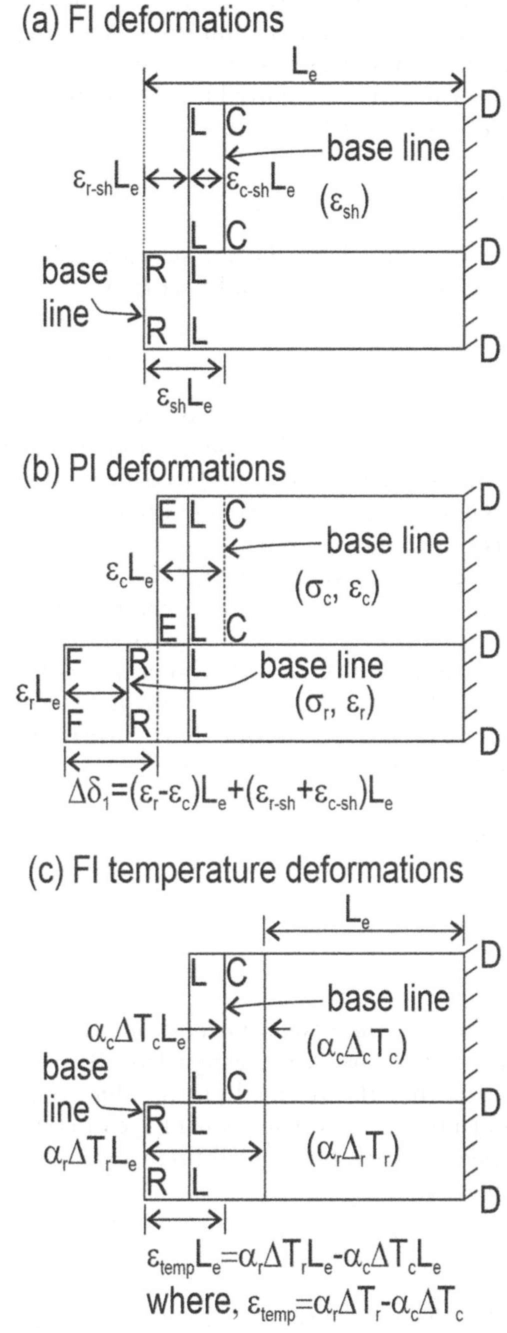

Figure 4.6 Residual strains (a) full interaction deformations with shrinkage; (b) partial interaction deformations with shrinkage; (c) full interaction deformations with thermal expansion.

the concrete. It can be seen that this expansion is entirely independent of the shrinkage strain and must simply occur to maintain equilibrium. Applying the same logic for the reinforcement, the reinforcement must expand from its base-line *R–R* by $\varepsilon_r L_e$ such that the stress in the reinforcement is σ_r, as required by the *PI* analyses.

From the deformations in Figure 4.6a and b, the change in slip $\Delta\delta$ is given by

$$\Delta\delta = (\varepsilon_r - \varepsilon_c) L_e + (\varepsilon_{r-sh} + \varepsilon_{c-sh}) L_e \tag{4.10}$$

However, as can be seen in Figure 4.6b, the second term is equal to $\varepsilon_{sh} L_e$. Hence, the change in slip under shrinkage conditions can be written

$$\Delta\delta = (\varepsilon_r - \varepsilon_c + \varepsilon_{sh}) L_e \tag{4.11}$$

It can be seen that the *PI* strain components in Eq. 4.5, that is $\varepsilon_r - \varepsilon_c$, have been increased by the shrinkage strain ε_{sh} and that the strains derived from the *FI* analyses in Figure 4.6a, that is $\varepsilon_{r\text{-}sh}$ and $\varepsilon_{c\text{-}sh}$ are not required to determine the change in slip. They were included in the discussion for visualisation purposes.

Concrete expansion is simply the opposite of concrete shrinkage, so the above analyses will apply, provided a concrete expansion strain ε_{exp} having an opposite sign to the shrinkage strain ε_{sh} is used. FI differential thermal expansion is illustrated in Figure 4.6c and shares the same mechanics as the *FI* shrinkage or expansion in Figure 4.6a. The concrete in Figure 4.6c expands by $\alpha_c \Delta T_c L_e$, where α_c is the coefficient of thermal expansion for the concrete and ΔT_c is the change in temperature of the concrete. Similarly, the reinforcement expands by $\alpha_r \Delta T_r L_e$. From these displacements, the displacement between *C–C* and *R–R*, labelled $\varepsilon_{temp} L_e$, is given by

$$\varepsilon_{temp} L_e = \alpha_r \Delta T_r L_e - \alpha_c \Delta T_c L_e \tag{4.12}$$

such that

$$\varepsilon_{temp} = \alpha_r \Delta T_r - \alpha_c \Delta T_c \tag{4.13}$$

Comparing Figure 4.6a and c, the displacement from *C–C* to *R–R* in Figure 4.6c, labelled $\varepsilon_{temp} L_e$, is equivalent to the displacement from *C–C* to *R–R* in Figure 4.6a, labelled $\varepsilon_{sh} L_e$. This means that the residual strains due to differential temperature changes can be treated identically to concrete shrinkage or concrete expansion (Knight et al. 2013). Hence, the generic equation for all three time-dependent material properties can be written as follows:

$$\Delta\delta = \left[(\varepsilon_r - \varepsilon_c) + \varepsilon_{time} \right] L_e \tag{4.14}$$

where ε_{time} is composed of the appropriate combinations of ε_{sh}, ε_{exp} and ε_{temp}. If ε_{sh} is assumed to be positive, that is when the concrete shrinks relative to the reinforcement (as in Figure 4.6a), then ε_{exp} is negative (as the concrete expands relative to the reinforcement), and ε_{temp} will be positive when the thermal expansion of the concrete is less than that of the reinforcement (as in Figure 4.6c).

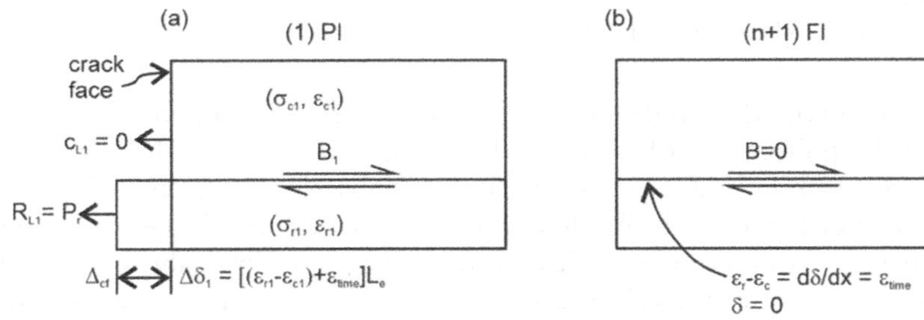

Figure 4.7 Inclusion of residual strains in elements (a) PI region; (b) FI region.

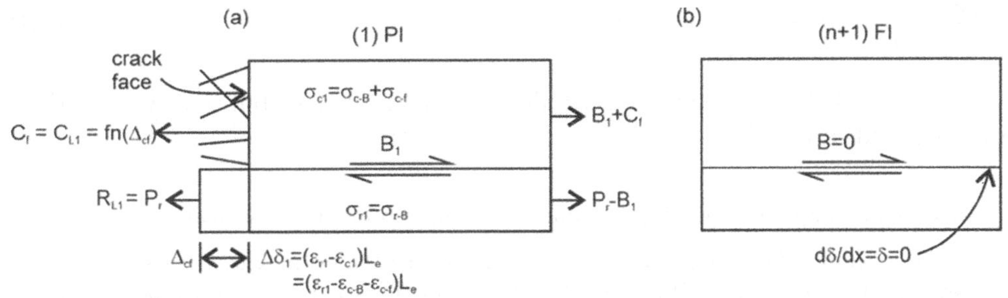

Figure 4.8 Inclusion of fibres in elements (a) PI region; (b) FI region.

Dividing both sides of Eq. 4.14 by L_e gives the change in slip per unit length, which is the slip–strain, as

$$\frac{\Delta\delta}{L_e} = \frac{d\delta}{dx} = \left(\varepsilon_r - \varepsilon_c\right) + \varepsilon_{time} \tag{4.15}$$

Rearranging Eq. 4.15 gives

$$\Delta\delta = \left(\varepsilon_r - \varepsilon_c\right) L_e + \varepsilon_{time} L_e \tag{4.16}$$

Hence, the effects of concrete shrinkage or expansion and thermal gradients can be included in the numerical analysis (Figure 4.4d) by simply adding an additional term $\varepsilon_{time}\, L_e$ to the increase in the element slip, as shown in Figure 4.7a.

4.2.1.3 Inclusion of fibres

The inclusion of fibres for the first element in Figure 4.4d is shown in Figure 4.8a (Visintin and Oehlers 2017; Sturm et al. 2018b). For a given half-crack width Δ_{cf}, the equivalent concrete stress across the crack due to the fibres, $\sigma_{c\text{-}f}$, is determined by the material properties shown in Figure 4.3b. Consequently, the force on the concrete across the initial crack, $A_c\sigma_{c\text{-}f}$, is no longer zero. The bond force B_1 in Figure 4.8a produces a stress in the concrete, $\sigma_{c\text{-}B}$, in the usual way, as shown in Figure 4.4d. Hence, from Figure 4.8a, the effect of fibres is to reduce the slip (as shown), and the slip–strain is consequently reduced by dividing the slip by L_e, as follows:

$$\frac{\Delta\delta}{L_e} = \frac{d\delta}{dx} = \left[\varepsilon_r - \left(\varepsilon_{c\text{-}B} + \varepsilon_{inet}\right)\right] + \varepsilon_{time} \tag{4.17}$$

It is important to note that to differentiate between the constant stress in the concrete arising from the pulling out of fibres at the crack face, $\sigma_{c\text{-}f}$, and the varying force in the concrete due to bond slip, $\sigma_{c\text{-}B}$, the additional subscript notations '-*f*' and '-*B*' have been introduced in Figure 4.8b. While the notation has changed, the way in which the concrete force $\sigma_{c\text{-}B}$ is determined is identical to that described above without fibres. Further, because strain in the reinforcement, ε_r, and in the concrete due to bond, $\varepsilon_{c\text{-}B}$, vary along the element, the concrete strain due to the fibres, $\varepsilon_{c\text{-}f}$, is uniform along the element, much like ε_{time}. Hence, it is possible to group $\varepsilon_{c\text{-}f}$ with ε_{time} such that

$$\frac{\Delta\delta}{L_e} = \frac{d\delta}{dx} = \left(\varepsilon_r - \varepsilon_{c\text{-}B}\right) + \left(\varepsilon_{time} - \varepsilon_{c\text{-}f}\right) \tag{4.18}$$

bearing in mind that $\varepsilon_{c\text{-}f}$ may not vary along the prism length but does vary with the half-crack width Δ_{cf}.

Worked Example WE-4.1: Equilibrium and compatibility at an element

As an example of the analysis of an element, consider the beam with a cross-section shown in Figure E4.1a. The tension stiffening prism is extracted from the beam; it has a single 20 mm diameter reinforcing bar and an overall height and width of 100 mm, thereby maintaining axial symmetry.

The concrete used to construct the element is *UHPFRC* and has tensile material properties as shown in Figure E4.1b. The concrete begins to strain harden due to micro cracking at a stress f_{SH} of 4 MPa, and cracks localise at f_{ct} of 5 MPa. After crack

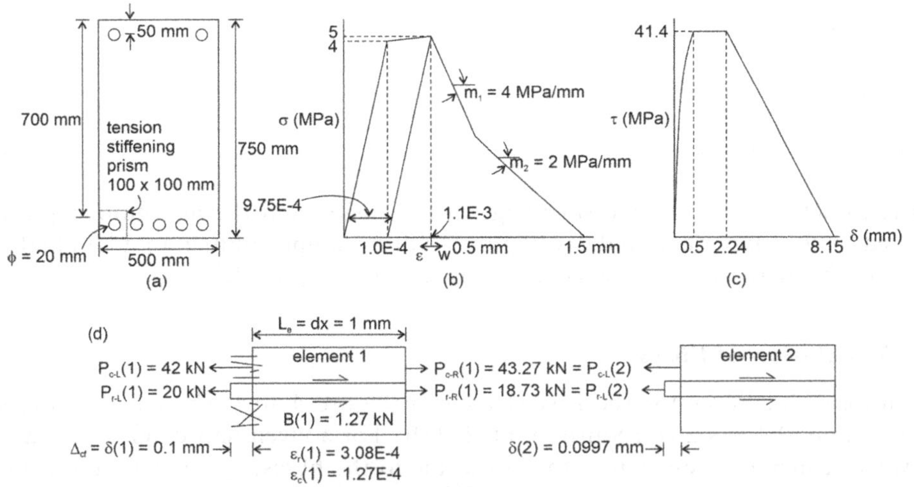

Figure E4.1 Worked example data: (a) cross-section; (b) tensile stress–strain and tensile stress–crack width properties; (c) reinforcement bond–stress–slip properties; (d) first and second elements in analyses.

localisation, the tensile stress–crack width relationship is bilinear and descending, defined by the slopes of the descending branches $m_1=4$ MPa/mm and $m_2=2$ MPa/mm. At the time of analysis, the concrete shrinkage strain is $1000\mu\varepsilon$. The elastic modulus of 40,000 MPa includes the impact of concrete creep. After cracking, the concrete exhibits an inelastic strain due to micro-cracking, $\varepsilon_{c\text{-}f}$, of 9.75E–4, as depicted in Figure E4.1b. The reinforcement bond stress–slip relationship is given in Figure E4.1c, with its magnitudes determined using the approach of Sturm and Visintin (2019) in Section 3.2.4.2.

For analysis, consider the first element of the tension stiffening prism, which is located at the crack face and shown in Figure E4.1d. Let us consider the scenario where a 0.1 mm slip of the reinforcement is imposed, such that the total crack width is 0.2 mm. Also assume that the material stress–crack width relationship in Figure 4.9b is given by

$$\sigma_{c\text{-}f} = f_{ct} - m_i 2\Delta_{cf} = 5 - 4\times 2\times 0.1 = 4.2 \text{ MPa} \quad \text{(E4.1.1)}$$

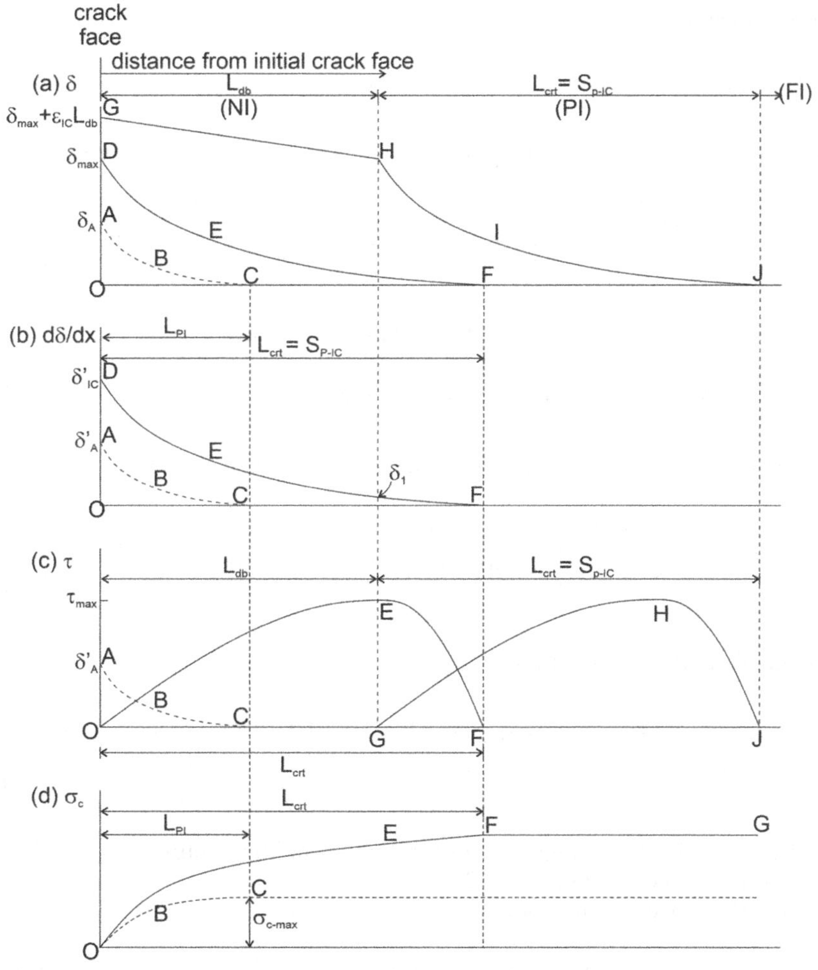

Figure 4.9 Initial crack behaviour (a) distribution of slip; (b) distribution of slip-strain; (c) distribution of bond stress; (d) distribution of concrete stress.

where, in Eq. E4.1.1, the half-crack width is multiplied by 2 to obtain the full crack width because the stress–crack width relationship in Figure E4.1b was determined using measurements of the full crack width.

By integrating the concrete stress over the concrete area, the force in the concrete at the crack face is $P_{c\text{-}L}(1)=42$ kN.

The *UHPFRC* bond stress–slip relationship is given by Eq. 3.42 in Section 3.2.4.2:

$$\tau(1) = \tau_{max}\left(\frac{\delta}{\delta_1}\right)^{\alpha} = 41.4\left(\frac{0.1}{0.5}\right)^{0.44} = 20.29 \text{ MPa} \tag{E4.1.2}$$

Integrating the bond stress over the bonded area yields the bond force:

$$B(1) = \tau(1)\, L_{per} L_e = 20.29 \times \pi \times 20 \times 1 = 1.27 \text{ kN} \tag{E4.1.3}$$

From equilibrium, the force in the reinforcement at the right-hand side of the first element in Figure E4.1d is

$$P_{r\text{-}R}(1) = P_{r\text{-}L}(1) - B(1) = 20 - 1.27 = 18.73 \text{ kN} \tag{E4.1.4}$$

From which, the average strain in the reinforcement in the first element is

$$\varepsilon_r(1) = \frac{P_{r\text{-}L}(1) + P_{r\text{-}R}(1)}{2E_r A_r} = \frac{20 + 18.73}{2 \times 40{,}000 \times 0.25 \times \pi \times 20 \times 20} = 3.08 \times 10^{-4} \tag{E4.1.5}$$

Similarly, the force in the concrete at the RHS of the first element is

$$P_{r\text{-}R}(1) = P_{r\text{-}L}(1) + B(1) = 42 + 1.27 = 43.27 \text{ kN} \tag{E4.1.6}$$

From which, the average strain in the concrete over the first element is

$$\varepsilon_c(1) = \frac{P_{c\text{-}L}(1) + P_{c\text{-}R}(1)}{E_c A_c} = \frac{42 + 43.27}{2 \times 40{,}000 \times 0.25 \times \pi \times 20 \times 20} = 1.07 \times 10^{-4} \tag{E4.1.7}$$

From Eq. 4.18, the slip–strain is

$$\frac{d\delta}{dx}(1) = (\varepsilon_r - \varepsilon_{c\text{-}B}) + (\varepsilon_{time} - \varepsilon_{c\text{-}f}) = [(3.08 - 1.07) + (10.0 - 9.75)] \times 10^{-4} = 2.26 \times 10^{-4} \tag{E4.1.8}$$

and the change in slip is simply the integration of the slip–strain along the element length of 1 mm, such that $\delta\Delta(1) = 2.26 \times 10^{-4}$.

Hence, the slip at the start of the second element is

$$\delta_2 = \delta_1 - \Delta\delta_1 = 0.1 - 2.26 \times 10^{-4} = 0.0997 \tag{E4.1.9}$$

4.2.2 Solving the numerical model

To this point we have determined the forces and deformations that occur within a small element of the tension stiffening prism, using Figure 4.4. It is now a matter of applying appropriate boundary conditions so that we can determine the relationship between the force developed in the reinforcement and the slip of the reinforcement relative to the surrounding concrete. In doing this, it will be shown that tension stiffening can be used to analyse not only infinite prisms but also short prisms, and also the development of cracks and intermediate cracking (the formation of cracks between adjacent cracks).

4.2.2.1 *Infinite prism*

The numerical analyses described in Figure 4.4 of the infinitely long prism without fibres or shrinkage can be used to determine the variation in slip and slip–strain along the prism for a fixed displacement at the crack, Δ_{cf}, and an assumed applied load, P_r. The question is whether the initial assumed load, P_r, is correct; to verify this, the boundary conditions for various scenarios must be satisfied.

4.2.2.1.1 Boundary conditions for infinite prism

Starting with the infinite prism shown in Figure 4.4, which has neither fibres nor residual strains (ε_{time}). At the crack face, the slip is Δ_{cf} and the strain in the concrete is zero, so the slip–strain is simply the strain in the reinforcement at the crack face, which is P_r/A_rE_r. Hence, the boundary condition at the crack face, $[BC]_{CF}$, is (Haskett et al. 2008)

$$[BC]_{CF} \quad \frac{d\delta}{dx} = \frac{P_r}{A_rE_r} \quad \text{and} \quad \delta = \Delta_{cf} \tag{4.19}$$

That is, the boundary condition at the crack face dictates that the slip–strain equals the strain in the reinforcement, and the slip equals Δ_{cf}.

As slip progresses along the prism shown in Figure 4.4, the only available boundary condition is that, at some point, the slip ceases at a *FI* region, depicted in Figure 4.4b as the slice labelled $(n+1)$. In this *FI* region, there is no slip (i.e. δ is zero), so the bond force B is zero, and the strains in the concrete and reinforcement are equal. Consequently, the slip–strain $d\delta/dx$ is also zero. To ensure that we are entering a *FI* region from a *PI region* (i.e. a region where there is no slip), it is not sufficient for the slip to merely be zero. It is also essential that the slip–strain tends to zero, such that the slip asymptotes to zero at the *FI* boundary. Hence, the *FI* boundary condition for an infinitely long prism, $[BC]_{FI\text{-}INF}$, is given by (Haskett et al. 2008)

$$[BC]_{FI\text{-}INF} \; \delta = \frac{d\delta}{dx} = 0 \tag{4.20}$$

The boundary conditions for the case with time-dependent strains are shown in Figure 4.7. At the crack face (Figure 4.7a), the change in slip, $\Delta\delta$, includes a residual strain component, $\varepsilon_{time}L_e$. Thus, the boundary condition becomes (Knight et al. 2013; Visintin et al. 2013)

$$[BC]_{CF\text{-}tim} \quad \frac{d\delta}{dx} = \frac{P_r}{A_r E_r} + \varepsilon_{time} \quad \text{and} \quad \delta =_{cf} \tag{4.21}$$

Furthermore, at the other boundary – the *FI* boundary in Figure 4.7b – the boundary condition for time-dependent residual strains is given by Knight et al. (2013) and Visintin et al. (2013)

$$[BC]_{FI\text{-}INF\text{-}tim} \quad \delta = 0 \quad \text{and} \quad \frac{d\delta}{dx} = \varepsilon_{time} \tag{4.22}$$

For concrete with fibres, the boundary conditions are shown in Figure 4.8. At the crack face, the concrete now experiences a stress due to the fibres, $\sigma_{c\text{-}f}$, which corresponds to a strain, $\varepsilon_{c\text{-}f}=\sigma_{c\text{-}f}/E_c$. This strain depends on the half-crack width, Δ_{cf}. Hence, the force at the crack face on the concrete due to the fibres, C_f, shown on the *LHS* of Figure 4.8a, is now $A_c\sigma_{c\text{-}f}$. The boundary condition at the crack face when fibres are present is now determined by both the strain in the reinforcement and the strain in the concrete (Visintin et al. 2017):

$$[BC]_{CF\text{-}fib} \quad \frac{d\delta}{dx} = \frac{P_r}{A_r E_r} - \frac{C_f}{A_c E_c} \quad \text{and} \quad \delta = \Delta_{cf} \tag{4.23}$$

and at the *FI* boundary in Figure 4.8b, the boundary condition is as in Eq. 4.20.

Bringing all the scenarios together, the generic boundary condition at the crack face is given by

$$[BC]_{CF\text{-}gen} \quad \frac{d\delta}{dx} = \frac{P_r}{A_r E_r} - \frac{C_r}{A_c E_c} + \varepsilon_{time} \quad \text{and} \quad \delta = \Delta_{cf} \tag{4.24}$$

and the generic boundary condition at the *FI* boundary is given by

$$[BC]_{FI\text{-}gen} \quad \delta \to 0 \quad \text{and} \quad \frac{d\delta}{dx} \to \varepsilon_{time} \tag{4.25}$$

where the reader is reminded that both slip and slip–strain cannot be zero, but tend towards zero.

4.2.2.1.2 Infinite prism behaviour

Continuing with the analysis in Figure 4.4 for the infinitely long prism without fibres and residual strains. It is necessary to vary P_r in the analysis in Figure 4.4 to ensure that both the slip δ and the slip–strain $d\delta/dx$ tend to zero at the same position along the beam. This is illustrated in Figure 4.9a where the ordinate is in line with the initial crack and the abscissa is the distance from the crack face. Applying a relatively small load P_r in Figure 4.4a, such that the interface shear lies on the ascending branch *O–A* or *O–D* in Figure 4.2, the numerical analysis must show a gradual reduction in slip in Figure 4.9a (e.g. *A–B–C*), where C is asymptotic to zero. This condition is met only when the slip–strain *A–B–C* in Figure 4.9b is also asymptotic to zero at C. Hence vary P_r until this occurs, bearing in mind that zero cannot be achieved.

Once a solution for a specific half-crack width Δ_{cf} is found, we have the extent of the *PI* region, labelled L_{PI} in Figure 4.9a; the variation in slip *A–B–C* in the *PI* region; and consequently the variation in the interface bond shear stress τ in Figure 4.9c. We also know the variation in the axial stress σ_c in the concrete, which is *O–B–C* in Figure 4.9d, where it can be seen to gradually increase in the *PI* region to $\sigma_{c\text{-}max}$ and remain constant thereafter in the FI region. Should the maximum stress $\sigma_{c\text{-}max}$ reach the tensile strength of the concrete f_{ct}, then a new crack will form at a distance L_{PI} from the initial crack. This is referred to as a primary crack at a spacing S_p equal to $L_{PI.}$

If the displacement Δ_{cf} in Figure 4.4a is increased to δ_{max}, which is the interface slip beyond which the shear stress can be assumed to be zero as shown in Figure 4.2. This will result in the slip variation *D–E–F* in Figure 4.9a over the critical length L_{crt}, where *D* corresponds to Point C in Figure 4.2 and F corresponds to Point O in Figure 4.2. Hence, the full shear stress distribution in Figure 4.2 is incorporated into the interface shear stresses shown as *O–E–F* in Figure 4.9c; consequently, the interface shear force cannot be increased. This interface shear force is equal to the force in the reinforcement at the crack and is referred to as the intermediate-crack (*IC*) debonding force P_{IC}. This is the maximum force that the reinforcement can sustain for that particular bond-slip τ/δ configuration, and it can be seen that the reinforcement length must be greater than $L_{crt}=S_{p\text{-}IC}$ (Figure 4.9) for this to occur; it will be shown that $S_{p\text{-}IC}$ is the crack spacing when controlled by *IC* debonding, as explained later.

Should the half-crack width Δ_{cf} be increased beyond δ_{max} in Figure 4.9a, this simply transposes the stress profile *O–E–F* in Figure 4.9c to *G–H–J*, also of length L_{crt}. There is, therefore, a debonded length of reinforcement labelled L_{db}, comprising an *NI* region, followed by a *PI* region of length L_{crt}, and then a *FI* region. Another way of visualising this behaviour is illustrated by *G–H–I–J* in Figure 4.9a, where the slip distribution *D–E–F* is simply transposed to *H–I–J*, and the increase in slip from H to J is simple $\varepsilon_{IC}L_{db}$, where ε_{IC} is the strain in the reinforcement when the force is P_{IC}. The transposition of the P_{IC} stress block *G–H–J* in Figure 4.9c constitutes global debonding of the reinforcement and can continue to the end of the reinforcement, causing it to pull out. All of this can be followed using the numerical approach.

4.2.2.1.3 Infinite prism crack development

Consider further the development of cracks along the infinite prism. Let the infinite tension stiffening prism in Figure 4.4a be part of a simply supported beam whose half-span length is shown in Figure 4.10b. The *RC* beam has a half-span length of *L* and is subjected to a concentrated load at mid-span such that there is a linear variation in moment, as shown in Figure 4.10a. The beam is initially uncracked, and the cracking moment M_{cr} can be determined from an elementary *FI* elastic analysis.

The beam in Figure 4.10b is gradually loaded until the maximum moment at mid-span, M_1 in Figure 4.10a, reaches the cracking moment M_{cr}, at which point the initial flexural crack forms, as shown in Figure 4.10b. Once the flexural crack forms, it widens, stretching the reinforcement spanning it until the force in the reinforcement is sufficient to accommodate M_{cr}. This mechanism can be simulated and quantified using the mechanics in Figure 1.29 in Section 1.5.2 but is not necessary for this explanation. Once the system has stabilised after cracking, the force in the reinforcement is P_{r1} (Figure 4.10b, and the slip at the crack face is Δ_1(Figure 4.10c). The Δ_1-A variation is the distribution

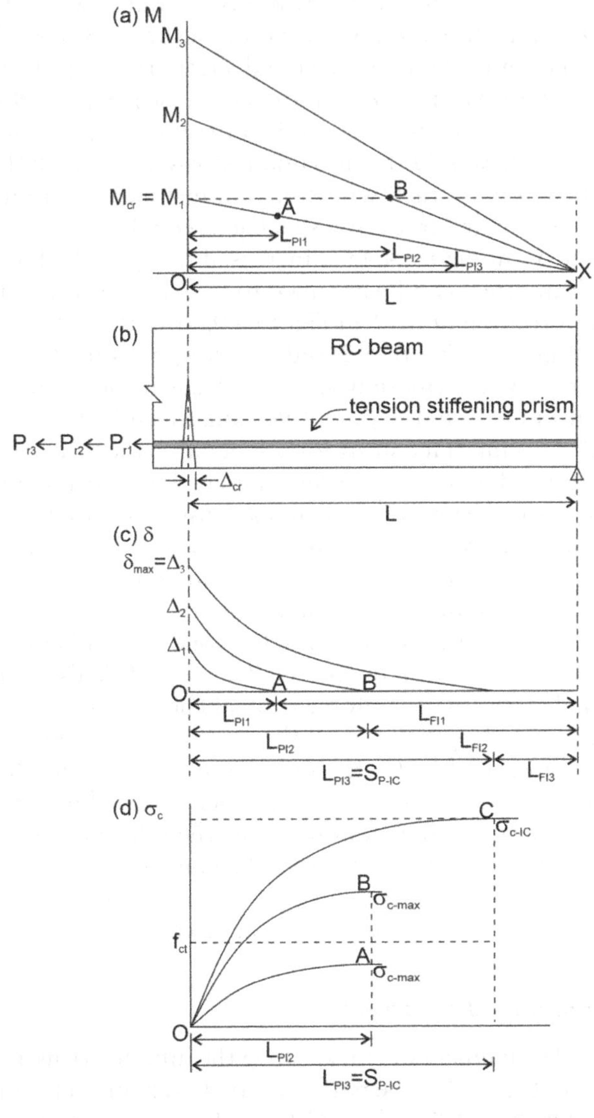

Figure 4.10 Primary crack propagation (a) distribution of moment between crack face and support; (b) beam elevation; (c) distribution of slip; (d) distribution of concrete stress.

of interface slip, with the *PI* region having a length of L_{PI1}. Hence, once a flexural crack forms, the tensile stresses in the concrete adjacent to the crack immediately drop to the post-cracking tensile stress, which, for concrete without fibres, is zero. Because a crack has now formed, the stresses in the concrete must build up through the tension stiffening mechanism over a distance L_{PI1} from the crack face. Consequently, the moment M_{cr} will not cause flexural cracking within this region, but only beyond it, where *FI* exists. If this were not the case, there would be an infinite number of flexural cracks within L_{PI1}. In the scenario shown in Figure 4.10a, the maximum moment beyond L_{PI1} in the *FI* region is at *A*, where the moment is less than M_{cr}. Therefore, a further flexural crack will not occur. The applied load is now increased gradually to a maximum moment M_2 (Figure 4.10a) such that the slip distribution $\Delta_2 - B$ (Figure 4.10c) and the applied moment distribution $M_2 - X$ (Figure 4.10a) are established. As the applied moment at B is M_{cr}, a flexural crack could form at B, which is located at a distance of L_{PI2} from mid-span. Although cracking could also occur within L_{PI2}, where the moment is higher, the tensile stresses are relieved through interface bond slip.

The concrete stress from a tension stiffening prism analysis is shown in Figure 4.10d. Let the concrete stress vary along the distribution $O-A$ such that the maximum concrete stress from tension stiffening mechanics $\sigma_{c\text{-}max}$ at A is less than the concrete tensile strength f_{ct}. In this case, cracking would not occur through the tension stiffening mechanism; that is it would not occur within L_{PI2}. It would occur through flexure in the *FI* region beyond L_{PI2} where M_{cr} is equal to or exceeded. However, as there is a moment gradient, it will probably occur where the moment is highest in the *FI* region, specifically, in the *FI* region adjacent to the *PI* region. Hence, in this case, the primary crack spacing $S_p = L_{PI2}$. In contrast, should the concrete stress vary due to tension stiffening along $O-B$, then cracking would occur within L_{PI2}. It can be seen that there are two mechanisms that control the onset of cracking, one being *FI* mechanics and the other *PI* mechanics. This is one of the reasons that cracking is considered to be random or at least a semi-random phenomenon.

Let the slip be increased to $\Delta_3 = \delta_{max}$ in Figure 4.10c. The reinforcement force P_{r3} is now the maximum force the reinforcement can take due to bond limitations; that is the IC debonding resistance P_{IC}. The *PI* length L_{PI3} has been labelled $S_{p\text{-}IC}$, which is L_{crit}. The buildup of stress in the concrete prism is shown as $O-C$ in Figure 4.10d, reaching a maximum of $\sigma_{c\text{-}IC}$. Should $\sigma_{c\text{-}IC} > f_{ct}$ as shown, then the primary crack spacing S_p will be less than L_{PI3}. However, should $\sigma_{c\text{-}IC} < f_{ct}$, then, as there is a moment gradient, the crack spacing $S_{p\text{-}IC} = L_{crit}$ is an upper bound to the primary crack spacing.

After the onset of the initial crack due to flexure, the infinite prism analysis in Figure 4.4 can be used to determine the second crack and its position, S_p, relative to the initial crack. The analysis can then be applied to the second crack in exactly the same way as for the initial crack and, unless some properties have changed, will give the same S_p for the next crack and so on. These cracks are called primary cracks, with a spacing of S_p.

4.2.2.2 Short prism

It is also worth noting that if the reinforcement length is less than L_{crt} in Figure 4.9, then we are now dealing with a prism of finite length. The analysis procedure in Figure 4.4 can still be used. The boundary condition at the crack face remains as Eq. 4.24, but for the other end, the reinforcement stress in the bar, σ_r, is now zero. This means that for short embedded bars less than L_{crt} long, the boundary condition is (Haskett et al. 2008)

$$[BC]_{SHT}\, \sigma_r = 0 \tag{4.26}$$

This scenario often occurs when short prisms are used in experiments to extract the bond-slip properties as explained in Section 3.2.

Short prisms (i.e. short lengths of bar) are also used in experiments to extract the shear friction properties using Z tests, as explained in Section 3.1.3.1. In this case, the ends of the bars are not unrestrained, as described above, but can be assumed to be fully restrained as they are anchored by wrapping around a transverse bar. Once again, the analysis procedure in Figure 4.4 can still be used. The boundary condition at the crack face remains as in Eq. 4.24; however, at the anchored end, where the reinforcement is fully fixed, the slip is zero, such that

$$[BC]_{ANC}\,\delta = 0 \tag{4.27}$$

4.2.2.3 Intermediate cracking

Once the primary cracks have formed at a spacing S_p, there is no longer an infinitely long prism, but rather a prism of length S_p between them. The mechanism in Figure 4.11a now applies. Through symmetry, only one half of this prism needs to be analysed using the numerical approach in Figure 4.4. In this case, by symmetry, it can be seen that the slip at mid-length (i.e. at $S_p/2$ from the primary crack faces) is zero, but not the slip–strain. Hence, the analysis in Figure 4.4 can be used to determine the behaviour of this mechanism between adjacent primary cracks with the following boundary condition. At the crack face, the boundary condition in Eq. 4.24 still applies. At the mid-length of the prism (i.e. at a distance of $S_p/2$ from the crack face), the boundary condition is now (Muhamad et al. 2012)

$$[BC]_{IC}\,\delta = 0 \quad \text{and} \quad \frac{d\delta}{dx} \neq 0 \tag{4.28}$$

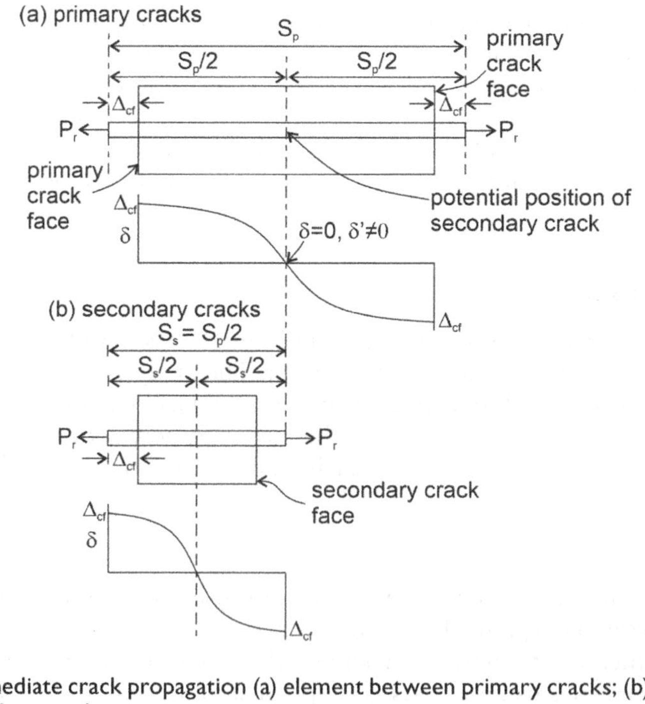

Figure 4.11 Intermediate crack propagation (a) element between primary cracks; (b) element between secondary cracks.

where in this case, the slip is zero as shown in the slip distribution in Figure 4.11a.

The above analysis between primary cracks can also be used to determine whether a secondary crack occurs midway between the primary cracks. In this case, the elementary moment to cause cracking M_{cr} used in flexural analyses cannot be used as there is no *FI* region; cracking will therefore occur when P_r from the tension stiffening analysis in Figure 4.11a causes the concrete to reach its tensile capacity. Whether a secondary crack forms [such as those in the lower prism Figure 1.10, the single short crack to the left in Figure 1.27a and the short cracks in Figure 1.30b] depends not only on the increase in moment but also on whether there is sufficient length between the cracks for the stress in the concrete to increase to f_{ct}. For secondary cracks to occur, there must either be a very large bonded perimeter of reinforcement L_{per}; the concrete stress at the crack face must be high, as in the case of fibre-reinforced concrete; or the bond stiffness must be very high.

If secondary cracks form, the crack spacing S_s is now $S_p/2$ and it is then possible to analyse their behaviour as in Figure 4.11b and determine whether tertiary cracks will occur at a spacing of $S_p/4$ using the same boundary conditions, namely, those in Eqs. 4.24 and 4.28 (Muhamad et al. 2012). The smaller the spacing of the cracks, the less likely new cracks are to form because the bonded length reduces and consequently, the available bond force and its associated concrete force required to crack the concrete diminish. To encourage further research, a flow chart showing the algorithm for tension stiffening is given in the Appendix. The reader may find it worthwhile to re-read this section with the accompanying flow chart, as the applicability of tension stiffening mechanics to many scenarios results in multiple derivations of slip–strain and boundary conditions. Despite this, the algorithm, once understood, is straightforward and the changes associated with residual strains and boundary conditions are minor. For clarity in this book, once secondary cracks form between primary cracks, all cracks will be referred to as secondary cracks; the same applies to tertiary cracks.

4.2.3 Tension stiffening behaviour

Of importance in the tension stiffening analysis is the determination of the crack spacing S_p and the crack opening stiffness, that is, the reinforcement load–slip (P_r/Δ_{cf}) behaviour at the crack face. Both of these are required in the flexural segmental analysis described in Section 1.5.2 using Figure 1.29 and in the shear flexural segmental analysis described in Section 1.5.3. The numerical model can not only quantify behaviours that can be used in design, but can also be used in parametric studies to help the engineer provide better designs. For example, parametric studies using the numerical model (Visintin et al. 2013) have shown that crack spacing reduces as the shrinkage strain increases and as the concrete creep increases. However, the importance of these time effects needs to be considered carefully. For example, flexural cracking at serviceability loads will in all probability occur at the very early stages of a member's life when the effects of creep and shrinkage are minimal. Hence crack spacing could be assumed to be independent of time effects. However, in contrast, the crack width will be highly dependent on time effects.

Parametric studies have also been used to aid understanding of fibre concrete (Visintin and Oehlers 2017). Consider, for example, the results in Figure 4.12 from a numerical

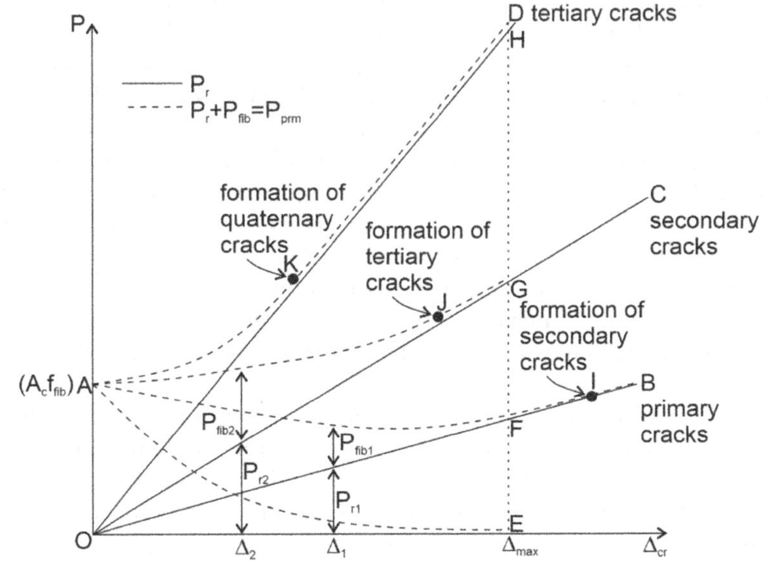

Figure 4.12 Tension stiffening with fibre concrete.

analysis of intermediate cracking in tension stiffening prisms made with fibre concrete. The ordinate represents the total force at the ends of the prism, P_{prm}, which comprises the force in the reinforcement, P_r, and the force in the fibres, P_{fib}, while the abscissa represents the half-crack width Δ_{cf}. The dashed lines represent the results for P_{prm} from a numerical simulation, and the unbroken lines represent the force in the reinforcement, P_r. The force at Point *A* in Figure 4.12 is the equivalent concrete force due to the fibres, at which a dominant crack begins to widen. This force is represented by $A_c f_{fib}$, where f_{fib} is the stress in the fibres as the crack width tends to zero, a value found at either Point *D* or Point *E* along the ordinate in Figure 4.3b.

The dashed line variation *A–E* in Figure 4.12 represents the force in the fibres, P_{fib}, alone, calculated using the strain softening material properties *E–F–G* presented in Figure 4.3. Here, Δ_{max} is defined as the half-crack width beyond which the fibre stress can be assumed to be zero. Hence, Point *E* in Figure 4.12 corresponds to Δ_{max} in Figure 4.3b. The dashed line *A–F–B* in Figure 4.12 is derived from a primary crack spacing analysis. The line *O–B* represents the force in the reinforcement at the crack, P_r, from the numerical analysis, shown here as a straight line for explanatory purposes. Hence, at a half-crack width of Δ_1, the reinforcement force is P_{r1}, and the fibre force is P_{fib1}. As can be seen, the fibre force is at its maximum when the crack width is close to zero, where the reinforcement force tends to zero. As the half-crack width increases, the contribution of the fibres diminishes, and that of the reinforcement increases until a half-crack width of Δ_{max} is reached, where the fibre contribution becomes zero. From the numerical analysis, the secondary crack occurred at Point *I*, at which point the fibres were ineffective.

The behaviour of secondary cracks (i.e. cracks spaced at $S_p/2$) is depicted by curve *A–G–C* in Figure 4.12, where *O–C* represents the force in the reinforcement. At Δ_2, the reinforcement force is P_{r2} and the fibre force is P_{fib2}. For a given value of Δ, such as

Δ_2, the reinforcement force for secondary cracks is larger than that for primary cracks, indicating that the reinforcement is more effective at smaller cracks, where the fibre forces are also more effective. Therefore, in general, the contribution of the fibres is greater in the presence of secondary cracks. It can be seen that fibres are more effective with smaller crack spacings, as smaller spacings result in smaller crack widths, a condition under which fibres are more effective. The formation of tertiary cracks occurred at Point *J*, where fibres were still effective. The analysis for tertiary cracks is represented by *A–K–D*, and the formation of quaternary cracks occurred at Point *K*, where the fibre contribution is significant. It can be seen that the benefits of fibre concrete are greatest when members are designed for small crack spacings.

The variation of bond stress from a tension stiffening analysis of a finite prism longer than the critical length L_{crt} is shown in Figure 4.13 (Muhamad et al. 2011). As the length of the prism was greater than L_{crt}, it can be analysed as an infinite prism up to global debonding. The bilinear bond-slip relationship *O–A–C* shown in Figure 4.2 was used. The distance is measured from the crack face on the left in Figure 4.13. The results are obtained when the slip at the crack face is δ_{max}, such that the *IC* debonding resistance is achieved. It can be seen that both the primary crack spacing L_{crit} and the *IC* debonding resistance P_{IC} have been determined. *PIC* is the force in the reinforcement when this debonding occurred and is equivalent to the integral of the bond stress over the surface of the bar. The rising branch in Figure 4.13 corresponds to the bond stresses following the falling branch *A–C* in Figure 4.2, and the falling branch in Figure 4.13 corresponds to the ascending bond-slip properties *O–A* in Figure 4.2.

Muhamad et al. (2011) also analysed the tension stiffening of infinite prisms using the linear descending bond slip *E–C* (Figure 4.2) to study the behaviour at the ultimate limit state. The reinforcement was assumed to initially have an elastic modulus of E_r, which then, at yield, reduced considerably to $E_{r\text{-}sh}$. The results are shown in Figure 4.14. The variation *O–A–B* occurs when the reinforcement remains elastic with a modulus of E_r. The variation *O–A–D* accounts for both E_r and $E_{r\text{-}sh}$, with yielding occurring at P_y, followed by strain hardening along *A–D*. They also analysed a tension stiffening prism

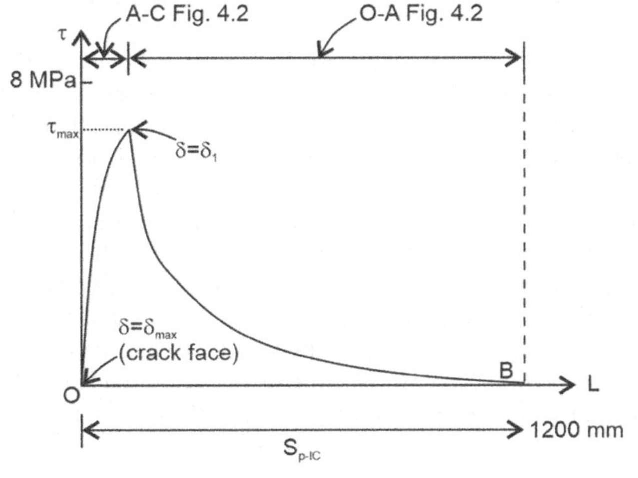

Figure 4.13 IC debonding.

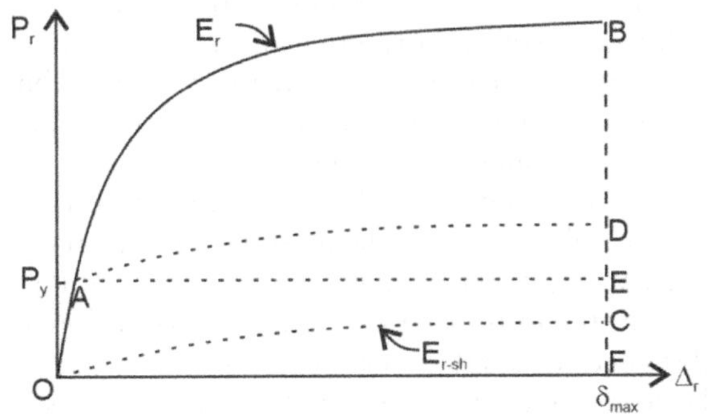

Figure 4.14 The effect of strain hardening.

in which the reinforcement had a modulus of $E_{r\text{-}sh}$ throughout, represented by *O–C*. From this analysis, they found that the distribution *O–A–D* was simply a transposition of the distribution *O–C*, meaning that *D–E* is equal to *C–F*. This may sound trivial, but it meant that strain hardening could be very easily included in closed-form solutions, as the closed-form solution for *O–A–B* could be used for *O–C* by simply changing the modulus. This is an example where numerical modelling, similar to experimental testing, can be a great help in developing simple closed-form solutions.

The numerical approach is very useful as it can accommodate any complex non-linear material property and can be used in parametric studies to isolate the important parameters that control the behaviour. The alternative involves the following closed-form solutions. The more complex the variation of the material properties, the more difficult it becomes to find a solution, often necessitating simplification of the material properties and a consequent loss of accuracy in the solution. However, closed-form solutions do identify the governing parameters so that the engineer can visualise and intuitively understand what is happening.

4.3 CLOSED-FORM SOLUTIONS FOR TENSION STIFFENING

The tension stiffening mechanics introduced through the above visual numerical model are now used to develop closed-form solutions using a more mathematical approach (Muhamad et al. 2011, 2012; Sturm et al. ; Vaculik et al. 2018). Invariably, doing this requires a simplification of the material properties and often leads to an associated reduction in accuracy compared with the numerical model. However, unlike the numerical approach, closed-form solutions are very useful for directly identifying the parameters that control tension stiffening. For example, it is shown that the total crack width in a hogging or sagging region of a beam at serviceability, that is, the sum of all the crack widths within a region, is independent of the bond stiffness when the linear ascending bond–stress–slip relationship in Figure 4.2 is assumed. At first glance, this would suggest that the engineer cannot control the crack widths when designing for durability. However, this is certainly not the case, as the crack spacing can be reduced, leading to finer cracks even though the sum of all the crack widths remains the same.

Fibres are then introduced, showing their enormous benefit in further reducing crack widths and consequently enhancing durability.

Having visually explained the fundamental mechanics to the reader, closed-form solutions are introduced, beginning with the derivation of the fundamental governing equation. To help the reader in understanding this derivation, the mathematics of the closed-form mechanics are directly linked to the visual mechanics of the numerical model. Based on the mechanics or principles postulated in Section 4.2 for developing the numerical approach to tension stiffening, the fundamental governing equation is first developed assuming that both the reinforcement and concrete materials remain linear–elastic; solutions are then found for the various boundary conditions. It must be emphasised that most solutions are limited to the materials remaining linear–elastic, in contrast to the numerical model, which does not require this caveat.

The bond-slip relationships in Figure 4.2 are considered. These include the linear ascending bond-slip relationship *O–D* in Figure 4.2, the linear descending relationship *E–C*, and the ascending region of the non-linear FIB relationship *O–A*. The FIB model for internal reinforcing bars has been found to give accurate predictions of the primary crack spacing S_p, in contrast to the linear ascending model, which gives poor correlation. However, it has been found that using the primary crack spacing from the CEB–FIB model in the linear ascending analysis gives good predictions of P/Δ. It has also been found that the linear descending bond-slip relationship gives good correlation with test results when used in the analysis of external reinforcement, such as adhesively bonded *FRP* plates.

4.3.1 Fundamental governing equation

Having considered an element in the numerical model (Figure 4.4d), the change in stress in the reinforcement due to interface slip is given by Eq. 4.3:

$$\frac{d\sigma_r}{dx} = \frac{L_{per}\tau}{A_r} \tag{4.29}$$

and that in the concrete by Eq. 4.2:

$$\frac{d\sigma_c}{dx} = -\frac{L_{per}\tau}{A_c} \tag{4.30}$$

Furthermore, from the numerical model, the slip–strain is given by Eq. 4.6; when accounting for time-residual strains and fibre strains, it is also expressed by Eq. 4.18 as

$$\frac{d\delta}{dx} = \left(\varepsilon_r - \varepsilon_{c\text{-}B}\right) + \left(\varepsilon_{time} - \varepsilon_{c\text{-}f}\right) \tag{4.31}$$

which can be written in terms of stress when both materials are linear elastic as

$$\frac{d\delta}{dx} = \left(\frac{\sigma_r}{E_r} - \frac{\sigma_c}{E_c}\right) + \left(\frac{\varepsilon_{time} - \varepsilon_{c\text{-}f}}{E_c}\right) \tag{4.32}$$

Bearing in mind that the second parameter in Eq. 4.32 is constant over the length of the segment, differentiating Eq. 4.32 gives

$$\frac{d^2\delta}{dx^2}=\frac{1}{E_r}\left(\frac{d\sigma_r}{dx}\right)-\frac{1}{E_c}\left(\frac{d\sigma_c}{dx}\right) \tag{4.33}$$

and substituting Eqs. 4.29 and 4.30 gives the following fundamental governing equation, first defined by Volkersen (1938) and applied in the derivation of analytical solutions (Muhamad et al. 2011, 2012; Sturm et al. 2018a,b)

$$\frac{d^2\delta}{dx^2}-\beta\tau=0 \tag{4.34}$$

where

$$\beta=L_{per}\left(\frac{1}{E_rA_r}+\frac{1}{E_cA_c}\right) \tag{4.35}$$

It can be seen from Eq. 4.34 that the governing equation is neither dependent on the fibre-induced strains $\varepsilon_{c\text{-}f}$ nor the time-dependent strains ε_{time}. These effects are incorporated through the boundary conditions as follows. The governing equation is now solved for various forms of the bond/slip (τ/δ) properties.

4.3.2 Solving for linear ascending bond slip

Starting with the simplest case of the linear ascending bond-slip variation *O–D* in Figure 4.2, which has a stiffness k_e (Muhamad et al. 2012; Sturm et al. 2018a,b), the bond stress–slip variation is given by

$$\tau(x)=k_e\delta(x) \tag{4.36}$$

Substituting Eq. 4.36 into the governing equation (Eq. 4.34) gives

$$\frac{d^2\delta}{dx^2}=\beta k_e\delta(x) \tag{4.37}$$

and solving gives the slip variation as

$$\delta(x)=c_1\cosh(\lambda_1x)+c_2\sinh(\lambda_1x) \tag{4.38}$$

in which

$$\lambda_1=\sqrt{k_e\beta} \tag{4.39}$$

where x is measured from the crack-face boundary.

Differentiating Eq. 4.38 yields

$$\frac{d\delta}{dx}=\lambda_1c_1\sinh(\lambda_1x)+\lambda_1c_2\cosh(\lambda_1x) \tag{4.40}$$

and substituting Eq. 4.36 into Eq. 4.38 gives

$$\tau(x) = k_e\left[c_1 \cosh(\lambda_1 x) + c_2 \sinh(\lambda_1 x)\right] \tag{4.41}$$

The constants of integration c_1 and c_2 can now be determined by applying the boundary conditions for the infinite prism and for intermediate cracking, as follows.

4.3.2.1 *Infinite prism*

4.3.2.1.1 Solving governing equation

The slip–strain boundary condition in Eq. 4.19 applies to the crack-face boundary in Figure 4.4a. Applying this boundary condition to Eq. 4.40 gives

$$c_2 = \frac{P_r}{A_r E_r \lambda_1} \tag{4.42}$$

The boundary condition at the *FI* boundary at $x=S_p$ from the crack face is given by Eq. 4.20. Applying this boundary condition, along with Eqs. 4.42–4.38 and 4.40 gives

$$c_1 \cosh(\lambda_1 S_p) + \frac{P_r}{A_r E_r \lambda_1} \sinh(\lambda_1 S_p) = 0 \tag{4.43}$$

and

$$\lambda_1 c_1 \sinh(\lambda_1 S_p) + \frac{P_r}{A_r E_r} \cosh(\lambda_1 S_p) = 0 \tag{4.44}$$

Substituting Eq. 4.43 into Eq. 4.44 to eliminate c_1 gives

$$\tanh^2(\lambda_1 S_p) = 1 \tag{4.45}$$

A simple solution of Eq. 4.45 is

$$S_p = \frac{2}{\lambda_1} \tag{4.46}$$

which is based on an assumption by Yuan et al. (2004) that the bond stress is resisted by $0.97P_r$ bearing in mind that the slip and slip–strain asymptote to zero such that the total bond force can only approach P_r. The need for this assumption is a good example of the challenge in developing analytical solutions even when simple material properties are assumed. The assumption that the bond stress is resisted by $0.97P_r$ was later shown by Sturm et al. (2018b) to yield poor estimates of the crack spacing when considering internal steel reinforcement. This solution is nevertheless important in identifying the parameters and proportionalities that define crack spacing and hence can be useful in developing an intuitive approach or rules of thumb for design.

Substituting Eq. 4.46 into Eq. 4.44 gives the other constant of integration as

$$c_1 = -\frac{P_r}{A_r E_r \lambda_1 \tanh 2} \tag{4.47}$$

Hence, we now know the variation of slip, slip–strain and bond along the prism, starting from the initial crack face, as derived from Eqs. 4.38, 4.40 and 4.41.

4.3.2.1.2 *Load to cause primary cracks*

From Figure 4.4 and Eq. 4.4, the relationship between the reinforcement force, P_r, at the crack-face boundary (Figure 4.4a) and the forces in the reinforcement ($P_{r\text{-}fi}$) and concrete ($P_{c\text{-}fi}$) at the *FI* boundary, located at a distance S_p from the crack-face boundary, is given by

$$P_r = P_{r\text{-}fi} + P_{c\text{-}fi} \tag{4.48}$$

Furthermore, $P_{c\text{-}fi}$ is the bond force B_{Sp} that acts over the length S_p between the crack-face boundary and the *FI* boundary, that is

$$B_{S_p} = \int_{x=0}^{x=S_p} \tau L_{per} dx \tag{4.49}$$

Hence Eq. 4.48 can be written as

$$P_r = P_{r\text{-}fi} + \int_{x=0}^{x=S_p} \tau L_{per} dx \tag{4.50}$$

For the concrete at the *FI* boundary to crack in tension at a stress f_{ct}, the bond force is given by

$$B_{S_p} = \int_{x=0}^{x=S_p} \tau L_{per} dx = f_{ct} A_c \tag{4.51}$$

Within the *FI* region adjacent to the *FI* boundary, the strains in the concrete and reinforcement are the same. Hence, an *FI* analysis, such as one using transformed sections, gives the reinforcement load required to cause the concrete to crack as

$$P_{r\text{-}fi} = \frac{f_{ct}}{E_c} A_r E_r \tag{4.52}$$

Substituting Eqs. 4.51 and 4.52 into Eq. 4.50 gives the load required to cause a primary crack in the *FI* region:

$$P_{r\text{-}cr\text{-}fi} = \frac{f_{ct}}{E_c} A_r E_r + f_{ct} A_c \tag{4.53}$$

The above force from an *FI* analysis can also be obtained from a *PI* analysis by solving Eq. 4.51. Substituting the bond stress from Eq. 4.41 into Eq. 4.51, where the constants of integration c_1 and c_2 are given by Eqs. 4.47 and 4.42, gives the load to cause a primary crack as

$$P_{r\text{-}cr\text{-}p} = \frac{A_r E_r \lambda_1^2 f_{ct} A_c}{k_e L_{per}} \tag{4.54}$$

It should be noted that Eq. 4.53, based on an *FI* analysis, and Eq. 4.54, based on a *PI* analysis, give the same results. It is also worth noting that this analysis, with its linear ascending bond slip, does not give an *IC* debonding resistance as described in Section 4.2.2.1 because the bond-slip properties, such as those shown in Figure 4.2, do not include the limit δ_{max}. It should also be remembered that the axial loads in the prism that cause cracking (Eqs. 4.53 and 4.54) may not necessarily govern initial crack development, as the prism is part of a flexural member where flexural forces might induce cracking.

4.3.2.1.3 *P/Δ for infinite prism*

For the infinite prism, the load–slip relationship at the crack face $P_{r\text{-}inf}/\Delta_{inf}$ can be obtained by substituting the constants of integration from Eqs. 4.42 and 4.47 into the slip variation of Eq. 4.38 to yield the relationship at $x=0$ (i.e. at the crack face):

$$\Delta_{inf} = \frac{w_{inf}}{2} = \frac{P_{r\text{-}inf}}{A_r E_r \lambda_1} \tag{4.55}$$

It should be borne in mind that Δ represents the half-crack width w.

4.3.2.2 Intermediate cracking

The above analysis of the infinite prism (Figure 4.4) determines the spacing (S_p) and occurrence of primary cracks. Once these cracks have formed, as shown in Figure 4.11a, the analysis shifts to finite prisms of length S_p.

4.3.2.2.1 *Prisms of length S_p*

Consider the LHS of the prism in Figure 4.11a; by symmetry, only half the prism needs to be analysed. The boundary conditions on the LHS, that is at $x = 0$, remain unchanged, as in Eq. 4.19; thus, following the same procedure as for the finite prism, the constant c_2 is as in Eq. 4.42. At the other boundary at mid-length of the prism in Figure 4.11a, the boundary condition is given by Eq. 4.28 at $x=S_p/2$. Substituting Eqs. 4.42 and 4.28 into Eq. 4.38, evaluated at $x=S_p/2$, gives the constant as

$$c_{1-s} = -\frac{P_r}{A_r E_r \lambda_1} \tanh(1) \tag{4.56}$$

Substituting Eqs. 4.56 and 4.42 into Eq. 4.40 for the boundary limits $x=0$ and $x=S_p/2$ gives the load to cause secondary cracks as

$$P_{r\text{-}cr\text{-}s} = \frac{A_r E_r \lambda_1^2 f_{ct} A_c}{0.35 k_e L_{per}} \tag{4.57}$$

The load P_r and slip Δ_{rs} at the crack face can be obtained by substituting c_2 from Eq. 4.42 and c_{1-s} from Eq. 4.56 into Eq. 4.38, evaluated at $x=0$, to give

$$\Delta_p = \frac{P_r \tanh(1)}{A_r E_r \lambda_1} \tag{4.58}$$

where Δ_p is the half-crack width when there are adjacent primary cracks.

4.3.2.2.2 *Prisms of length $S_p/2$*

When secondary cracks form at a spacing of $S_p/2$, the same analysis applies as in Section 4.3.2.2.1 for prisms between primary cracks, except that the second boundary condition is at $S_p/4$. This gives the constant as

$$c_{1-t} = -\frac{P_r}{A_r E_r \lambda_1} \tanh(0.5) \tag{4.59}$$

and, consequently, the load to cause tertiary cracks as

$$P_{r\text{-}cr\text{-}t} = \frac{A_r E_r \lambda_1^2 f_{ct} A_c}{0.113 k_e L_{per}} \tag{4.60}$$

and P/Δ as

$$\Delta_s = \frac{P_r \tanh(0.5)}{A_r E_r \lambda_1} \tag{4.61}$$

The same procedure can be applied to determine whether quaternary or further intermediate cracks can occur. However, a limit exists: as the crack spacing reduces, so too do the available bond force and, consequently, the concrete force available to crack the concrete.

Worked Example WE-4.2: Linear ascending bond stiffness

As an example of the application of analytical solutions to determine the crack spacing and crack opening stiffness, consider the same beam cross-section as in Figure E4.1a, now shown in Figure E4.2a. For this example, consider the beam to be constructed from a 40 MPa concrete without fibres, with an elastic modulus of 32,000 MPa and a tensile cracking stress of 4 MPa. The linear ascending bond properties used in the analysis are shown in Figure E4.2b and have a stiffness k_e of 15.8 MPa/mm. This bond stiffness has been chosen because it is approximately equal to τ_{max}/δ_1 for a 40 MPa concrete determined using the FIB approach in Section 3.2.4.1.

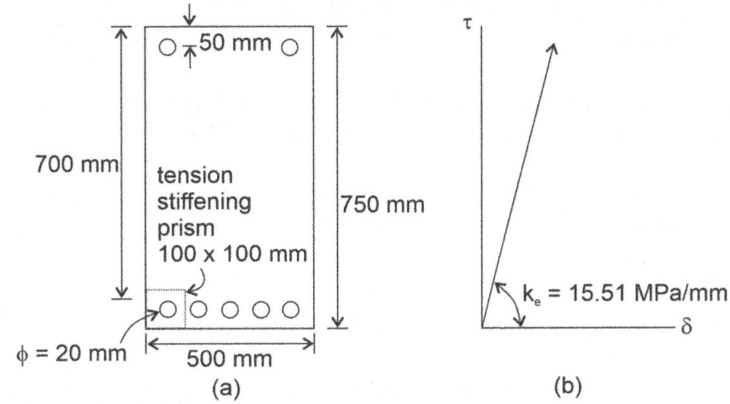

Figure E4.2 Section and material properties (a) cross section; (b) bond properties.

From Eq. 4.46 the primary crack spacing is given by

$$S_p = \frac{2}{\lambda_1} \tag{E4.2.1}$$

in which, from Eq. 4.35,

$$\beta = L_{per}\left(\frac{1}{E_r A_r} + \frac{1}{E_c A_c}\right) = \pi \times 20\left(\frac{1}{200{,}000 \times 0.25 \times \pi \times 20^2} + \frac{1}{32{,}000 \times 100 \times 100}\right)$$
$$= 1.196 \times 10^{-6} \tag{E4.2.2}$$

and from Eq. 4.39,

$$\lambda_1 = \sqrt{k_e \beta} = \sqrt{15.81 \times 1.196 \times 10^{-6}} = 0.00435 \tag{E4.2.3}$$

Substituting Eq. E4.2.3 into Eq. E4.2.1 gives the primary crack spacing

$$S_p = \frac{2}{0.00435} = 459.8 \text{ mm} \tag{E4.2.4}$$

From Eq. 4.54, the corresponding load to cause primary cracking is

$$P_{r\text{-}cr\text{-}p} = \frac{A_r E_r \lambda_1^2 f_{ct} A_c}{k_e L_{per}} = \frac{0.25 \times \pi \times 20^2 \times 200{,}000 \times 0.00435^2 \times 4 \times 100 \times 100}{15.81 \times \pi \times 20}$$
$$= 47.87 \text{ kN} \tag{E4.2.5}$$

The relationship between the load and slip of the reinforcement for primary cracking is given by Eq. 4.58 such that the crack opening stiffness, P_r/Δ_p, is

$$\frac{P_r}{\Delta_p} = \frac{A_r E_r \lambda_1}{\tanh(1)} = \frac{0.25 \times \pi \times 20^2 \times 200{,}000 \times 0.00435}{\tanh(1)} = 358.9 \text{ kN/mm} \quad \text{(E4.2.6)}$$

The crack opening stiffness given by Eq. E2.6 is applied until the load causing secondary cracking is reached, which is given by Eq. 4.57, that is

$$P_{r\text{-}cr\text{-}s} = \frac{A_r E_r \lambda_1^2 f_{ct} A_c}{0.35 k_e L_{per}} = \frac{0.25 \times \pi \times 20^2 \times 200{,}000 \times 0.00435^2 \times 4 \times 100 \times 100}{0.35 \times 15.81 \times \pi \times 20} \quad \text{(E4.2.7)}$$

$$= 136.8 \text{ kN}$$

after which the crack opening stiffness can be quantified using

$$\frac{P_r}{\Delta_s} = \frac{A_r E_r \lambda_1}{\tanh(0.5)} = \frac{0.25 \times \pi \times 20^2 \times 200{,}000 \times 0.00435}{\tanh(0.5)} = 591.4 \text{ kN/mm} \quad \text{(E4.2.8)}$$

Comparing the crack opening stiffness for only primary cracks with that for primary and secondary cracks shows that the formation of a new crack results in an increase in stiffness of 64%, or conversely, a reduction in crack width of approximately 60% for the same applied load.

4.3.2.3 Parametric study for linear ascending

Unlike numerical solutions which do not directly identify the parameters that affect behaviour, closed-form solutions do. Let us first consider the IC mechanics when the primary cracks have formed. The half-crack width is given by Eq. 4.58 and the crack spacing by Eq. 4.46. In a similar way to Sturm et al. (2018a), dividing the half-crack width by the half-crack spacing gives the crack width density (*CWD*), which represents the crack width per unit length. Between primary cracks, the CWD is given by

$$CWD_p = \frac{w}{S_p} = \frac{\Delta}{S_p/2} = \frac{P_r \tanh(1)}{E_r A_r} = \frac{0.76 P_r}{E_r A_r} = 0.76 \varepsilon_{r\text{-}p} \quad (4.62)$$

where $\varepsilon_{r\text{-}p}$ is the strain in the reinforcement at a primary crack face.

It can be seen in Eq. 4.62 that the *CWD* is totally independent of the bond stiffness k_e. It would appear at first sight that there is no point in trying to improve the performance of a structure by using reinforcement with high bond. For example, doubling the bond stiffness would give exactly the same total crack width. However, it is not the total crack width that is important in designing for durability, but rather the actual crack width. Doubling the bond stiffness will give more cracks but finer cracks. In contrast, reducing the bond stiffness will reduce the number of cracks but lead to wider cracks. For example, using reinforcement such as some profiled sheets which can have very low bond stiffnesses may lead to just one crack at the position of maximum moment but this will be a very wide crack. It can also be seen from Eq. 4.62 that the *CWD* is independent of the concrete properties and is only dependent on the strain in the reinforcement at the crack.

When secondary cracks form between primary cracks as in Figures 1.10 and 1.30b, the width of the adjacent primary cracks reduces to accommodate the secondary cracks. It should be noted that once secondary cracks have formed between primary cracks, all the cracks will be referred to as secondary cracks. For secondary cracks, the *CWD* is as follows:

$$CWD_s = \frac{w}{S_p/2} = \frac{\Delta}{S_p/4} = \frac{P_r \tanh(0.5)2}{E_r A_r} = \frac{0.92P_r}{E_r A_r} = 0.92\varepsilon_{r\text{-}s} \tag{4.63}$$

which is a slightly higher *CWD* than for primary cracks, but it should be noted that the crack spacing has halved, which will lead to finer cracks, 61% of the width of the primary crack.

The primary crack spacing, as in Eq. 4.46, is inversely proportional to λ_1, the parameters of which are all shown below. Hence it can be seen that increasing k_e and L_{per} increases λ_1 and, therefore, reduces the crack spacing S_p. For example, reducing the bar diameter from 20to 10 mm will double L_{per} and, hence, the crack spacing will reduce by the inverse of the square root of 2, that is by 29%. Furthermore, reducing E_c reduces the crack spacing but the significance of this is not obvious.

$$\lambda_1 = \sqrt{k_e L_{per}\left(\frac{1}{E_r A_r} + \frac{1}{E_c A_c}\right)} \tag{4.64}$$

All of this suggests that the engineer should and can design for as many cracks as possible, as these will lead to finer cracks and longer durability. The linear ascending bond stiffness is probably appropriate at serviceability. It probably does not accurately resemble the experimental rising branch and, therefore, may not be as precise as non-linear modelling when used to estimate crack spacing and crack widths, but it does give a clear indication to the engineer what the important parameters are.

In the previous section it was shown that the fundamental governing equation is independent of the residual strains and the bond-slip properties. The reader has then been taken through all the steps in deriving a closed-form solution for the tension stiffening behaviour using the linear ascending bond-slip model. The same procedure is used for bond-slip characteristics that more closely simulate the actual experimentally derived behaviour. Invariably the mathematics becomes more complex but the procedure is the same. From here onwards, only the solutions will be given with reference to the papers from which they were derived.

4.3.2.4 Intermediate cracking – linear model

The complexity of the CEB–FIB model (FIB 2013) means that it does not yield closed-form solutions simple enough for application; consequently, a linear model allowing for residual strains and fibres will be used instead. As shown by Sturm et al. (2018a), this approach does not significantly impact the accuracy of the model over the practical range of reinforcement slips. Applying the linear ascending bond-slip model of Eq. 4.36 gives the *P*/Δ relationship for primary cracks as

$$\Delta_p = \frac{\dfrac{P_r}{E_r A_r} + \varepsilon_{time} - \dfrac{f_i}{E_c} - \varepsilon_{inel}}{\dfrac{\lambda_1}{\tanh(\lambda_1 S_p)} - \dfrac{m_i}{E_c}} \tag{4.65}$$

where ε_{inel} is defined in Figure 4a, and f_i and m_i can have the subscripts 1 or 2, as shown in Figure 4.3b.

4.3.3 Solved for linear descending

4.3.3.1 Material properties

The linear descending bond-slip property is shown as E–C in Figure 4.2 and is given by

$$\tau = \frac{\tau_{max}}{\delta_{max}}(\delta_{max} - \delta) \tag{4.66}$$

By substitution into the governing Eq. 4.34 and integrating the distributions of slip, slip–strain and bond stress along the bonded length, Muhamad et al. (2012) provide the following expressions:

$$\delta(x) = c_1 \sin(\lambda_3 x) + c_2 \cos(\lambda_3 x) + \delta_{max} \tag{4.67}$$

$$\frac{d\delta}{dx} = \lambda_3 c_1 \cos(\lambda_3 x) - \lambda_2 c_2 \sin(\lambda_3 x) \tag{4.68}$$

$$\tau(x) = -\frac{\tau_{max}}{\delta_{max}}\left[c_1 \sin(\lambda_3 x) + c_2 \cos(\lambda_3 x)\right] \tag{4.69}$$

where x is measured from the crack-face boundary and λ_3 is given by

$$\lambda_3 = \sqrt{\frac{\tau_{max}}{\delta_{max}}\beta} \tag{4.70}$$

Similarly to the linear ascending case, Eqs. 4.66–4.69 can be used to determine the crack spacing and load-slip relationship by using the boundary conditions in Eqs. 4.19 and 4.20 to obtain specific solutions for the coefficients c_1 and c_2.

4.3.3.2 Infinitely long prism

The load in the reinforcement required to cause a primary crack is given by Muhamad et al. (2012) as

$$P_{r\text{-}cr\text{-}p} = \frac{A_r E_r \lambda_3^2 \delta_{max} f_{ct} A_c}{\tau_{max} L_{per}} \tag{4.71}$$

and the primary crack spacing is

$$S_p = \frac{\arcsin\left(\dfrac{\lambda_3 f_{ct} A_c}{\tau_{max} L_{per}}\right)}{\lambda_3} \tag{4.72}$$

in which

$$\lambda_3 = \sqrt{\frac{\tau_{max}}{\delta_{max}}\beta} \tag{4.73}$$

The P/Δ relationship for the initial crack is given by

$$\Delta_p = \delta_{max} - \delta_{max}\sqrt{1 - \frac{P_{r\text{-}inf}^2}{A_r^2 E_r^2 \lambda_3^2 \delta_{max}^2}} \tag{4.74}$$

4.3.3.3 Intermediate cracking

The P/Δ relationship between primary cracks with spacing S_p is given by Muhamad et al. (2012)

$$\Delta_p = \delta_{max}\left(1 - \frac{1}{\cos(\lambda_3 S_p / 2)}\right) - \frac{P_r}{A_r E_r \lambda_3}\tan(\lambda_3 S_p / 2) \tag{4.75}$$

4.3.3.4 Reinforcement debonding in an infinitely long prism

If the reinforcement continues to be pulled such that Δ_p is equal to δ_{max} in Eq. 4.74, the IC debonding condition, given by line *D–E–F* in Figure 4.9a, is reached, and the force P_r is equal to the IC debonding force P_{IC}.

Substituting Eqs. 4.35 and 4.70 into Eq. 4.74 and neglecting the concrete area, the *IC* debonding force P_{IC} is given in the well-known form commonly applied to *FRP* plate debonding scenarios (Yuan et al. 2004; Haskett et al. 2008).

$$P_{IC} = \sqrt{\tau_{max}\delta_{max}}\sqrt{L_{per} A_r E_r} = \sqrt{2G_f}\sqrt{L_{per} A_r E_r} \tag{4.76}$$

Applying the boundary condition that the slip–strain is equal to the strain in the reinforcement at the loaded end and the slip is equal to zero at the free end, Eq. 4.19 allows Eq. 4.74 to be substituted into Eq. 4.68 to yield the length required to achieve the *IC* debonding force, that is

$$L_{crt} = \frac{\pi}{2\sqrt{\dfrac{\tau_{max} L_{per}}{\delta_{max} A_r E_r}}} \tag{4.77}$$

4.3.4 Solved for constant bond-slip relationship

When only friction is present, such as with smooth reinforcement, or if the bar has been pulled until it debonds, the bond-slip relationship can be taken as a constant, such as branch H–I in Figure 4.2. In this scenario, the somewhat trivial solution is obtained, in which the reinforcement force P_r can be directly related to the bonded length by Vaculik et al. (2018):

$$P_r = \tau_r L_{per}\left(L - \Delta_p\right) \tag{4.78}$$

where the bonded length L is reduced as it is pulled out of the concrete.

4.3.5 Solved for residual strains and fibres

4.3.5.1 Material properties

Full details of the fibre concrete properties are given in Section 2.4.1.2 and in Section 3.3.1. These are idealised in Figure 4.3 (Sturm et al. 2018b). The path followed depends on the quantity of fibre. For concrete with small amounts of fibre, a micro-crack will immediately lead to a macro-crack and this is referred to as a non-strain hardening concrete. The uncracked concrete will be assumed to follow a loading or unloading path O–A–B in Figure 4.3a with modulus E_c and strength f_{ct}. If there are sufficient fibres to allow an increase in strength after micro-cracking, then it is referred to as a strain hardening concrete. On the formation of a micro-crack at A with strength f_{SH}, further micro-cracks will form until one of the micro-cracks forms a macro-crack at C with strength f_{ct}. The concrete will follow the path O–A with modulus E_c, then A–C with the strain hardening modulus E_{SH}, and the uncracked concrete then unloads along C–I with modulus E_c where the yield plateau corresponds to the strain offset ε_{inel}. The properties of the macro-crack are shown in Figure 4.3b. Point D corresponds to either Point B or Point C and occurs at f_{ct} in Figure 4.3a. Following this, the descending branch can be linearized into segments D–H and H–G where each linear segment is defined by m_i and f_i and the intercept f_{pc}.

The FIB bond slip (FIB 2013) is shown in Figure 4.2 as the non-linear path O–A, where the peak stress τ_{max} at A occurs at a slip δ_1, followed by the path A–G–H–I. The non-linear path O–A used in this analysis is given by

$$\tau = \tau_{max}\left(\frac{\delta}{\delta_1}\right)^{\alpha} \tag{4.79}$$

where α is a constant to fit the experimental data and typically has a value of 0.4. The FIB bond model is selected here as it has been shown for both concrete with and without fibres to give the most reliable estimates of crack spacing. That is, although a linear ascending bond property can be identified which leads to a reasonable approximation of the crack opening stiffness, it leads to a poor prediction of the crack width (Sturm et al. 2018b). This difficulty arises because of the approximation required for Eq. 4.45 to avoid an asymptotic solution.

4.3.5.2 *Infinite prism – FIB model*

To determine a closed-form solution, Sturm et al. (2018b) assumed that the stress at which a new primary crack occurs is f_{ct} (Figure 4.3a) and the stress at the initial crack face, f_{pc} (Figure 4.3b), is at the intercept between two linear descending paths at Point H (Sturm et al. 2018b). Taking f_{pc} as the intercept between the two falling branches is an approximation required to create the analytical solution and is conservative, as it assumes a significant crack opening occurs at the initial crack before primary cracks form. This assumption has been validated and shown to correlate well with test results (Sturm et al. 2018b).

The primary crack spacing is therefore

$$S_p = \left[\frac{2^{\alpha}(1+\alpha)}{\lambda_2(1-\alpha)^{1+\alpha}}\right]^{\frac{1}{1+\alpha}}\left[\frac{f_{ct}-f_{pc}}{E_c}\left(\frac{E_cA_c}{E_rA_r}+1\right)\right]^{\frac{1-\alpha}{1+\alpha}} \tag{4.80}$$

where

$$\lambda_2 = \frac{\tau_{max}}{\delta_1^{\alpha}}\beta \tag{4.81}$$

4.3.5.3 *Intermediate cracking – linear model*

Although leading to a simple and accurate solution for the primary crack spacing, as described in Section 4.3.2.4, the FIB model does not yield simple closed-form solutions for the relationship between P_r and Δ. It has, however, been shown that using the FIB model to predict S_p and then a linear ascending model to predict the P_r/Δ relationship yields accurate predictions for tension stiffening and crack widths, including fibre-reinforced concrete and concrete subjected to shrinkage and creep. Applying the linear ascending bond-slip model of Eq. 4.36 gives the P/Δ relationship for primary cracks (Sturm et al. 2018b):

$$\Delta_p = \frac{\dfrac{P_r}{E_rA_r}+\varepsilon_{time}-\dfrac{f_i}{E_c}-\varepsilon_{inel}}{\dfrac{\lambda_1}{\tanh(\lambda_1 S_p)}-\dfrac{m_i}{E_c}} \tag{4.82}$$

where it can be seen that the residual strains, fibre strains and micro-cracking or yield strains act in a similar way and, as would be expected, shrinkage will increase the crack width.

4.3.5.4 *Parametric study*

Rearranging Eq. 4.82 as a function of the load on the reinforcement, P_r, gives

$$P_r = k_{pi}(\Delta - \Delta_o) \tag{4.83}$$

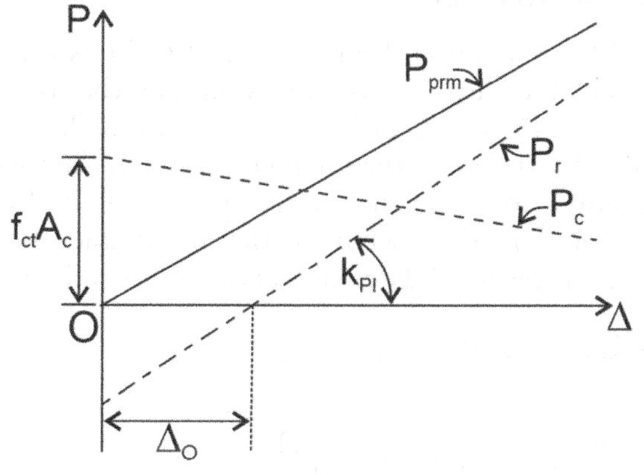

Figure 4.15 Effect of residual strains on crack opening stiffness.

as shown in Figure 4.15, where P_{prm} is the total force in the prism (P_r+P_c), and the crack opening stiffness k_{pi} is given by

$$k_{pi} = E_r A_r \left(\frac{\lambda_1}{\tanh(\lambda_1 S_p / 2)} - \frac{m_i}{E_c} \right) \tag{4.84}$$

and the residual strain offset Δ_o is given by

$$\Delta_o = \frac{\varepsilon_{time} - \dfrac{f_i}{E_c} - \varepsilon_{inel}}{\dfrac{\lambda_1}{\tanh(\lambda_1 S_p/2)} - \dfrac{m_i}{E_c}} \tag{4.85}$$

The average axial strain in the tension prism is $\Delta/(S_p/2)$, such that, using Eq. 4.84, the effective axial rigidity is $k_{pi}S_p/2$. Using the effective axial rigidity, the effective modular ratio, n_{PI}, from *PI* theory can be defined as

$$n_{PI} = \frac{K_{PI} S_p}{E_c A_r} = \left(\frac{\lambda_1 S_p}{\tanh(\lambda_1 S_p)} - \frac{m_i}{E_c} S_p \right) n_{FI} \tag{4.86}$$

where n_{FI} is the *FI* modular ratio (E_r/E_c). Equation 4.86 is important as it shows which parameters can be changed to increase the stiffness due to tension stiffening.

The total load in the tension prism P_{prm}, (P_c+P_r), is as follows:

$$P_{prm} = K_{pi}(\Delta - \Delta_o) + f_i A_c - m_i A_c \Delta \tag{4.87}$$

which shows the contribution of the fibres particularly for small deformations.

From the above equations, it can be seen that fibres reduce the crack opening stiffness (Eq. 4.84) because the uncracked concrete between cracks resists a smaller proportion of the overall load; fibres reduce the initial deformation due to shrinkage (Eq. 4.85), and this is particularly significant in concretes that strain harden; from Eq. 4.80, strain hardening does not affect S_p; furthermore, strain hardening delays the formation of cracks, and therefore, for a given load level, crack widths are reduced in concretes that strain harden compared to those that do not.

Worked Example WE-4.3: FIB crack spacing and linear ascending crack opening stiffness

Let us now return to Example WE-4.2 and apply the analytical solutions in Sections 4.3.5.2 and 4.3.5.4 to determine the crack spacing and opening stiffness. Recall that in this example the beam cross-section is as shown in Figure E4.3a, with concrete tension properties in Figure E4.3b and reinforcement bond properties in Figure E4.3c.

The primary crack spacing is given by Eq. 4.80, which is dependent on the non-linearity of the rising branch of the τ/δ bond properties, defined by α (found to be 0.44 from Eq. E1.2). In Eq. 4.80, the parameter f_{pc} corresponds to the stress at the inflection point of the stress–crack width relationship in Figure E4.3b, which is equal to 2 MPa. Finally, the bond stress parameter λ_2 is given by Eq. 4.81, where the parameter β is defined in Eq. 4.35. That is

$$\lambda_2 = \frac{\tau_{max}}{\delta_1^{\alpha}}\beta = \frac{41.4}{0.5^{0.44}} \times 1.196 \times 10^{-6} = 6.72 \times 10^{-5} \tag{E4.3.1}$$

$$S_p = \left[\frac{2^{0.44}(1+0.44)}{6.72\times10^{-5}(1-0.44)^{1+0.44}}\right]^{\frac{1}{1+0.44}} \left[\frac{5-2}{40,000}\left(\frac{40,000\times100\times100}{200,000\times0.25\times\pi\times20\times20}+1\right)\right]^{\frac{1-0.44}{1+0.44}}$$

$$= 121.5 \text{ mm} \tag{E4.3.2}$$

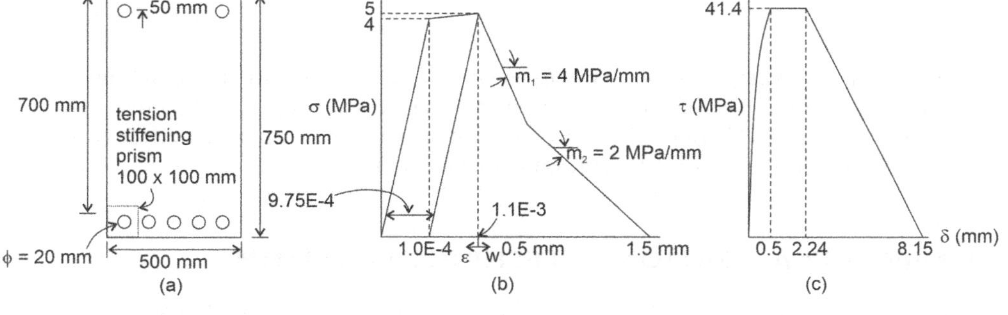

Figure E4.3 Cross-section and material properties (a) cross section; (b) concrete tensile stress-strain and tensile crack width properties; (c) reinforcement bond properties.

Now determine the crack opening stiffness when on Branch 1 of the stress–crack width relationship in Figure E4.3b, with m_1=4 MPa/mm. From Eq. 4.84, the crack opening stiffness is dependent on the crack spacing given by Eq. E2.1 and the λ_1 parameter given by Eq. 4.39. To determine the λ_1 parameter, the stiffness of the rising branch of the τ/δ relationship must be linearised and is taken as τ_{max}/δ_1

$$\lambda_1 = \sqrt{k_e\beta} = \sqrt{\frac{41.4}{0.5}6.72\times10^{-5}} = 0.0750 \tag{E4.3.3}$$

Substituting λ_1, S_p and the material properties into Eq. 4.84 yields

$$k_{pi} = E_r A_r\left(\frac{\lambda_1}{\tanh(\lambda_1 S_p/2)} - \frac{m_i}{E_c}\right) = 200,000\times0.25\times\pi\times20\times20$$

$$\left(\frac{0.0750}{\tanh(0.0750\times121.5)} - \frac{4}{40,000}\right) = 4706 \text{ kN/mm} \tag{E4.3.4}$$

As shown in Figure 4.15, this stiffness is offset from the origin by the slip Δ_o, which is given by Eq. 4.85. As an example of the magnitude of this offset, first, consider the shrinkage strain to be very small such that ε_{time}=100$\mu\varepsilon$. From Eq. 4.85, if slips are assumed to be restricted to Branch 1 (with m_i=4 MPa/mm) and the maximum fibre stress for Branch 1 is taken as f_1=5 MPa, then

$$\Delta_o = \frac{\varepsilon_{time} - \frac{f_i}{E_c} - \varepsilon_{inel}}{\frac{\lambda_1}{\tanh(\lambda_1 S_{p/2})} - \frac{m_i}{E_c}} = \frac{100\times10^{-6} - \frac{5}{40,000} - 1.09\times10^{-3}}{\frac{0.0750}{\tanh(0.0750\times121.5)} - \frac{4}{40,000}} = -0.0149 \text{ mm} \tag{E4.3.5}$$

If the shrinkage strain is now taken as very large such that ε_{time}=1000$\mu\varepsilon$

$$\Delta_o = \frac{\varepsilon_{time} - \frac{f_i}{E_c} - \varepsilon_{inel}}{\frac{\lambda_1}{\tanh(\lambda_1 S_{p/2})} - \frac{m_i}{E_c}} = \frac{1000\times10^{-6} - \frac{5}{40,000} - 1.09\times10^{-3}}{\frac{0.0750}{\tanh(0.0750\times121.5)} - \frac{4}{40,000}} = -0.00287 \text{ mm} \tag{E4.3.6}$$

where, in the example above, a negative value of Δ_o increases the force that must be applied to the bar to generate a given total half-crack width. That is, the high tensile strength and high tensile strains generated by micro-cracking are effective in minimising the impact of shrinkage on tensile cracking.

4.4 APPENDIX A: TENSION STIFFENING ALGORITHM

The flow chart in Figure 4.16 can be used to develop numerical models for primary crack spacing, and the flow chart in Figure 4.17 for tension stiffening crack opening stiffness.

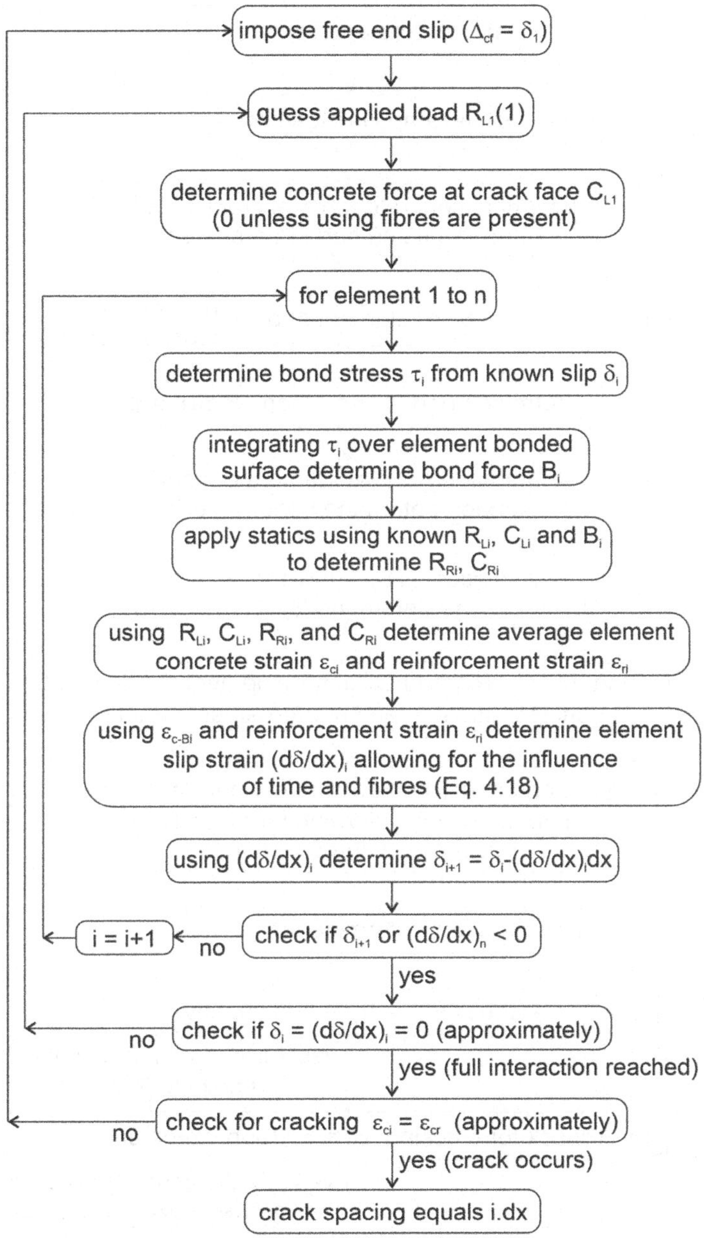

Figure 4.16 Algorithm for determining primary crack spacing.

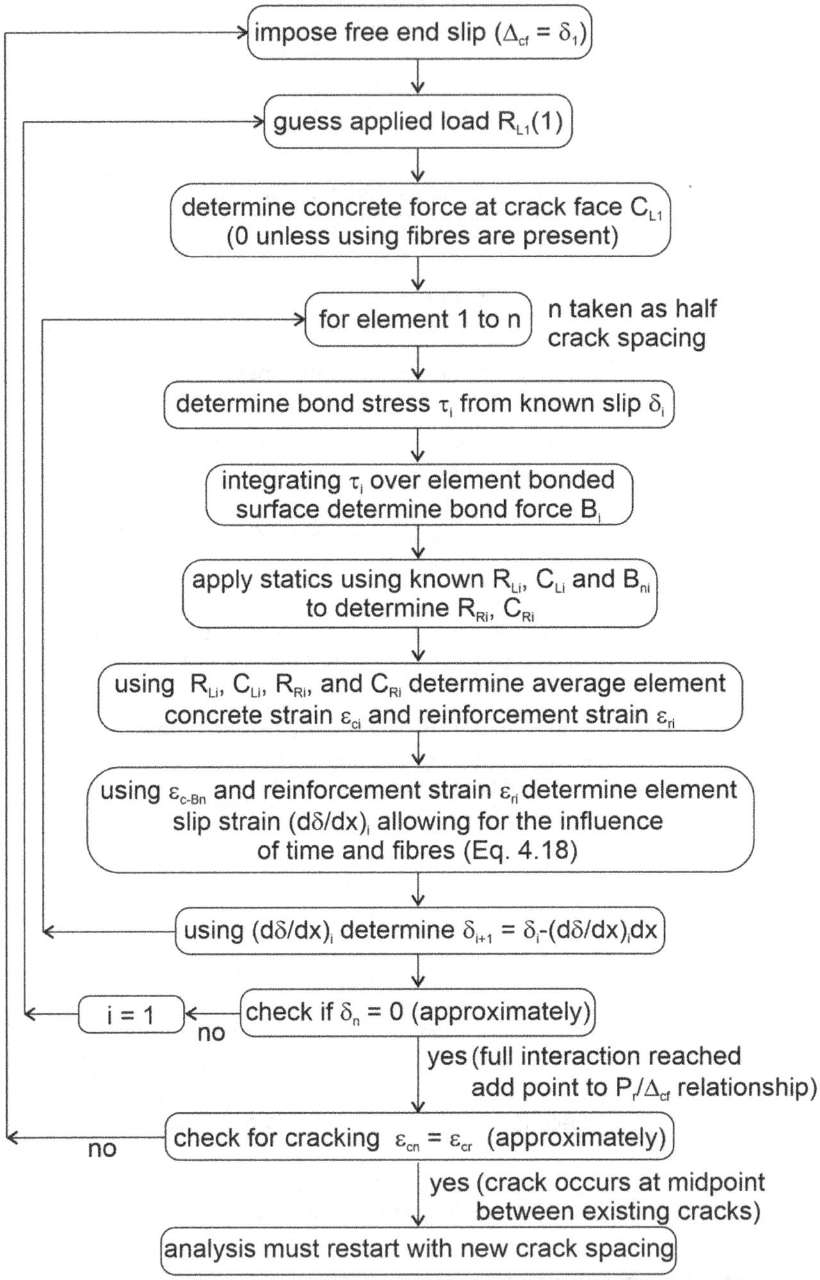

Figure 4.17 Algorithm for determining tension stiffening crack opening stiffness.

4.5 NOTATION AND ABBREVIATIONS

A_c	cross-sectional area of concrete in tension stiffening prism
A_r	cross-sectional area of tension reinforcement in tension stiffening prism
B	bond force
B_{Sp}	bond force acting over length S_p
$[BC]_{CF\text{-}fib}$	boundary condition at crack-face boundary with fibres
$[BC]_{CF\text{-}tim}$	boundary condition at crack-face boundary with residual strains
$[BC]_{CF\text{-}gen}$	generic boundary condition at crack-face boundary
$[BC]_{FI\text{-}gen}$	generic boundary condition at *FI* boundary of infinite prism
$[BC]_{FI\text{-}INF}$	boundary condition at *FI* boundary of infinite prism
$[BC]_{FI\text{-}INF\text{-}tim}$	boundary condition at *FI* boundary of infinite prism with residual strains
$[BC]_{IC}$	boundary condition between intermediate cracks
$[BC]_{SHT}$	opposing boundary condition to the crack face for a short prism
b	width of member
C	force in concrete in prism
C_L	force in concrete on the left of prism
C_R	force in concrete on the right of prism
CWD_p	crack width density with primary cracks
CWD_s	crack width density with secondary cracks
c	constant of integration
d_r	distance of tension reinforcement from tension face of member
$d\delta/dx$	slip–strain
E_c	concrete modulus; concrete elastic modulus
$E_{c\text{-}mi}$	effective concrete modulus to allow for strain hardening in fibre concrete
E_r	reinforcement modulus
$E_{r\text{-}sh}$	reinforcement modulus whilst strain hardening
E_{SH}	strain hardening modulus of concrete
FI	full interaction
f_{ct}	tensile strength of concrete
f_{fib}	equivalent concrete stress due to fibres spanning crack
f_i	fibre stress at zero crack width
fn	function
f_{pc}	fibre stress at transition point of m_i to m_{i+1}
f_{SH}	stress at which micro-cracks start to develop in fibre concrete
G_f	area of τ/δ
IC	intermediate crack
k_e	bond-slip stiffness; τ/δ
k_{pi}	crack opening stiffness
$k_{pi}S_p/2$	effective axial rigidity
L	span of beam; bonded length
L_{crt}	prism length required to develop *IC* debonding resistance; $S_{p\text{-}IC}$
L_{db}	debonded length
L_e	element length
LHS	left-hand side
L_{per}	perimeter length of bonded region; circumference of bar
L_{PI}	length of which there is *PI*

M	moment
M_{cr}	moment to cause a flexural crack in an uncracked *FI* region
m_i	rate of change of fibre stress with Δ
NI	no interaction
n_{fi}	*FI* modular ratio; E_r/E_c
n_{pi}	*PI* modular ratio
P	force
P_c	force in concrete at crack face
$P_{c\text{-}fi}$	force in concrete in *FI* region
P_{fib}	force in fibre across crack
P_{IC}	*IC* debonding resistance
PI	partial interaction
P_{prm}	total force applied to prism at crack; P_r+P_{fib}
P_r	force in reinforcement at crack face
$P_{r\text{-}cr\text{-}fi}$	P_r to cause a crack in the *FI* region
$P_{r\text{-}cr\text{-}p}$	P_r to cause a primary crack
$P_{r\text{-}cr\text{-}s}$	P_r to cause a secondary crack
$P_{r\text{-}fi}$	force in reinforcement in *FI* region
$P_{r\text{-}inf}$	P_r for an infinite prism
P_y	force in reinforcement at start of yield
P/Δ	at crack, force-in-reinforcement/slip-of-reinforcement at crack face; crack opening stiffness
R	force in reinforcement in prism
RC	reinforced concrete
R_L	force in reinforcement on the left of prism
R_R	force in reinforcement on the right of prism
RHS	right-hand side
S_p	primary crack spacing
$S_{p\text{-}IC}$	maximum primary crack spacing
S_s	crack spacing after secondary cracks developed
w	crack width; $2\Delta_{cf}$
w_{inf}	crack width in infinite prism analysis
α	bond stiffness coefficient
α_c	coefficient of thermal expansion of concrete
α_r	coefficient of thermal expansion of reinforcement
β	tension stiffening axial stiffness parameter
Δ	half-crack width
Δ_{cf}	slip at crack -face
Δ_{inf}	Δ for infinite prism
Δ_{max}	slip at crack face beyond which fibre stresses are zero
Δ_o	Δ offset due to residual strains
Δ_p	Δ when there are adjacent primary cracks
Δ_s	Δ when there are adjacent secondary cracks
ΔT_c	change in concrete temperature
ΔT_r	change in reinforcement temperature
$\Delta\delta$	change in slip in an element
$\Delta\sigma_c$	change in concrete stress in an element

$\Delta\sigma_r$	change in reinforcement stress in an element
δ	interface slip
δ_{max}	interface slip beyond which the shear stress is zero
δ_1	interface slip at τ_{max}
δ_2	interface slip at end of plateau
δ_3	interface slip at commencement of friction
δ'	slip–strain
δ'_{IC}	slip–strain at crack when P_{IC} fully developed
ε	strain
ε_c	concrete strain
$\varepsilon_{c\text{-}B}$	strain in concrete due to bond transfer
$\varepsilon_{c\text{-}f}$	strain in concrete due to fibre forces
$\varepsilon_{c\text{-}sh}$	concrete strain due to concrete shrinkage
ε_{exp}	residual strain due to concrete expansion
ε_{inel}	strain offset due to strain hardening of fibre concrete
ε_{IC}	reinforcement strain at P_{IC} in debonded region
ε_r	reinforcement strain
$\varepsilon_{r\text{-}p}$	strain in reinforcement at primary crack face
$\varepsilon_{r\text{-}s}$	strain in reinforcement at secondary crack face
$\varepsilon_{r\text{-}sh}$	reinforcement strain due to concrete shrinkage
ε_{sh}	shrinkage strain
ε_{temp}	residual strains due to thermal deformations
ε_{time}	residual strain due to material deformations with time; $\varepsilon_{sh}+\varepsilon_{exp}+\varepsilon_{temp}$
ϕ	diameter
λ	tension stiffening parameter
σ	stress
σ_c	concrete stress
$\sigma_{c\text{-}B}$	stress in concrete due to bond transfer
$\sigma_{c\text{-}f}$	stress in concrete due to fibres
$\sigma_{c\text{-}IC}$	maximum stress in concrete when P_{IC} attained
$\sigma_{c\text{-}max}$	maximum stress in concrete at *FI* boundary
σ_r	reinforcement stress
τ	interface shear stress
τ/δ	bond-slip properties
τ_{max}	maximum interface shear stress

REFERENCES AND DEVELOPMENT

Choi, C.K. and Cheung, S.H., 1996. Tension-stiffening model for planar reinforced concrete members. *Computers & Structures*, *59*(1), 179–190.

Gupta, A.K. and Maestrini, S.R., 1990. Tension-stiffness model for reinforced concrete bars. *Journal of Structural Engineering*, *116*(3), 769–790.

Haskett, M., Oehlers, D.J. and Ali, M.M., 2008. Local and global bond characteristics of steel reinforcing bars. *Engineering Structures*, *30*(2), 376–383.

International Federation for Structural Concrete (FIB), 2013. FIB model code for concrete structures 2010. Lausanne, Switzerland

Knight, D., Visintin, P., Oehlers, D.J. and Ali, M., 2014. Short-term partial-interaction behavior of RC beams with prestressed FRP and steel. *Journal of Composites for Construction ASCE*, *18*(1), 04013029.

Knight, D., Visintin, P., Oehlers, D.J. and Jumaat, M.Z., 2013. Incorporating residual strains in the flexural rigidity of RC members with varying degrees of prestress and cracking. *Advances in Structural Engineering*, *16*(10), 1701–1718.

Marti, P., Alvarez, M., Kaufmann, W. and Sigrist, V., 1998. Tension chord model for structural concrete. *Structural Engineering International*, *8*(4), 287–298.

Muhamad, R., Ali, M., Oehlers, D.J. and Griffith, A.H., 2012. The tension stiffening mechanism in reinforced concrete prisms. *Advances in Structural Engineering*, *15*(12), 20532069.

Muhamad, R., Ali, M., Oehlers, D.J. and Sheikh, A.H., 2011. Load-slip relationship of tension reinforcement in reinforced concrete members. *Engineering Structures*, *33*, 1098–1106.

Sturm, A.B., Visintin, P. and Oehlers, D.J., 2018a. Time-dependent serviceability behavior of reinforced concrete beams: Partial interaction tension stiffening mechanics. *Structural Concrete*, *19*, 508–523.

Sturm, A.B., Visintin, P., Oehlers, D.J. and Seracino, R., 2018b. Time-dependent tension-stiffening mechanics of fibre-reinforced and ultra-high-performance fibre-reinforced concrete. *ASCE Journal of Structural Engineering*, *144*(8), 04018122.

Sturm, A.B. and Visintin, P., 2019. Local bond slip behavior of steel reinforcing bars embedded in ultra high performance fibre reinforced concrete. *Structural Concrete*, *20*(1), 108–122.

Vaculik, J., Sturm, A.B., Visintin, P. and Griffith, M.C., 2018. Modelling FRP-to-substrate joints using the bilinear bond-slip rule with allowance for friction – full-range analytical solutions for long and short bonded lengths. *International Journal of Solids and Structures*, *135*, 245–260.

Visintin, P. and Oehlers, D.J., 2017. Fundamental mechanics that govern the flexural behaviour of reinforced concrete beams with fibre-reinforced concrete. *Advances in Structural Engineering*, *21*(4), 1–15.

Visintin, P., Oehlers, D.J. and Haskett, M., 2013. Partial-interaction time dependent behaviour of reinforced concrete beams. *Engineering Structures*, *49*, 408–420.

Volkersen, O., 1938. Die Nietkraftverteilung in zugbeanspruchten Nietverbindungen mit konstanten Laschenquerschnitten. *Luftfahrtfor schung*, *15*, 41–47.

Wu, Z., Yoshikawa, H. and Tanabe, T.A., 1991. Tension stiffness model for cracked reinforced concrete. *Journal of Structural Engineering*, *117*(3), 715–732.

Yuan, H., Teng, J.G., Seracino, R. and Wu, Z.S., 2004. Full range behaviour of FRP-to-concrete bonded joints. *Engineering Structures*, *26*(5), 543–691.

Chapter 5

Concrete compressive softening mechanism

Testing concrete from the same batch in specimens of the same shape and size can show a wide range of properties, such as those shown in Figure 5.1, where σ_c is the axial stress in the specimen and ε_c is the axial strain. Of importance to engineers and researchers is the concrete peak strength f_c. However, of much greater importance, as will be illustrated in Chapters 6 and 7, is the concrete ductility. Concrete ductility is not easy to quantify, but will, for the time being, be considered to depend to a large extent on the area under the stress/strain curve, such as under *O–B–C–D–E* for the unconfined concrete; under *O–G–H–L* for concrete encased in stirrups; or under *O–B–G–K–I–J* for the FRP-wrapped concrete or concrete encased in steel tubes. It can be seen in Figure 5.1 that there is a very wide range of stress–strain shapes and sizes. The question is what do we use in design and analysis, which is the subject of this chapter. The apparently incredibly complex behaviour of concrete under compression, as illustrated in Figure 5.1, is simply due to the formation of cracks that start to develop at the early stages of compressive loading and consequently affect both the ascending and descending branches of the stress–strain relationship.

A typical stress–strain property for unconfined concrete is shown in Figure 5.1. Initially, the material can be considered to be linear–elastic at a modulus E_c over the

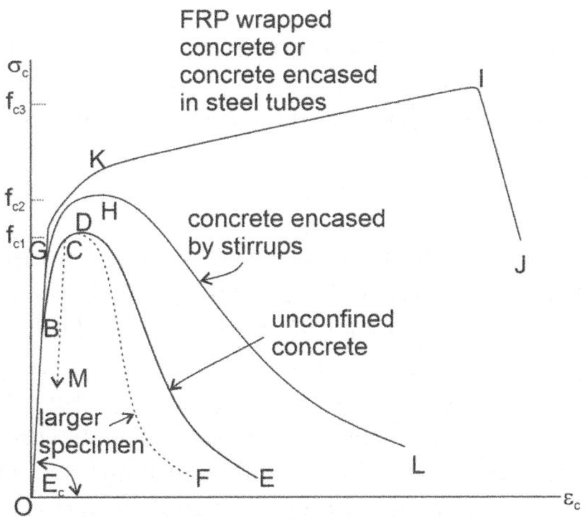

Figure 5.1 Typical concrete compressive stress–strain.

DOI: 10.1201/9781003711407-8

region *O–B*. At Point *B*, micro-cracks begin to exhibit a quantifiable effect such that should unloading occur at Point *C*, the descending branch *C–M* will be offset from the ascending branch. Upon further loading, which causes displacement, at the peak at Point *D*, the micro-cracks coalesce to form a sliding plane that controls the descending branch and makes it dependent on the size and shape of the specimen being tested. For example, if the specimen is increased in length, then the descending branch will decline more rapidly, changing from *D–E* to *D–F*. Should the concrete specimen be encased by a ductile material such as a steel tube or by steel stirrups, this confinement also affects micro-cracking and consequently the peak strength f_c. However, and much more importantly, the ductility has increased considerably from the enclosed area *O–B–C–D–E* to *O–G–K–I–J*. Even confinement using a brittle *FRP* wrap can substantially improve the ductility to *O–G–K–I–J*. It can be seen that the benefits of confinement are simply too significant to be ignored.

In this chapter, the mechanics of concrete softening are considered for members with rectangular cross-sections, after which any differences that might occur in circular sections are explained. The fundamental material behaviour observed in tests is first described. It is then shown how the Poisson ratio can vary with orientation due to interface sliding. Shear friction (SF) theory is then used to determine the softening σ_c/ε_c properties. The simplest problem of axially loaded members is first considered. The theory is then expanded to include the more complex problem of regions where flexure dominates, followed by the even more complex problem of regions where transverse shear dominates. This is then followed by a consideration of the minor differences that occur with circular sections, and alternative methods of analysis are examined. It is shown how passive confinement is beneficial and can be quantified through mechanics, and how it affects both the strength and the ductility of plain and fibre-reinforced concrete. It is also demonstrated how transverse shear is detrimental and reduces both strength and ductility, and importantly, how this reduction can also be quantified through SF theory.

5.1 BASIC FUNDAMENTAL MATERIAL BEHAVIOUR

Because of cracking and micro-cracking, concrete cannot be considered a truly homogeneous material. This is exhibited through the development of sliding planes when concrete softens and in the size dependency of the stress–strain properties during softening.

5.1.1 Sliding plane

Five examples of the concrete softening mechanism are shown in Figures 5.2 and 5.3. Except for the cube in Figure 5.2a, the height was twice the diameter. In Figure 5.2a, a circumferential wedge has formed such that the sliding planes are composed of two cones that are encased by the outer concrete. In Figure 5.2b, only one circumferential wedge has formed in the weaker of the two ends. The circumferential wedge often occurs in specimens where the slenderness ratio, that is, the height-to-width ratio, is less than or equal to two, and also occurs in the cubes in Figures 1.19c and 5.2a. Chapter 3 demonstrates how the SF properties can be extracted from this circumferential wedge mechanism. However, the application of this mechanism provides an upper bound

(a) double pyramid (b) single cone

Figure 5.2 Softening mechanisms: circumferential wedge. (Photos courtesy of Lloyd Dadd.) (a) double pyramid shape; (b) single cone shape.

(a) unconfined (b) confined (c) confined

Figure 5.3 Softening mechanisms: single sliding plane (a) unconfined concrete; (b) and (c) confined concrete. ((a) Courtesy of Lloyd Dadd. (b and c) From Oehlers et al. 2017.)

to dilation and consequently passive confinement, and is therefore unconservative. Furthermore, it rarely, if ever, occurs in columns or beams and so will not be considered further. A single sliding plane is shown in the unconfined specimen in Figure 5.3a and in the passively confined specimens in Figure 5.3b and c. This single-plane mechanism typically occurs in specimens with a slenderness ratio of two or more, where the end platens, as in Figure 1.19, have a reduced effect, and where the angle of the sliding plane, often taken as 26°, can accommodate a sliding plane throughout the length of the specimen. Furthermore, this type of sliding plane is associated with flexural failure, as in Figure 1.7, and shear failure, as in Figure 1.8. Hence, this chapter will deal with single sliding planes.

5.1.2 Size-dependent stress/strain

Consider the axially loaded specimen in Figure 5.4. The concrete core is of length L_{def}, depth D and width B, the elastic modulus is E_c with a material Poisson ratio ν_{mat}. The tube has a thickness t, and let us assume that $t = D$ such that any axial stress in the tube is negligible. Alternatively, let us assume that the interface bond between the tube and the concrete is zero so that the axial stress in the tube is zero. Either way, the tube only resists lateral expansion; that is, it provides passive confinement. A sliding plane forms at an angle α, which is often assumed to be 26°, such that when the interface slip is S, as shown, the longitudinal component of the slip is $S \cos \alpha$ and the transverse component is $S \sin \alpha$.

A load N is applied to the concrete component of the specimen in Figure 5.4, which causes a stress σ_c. The specimen contracts by δ such that the overall or global strain ε_c is δ / L_{def}. Prior to the sliding plane forming, the global strain ε_c is simply that of the material strain ε_{mat}. However, once sliding occurs, the axial deformation δ increases by $S \cos \alpha$, which is an effective strain due to slip ε_{slip} of

$$\varepsilon_{slip} = \frac{S \cos \alpha}{L_{def}} \tag{5.1}$$

such that the global or overall strain is given by

$$\varepsilon_c = \varepsilon_{mat} + \varepsilon_{slip} \tag{5.2}$$

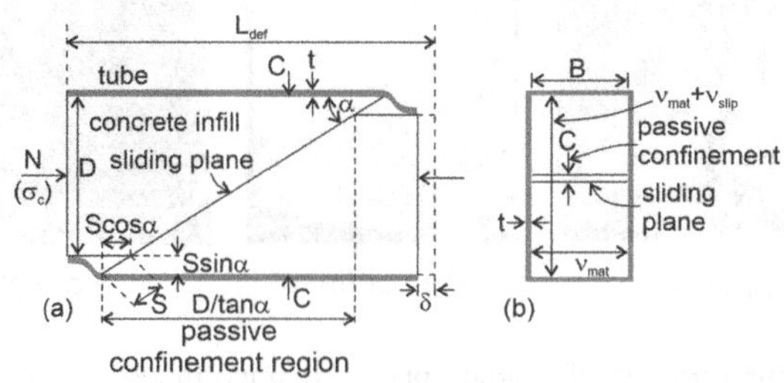

Figure 5.4 Encased axially loaded member (a) passively confined segment elevation; (b) passively confined cross section.

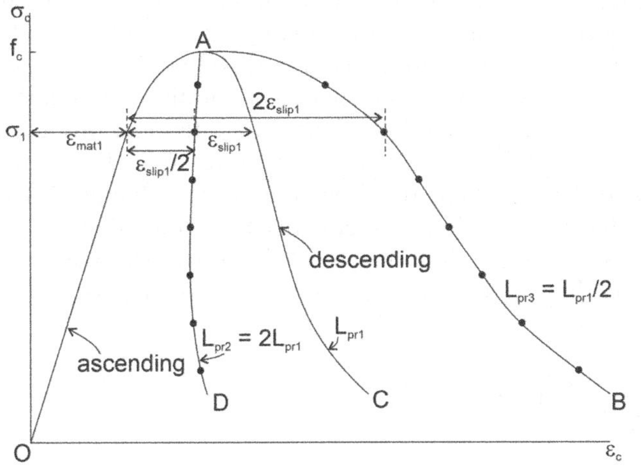

Figure 5.5 Size-dependent stress–strain.

From Eqs. 5.1 and 5.2, it can be seen that the global strain is size-dependent once the sliding plane occurs (Chen et al. 2014, 2016; Visintin et al. 2015a).

Let us assume that when testing the specimen in Figure 5.4, the stress–strain results are represented by *O–A–C* in Figure 5.5, where the length of the prism is labelled as L_{pr1}; a commonly used length is 200 mm. The peak stress at Point *A* is where the sliding plane begins to slip. In the ascending portion of the curve from Point O–*A*, the global strain in Eq. 5.2 is simply due to the material contraction $\varepsilon_{mat,}$ but beyond Point *A*, that is from Point *A–C*, the global strain in Eq. 5.2 comprises both the material strain ε_{mat} and the strain due to interface slip ε_{slip}. For example, at the stress level σ_1, the material strain is ε_{mat1} and the effective strain due to slip is ε_{slip1}, and this applies at all stress levels.

Let us double the length of the prism from L_{pr1}, which applies to O–*A–C* in Figure 5.5, to $2L_{pr1}$. At the stress level σ_1, the material strain ε_{mat} remains the same, but the effective strain due to slip would have halved. Instead of an effective strain due to slip of $S \cos \alpha / L_{pr1}$, we now have $S \cos \alpha / 2L_{pr1}$. It can be seen that the descending branch can be constructed for any prism length L_{pr2} from the test results of one prism of length _Lpr_1, such that for a stress level σ_n, the overall strain in a prism of length L_{pr2}, denoted as ε_{cgl2}, is given by (Chen et al. 2014, 2016)

$$\left(\varepsilon_{cgl2}\right)_n = \left[\left(\varepsilon_{cgl1}\right)_n - \left(\varepsilon_{mat}\right)_n\right]\frac{L_{pr1}}{L_{pr2}} + \left(\varepsilon_{mat}\right)_n \tag{5.3}$$

5.1.3 Poisson ratio

Prior to the sliding plane forming in Figure 5.4, the concrete material's Poisson ratio ν_{mat}, which is the lateral strain induced by a longitudinal strain as a proportion of the longitudinal strain, is about

$$\nu_{mat} = 0.2 \tag{5.4}$$

As shown in Figure 5.4a, when the longitudinal axial load N is applied, the concrete would expand laterally, in this case vertically, by a strain of 20% of the longitudinal strain, assuming the concrete is unrestrained in this direction.

When a slip S occurs in Figure 5.4a, there is an additional lateral expansion represented as $S \sin \alpha$, and hence, an additional effective lateral strain of $S \sin \alpha/D$, where D is the lateral dimension of the concrete prism (Chen et al. 2014, 2016). This additional lateral strain due to slip is accompanied by an additional longitudinal contraction due to slip of $S \cos \alpha$; this is in an additional effective longitudinal strain due to slip of $S \cos \alpha/L_{def}$, where L_{def} is the longitudinal length of the prism over which the strain is being determined. Hence, the Poisson ratio due to interface slip ν_{slip} is given by

$$\nu_{slip} = \frac{\dfrac{S \sin \alpha}{D}}{\dfrac{S \cos \alpha}{L_{def}}} = \frac{L_{def}}{D} \tan \alpha \tag{5.5}$$

The longitudinal length of the sliding plane in Figure 5.4 is $D/\tan \alpha$ as shown. It is this longitudinal region that is passively restrained by interface slip and, hence, of most interest. Replacing L_{def} in Eq. 5.5 with the length of the passively restrained region $D/\tan \alpha$ gives the Poisson ratio over the passively restrained sliding region as

$$\nu_{slip} = 1 \tag{5.6}$$

From the above discussion, the encased tube in Figure 5.4b has, after sliding commences, a vertical Poisson effect of ν_{mat+} ν_{slip} and a horizontal Poisson effect of ν_{mat}. Hence, the Poisson ratio varies in direction or orientation.

It can be seen that ν_{slip}, equal to 1 from Eq. 5.6, is much greater than the material value ν_{mat}, equal to 0.2 from Eq. 5.4; that is, the lateral expansion due to slip is much greater than that due to material expansion. It will be shown later, through SF analyses, that the effect of ν_{mat}, is to allow a gradual buildup of the passive constraint and hence to ensure that the peak strength increases due to passive confinement. However, it will also be shown through SF analyses that it is the much greater Poisson effect of ν_{slip} that brings out the greatest benefits of passive confinement, as can be seen by the very large distortion of the steel tube in Figure 1.20. It can be seen in Figure 5.4a that the lateral passive confinement force C induced by the slip has a vertical component and consequently has a component normal to the sliding plane; hence, it is of great benefit in enhancing the SF resistance to sliding.

5.1.4 Angle of sliding plane α

It has been suggested in Section 1.1 that the angle of the sliding plane α is reasonably close to 26°. However, this angle does vary and can be determined from the material SF properties, as in Section 3.1.4 and in Figure 1.4, and in particular the shear stress to start sliding V_u in that figure. Analysis of the SF properties by Chen et al. (2015) showed that V_u could be described as a quadratic, which is difficult to manipulate but suitable for a simple iterative numerical approach. However, for ease of explanation, it can be simplified to the following form:

$$V_u = C_1\sigma_N + C_2 f_{co} \tag{5.7}$$

where, for conventional concrete, the shear capacity at zero slip coefficient C_1 is approximately 0.86 and C_2 is approximately 0.16, and where f_{co} is the unconfined compressive strength of the concrete and σ_N the compressive stress normal to the sliding plane.

Consider the free-body diagrams of a prism in Figure 5.6 that are subjected to various stress resultants. The prism has a depth of D, is of any dimension into the sheet which, for ease of explanation, may be considered as 1 mm, and the sliding plane is at an angle α as shown. The prism in Figure 5.6a is subjected to an axial load N which, divided by the cross-sectional area of the prism, induces an axial compressive stress σ_c shown in brackets. The axial force N can be resolved along the sliding plane to give a shear force $N \cos \alpha$, which, divided by the cross-sectional area of the sliding plane, induces a shear stress τ_N shown in brackets, and a force normal to the sliding plane of

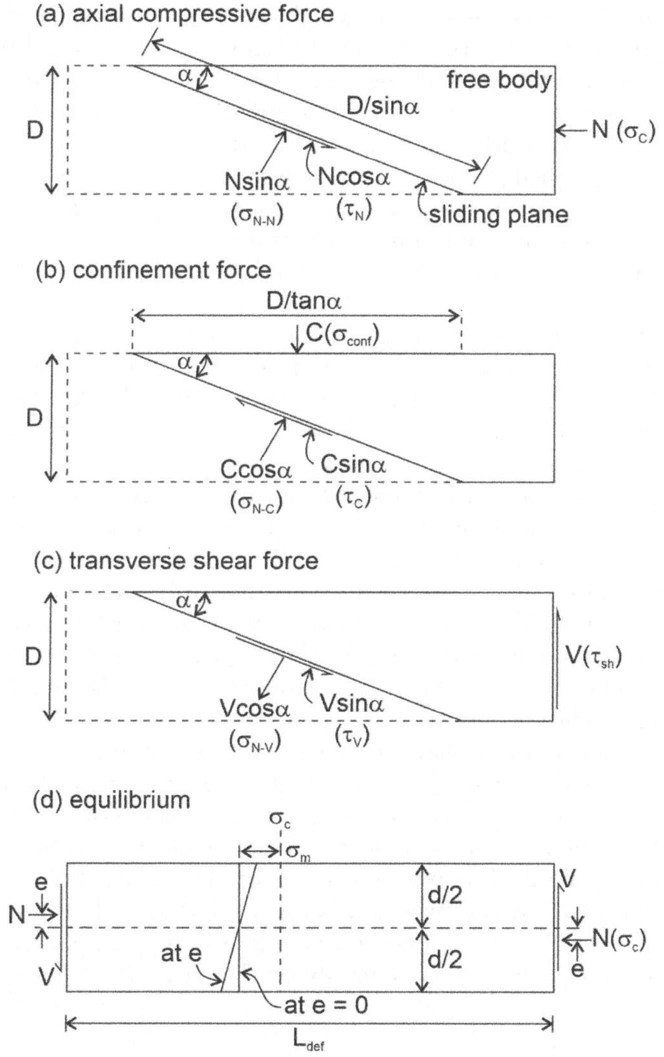

Figure 5.6 Free-body diagrams of concrete prism (a) axial compressive force; (b) confinement force; (c) transverse shear force; (d) equilibrium.

$N \sin \alpha$ that induces a normal stress to the sliding plane of $\sigma_{N\text{-}N}$, also shown in brackets. Please bear in mind that $\sigma_{N\text{-}N}$ and τ_N govern V_u in the SF properties in Figure 1.4 and Section 3.1.4.

The same prism is now subjected to a confining force in Figure 5.6b of magnitude C over the width of the prism that encompasses the sliding plane, which is over the length $D/\tan \alpha$ as shown. The resultants along the sliding plane are the shear force $C \sin \alpha$ (stress τ_C) and the normal force $C \cos \alpha$ (stress $\sigma_{N\text{-}C}$). The shear force $C \sin \alpha$ acts in the opposite direction to the shear force $N \cos \alpha$ in Figure 5.6a and hence reduces the shear along the sliding plane, consequently allowing for a larger axial force N (that is, σ_c) at failure. The normal stress due to confinement $\sigma_{N\text{-}C}$ in Figure 5.6b is compressive and will act in addition to $\sigma_{N\text{-}N}$ in Figure 5.6a to enhance the shear frictional resistance V_u.

The prisms in Figure 5.6a and b have both translational equilibrium and rotational equilibrium. The same prism is now subjected to a transverse shear force V in Figure 5.6c. It can be seen that the prism is still in translational equilibrium but is no longer in rotational equilibrium. This means that, when combined with the axial forces N in Figure 5.6a, these axial forces have to be slightly eccentric, as in Figure 5.6d, to provide a moment of magnitude $N2e$ to counteract that induced by the transverse shear V that is VL_{def}. As explained previously, in research to find solutions that can be used in practice and which can also be visualised, it is often necessary to make simplifications. In this case, rotational equilibrium due to the transverse shear VL_{def} will be ignored. It is suggested that this is a minor effect for the following mechanical interactions.

1. The *SF* properties of the sliding plane are illustrated in Figure 3.4, where it can be seen that the SF capacities V_u depend on σ_N. It can also be seen that the *SF* capacities for a specific σ_N are maintained over a small slip Δ_{SF}, that is, it does not suddenly drop to zero when the capacity is reached but maintains its capacity over a small slip and then gradually reduces.
2. Failure of the inclined sliding planes in Figure 5.6a–c consists of a rigid body displacement of the member on either side of the sliding plane; that is, slip across the sliding plane is at a constant Δ_{SF}. Hence, should a part of the sliding plane reach its capacity V_u, the shear V_u at $\Delta_{SF} = 0$ is maintained, and failure occurs when all parts of the sliding plane reach their capacities; that is, it fails as a whole and not when an individual part fails along the sliding plane. Consequently, it depends on the mean axial stress along the sliding plane and not on the variation of the axial stress along the sliding plane.
3. Now, the eccentric axial loads N in Figure 5.6d cause a mean axial stress σ_m (labelled $e=0$) and a variation σ_c (labelled e). Hence, the variation in the axial stress σ_c (labelled e) does not change the mean of the axial stresses σ_m, to state the obvious. Consequently, by ignoring the moment $N2e$ required for rotational equilibrium, we are dealing with mean stresses over the member of depth D and not the variation in stress. This is fine because it has been shown in Point (2) above that the sliding plane fails as a whole and not at individual parts along the sliding plane; thus, its capacity depends mainly on the mean axial stress and not on the variation.

Consequently, it would appear that ignoring the required eccentricity e of N shown in Figure 5.6d will not be significant in quantifying the failure, such that the free body diagram in Figure 5.6c can be used to simplify the analysis. It will be shown that this leads to relatively simple solutions. No doubt it can be improved over time.

The prism in Figure 5.6c is subjected to a transverse shear force V (shear stress τ_{sh} along the transverse plane). The resultant along the sliding plane gives the shear force $V \sin \alpha$ (stress τ_V) and normal force $V \cos \alpha$ (stress $\sigma_{N\text{-}V}$). It can be seen that the normal stress $\sigma_{N\text{-}V}$ is tensile; that is, it reduces the confinement along the sliding plane, thereby reducing the SF shear strength along the sliding plane and consequently the axial strength of the concrete. Furthermore, the shear force $V \sin \alpha$ is in the same direction as that in Figure 5.6a, which will reduce the axial strength.

Let us assume that a prism is subjected to all three stress resultants in Figure 5.6a–c, namely an axial stress σ_c, a confining stress σ_{conf} and a shear stress τ_{sh}. From the equilibrium of the forces induced by these stresses, the normal confinement along the sliding plane is given by

$$\sigma_N = \sigma_{N\text{-}N} + \sigma_{N\text{-}C} - \sigma_{N\text{-}V} = \sigma_c \sin^2 \alpha + \sigma_{conf} \cos^2 \alpha - \tau_{sh} \sin \alpha \cos \alpha \tag{5.8}$$

where the first component $\sigma_{N\text{-}N}$ is induced by the axial stress σ_c in Figure 5.6a and is equal to $\sigma_c \sin^2 \alpha$, the second component $\sigma_{N\text{-}C}$ is induced by the confining stress σ_{conf} in Figure 5.6b and is equal to $\sigma_{conf} \cos^2 \alpha$, and the third component $\sigma_{N\text{-}V}$ is induced by the shear component τ_{sh} in Figure 5.6c and is equal to $\tau_{sh} \sin \alpha \cos \alpha$. Furthermore, the shear stress along the sliding plane is given by

$$\tau_{SP} = \tau_N - \tau_C + \tau_V = \sigma_c \sin \alpha \cos \alpha - \sigma_{conf} \cos \alpha + \tau_{sh} \sin^2 \alpha \tag{5.9}$$

where τ_N is due to N and is equal to $\sigma_c \sin \alpha \cos \alpha$, τ_c is due to confinement C and is equal to $\sigma_{conf} \cos \alpha$ and τ_V is due to V and is equal to $\tau_{sh} \sin^2 \alpha$.

Equating Eq. 5.9, which is derived from forces, with Eq. 5.7, which represents the material strengths, that is, applying the equation when sliding is about to occur, and substituting for σ_N from Eq. 5.8 gives the following dependence of the axial strength on α:

$$f_{c-\alpha} = \frac{C_2 f_{co} + \sigma_{conf} \cos \alpha (1 + C_1) - \tau_{sh} \left(\sin^2 \alpha + C_1 \sin \alpha \cos \alpha \right)}{\sin \alpha \cos \alpha - C_1 \sin^2 \alpha} \tag{5.10}$$

where it can be seen that the effect of lateral confinement is given by the second term in the numerator and the effect of transverse shear stresses by the third term.

Purely as an example to illustrate the dependence on α, Eq. 5.10 has been plotted for a laterally unconfined prism with $\sigma_{conf} = 0$ in Figure 5.7, using the SF material properties for conventional concrete without fibres from Chen et al. (2015). The minimum value of $f_{c\text{-}\alpha}$ is approximately 27 MPa, where α is about 26°. As shown on the sliding plane, it has already been established in Eq. 5.8 that σ_N is equal to $\sigma_c \sin^2 \alpha$ normal confinement σ_N of 5.2 MPa. Furthermore, as demonstrated on the sliding plane for a confinement of σ_{conf}, it has also been shown in Eq. 5.8 that the normal stress σ_N is equal to $\sigma_{conf} \cos^2 \alpha$. To achieve a σ_N of 5.2 MPa would require a σ_{conf} of 6.4 MPa. If we apply a confinement stress of 25% of this value, which is 1.6 MPa, then the variation of strength with α is

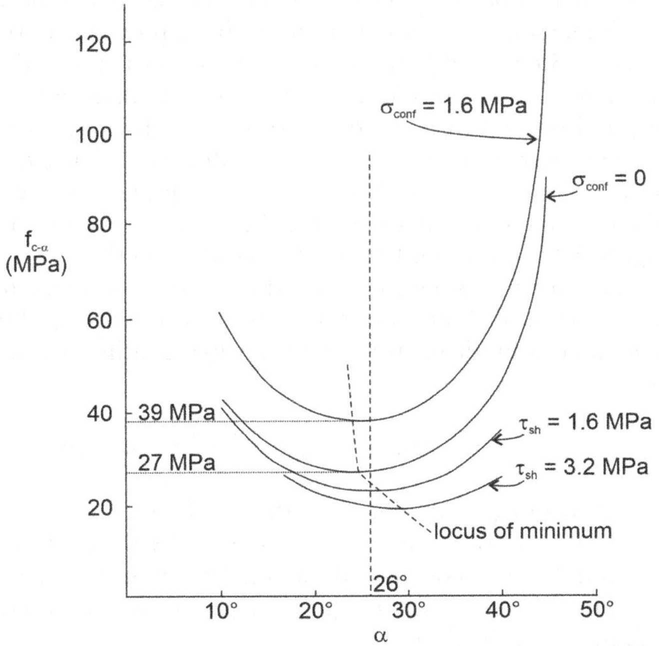

Figure 5.7 Variation of compressive strength with α.

also shown in Figure 5.7, where the strength has increased to 39 MPa. It can be seen that only a small additional lateral confining stress of 1.6 *MPa* can produce substantial increases in axial strength, increasing from 27 to 39 MPa. This is in addition to the increase in ductility, which will be explained later. It can also be seen in Figure 5.7 that α of 26° is a good estimate and that it does not change much with confinement. However, this may not necessarily be the case for other types of concrete, such as high-strength concretes or those with fibres.

The effect of applying a transverse shear stress τ_{sh} of 1.6 and 3.2 MPa to the unconfined prism using Eq. 5.10 is also shown in Figure 5.7. It can be seen through the locus of the minima that a transverse shear stress not only reduces the strength of the concrete but also increases the angle of the sliding plane α.

It may be worth emphasising that the first term in Eq. 5.10, which is shown in the following equation, is the concrete compressive strength as might be obtained from conventional tests:

$$f_{c-\alpha} = \frac{C_2 f_{co}}{\sin\alpha\ \cos\alpha - C_1 \sin^2\alpha} \tag{5.11}$$

The second term in Eq. 5.10 is the increase in the axial strength due to confinement and can be written as

$$Q_{conf} = \frac{\sigma_{conf} \cos\alpha(1 + C_1)}{\sin\alpha\ \cos\alpha - C_1 \sin^2\alpha} \tag{5.12}$$

The third term is the reduction in axial strength due to transverse or lateral shear forces and can be written as

$$Q_{sh} = \frac{\tau_{sh}\left(\sin^2\alpha + C_1 \sin\alpha\cos\alpha\right)}{\sin\alpha\ \cos\alpha - C_1\sin^2\alpha} \tag{5.13}$$

If, for example, we fix α at 26° and use a value of 0.86 for C_1, then Q_{sh} becomes $2.3\tau_{sh}$, indicating that the transverse shear stresses τ_{sh} reduce the axial strength of the concrete by $2.3\tau_{sh}$. It can be seen that a relatively small transverse shear stress can significantly reduce the compressive strength of the concrete. It is suggested that determining experimentally the compressive strength of concrete cylinders when the cylinders are also subjected to applied transverse shear would be a good avenue for research.

5.1.5 Preliminary mechanics of the softening branch

The following procedures are an introduction to concrete softening, which are elaborated upon in the remainder of this chapter. The approach taken here is similar to that of Ali et al. (2010), Visintin et al. (2015a,b) and Hao et al. (2022a,b), but it is presented in a generalised way by considering all combinations of actions that can be applied to the sliding plane. These references are nevertheless very useful in developing an understanding of the approach.

The shear sliding plane in Figure 5.4 has the SF properties illustrated in Figure 1.4. Prior to the formation of the sliding plane, and as the load N in Figure 5.4 is gradually applied to the prism, the contraction is due to the material contraction of the concrete, such that it follows the material stress–strain ascending branch O–A in Figure 5.5. At Point A, a sliding plane forms, and when this occurs can be derived from the SF properties in Figure 1.4. The strain due to material contraction is shown in Figure 5.5 as ε_{mat}, and the effective strain due to sliding within a prism of length L is shown as ε_{slip}.

The SF properties in Figure 1.4 are shown in Figure 5.8 without the aggregate interlock component w/Δ_{SF} in Figure 1.4. It will be assumed that their effects are an order of magnitude smaller. These material properties apply to a sliding plane such as that in Figure 5.4a. When the shear stress along the sliding plane is some value τ_{sp}, then the axial stress σ_c can be derived from the corresponding component in Eq. 5.9, which is $\tau_{sp}=\tau_N=\sigma_c \sin\alpha \cos\alpha$. Rearranging this equation gives

$$\sigma_c = \frac{\tau_{sp}}{\sin\alpha\cos\alpha} \tag{5.14}$$

which is also shown in the square bracket on the ordinate in Figure 5.8, just to emphasise the fact that the *SF* properties can be converted directly to stresses. It is also worth noting that the subscript *sp* has been used to remind the reader that we are dealing with shear stresses along the sliding plane. For a given value of shear stress τ_{sp} along the sliding plane, the axial stress σ_c can be determined. Furthermore, when the slip along the sliding plane is Δ, then from Figure 5.4a where S is Δ, the axial contraction is $\Delta\cos\alpha$, such that the effective axial strain in a prism of length L due to sliding is

$$\varepsilon_s = \frac{\Delta_{SF}\cos\alpha}{L} \tag{5.15}$$

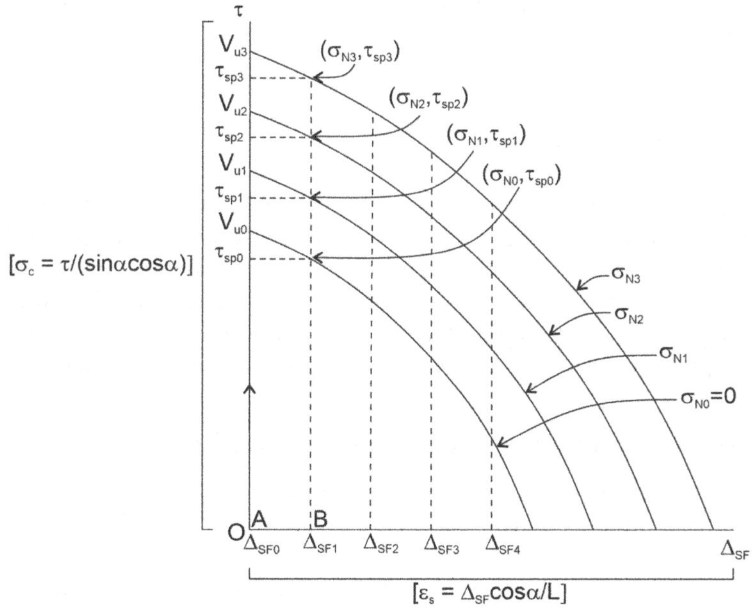

Figure 5.8 Shear-friction properties along compressive sliding plane.

which is also shown in square brackets along the abscissa in Figure 5.8. Hence, the SF properties have been converted to compressive axial stresses and strains for the descending branch of the compressive concrete stress/strain relationship.

We will derive the stress/strain relationship of the descending branch at the five interface slips Δ_{SF0} to Δ_{SF4} in Figure 5.8. One way of visualising the problem is to start at a fixed slip Δ_x and then determine the axial stress σ_{cx} at which further slip commences at this slip Δ_x. Let us illustrate this procedure by starting at Point *A*, where the interface slip Δ_{SF} is zero, that is, at the commencement of sliding, and the interface shear stress τ_{sp} is also zero. Moving upwards from Point *A*, that is, the load path is vertical and upwards, we gradually increase the axial stress σ_{cx} in Figure 5.6a, which will gradually increase both the shear stress τ_{Nx} and the normal stress $\sigma_{N\text{-}Nx}$. These are the combinations of τ_{Nx} and $\sigma_{N\text{-}Nx}$ required for equilibrium along this load path. Along the same load path, we also know the combination of material properties such as V_{u1} and σ_{n1} in Figure 5.8 that are required to cause slip. When the combination of stresses required for equilibrium, that is, τ_{Nx} and $\sigma_{N\text{-}Nx}$, is equal to the combination of material properties V_{u1} and σ_{n1} required for slip to increase, then this is the point at which slip increases. From the shear stress τ_{Nx} at which this occurs, σc can be derived σ_c from Eq. 5.14. The same procedure can be used for the loading path starting at Point *B*, where the material properties such as (σ_{N2}, τ_{sp2}) are shown. In this case, the interface slip is Δ_{SF1}, from which the effective strain due to sliding can be derived from Eq. 5.15.

To further aid in understanding this approach, let us view the problem from a slightly different perspective, which will yield exactly the same solution. The SF properties in Figure 5.8 are plotted in Figure 5.9, but now with σ_N as the abscissa. The plot *A*–*B* in Figure 5.9 represents the material properties at $\Delta_{SF0,}$ such as those in Eq. 5.7. The plot *C*–*D* in Figure 5.9 is at a slip Δ_{SF1}, and so on. These plots represent the material properties.

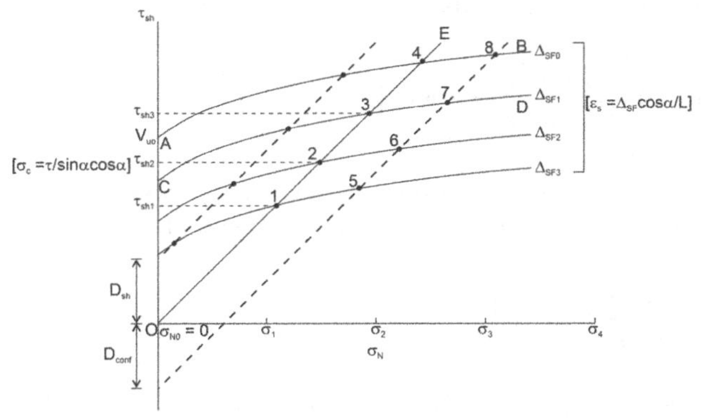

Figure 5.9 Derivation of softening branch.

We now need to plot the variation of τ and σ_N with applied load. Where these plots intersect is where sliding occurs or resumes; that is, they are points on the stress–strain graph.

Starting with the prism in Figure 5.6a, which has only an axial compressive force. From Eq. 5.8,

$$\sigma_N = \sigma_c \sin^2 \alpha \tag{5.16}$$

and from Eq. 5.9,

$$\tau_{sp} = \sigma_c \sin\alpha \, \cos\alpha \tag{5.17}$$

Rearranging the above two equations to eliminate σ_c gives

$$\tau_{sp} = \sigma_N \; cot\alpha \tag{5.18}$$

which is plotted as the line *O–E* in Figure 5.9. Line *O–E* is the loading line, and where this loading line intercepts the material properties at Points 1–4 are the shear stresses required to initiate sliding. For example, at Point 2, the shear stress along the sliding plane τ_{sp2} is the shear stress necessary to cause sliding when the slip is Δ_{SF2}; these values can be converted to an axial stress σ_c and axial strain ε_{SL} as explained previously and shown in square brackets in the figure. This procedure can be used to plot the descending branch *A–C* of the unconfined concrete in Figure 5.5, and if the length of the prism is doubled, the descending branch *A–D*.

Now let us consider a prism subjected to both an axial force and a confinement force, as shown by combining Figure 5.6a and b. From Eq. 5.8,

$$\sigma_{conf} = \sigma_c \sin^2 \alpha +_{conf} \cos^2 \tag{5.19}$$

and from Eq. 5.9

$$\tau_{sp} = \sigma_c \sin\alpha \, \cos\alpha - \sigma_{conf} \cos\alpha \tag{5.20}$$

rearranging the above two equations to eliminate σ_c gives

$$\tau_{sp} = \sigma_N \cot\alpha - \sigma_{conf}\left(\frac{\cos^3\alpha}{\sin\alpha} + \cos\alpha\right) \tag{5.21}$$

Hence, confinement is beneficial as it reduces the shear stress by the second component in Eq. 5.21, D_{conf}, thus the beneficial effect of confinement is

$$D_{conf} = \sigma_{conf}\left(\frac{\cos^3\alpha}{\sin\alpha} + \cos\alpha\right) \tag{5.22}$$

which is shown in Figure 5.9. It can be seen that the intercepts 5–8 have been increased in comparison to the unconfined intercepts 1–4; that is, the axial stresses σ_c at sliding have increased. This procedure can be used to plot the descending branch of confined concrete.

The above procedure can be applied to determine the effect of transverse shear by combining the stress resultants in Figure 5.6a and c. This gives

$$\tau_{sp} = \sigma_N \cot\alpha + \tau_{sh}\left(\sin\alpha\cos\alpha + \sin^2\alpha\right) \tag{5.23}$$

where the transverse shear can be seen to be detrimental by increasing the shear through the second component D_{sh}, which is

$$D_{sh} = \tau_{sh}\left(\sin\alpha\cos\alpha + \sin^2\alpha\right) \tag{5.24}$$

which is also plotted in Figure 5.9, where it can be seen to be detrimental. Of course, all three components in Figure 5.6 can be combined directly.

5.2 AXIALLY LOADED MEMBERS

In this section, the simplest case of concrete softening in an axially loaded member, as shown in Figures 1.17 and 1.19, is considered. This is followed in Sections 5.3 and 5.4 by softening in regions dominated by flexure, as illustrated in Figure 1.7, and then in regions dominated by shear, as depicted in Figures 1.8 and 1.30.

The SF properties that apply to a sliding plane are shown in Figure 5.10, where it can be seen that three variables are required to quantify the properties. These are the shear stress along the sliding plane τ_{SF}; the stress normal to the sliding plane $\sigma_{N\text{-}SF}$; and the slip along the sliding plane Δ_{SF}. The aim is to find the τ_{SF}/Δ_{SF} path, such as *A–B–C* for unconfined concrete or *A–E–F* for passively confined concrete, that the concrete follows whilst softening; from which, as will be shown later, the σ_c/ε_c path during softening can be determined. Hence, the first step is to eliminate $\sigma_{N\text{-}SF}$ from the three material properties, leaving τ_{SF}/Δ_{SF} for the following analyses.

5.2.1 Unconfined concrete stress/strain

The unconfined concrete stress/strain σ_c/ε_c is the simplest case to start with.

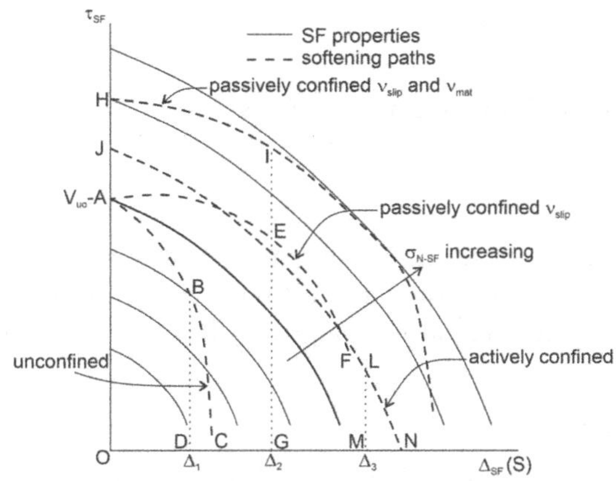

Figure 5.10 Softening path.

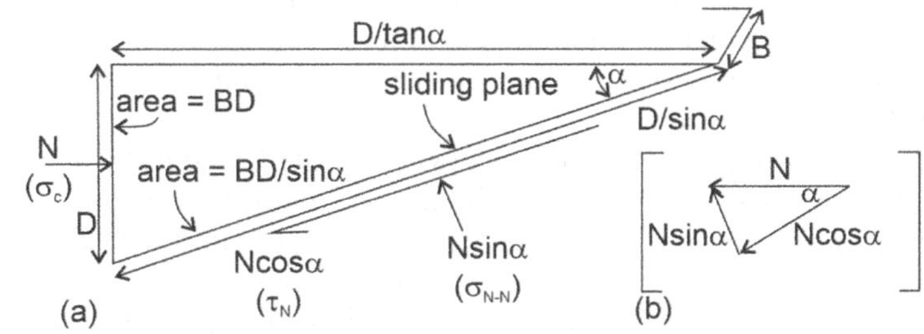

Figure 5.11 Unconfined free-body diagram (a) unconfined free body; (b) resolution of axial force.

5.2.1.1 Equilibrium

Consider the concrete prism in Figure 5.4 without the encasing tube and examine the softening branch of the unconfined concrete. The equilibrium of the free body of an element adjacent to the sliding plane is shown in Figure 5.11a. The free body is only subjected to an axial force N on the left-hand side (LHS), which when divided by the cross-sectional area of the prism BD gives the axial stress:

$$\sigma_c = \frac{N}{BD} \tag{5.25}$$

The components of the axial force on the right-hand side (RHS) along the sliding plane are shown in Figure 5.11b, which comprise a shear component $N \cos \alpha$ along the sliding plane, which when divided by the cross-sectional area of the sliding plane $BD/\sin \alpha$ gives the shear stress τ_N as

$$\tau_N = \frac{N \cos \alpha}{BD/\sin \alpha} = \sigma_c \sin \alpha \cos \alpha \tag{5.26}$$

and a normal component to the sliding plane $N \sin \alpha$, which gives a normal stress $\sigma_{N\text{-}N}$ of

$$\sigma_{N\text{-}N} = \frac{N \sin\alpha}{BD/\sin\alpha} = \sigma_c \sin^2\alpha \tag{5.27}$$

Equating Eqs. 5.26 and 5.27 to eliminate σ_c gives

$$\sigma_{N\text{-}N} = \tau_N \tan\alpha \tag{5.28}$$

This relationship can be visualised as a load path, as it depends on the forces acting on the free body in Figure 5.11. We now have a direct relationship between $\sigma_{N\text{-}N}$ and τ_N in Eq. 5.28, allowing the softening paths in Figure 5.10 to be determined as follows.

5.2.1.2 Numerical solution

It can be seen from Eq. 5.28 that there is a direct relationship between the normal confinement to the sliding plane, that is, the normal stress $\sigma_{N\text{-}N}$, and the shear along the sliding plane τ_N. To understand the relevance of this, consider the SF properties (which could be obtained from Eqs. 3.17 and 3.18) in Figure 5.10; once again, we are ignoring the crack widening component h_{cr} due to aggregate interlock, as in Section 3.1.4, since this is an order of magnitude smaller than the crack widening due to slip, but it could be included if deemed necessary.

To derive the softening stress–strain relationship of the laterally unconfined concrete, that is, σ_c/ε_c, we first need to derive the unconfined softening τ_{SF}/Δ_{SF} relationship sketched as the variation A–B–C in Figure 5.10, bearing in mind that Δ_{SF} in Figure 5.10 is the slip S in Figure 5.4. For a specific value of slip Δ_1 in Figure 5.10, it is simply a question of moving along the vertical load path D–B until Eq. 5.28 is satisfied; that is, Point B is found where the material property $\sigma_{N\text{-}N}$ is equal to $\tau_N \tan\alpha$ from Eq. 5.28. The slip Δ in Figure 5.10 can be varied to construct the Path A–B–C for the unconfined concrete, where Point A is V_{u0}, which is the shear strength for unconfined concrete in Sections 3.1.2.2 and 3.1.5. Having quantified the softening τ_{SF}/Δ_{SF} relationship such as A–B–C in Figure 5.10, this can then be converted into the softening σ_c/ε_c as follows.

5.2.1.3 Laterally unconfined σ_c/ε_c from τ_{SF}/Δ_{SF}

From the variation A–B–C in Figure 5.10 that has now been determined as above, for a specific slip $\Delta_{SF\text{-}x}$, we know the shear stress $\tau_{SF\text{-}x}$. From Eq. 5.26, the axial stress is

$$\sigma_{c\text{-}x} = \frac{\tau_{SF\text{-}x}}{\sin\alpha \cos\alpha} \tag{5.29}$$

which is the axial stress for the specific slip $\Delta_{SF\text{-}x}$ and in which α can be assumed to be 26° or derived from Section 3.1.2.2 (Chen et al. 2015).

The material strain $\varepsilon_{mat\text{-}x}$ for that axial stress $\sigma_{c\text{-}x}$ can now be derived from the strain-based material properties in Section 2.3. Furthermore, the specific slip $\Delta_{SF\text{-}x}$ induces an effective longitudinal strain:

$$\varepsilon_{slip\text{-}x} = \frac{S\cos\alpha}{L_{def}} = \frac{\Delta_{SF\text{-}x}\cos\alpha}{L_{def}} \tag{5.30}$$

where L_{def} is the assumed length of the concrete prism L_{pr1} in Eq. 5.3. The procedure is repeated for a range of $\Delta_{SF\text{-}x}$ from zero to Point A in Figure 5.10 to obtain the variation of stress with strain, after which Eq. 5.3 can be used to plot the unconfined stress–strain relationship for the concrete prism of any length L_{def}.

5.2.1.4 Closed-form solution for plain concrete

The upper variation of Eq. 3.17 gives the SF properties for unconfined plain concrete, which can be used to derive a closed-form solution for the softening of plain concrete; the units are N and mm. Substituting Eq. 5.28 into the SF expressions of Hao et al. (2020) in Eq. 3.17 gives

$$\tau_{SF} = \begin{bmatrix} (1.50\tau_{SF}\tan\alpha + 0.00112f_c - 0.337)\tau_{SF}\tan\alpha \\ -0.000784f_c^2 + 0.0152f_c + 0.556 \end{bmatrix}\Delta_{SF} + 0.105f_c \tag{5.31}$$

which can be used to plot the unconfined softening path A–B–C in Figure 5.10.

As an example, for a concrete strength f_c of 50 MPa and assuming α is 26°, and substituting into Eq. 5.31 gives

$$\tau_{SF} = \left[0.357\tau_{SF}^2 - 0.137\tau_{SF} - 0.644\right]\Delta_{SF} + 5.25 \tag{5.32}$$

For a specific slip Δ_{SF} of 0.1 mm, Eq. 5.32 can be rearranged into a quadratic to give a value of τ_{SF} of 3.26 MPa. From this, using Eq. 5.26, the axial stress σ_c is equal to 8.27 MPa. From Eq. 5.30, the effective strain due to slip ε_{slip} is 449×10^{-6}, assuming the length of the prism is 200 mm.

5.2.1.5 Summary of procedure for unconfined concrete

First, construct the τ_{SF}/Δ_{SF} relationship of the unconfined softening concrete:

Such as A–B–C in Figure 5.10. To do this, at specific values of Δ_{SF} and loading vertically, determine where the fundamental equation:

$$\sigma_{N\text{-}N} = \tau_N \tan\alpha \qquad \text{(from 5.28)}$$

in which $\sigma_{N\text{-}N}$ is the normal stress on the sliding plane due to the applied axial load and τ_N is the shear stress along the sliding plane due to the applied axial load, occurs.

Now determine the stress–strain relationship of the softening concrete:

From the τ_{SF}/Δ_{SF} relationship derived above, for each specific point $\tau_{SF\text{-}x}$ and $\Delta_{SF\text{-}x}$ on this relationship:

$$\sigma_{c\text{-}x} = \tau_{SF\text{-}x} / (\sin\alpha\cos\alpha) \qquad \text{(from 5.29)}$$

$$\varepsilon_{slip\text{-}x} = \Delta_{sh\text{-}x}\cos\alpha / L_{def} \qquad \text{(from 5.30)}$$

where $\sigma_{c\text{-}x}$ is the concrete axial stress and $\varepsilon_{slip\text{-}x}$ is the effective axial strain due to slip for a specific slip $\Delta_{SF\text{-}x}$.

Then determine the global stress–strain relationship:

The slip-based softening branch derived above can now be added to the strain-based material property *O–A* in Figure 5.5 from Chapter 2, to give *A–C* in Figure 5.5, where L_{pr1} is equal to L_{def} above. From this, the global stress/strain relationship for any length of prism can be derived from Eq. 5.3.

Alternative closed-form solution:

An alternative way of solving the problem is to substitute the above fundamental Eq. 5.29 into the equation of the *SF* material properties and solve for τ_{SF} at specific values of Δ_{SF} as in Section 5.2.1.4.

5.2.2 Actively confined concrete stress–strain

Having dealt with unconfined concrete above, let us now move on to laterally confined concrete where the lateral confinement is held constant; that is, we are dealing with laterally actively confined concrete.

5.2.2.1 Equilibrium

A lateral confinement force C has been added to the free-body diagram in Figure 5.11 as shown in Figure 5.12a. The confinement force is held constant; hence, the concrete is actively confined. The components of the confinement force acting on the sliding plane are shown in Figure 5.12b, where C cos α is the normal component to the sliding plane, and when divided by the cross-sectional area of the sliding plane *BD*/sin α gives the normal stress $\sigma_{N\text{-}C}$; and where the shear component C sin α, also divided by the area of the sliding plane, gives a shear stress τ_C.

It can be seen in Figure 5.12a that the confinement force reduces the shear along the sliding plane, such that the resulting shear stress τ^* is given by

$$\tau^* = \tau_N - \tau_C \tag{5.33}$$

and also increases the normal confinement force, such that the resulting normal force $\sigma_N^*$is given by

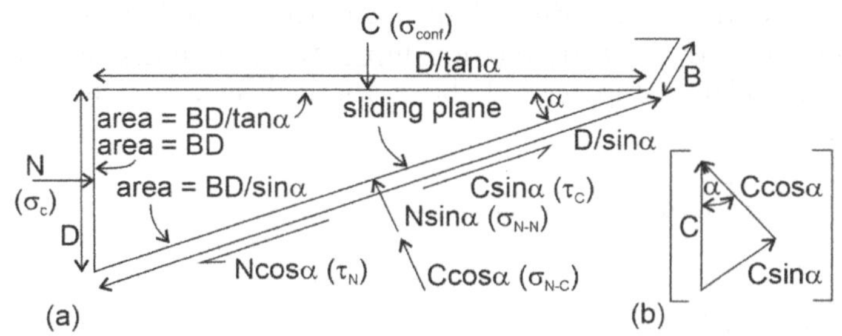

Figure 5.12 Confined free-body diagram (a) confined free body diagram; (b) resolution of confinement force.

$$\sigma_N^* = \sigma_{N\text{-}N} + \sigma_{N\text{-}C} \tag{5.34}$$

Both of these confinement effects in Eqs. 5.33 and 5.34 improve the concrete's ability to resist shear through SF.

Inserting τ_N, that is, the shear induced by the axial force, from Eq. 5.26 into Eq. 5.33 gives

$$\sigma_c = \frac{\tau^* + \tau_C}{\sin\alpha\cos\alpha} \tag{5.35}$$

and similarly, inserting $\sigma_{N\text{-}N}$ from Eq. 5.27 into Eq. 5.34 gives

$$\sigma_c = \frac{\sigma_N^* - \sigma_{N\text{-}C}}{\sin^2\alpha} \tag{5.36}$$

Eliminating σ_c from Eqs. 5.35 and 5.36 gives the resulting normal force as

$$\sigma_N^* = \tau^* \tan\alpha + \tau_C \tan\alpha + \sigma_{N\text{-}C} = \tau^* \tan\alpha + K_C \tag{5.37}$$

From Eq. 5.37, it can be seen that the second and third terms on the *RHS* of Eq. 5.37, that is $\tau_C \tan\alpha$ and $\sigma_{N\text{-}C}$ respectively, are the benefits due to confinement, which will be referred to as the active confinement coefficient:

$$K_C = \tau_C \tan\alpha + \sigma_{N\text{-}C} \tag{5.38}$$

where the first term on the *RHS* is the increase in the normal stress or the confinement normal to the sliding plane due to the shear stress induced by the lateral confinement, and the second term is this normal stress directly induced by the lateral confinement. For active confinement, as C is constant, K_C is constant.

5.2.2.2 Numerical solution

As with the unconfined concrete in Section 5.21, it is a matter of following the vertical path such as *M–L* in Figure 5.10 until, from Eq. 5.37, the normal confining stress $\sigma_N^*$is equal to the shear stress component ($\tau^* \tan\alpha + K_C$). This will yield τ^* and, from Eq. 5.33, τ_N can be determined as τ_C is known for a specific C. Consequently, from Eq. 5.26, σ_c can be derived from the now known value of τ_N.

5.2.2.3 Summary of procedure for actively confined concrete

Using the same steps as in Section 5.2.1.5.

First, construct the τ_{SF}/Δ_{SF} relationship of the actively confined softening concrete:

Such as *J–L–N* in Figure 5.10. To do this, at specific values of $\Delta_{SF\text{-}x}$, and loading vertically, determine where the following fundamental equation applies. From Eq. 5.37,

$$\sigma_N^* = \tau^* \tan\alpha + K_C \tag{5.39}$$

where from Eq. 5.38,

$$K_C = \tau_C \tan\alpha + \sigma_{N\text{-}C} \tag{5.40}$$

and in which τ_C can be derived from Eq. 5.9 as

$$\tau_C = \sigma_{conf} \cos\alpha \tag{5.41}$$

and $\sigma_{N\text{-}C}$ from Eq. 5.8 as

$$\sigma_{N\text{-}C} = \sigma_{conf} \cos^2\alpha \tag{5.42}$$

The *LHS* of Eq. 5.39 is the normal component, which is equal to the shear component, represented by the *RHS*, bearing in mind that for a specific lateral confinement C, the parameter K_C is constant. This will give τ^* from Eq. 5.33:

$$\tau^* = \tau_N - \tau_C \tag{5.43}$$

from which τ_N, the shear force on the sliding plane, can be derived. Sliding will occur when it is equal to the *SF* material strength $\tau_{SF\text{-}x}$ for the specific slip $\Delta_{SF\text{-}x}$. The procedure is repeated for varying $\Delta_{SF\text{-}x}$ to determine the τ_{SF}/Δ_{SF} relationship for the softening branch.

Now determine the stress–strain of the softening concrete:

From the τ_{SF}/Δ_{SF} relationship derived above, for each specific point $\tau_{SF\text{-}x}$ and $\Delta_{SF\text{-}x}$ on this relationship from Eqs. 5.29 and 5.30:

$$\sigma_{c\text{-}x} = \tau_{SF\text{-}x} / (\sin\alpha \cos\alpha) \tag{5.44}$$

$$\varepsilon_{slip\text{-}x} = \Delta_{sh\text{-}x} \cos\alpha \,/\, L_{def} \tag{5.45}$$

where $\sigma_{c\text{-}x}$ is the concrete axial stress and $\varepsilon_{slip\text{-}x}$ is the effective strain due to slip for a specific slip $\Delta_{SF\text{-}x}$. This step is the same as that for unconfined concrete in Section 5.2.1.5.

Determining the global stress–strain and the alternative approach are the same as for unconfined concrete, as discussed in Section 5.2.1.5.

Alternative closed-form solution:

An alternative way of solving the problem is to substitute the above fundamental Eq. 5.39 into the equation of the *SF* material properties and solve for τ_{SF} at specific values of Δ_{SF}, as outlined in Section 5.2.1.4. Since the confinement parameter K_C in Eq. 5.40 is constant for a specific active confinement force C, a direct solution can be achieved.

5.2.3 Passively confined concrete

It is now a question of determining K_C from the mechanics of passive confinement, where the confinement force C varies as follows.

Consider the encased prism in Figure 5.4. The lateral expansion within the passive-confinement region of length $D/\tan\alpha$ is due to both the slip component $S \sin\alpha$ and the concrete material Poisson expansion ν_{mat}. Consider first the simplest problem by assuming that the interface bond between the tube and the concrete is zero; the effect of the interface bond will be considered in detail later.

Within the passively constrained region in Figure 5.4a of length $D/\tan\alpha$, the lateral expansion, that is the vertical expansion in Figure 5.4b, is due to both the slip $S \sin\alpha$ and the Poisson material expansion ν_{mat}. The effective lateral strain due to slip is

$$\varepsilon_{lat\text{-}slip} = \frac{S\sin\alpha}{D} \tag{5.46}$$

The lateral strain due to the Poisson expansion of the concrete is affected by the lateral restraints on the concrete. The elements of the concrete-filled tube are shown separately in Figure 5.13. The unstressed concrete has the size A–B–C–D, and the unstressed tube has the size E–F–G–H. The Poisson expansion due to a longitudinal or axial stress σ_c is $\nu_{mat}\sigma_c D/E_c$, as shown. Hence, I–J is the position when a longitudinal stress is applied, but the concrete is free to expand laterally, that is, it is laterally unrestrained and consequently unstressed laterally. Of course, the tube restricts this lateral expansion such that the tube expands from F–G to the level K–K, and the concrete contracts from the unrestrained level I–J to the level K–K. For equilibrium, the magnitude of the compressive force in the concrete is equal to the magnitude of the tensile force in the tube when they are both at the level K–K, and this force is the passive confinement force C_{pas}. If ε_{red} is the reduction in the concrete lateral strain due to the lateral confinement such that $\varepsilon_{red}D$ is the contraction of the concrete due to the passive confinement, as shown on the *RHS* of Figure 5.13b, then equating the concrete and tube forces,

$$C_{pas} = \frac{\varepsilon_{red}E_cBD}{\tan\alpha} = \left(\nu_{mat}\frac{\sigma_c}{E_c} - \varepsilon_{red}\right)\frac{E_r 2tD}{\tan\alpha} \tag{5.47}$$

from which

$$\varepsilon_{red} = \nu_{mat}\frac{\sigma_c}{E_c}\cdot\frac{E_r 2t}{E_cB + E_r 2t} \tag{5.48}$$

Hence, the lateral strain of the Poisson effect of the passively confined concrete is

$$\varepsilon_{lat\text{-}pois} = \nu_{mat}\frac{\sigma_c}{E_c} - \nu_{mat}\frac{\sigma_c}{E_c}\frac{E_r 2t}{E_cB + E_r 2t} = \nu_{mat}\frac{\sigma_c}{E_c}\left(1 - \frac{E_r 2t}{E_cB + E_r 2t}\right) \tag{5.49}$$

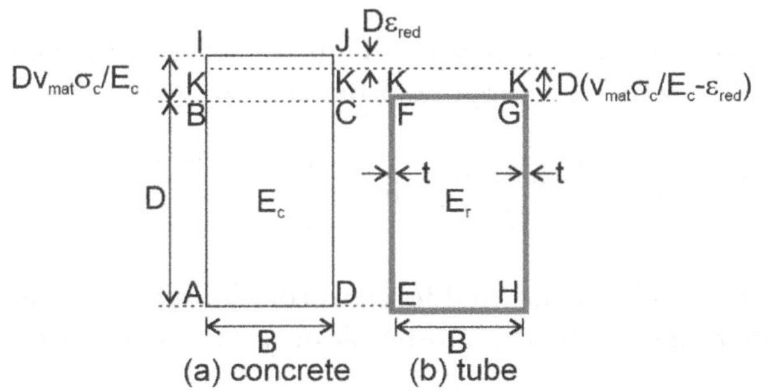

Figure 5.13 Passive-confinement rectangular section (a) concrete element; (b) steel tube element.

Consequently, from Eqs. 5.46 and 5.49, the total lateral strains that these expansions induce, ε_{lat}, are given by

$$\varepsilon_{lat} = \varepsilon_{lat\text{-}slip} + \varepsilon_{lat\text{-}pois} = \frac{S\sin\alpha}{D} + v_{mat}\frac{\sigma_c}{E_c}\left(1 - \frac{E_r 2t}{E_c B + E_r 2t}\right) \tag{5.50}$$

where the first component is the effective lateral strain due to slip and the second component is due to material expansion in which the axial stress in the concrete is σ_c, and it is assumed that the concrete has a modulus of E_c.

Applying these lateral strains to the tube in Figure 5.4 gives the following lateral confining force C_{pas} within the confinement zone of length $D/\tan\alpha$:

$$C_{pas} = \varepsilon_{lat} E_r 2t \frac{D}{\tan\alpha} \tag{5.51}$$

where E_r is the modulus of the tube and t is the tube thickness. Substituting the lateral strain in Eq. 5.50 gives the confining force:

$$C_{pas} = E_r 2tS\cos\alpha + E_r 2tv_{mat}\frac{\sigma_c}{E_c}\frac{D}{\tan\alpha}\left(1 - \frac{E_r 2t}{E_c B + E_r 2t}\right) \tag{5.52}$$

Dividing Eq. 5.52 by the cross-sectional area it acts on, that is, $BD/\tan\alpha$ in Figure 5.12, gives the confinement stress due to slip:

$$\sigma_{conf\text{-}slip} = \frac{E_r 2tS}{BD}\sin\alpha(5.53)$$

and that due to material expansion as

$$\sigma_{conf\text{-}mat} = E_r 2tv_{mat}\frac{\sigma_c}{BE_c}\left(1 - \frac{E_r 2t}{E_c B + E_r 2t}\right) \tag{5.54}$$

From Figure 5.12, the shear force due to confinement is $C\sin\alpha$; dividing by the area of the sliding plane gives the shear stress as

$$\tau_c = \frac{C\sin^2\alpha}{BD} \tag{5.55}$$

where B is the width of the tube as shown in Figure 5.4. Substituting for C_{pas} in Eq. 5.52 and multiplying by $\tan\alpha$ for the component in Eq. 5.40 gives

$$\tau_c\tan\alpha = \frac{E_r 2tS}{BD}\left(\sin^3\alpha\right) + \frac{E_r 2tv_{mat}\sigma_c}{BE_c}\left(\sin^2\alpha\right)\left(1 - \frac{E_r 2t}{E_c B + E_r 2t}\right) \tag{5.56}$$

Similarly, from Figure 5.12, the normal force to the sliding plane due to confinement is $C\cos\alpha$, which, when divided by the sliding plane area, gives a normal stress:

$$\sigma_{N\text{-}C} = \frac{C \sin\alpha \cos\alpha}{BD} \tag{5.57}$$

and substituting for C_{pas} in Eq. 5.52 gives

$$\sigma_{N\text{-}C} = \frac{E_r 2tS \sin\alpha \cos^2\alpha}{BD} + \frac{E_r 2tv_{mat}\sigma_c \cos^2\alpha}{E_c B}\left(1 - \frac{E_r 2t}{E_c B + E_r 2t}\right) \tag{5.58}$$

To form the confinement parameter required in Eq. 5.38, add Eqs. 5.56 and 5.58

$$\tau_c \tan\alpha + \sigma_{N\text{-}C} = K_{C\text{-}slip\text{-}1} + K_{C\text{-}mat\text{-}1} \tag{5.59}$$

where the slip parameter is

$$K_{C\text{-}slip\text{-}1} = \frac{E_r 2t \sin\alpha}{BD}\Delta_{SF} \tag{5.60}$$

and the material strain parameter is

$$K_{C\text{-}mat\text{-}1} = \frac{E_r 2tv_{mat}\sigma_c}{BE_c}\left(1 - \frac{E_r 2t}{E_c B + E_r 2t}\right) \tag{5.61}$$

and in which Δ_{SF} in Eq. 5.60 is S in Figure 5.4.

5.2.3.1 Passively confined stress–strain solution

Equation 5.39 applies whether the lateral confinement force is active or passive; therefore, K_C can be replaced as follows:

$$\sigma_N^* = \tau^* \tan\alpha + \tau_C \tan\alpha + \sigma_{N\text{-}C} = \tau^* \tan\alpha + K_{C\text{-}slip} + K_{C\text{-}mat} \tag{5.62}$$

where in this case the confinement parameters are given by Eqs. 5.60 and 5.61.

Equation 5.62 can now be solved using the procedure in Section 5.2.2.3. If the slip parameter $K_{C\text{-}slip}$ is used by itself, that is, not with the material parameter $K_{C\text{-}mat}$, then the passive confinement depends directly on the slip, providing a closed-form solution for a specific slip. However, if the material parameter $K_{C\text{-}mat}$ is included, then an iterative procedure is required. In this case, σ_c in Eq. 5.61 will have to be estimated, and if the analysis in Section 5.2.2.3 yields a different value, then the estimated value will have to be adjusted until the estimated and calculated values of σ_c converge.

5.3 PREDOMINANTLY FLEXURE REGION

Flexural members can have regions of pure flexure, such as at mid-span in a simply supported beam with a pair of symmetrical point loads, which is suggested to rarely occur in practice. However, flexural members never have regions of pure shear, as occurs in the Z test, in line with the applied loads in Figure 3.7a. Hence, for all intents and purposes, all regions are subjected to both flexure and transverse shear, and consequently,

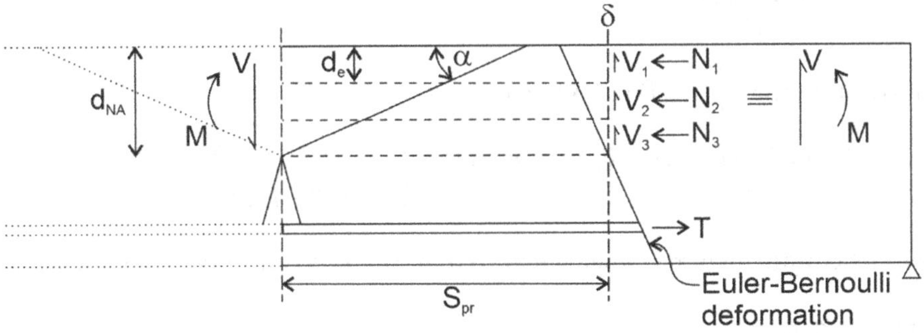

Figure 5.14 Flexural member softening.

all failures involve both flexure and transverse shear. The regions of a beam have been categorised below as 'predominantly flexural region' where convention assumes that flexure dominates the behaviour or design; and 'predominantly shear region' where convention assumes that transverse shear dominates.

5.3.1 Allowance for variable compressive stresses

Failure of a beam in a predominantly flexural region is shown in Figure 1.7. The beam is subjected to flexure, which causes the Euler–Bernoulli deformation, explained using Figure 1.29b, and also shown in Figure 5.14, such that there is a linear variation of deformations and strains in the compression region. The question is how to account for this variation. Consider slicing the compression region into small elements of depth d_e where the length of the elements $L_{def}=S_{pr}$ is sufficient to encompass the potential sliding plane. Each element will have exactly the same material stress/strain. Furthermore, whilst softening, each will have exactly the same effective stress/strain due to slip, as this does not depend on the position of each sliding plane in an element. Hence, the global stress/strain will be the same for each element. Consequently, the analyses in Section 5.2 can be used to determine the global stress/strain relationship required for the flexural analysis of a beam with varying strains.

5.3.2 Allowance for transverse shear

The beam in Figure 5.14 is subjected to a transverse shear force V where the free-body diagram of the sliding plane is shown in Figure 5.15. It can be seen that the direction of the component of the shear force $V \sin \alpha$ along the sliding plane is opposite to the component $N \cos \alpha$ and, hence, reduces the total shear along the sliding plane. Furthermore, the component normal to the sliding plane $V \cos \alpha$ increases the total confining stress σ_N normal to the sliding plane. Both effects enhance the stress/strain properties of the concrete, and so for flexural analyses of members in predominantly flexural regions, that is, with small amounts of transverse shear, the effect of shear on the concrete stress/strain properties can be ignored; that is, omitting transverse shear gives a safe analysis.

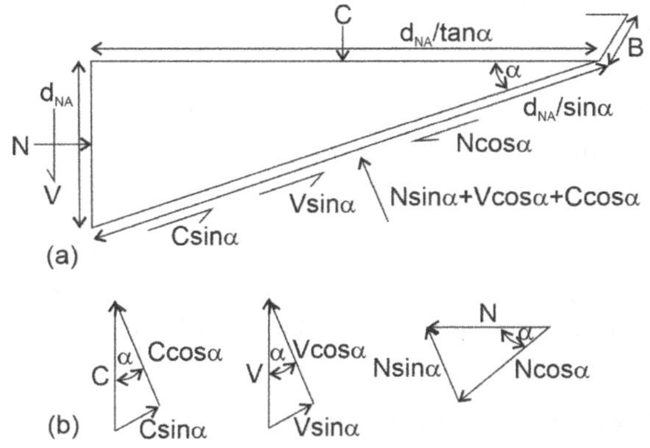

Figure 5.15 Transverse shear in predominantly flexural region (a) free body diagram; (b) resolution of applied forces along and perpendicular to sliding plane.

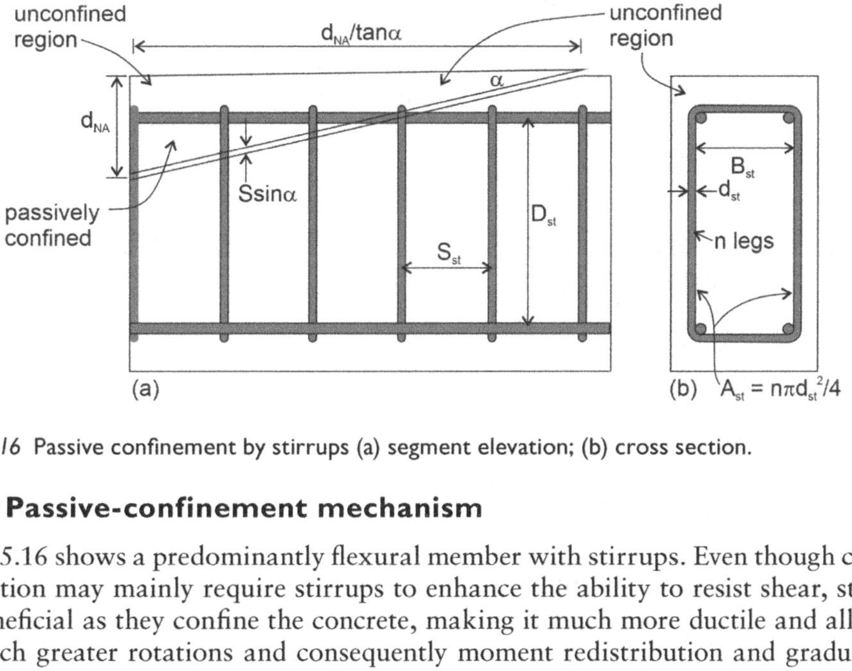

Figure 5.16 Passive confinement by stirrups (a) segment elevation; (b) cross section.

5.3.3 Passive-confinement mechanism

Figure 5.16 shows a predominantly flexural member with stirrups. Even though current convention may mainly require stirrups to enhance the ability to resist shear, stirrups are beneficial as they confine the concrete, making it much more ductile and allowing for much greater rotations and consequently moment redistribution and gradual collapse (Hao et al. 2020).

The beam in Figure 5.16 can be considered to consist of two concrete materials: the passively confined concrete encased by the stirrups, and the unconfined concrete outside the encasement. When the load is first applied to the beam, the section can be analysed as though the concrete is unconfined with the properties in Section 5.2.1. The concrete in the upper cover to the stirrups in Figure 5.16 will eventually start to weaken or soften and ultimately spall, leaving the passively confined concrete beam such that the top surface of the beam is at the level of the stirrups. Let us assume that a wedge forms at an angle α and for the convenience of analysis at depth d_{NA} as shown, such that the confinement zone is of length $d_{NA}/\tan\alpha$. The spacing of the groups of stirrups is S_{st}, the depth of the stirrups is D_{st}, the cross-sectional area of the legs of the stirrups in a group is A_{st} as shown in Figure 5.16b, the width of a group of stirrups is B_{st} and the modulus of the stirrups is E_{st}. Please note that each group of stirrups is spaced S_{st} in Figure 5.16a such that there are n legs per group as in Figure 5.16b.

The free-body diagram of the sliding plane in Figure 5.16 is the same as that in Figure 5.12 with D replaced by d_{NA}; any transverse shear forces in the beam have been omitted as they have been shown to be beneficial. Furthermore, we will ignore the passive confinement due to the material Poisson effect. This will be included, but the vertical variation in the compressive stress in Figure 5.14 will need to be accounted for.

5.3.3.1 Slip-based confinement of unbonded stirrups

First, consider the slip-based confinement when the stirrups are not bonded to the adjacent concrete but are simply anchored at their ends by the longitudinal reinforcement at a distance D_{st} apart, as shown in Figure 5.16.

The strain in the stirrups due to slip is $S\sin\alpha/D_{st}$; consequently, the stress is $E_{st}S\sin\alpha/D_{st}$ and the force in one group of stirrups is $A_{st}E_{st}S\sin\alpha/D_{st}$. The number of groups of stirrups confining the compression zone is the length of the compression zone divided by the group spacing S_{st}, which is $D_{NA}/(S_{st}\tan\alpha)$. Hence, the total confinement force is

$$C_{pas} = \frac{E_{st}A_{st}SD_{NA}\cos\alpha}{D_{st}S_{st}} \tag{5.63}$$

bearing in mind that there is a limit to the yield or fracture strength. Dividing Eq. 5.63 by the cross-sectional area $BD_{NA}/\tan\alpha$ over which it acts gives the confinement stress

$$\sigma_{conf\text{-}slip} = \frac{E_{st}A_{st}S\sin\alpha}{D_{st}S_{st}B_{st}} \tag{5.64}$$

From Figure 5.12a, τ_C is equal to $C\sin\alpha$ divided by the area of the sliding plane that is

$$\tau_C = \frac{E_{st}A_{st}S}{D_{st}S_{st}B_{st}}\sin^2\alpha\cos\alpha \tag{5.65}$$

Multiplying by $\tan\alpha$ and substituting for C_{pas} from Eq. 5.63 gives

$$\tau_C\tan\alpha = \frac{E_{st}A_{st}S}{D_{st}S_{st}B_{st}}\sin^3\alpha \tag{5.66}$$

Similarly, dividing $C\cos\alpha$ in Figure 5.12 by the area of the sliding plane gives

$$\sigma_{N\text{-}C} = \frac{E_{st}A_{st}S}{D_{st}S_{st}B_{st}}\sin\alpha\cos^2\alpha \tag{5.67}$$

Adding Eqs. 5.66 and 5.67 gives

$$K_{C\text{-}slip\text{-}2} = \tau_C\tan\alpha + \sigma_{N\text{-}C} = \frac{E_{st}A_{st}\sin\alpha}{D_{st}S_{st}B_{st}}\Delta_{SF} \tag{5.68}$$

The solutions for this are provided in Sections 5.2.2.3 and 5.2.3.1.

5.3.3.2 Slip-based confinement of bonded stirrups

Assuming that there is no bond between the stirrup and the adjacent concrete, as stated above, simplifies the problem. However, the tension stiffening mechanics of Chapter 4 can be used to incorporate bond for a more accurate analysis, using a similar approach to that developed by Hao et al. (2020, 2022a,b). A single group of stirrups crossing the sliding plane in Figure 5.16 is shown in Figure 5.17. A slip S across the sliding plane causes a vertical separation of _$S \sin \alpha$. Furthermore, from the SF properties in Section 3.1.2.2, the slip S also causes a transverse opening of the crack by h_{cr}, which has a vertical component of $h_{cr}/\cos \alpha$, as shown; please bear in mind that h_{cr} is also a function of S. Hence, the vertical component of the effective crack opening is

$$w_v = S \sin \alpha + h_{cr} / \cos \alpha \tag{5.69}$$

where both components are a function of S. Often, the aggregate interlock component h_{cr} is ignored.

Ignoring h_{cr} for the moment, the vertical component of the crack width w_v in Figure 5.17 causes the group of stirrups to go into tension with a force denoted as P_{gr} in Figure 5.17. Hence, at each crack face, the bar has slipped $w_v/2$. For each individual stirrup leg crossing the sliding plane, the tension stiffening mechanics in Chapter 4, and in particular in Section 4.3, give the relationship between the force in an individual bar P_{ind} at the crack and the slip at the crack face Δ, which is half the crack width w_v, that is

$$P_{ind} = K_{ts} \frac{w_v}{2} \tag{5.70}$$

that is the coefficient K_{ts} can be determined from a tension stiffening analysis and which primarily depends on the bond-slip property; however, for large slips, it depends on the anchorage at the ends of the stirrups. For this derivation, it will be assumed that K_{ts} is constant, but it would be easy to allow for variations in K_{ts} as the slip becomes larger if required.

Following the same procedure as in the previous section, the confining force is given by

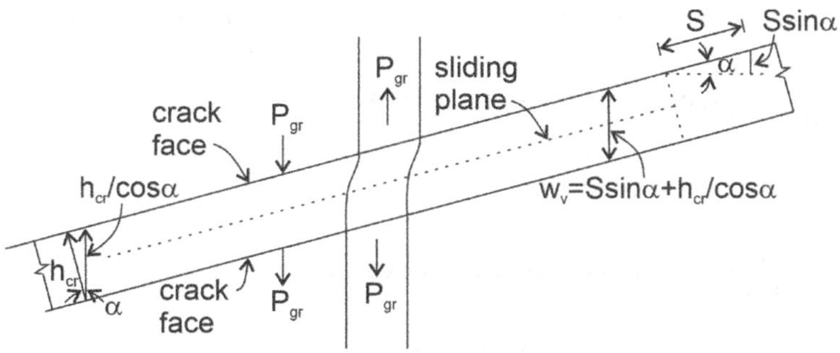

Figure 5.17 Sliding plane deformation.

$$C_{pas} = \frac{nK_{ts}SD_{NA}}{2S_{st}}\cos\alpha \tag{5.71}$$

where n is the number of stirrup legs in a group, as shown in Figure 5.16b. Dividing by the cross-sectional area $BD_{NA}/\tan\alpha$ gives the confinement stress

$$\sigma_{conf\text{-}slip} = \frac{nK_{ts}}{2S_{st}B_{st}}\sin\alpha\Delta_{SF} \tag{5.72}$$

and furthermore,

$$\tau_C \tan\alpha = \frac{nK_{ts}}{2S_{st}B_{st}}\sin^3\alpha\Delta_{SF} \tag{5.73}$$

and

$$K_{C\text{-}slip\text{-}3} = \frac{nK_{ts}}{2B_{st}S_{st}}\sin\alpha\left(1+\sin^2\alpha\right)\Delta_{SF} \tag{5.74}$$

which applies when the crack widening h_{cr} due to aggregate interlock is ignored.

Let us now include the aggregate interlock term h_{cr} in Eq. 5.69. From the SF properties in Section 3.1.2.2, crack widening due to aggregate interlock is a function of the slip S, so let

$$h_{cr} = K_{hcr}S \tag{5.75}$$

where the coefficient K_{hcr} can be determined from the material properties in Section 3.1.2.2. Following the same procedure as above, the passive-confinement force is

$$C_{pas} = \frac{nK_{ts}D_{NA}}{2S_{st}}\left(\cos\alpha + K_{hcr}\frac{\cos\alpha}{\tan\alpha}\right)\Delta_{SF} \tag{5.76}$$

the confinement stress is

$$\sigma_{conf\text{-}slip} = \frac{nK_{ts}}{2S_{st}B_{st}}\left(\sin\alpha + K_{hcr}\cos\alpha\right)\Delta_{SF} \tag{5.77}$$

and the shear stress along the sliding plane due to the confinement force is

$$\tau_C = \frac{nK_{ts}}{2S_{st}B_{st}}\left(\sin^2\alpha\cos\alpha + K_{hcr}\sin\alpha\cos^2\alpha\right)\Delta_{SF} \tag{5.78}$$

and the confinement coefficient comes to

$$K_{C\text{-}slip\text{-}4} = \frac{nK_{ts}}{2B_{st}S_{st}}\left(1+\sin^2\alpha\right)\left(\sin\alpha + K_{hcr}\cos\alpha\right)\Delta_{SF} \tag{5.79}$$

which now includes aggregate interlock widening.

Ignoring the bond between the stirrups in Figure 5.16 and the adjacent concrete is a conservative, that is, safe, analysis. This is because it assumes that the displacement due to crack widening, S sin α plus h_{cr}/cos α as in Eq. 5.69, is applied to the whole of the stirrup length D_{st} to induce a strain and consequently confining force. Allowing for the bond using a tension stiffening analysis causes the displacement to be applied over a smaller length than D_{st} and consequently a higher local strain and a larger confinement.

The solution of the above parameters is given in Sections 5.2.2.3 and 5.2.3.1.

5.4 PREDOMINANTLY TRANSVERSE SHEAR REGION

5.4.1 Sliding plane orientation

The orientation of the cracks associated with so-called shear failure can clearly be seen in Figures 1.8 and 1.30. The cracks are orientated from the support to the applied load, as illustrated in Figure 5.18, and are composed of two distinct slopes shown as β in the tension zone and α in the compression zone.

The free-body diagram of the crack in the compression zone in Figure 5.18 is shown in Figure 5.19, where it can be seen that the transverse shear V increases the total or resulting shear along the sliding plane τ^* and reduces the resulting confinement to the

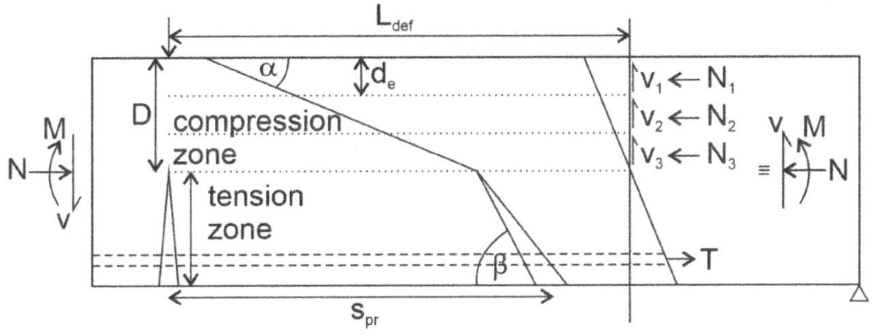

Figure 5.18 Predominantly shear region crack orientation.

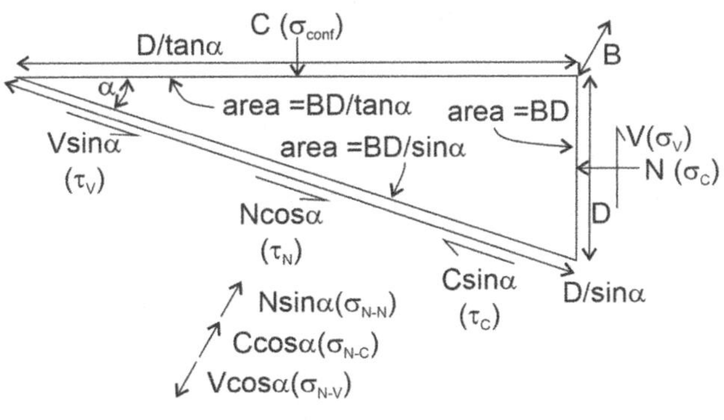

Figure 5.19 Detrimental effect of transverse shear.

sliding plane σ_N^*. Both effects are detrimental to the global concrete stress–strain. This is in total contrast to what occurs in predominantly flexural regions in Section 5.3, where it is safe to ignore the effects of transverse shear on the concrete properties, as it is beneficial. It would appear that the angle α and orientation of the sliding plane, as depicted in Figure 5.15 for predominantly flexure and in Figure 5.18 for predominantly shear, warrant more research.

5.4.2 Transverse shear on concrete stress–strain

Let us assume that an element of depth d_e in the compression zone in Figure 5.18 is subjected to an axial stress σ, a confining stress σ_{conf}, and a transverse shear stress σ_V. From the free-body diagram in Figure 5.19, the applied forces in terms of their stresses are given by

$$N = \sigma_c BD \tag{5.80}$$

$$C = \frac{\sigma_{conf} BD}{\tan\alpha} \tag{5.81}$$

$$V = \sigma_V BD \tag{5.82}$$

The components of these forces along the sliding plane divided by the area over which they act give the following shear stresses acting on the free body:

$$\tau_N = \sigma_c \sin\alpha\cos\alpha \tag{5.83}$$

$$\tau_C = \sigma_{conf} \sin\alpha\cos\alpha \tag{5.84}$$

$$\tau_V = \sigma_V \sin^2\alpha \tag{5.85}$$

and the components normal to the sliding plane:

$$\sigma_{N\text{-}N} = \sigma_c \sin^2\alpha \tag{5.86}$$

$$\sigma_{N\text{-}C} = \sigma_{conf} \cos^2\alpha \tag{5.87}$$

$$\sigma_{N\text{-}V} = \sigma_V \sin\alpha\cos\alpha \tag{5.88}$$

The total shear stress acting along the sliding plane is given by

$$\tau^* = \tau_N + \tau_V - \tau_C \tag{5.89}$$

and substituting τ_N from Eq. 5.83 gives

$$\sigma_c = \frac{\tau^* - \tau_V + \tau_C}{\sin\alpha\cos\alpha} \tag{5.90}$$

The total normal stress acting along the sliding plane is

$$\sigma_N^* = \sigma_{N\text{-}N} + \sigma_{N\text{-}C} - \sigma_{N\text{-}V} \tag{5.91}$$

and substituting σ_{NN} from Eq. 5.86 gives

$$\sigma_c = \frac{\sigma_N^* - \sigma_{N\text{-}C} + \sigma_{N\text{-}V}}{\sin^2\alpha} \tag{5.92}$$

Eliminating σ_c from Eqs. 5.90 and 5.92 gives

$$\sigma_N^* = \tau^* \tan\alpha + (\tau_C \tan\alpha + \sigma_{N\text{-}C}) - (\tau_V \tan\alpha + \sigma_{N\text{-}V}) \tag{5.93}$$

where the second term on the *RHS* gives the beneficial effects of confinement and the third term is the detrimental effect of transverse shear:

$$K_V = \tau_V \tan\alpha + \sigma_{N\text{-}V} \tag{5.94}$$

Substituting in Eq. 5.94, Eqs. 5.85 and 5.88 gives

$$K_V = \sigma_V \left(\sin^2\alpha \tan\alpha + \sin\alpha\cos\alpha\right) \tag{5.95}$$

5.5 SUMMARY OF GLOBAL STRESS–STRAIN DERIVATION

It is necessary to first construct the τ_{SF}/Δ_{SF} relationship of the concrete. To do this, at specific values of $\Delta_{SF\text{-}x}$ and loading vertically, as in Figure 5.10, determine where the following fundamental equation applies:

$$\sigma_N^* = \tau^* \tan\alpha + K_{C\text{-}slip} + K_{C\text{-}mat} - K_V \tag{5.96}$$

In which, for actively confined concrete, both the K_C values are replaced from Eq. 5.38 by

$$K_C = \tau_C \tan\alpha + \sigma_{N\text{-}C} \tag{5.97}$$

For concrete passively encased by a tube, the slip parameter from Eq. 5.60 is

$$K_{C\text{-}slip\text{-}1} = \frac{E_r 2t \sin\alpha}{BD} \Delta_{SF} \tag{5.98}$$

and from Eq. 5.61, the material strain parameter is

$$K_{C\text{-}mat\text{-}1} = \frac{E_r 2t v_{mat} \sigma_c}{BE_c}\left(1 - \frac{E_r 2t}{E_c B + E_r 2t}\right) \tag{5.99}$$

which can be seen to depend on the axial stress σ_c, and as such, when used, an estimate of σ_c has to be made and iterated until the estimated value is equal to the calculated value.

For concrete passively encased by stirrups and when the stirrup legs are only anchored at their ends, that is, there is no interface bond, then from Eq. 5.68,

$$K_{C\text{-}slip\text{-}2} = \tau_C \tan\alpha + \sigma_{N\text{-}C} = \frac{E_{st} A_{st} \sin\alpha}{D_{st} S_{st} B_{st}} \Delta_{SF} \tag{5.100}$$

However, should there be an interface bond, according to Eq. 5.74, then

$$K_{C\text{-}slip\text{-}3} = \frac{nK_{ts}}{2B_{st} S_{st}} \sin\alpha\left(1 + \sin^2\alpha\right)\Delta_{SF} \tag{5.101}$$

and should the crack widening variable h_{cr} be included in the analysis, as in Eq. 5.79, then

$$K_{C\text{-}slip\text{-}4} = \frac{nK_{ts}}{2B_{st} S_{st}}\left(1 + \sin^2\alpha\right)\left(\sin\alpha + K_{hcr}\cos\alpha\right)\Delta_{SF} \tag{5.102}$$

For concrete subjected to transverse shear, from Eq. 5.95,

$$K_V = \sigma_V\left(\sin^2\alpha \tan\alpha + \sin\alpha\cos\alpha\right) \tag{5.103}$$

In the above K parameters, σ_v is the transverse shear stress and τ_C can be derived from Eq. 5.9 as

$$\tau_C = \sigma_{conf}\cos\alpha \tag{5.104}$$

and $\sigma_{N\text{-}C}$ from Eq. 5.8 gives

$$\sigma_{N\text{-}C} = \sigma_{conf}\cos^2\alpha \tag{5.105}$$

Following a vertical path at a specific Δ in the *SF* properties, as shown in Figure 5.10, when Eq. 5.96 is satisfied, this will yield τ^* from Eq. 5.89, which gives

$$\tau^* = \tau_N - \tau_C + \tau_V \tag{5.106}$$

τ_N can be derived, which represents the shear force on the sliding plane due to the applied axial force; sliding will occur when it is equal to the *SF* material strength $\tau_{SF\text{-}x}$ for the specific applied slip $\Delta_{SF\text{-}x}$. The procedure is repeated for varying $\Delta_{SF\text{-}x}$ to determine the τ_{SF}/Δ_{SF} relationship for the softening branch.

From the τ_{SF}/Δ_{SF} relationship derived above, for each specific point $\tau_{SF\text{-}x}$ and $\Delta_{SF\text{-}x}$ on this relationship, the stress from Eq. 5.29 is

$$\sigma_{c\text{-}x} = \tau_{SF\text{-}x} / (\sin\alpha \cos\alpha) \tag{5.107}$$

and the effective strain due to slip from Eq. 5.30 can be determined as

$$\varepsilon_{slip\text{-}x} = \Delta_{sh\text{-}x} \cos\alpha / L_{def} \tag{5.108}$$

where $\sigma_{c\text{-}x}$ is the concrete axial stress and $\varepsilon_{slip\text{-}x}$ is the effective strain due to slip for the specific slip $\Delta_{SF\text{-}x}$. If the parameter depends on σ_v as in Eq. 5.95, the analysis can be conducted for a specific shear stress, meaning that σ_v must be fixed. If the parameter depends on σ_c as in Eq. 5.61, then this variable must be estimated, and the analysis completed to determine σ_c. If this calculated value does not equal the estimated value, the procedure is iterated until it does.

An alternative way of solving the problem is to substitute the fundamental Eq. 5.96 into an equation of the *SF* material properties and solve for τ_{SF} at specific values of Δ_{SF}. An example of this is given in Section 5.2.1.4.

The above analysis gives the increase in strain of the softening branch of the concrete, ε_{slip}. To determine the global stress–strain, the material strain ε_{mat} must be added to ε_{slip}, as illustrated in Figure 5.5. The material strain ε_{mat} is a strain-based property and can be obtained from Chapter 2. Having determined the global stress–strain for a specimen of length L_{def} used in Eq. 5.30, the global stress–strain for any length can be obtained from Eq. 5.3.

5.6 CIRCULAR SECTION MEMBERS

The analysis of circular sections follows all the steps for that of rectangular sections in Sections 5.2–5.5. The slight differences caused by the change in shape are explained in this section.

5.6.1 Free-body diagram

A circular tube filled with concrete is shown in Figure 5.20, where D is now the diameter of the tube, which has a thickness t and modulus E_r. The concrete has a modulus E_c and a material Poisson ratio ν_{mat}.

The free-body diagram of the sliding plane in Figure 5.20 is shown in Figure 5.21, where the only difference between this free body and that for the rectangular section in Figure 5.12 is the cross-sectional areas. Applying the steps in the analyses for the rectangular section to the circular section: from Eqs. 5.25 to 5.27 gives exactly the same relationship as in Eq. 5.28; and the steps from Eqs. 5.33 to 5.36 give Eq. 5.37. This is to be expected, as changes to the shape of a component do not affect the mechanics within the material. Hence, the only differences are the mechanics of confinement.

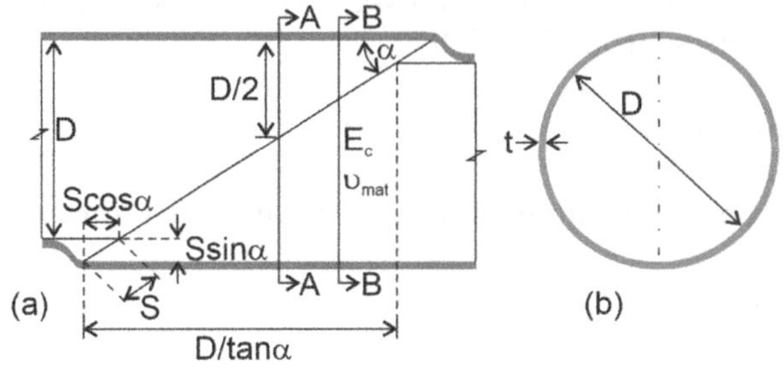

Figure 5.20 Concrete filled circular tube (a) segment elevation; (b) cross section.

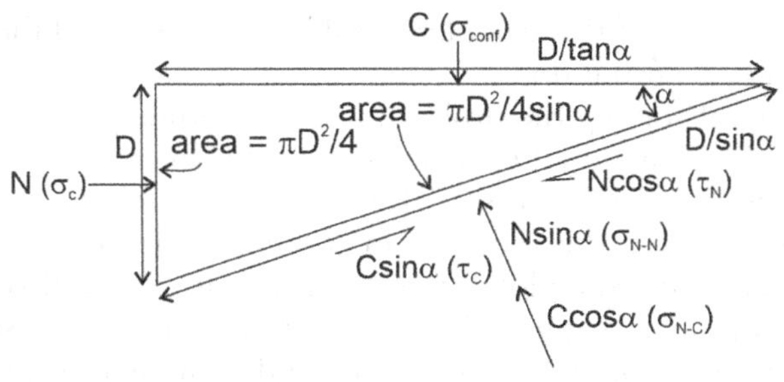

Figure 5.21 Free body of circular section.

5.6.2 Passive-confinement mechanism through sliding

5.6.2.1 *Deformation due to sliding*

Section *A–A* through the tube in Figure 5.20, which intercepts the sliding plane at mid-depth, is shown in Figure 5.22a as *A–A*. A slip at the interface of the sliding plane *S* in Figure 5.20 will cause an effective crack width $S \sin \alpha$ in Figure 5.22a. If it is assumed that the encasing tube and the encased concrete keep their circular shape, as shown in Figure 5.22b, then the half circumference of the tube of length $\pi D/2$ has extended by $S \sin \alpha$. Ignoring the bond between the concrete and the tube, the circumferential strain in the tube is $S \sin \alpha/(\pi D/2)$ and the circumferential stress is

$$\sigma_{circ} = \frac{2SE_r \sin \propto}{\pi D} \tag{5.109}$$

Consider Section *B–B* in Figure 5.20, shown as *B–B* in Figure 5.22a, where the sliding plane intercepts the upper part of the tube. If it is assumed again that a circular shape is maintained on either side of the crack, as in Figure 5.22c, then the extension of the half perimeter is still $S \sin \alpha$ and the circumferential stress is still that in Eq. 5.109. Hence, it can be assumed that if the circular shape is maintained, then the effective strain due to slip in Figure 5.20a is the same over the confinement length $D/\tan \alpha$ of the tube.

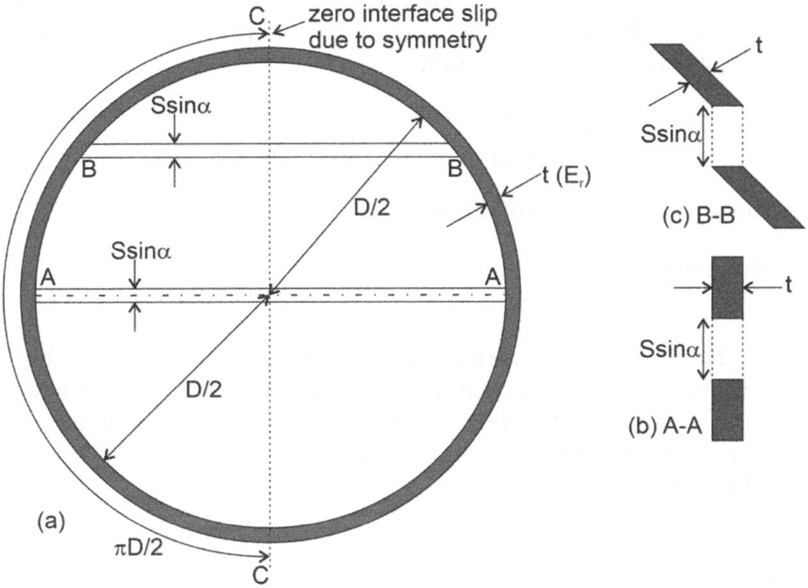

Figure 5.22 Vertical rigid body displacement (a) cross section deformation; (b) and (c) idealised deformation of tube.

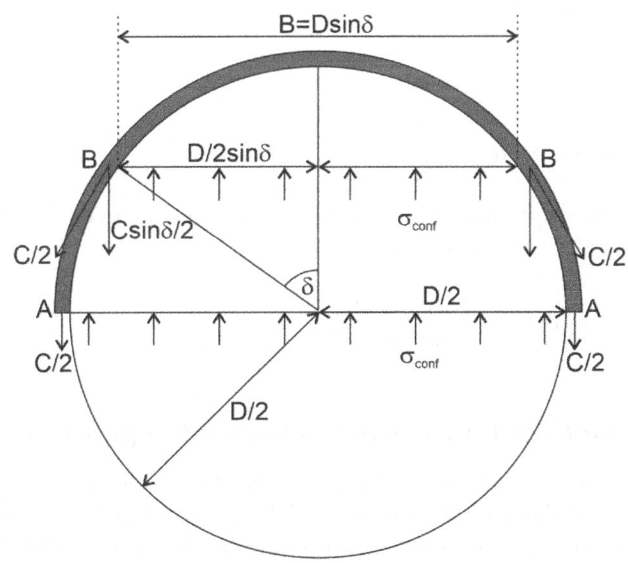

Figure 5.23 Confinement stresses.

5.6.2.2 Confinement stresses

A free-body diagram of half the tube in Figure 5.20 is shown in Figure 5.23, where the force in the tube is shown as *C*/2; that is, there is a total confinement force of *C*. At level *A–A*, the total confinement force *C* is vertical and acts over the length of the passive-confinement zone $D/\tan \alpha$ in Figure 5.20. Furthermore, the confinement force in Figure 5.23 is in equilibrium with the vertical stress in the concrete σ_{conf}. Hence, from the force in the tube and from equilibrium,

$$C_{pas} = \sigma_{conf} D \frac{D}{\tan\alpha} = 2t\sigma_{circ} \frac{D}{\tan\alpha} \tag{5.110}$$

and substituting σ_{circ} from Eq. 5.109 gives

$$\sigma_{conf} = \frac{4tSE_r \sin\alpha}{\pi D^2} \tag{5.111}$$

At level B–B in Figure 5.23, the vertical component of the force in the tube is $C \sin \delta$. Dividing this by the width over which it acts, that is $D \sin \delta$, gives exactly the same confinement stress as in Eq. 5.111. Hence, the passively restrained concrete is subjected to a uniform stress throughout of σ_{conf} in Eq. 5.111.

Following the same steps as in Eq. 5.55 onwards gives

$$\tau_C = \frac{16E_r tS}{\pi^2 D^2} \sin^2\alpha \cos\alpha \tag{5.112}$$

consequently

$$\tau_C \tan\alpha = \frac{16E_r tS}{\pi^2 D^2} \sin^3\alpha \tag{5.113}$$

and

$$\sigma_{N\text{-}C} = \frac{16E_r tS}{\pi^2 D^2} \sin\alpha \cos^2\alpha \tag{5.114}$$

Adding the above two equations gives the following confinement slip term for deriving the SF properties of the softening branch:

$$K_{C\text{-}slip\text{-}5} = \sigma_{N\text{-}C} + \tau_C \tan\alpha = \frac{16E_r t \sin\alpha}{\pi^2 D^2} \Delta_{SF} \tag{5.115}$$

5.6.3 Passive confinement through material expansion

The approach used for rectangular sections, as illustrated in Figure 5.13, will be applied to circular sections, as shown in Figure 5.24. When the concrete is subjected to an axial stress σ_c and is free to expand through the Poisson effect, the radius would increase, resulting in a strain of $\nu_{mat}\sigma_c/E_c$. However, the lateral concrete expansion is restrained by the tube, which reduces the concrete expansion by a strain ε_{red} as shown. This reduction in strain can be found from the equilibrium of the forces in the tube and the concrete.

The confinement stresses in the concrete cylinder are illustrated in Figure 5.24a. No matter where the section A–A through the diameter is taken, for example, if it is inclined, the confinement stress σ_{conf} is the same throughout. Hence, the confinement strain ε_{conf}

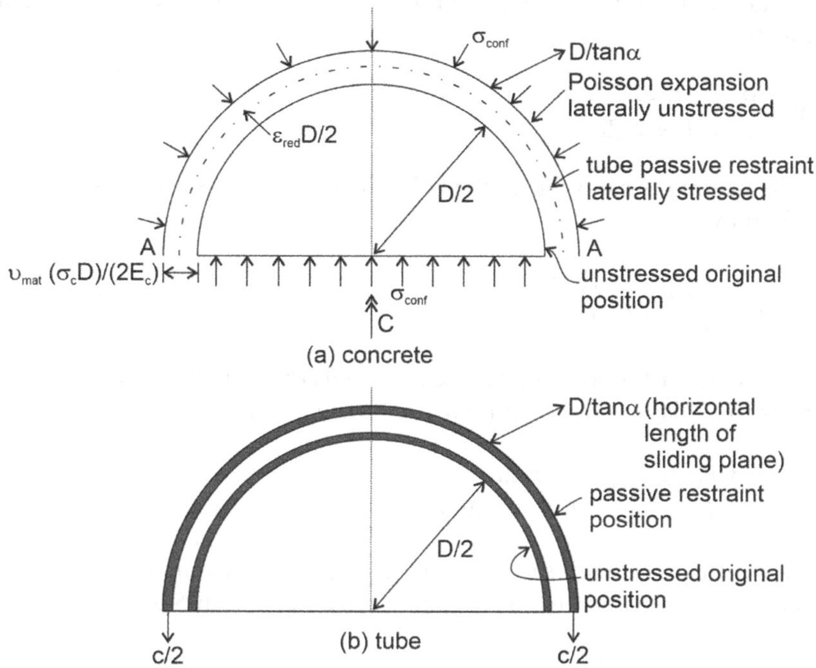

Figure 5.24 Passive confinement of Poisson expansion (a) concrete core; (b) confining tube.

is σ_{conf}/E_c. When that confinement strain is equal to ε_{red} then from that confinement stress σ_{conf}, which it induces, and from the equilibrium of the forces in Figure 5.24:

$$C_{pas} = \varepsilon_{red} E_c D \frac{D}{\tan\alpha} \tag{5.116}$$

The deformation of the tube is shown in Figure 5.24b. The expansion of the concrete causes the radius of the tube to increase to

$$R = \frac{D}{2} + \frac{D}{2}\left(\frac{\nu_{mat}\sigma_c}{E_c} - \varepsilon_{red}\right) \tag{5.117}$$

where the second term represents the increase in the radius, from which the increase in circumferential length of the tube can be derived, and consequently, the strain in the tube:

$$\varepsilon_{tube} = \frac{\nu_{mat}\sigma_c}{E_c} - \varepsilon_{red} \tag{5.118}$$

Therefore, the confining force from the tube is

$$C_{pas} = \left(\frac{\nu_{mat}\sigma_c}{E_c} - \varepsilon_{red}\right) E_r 2t \frac{D}{\tan\alpha} \tag{5.119}$$

From Eqs. 5.116 and 5.119, the reduction in strain due to the passive confinement can be determined:

$$\varepsilon_{red} = \nu_{mat} \frac{\sigma_c}{E_c} \frac{E_r 2t}{E_c D + E_r 2t} \tag{5.120}$$

From this, the lateral restraint due to the Poisson effect of a passively restrained circular section is given by

$$\varepsilon_{lat\text{-}pois} = \nu_{mat} \frac{\sigma_c}{E_c} \left(1 - \frac{E_r 2t}{E_c D + E_r 2t}\right) \tag{5.121}$$

Following the steps for the rectangular section yields the following passive-confinement parameter for the Poisson effect within a circular section:

$$K_{C\text{-}mat\text{-}2} = \frac{8\sigma_c E_r t \nu_{mat}}{\pi D E_c} \left(1 - \frac{E_r 2t}{E_c D + E_r 2t}\right) \tag{5.122}$$

This can be used to construct the softening τ_{SF}/Δ_{SF} relationship and consequently the stress/strain relationship, as previously summarised in Section 5.5.

5.7 ACTIVE CONFINEMENT APPROACH

Using the same mechanics described above in this chapter for a rectangular section, the passive stress–strain relationship can be derived from the actively confined concrete stress–strain relationship (Hao et al. 2022a,b); the latter can be derived experimentally from actively confined concrete or from the SF properties as follows. Passive confinement due to the material Poisson effect will not be considered here; only the passive confinement due to the slip of the sliding plane will be addressed.

5.7.1 Deriving passive stress–strain from active stress–strain

A sliding plane with transverse reinforcement is shown in Figure 5.25a, where for a given slip S, there is an effective crack width w of $S \sin \alpha$. From the tension stiffening mechanics of Chapter 4, the bar on each side of the crack slips relative to the crack face by Δ, which is equal to $w/2$, and this induces a tensile force in the bar across the crack P. The relationship between the tensile force P and the slip Δ, that is P/Δ, can be determined from the tension stiffening analyses in Section 4.3. For local equilibrium, this tensile force in the bar P induces a compressive force P across the sliding plane that is of equal magnitude. Hence, for a given slip S and the known P/Δ properties, the confining force P and consequently the confinement stress σ_{conf} are known. It is worth noting that in Figure 5.25a, the confinement σ_{conf} is applied internally to the sliding plane through the tension stiffening mechanism. In contrast, in an experimental test as shown in Figure 5.25b, the confinement is applied externally, but the effect on the sliding plane is the same as that in Figure 5.25a.

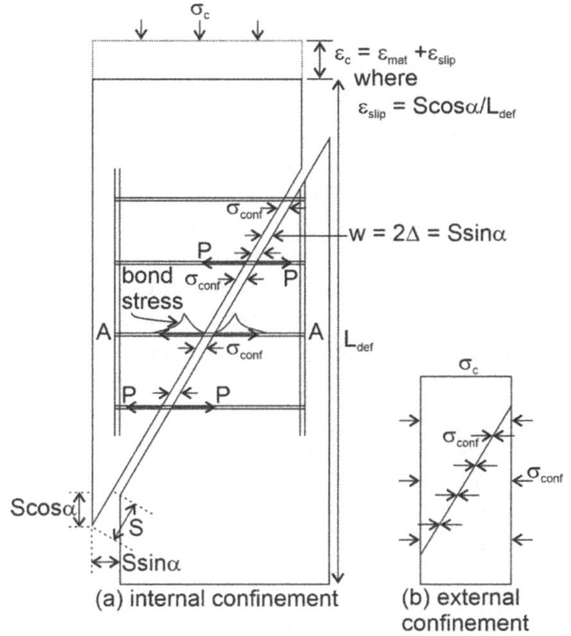

Figure 5.25 Passive internal confinement (a) internal confinement; (b) external confinement.

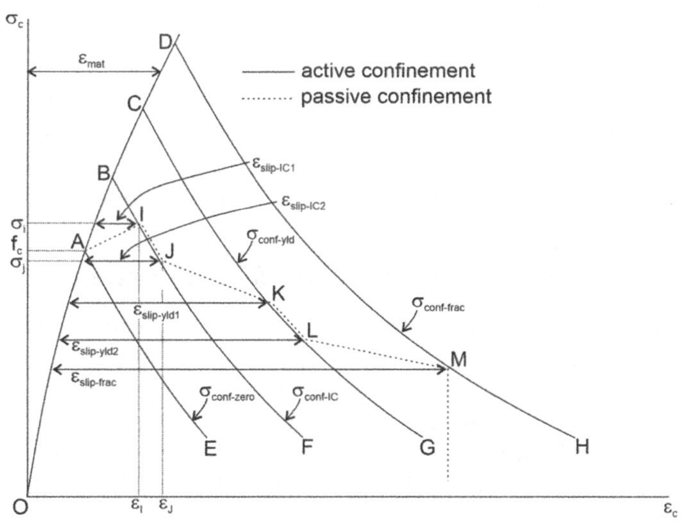

Figure 5.26 Deriving passive σ/ε from active σ/ε.

For a given slip S in Figure 5.25a, the axial contraction of the specimen of length L_{def} is $S\cos\alpha$. Therefore, the effective strain due to slip ε_{slip} is $S\cos\alpha/L_{def}$. It is now a question of finding the axial stress σ_c that causes this slip S.

The unbroken lines in Figure 5.26 represent the active stress/strain relationship of concrete when subjected to a variety of constant active lateral confining stresses σ_{conf}.

Curve O–A–E is for unconfined concrete, that is, $\sigma_{conf\text{-}zero}$ is zero. The remaining curves are for increasing confinements. However, to illustrate the behaviour of reinforcing bars in providing passive confinement, let us consider the different phases that the reinforcing bars in Figure 5.25a go through as the crack widens.

As the crack widens in Figure 5.25, the bond force between the concrete and the bar, as illustrated at level A–A, increases until it reaches its maximum value, referred to as the *IC* debonding resistance P_{IC}. When this occurs, P in Figure 5.25 is P_{IC}, from which the confining stress $\sigma_{conf\text{-}IC}$ can be determined. Concrete at this confinement is shown as O–B–F in Figure 5.26. Assuming that the reinforcing bars in Figure 5.25a are anchored at their ends by longitudinal reinforcement, further crack widening will cause debonding; that is, the bond stresses will move outwards towards their anchors. The anchors may eventually cause the reinforcing bars to yield; that is, P in Figure 5.25a is now P_{yield}, from which the confinement stress $\sigma_{conf\text{-}yield}$ can be determined. Concrete subjected to this confinement is shown as O–C–G in Figure 5.26. Eventually, fracture may occur at an increased stress, and this is represented by O–D–H.

On initial loading, the axial stress/axial strain in Figure 5.26 follows the path O–A until, at Point A, a sliding plane forms. The slip that causes *IC* debonding to start at S_{IC1} sin α is known from the P/Δ properties; therefore, the axial strain at which this occurs, $\varepsilon_{slip\text{-}IC1}$ is equal to S_{IC1} cos α/L_{def}, which is the axial contraction divided by the length of the prism. To find where this occurs, it is a simple matter of moving between the paths O–B and F–B until the gap between them is $\varepsilon_{slip\text{-}IC1} = S_{IC1}$ cos α/L_{def}. This is shown as Point I in Figure 5.26, which is one further point on the passive stress/strain relationship. As the crack widens, *IC* debonding will continue until, at a slip of S_{IC2}, the force starts to increase due to the end anchorage of the bar; the slip at which this occurs is given by the P/Δ relationship, and consequently, $\varepsilon_{slip\text{-}IC2}$ is equal to S_{IC2} cos α/L_{def}. This is shown to fit at Point J, which is therefore another point on the passive stress/strain graph. The same procedure can be applied to derive Points K, L and M. Hence, the passive confinement stress/strain is O–A–I–J–K–L–M, which can be used in analyses. The more active stress/strain relationships used, the smoother the curve. Exactly the same procedure can be employed when assuming the bars to be unbonded.

5.7.2 Deriving active stress/strain from *SF* properties

Procedures for deriving the active stress/strain have already been given in Section 5.2.2. Here is an alternative approach.

The active stress/strain relationship required to determine the passive response can be derived from the SF properties. The free-body diagram of the sliding plane in Figure 5.25 is shown in Figure 5.27 for a rectangular section. In this case, the confinement stress σ_{conf} is constant, so the confinement force C is also constant.

$$C = \frac{\sigma_{conf} BD}{\tan \alpha} \tag{5.123}$$

Following the same procedure used numerous times above in this chapter,

$$\tau_C = \frac{C \sin^2 \alpha}{BD} = \sigma_{conf} \sin \alpha \cos \alpha \tag{5.124}$$

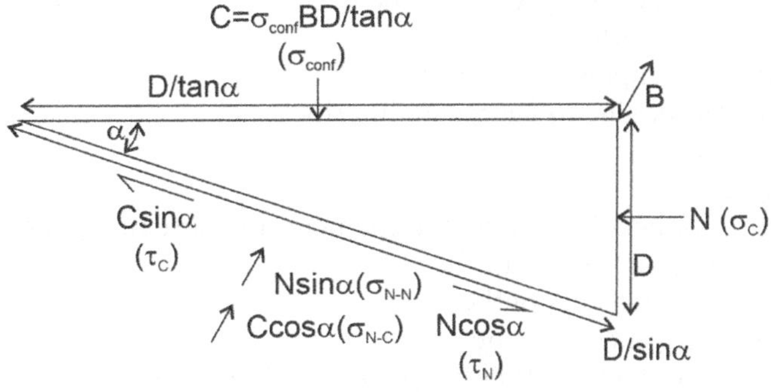

Figure 5.27 Constant confining stress.

$$\sigma_{N\text{-}C} = \frac{C \sin\alpha \cos\alpha}{BD} = \sigma_{conf} \cos^2 \alpha \tag{5.125}$$

$$\tau_C \tan\alpha + \sigma_{N\text{-}C} = \sigma_{conf} \tag{5.126}$$

that is

$$K_{C\text{-}slip\text{-}7} = \sigma_{conf} \tag{5.127}$$

from which the analysis can proceed as in Section 5.5.

5.8 NOTATION AND ABBREVIATIONS

A_{st}	total cross-sectional area of group of n stirrup legs
B	width of concrete prism; width
B_{st}	width of stirrup group
C	lateral confinement force; active lateral confinement force
C_{pas}	passive lateral confinement force
C_1	V_u coefficient
C_2	V_u coefficient
D	depth of concrete prism; diameter of tube; lateral dimension of Poisson expansion
D_{conf}	beneficial factor due to lateral confinement
D_{sh}	detrimental factor of transverse shear
D_{st}	depth of stirrup; length of stirrup leg
d_e	depth of elemental prism
d_{NA}	depth to neutral axis
d_{st}	diameter of stirrup leg
E_c	concrete modulus
E_r	reinforcement modulus; tube modulus
FRP	fibre reinforced polymer

f_c	cylinder strength of concrete; peak strength of concrete in compression
$f_{c\text{-}\alpha}$	variation of f_c with α
f_{c0}	unconfined compressive strength
h_{cr}	crack widening due to aggregate interlock
IC	intermediate crack
K	parameter
K_C	parameter of benefits due to confinement; increase in σ_N due to confinement
$K_{C\text{-}act}$	K_C for active confinement
$K_{C\text{-}mat}$	material confinement coefficient
$K_{C\text{-}slip}$	slip confinement coefficient
K_{hcr}	aggregate interlock coefficient; the h_{cr}/S relationship
K_{ts}	tension stiffening coefficient; the P/Δ relationship
K_V	detrimental effect on σ_N due to transverse shear
L	length; length of prism
L_{def}	length of specimen over which compressive strain determined; deformation length
L_{pr}	length of prism
LHS	left-hand side
M	moment
N	axial compressive force
n	number of stirrup legs in a transverse group of stirrups
P	tensile force in stirrup induced by crack widening
P_{gr}	tensile passive resistance of as group of n stirrup legs to sliding; compressive force across sliding plane for equilibrium
P_{IC}	IC debonding strength of reinforcing bar
P_{ind}	force in an individual stirrup leg
P_{yield}	yield strength of reinforcing bar
Q_{conf}	increase in axial strength f_c due to confinement
Q_{sh}	reduction in axial strength f_c due to transverse shear
R	radius of tube
RHS	right-hand side
S	slip along sliding plane; Δ_{SF}
S_{IC}	slip to cause IC debonding
S_p	flexural crack spacing; primary crack spacing
S_{st}	stirrup spacing
SF	shear friction
t	tube thickness
V	transverse shear force
V_u	τ_{SF} at zero Δ_{SF}
V_{u0}	V_u for unconfined concrete
w	crack width
w_v	vertical component of w when sliding plane inclined to reinforcement
α	angle of sliding plane in compression region
β	angle of sliding plane in tension region
Δ	slip along sliding plane
Δ_{SF}	slip along sliding plane as part of SF properties; S
δ	displacement; displacement profile

ε	strain
ε_c	concrete compressive strain; overall strain; $\varepsilon_{slip}+\varepsilon_{mat}$
ε_{cgl}	overall or global strain in specimen of length L_{pr2}
ε_{lat}	total lateral strain
$\varepsilon_{lat\text{-}pois}$	lateral strain induced by material Poisson effect of laterally restrained concrete
$\varepsilon_{lat\text{-}slip}$	effective lateral strain due to slip
ε_{mat}	concrete compressive strain in material
ε_{red}	reduced lateral strain in concrete due to passive confinement
ε_s	effective axial strain due to slip; ε_{slip}
ε_{slip}	effective concrete compressive strain due to interface slip
ν_{mat}	Poisson ratio of concrete material
ν_{slip}	effective Poisson ratio due to slip
$\nu_{slip\text{-}pass}$	effective Poisson ratio due to slip over passively restrained region
σ	stress
σ_c	concrete compressive stress
σ_{circ}	circumferential stress
σ_{conf}	concrete compressive lateral confinement stress
$\sigma_{conf\text{-}IC}$	σ_{conf} due to P_{IC}
$\sigma_{conf\text{-}mat}$	confinement stress due to Poisson material expansion
$\sigma_{conf\text{-}yield}$	σ_{conf} due to P_{yield}
$\sigma_{conf\text{-}slip}$	confinement stress due to slip
σ_N	compressive stress normal to sliding plane; total confinement normal to sliding plane
$\sigma_{N\text{-}C}$	σ_N induced by the confinement force C
$\sigma_{N\text{-}N}$	σ_N induced by the axial force N
$\sigma_{N\text{-}SF}$	compressive stress normal to sliding plane as part of SF properties
$\sigma_{N\text{-}V}$	σ_N induced by transverse shear force V
σ_V	lateral shear stress; transverse shear stress
σ_N^*	resulting normal stress to sliding plane; total normal stress
τ^*	resulting shear stress along sliding plane; total shear stress
τ_C	shear stress along sliding plane induced by confinement force C
τ_N	shear stress along sliding plane induced by the axial compressive force N
τ_{SF}	shear stress along sliding plane as part of SF properties
τ_{sh}	shear stress along transverse plane induced by transverse shear V
τ_{sp}	total shear stress along sliding plane
τ_V	shear stress along sliding plane induced by transverse shear force V

REFERENCES AND DEVELOPMENT

Ali, M.M., Oehlers, D.J. and Griffith, M.C., 2010. The residual strength of confined concrete. *Advances in Structural Engineering*, *13*(4), 603–618.

Chen, Y., Visintin, P. and Oehlers, D.J., 2015. Concrete shear-friction material properties: Derivation from actively confined compression cylinder tests. *Advances in Structural Engineering*, *18*(8), 1173–1185.

Chen, Y., Visintin, P. and Oehlers, D.J., 2016. Extracting size-dependent stress–strain relationships from FRP-confined concrete cylinders for varying diameters and heights. *Journal of Materials in Civil Engineering*, *28*(5), 04015182.

Chen, Y., Visintin, P., Oehlers, D.J. and Alengaram, U.J., 2014. Size-dependent stress/strain model for unconfined concrete. *Journal of Structural Engineering*, *140*(4), 04013088.

Hao, X., Visintin, P. and Oehlers, D.J., 2020. Closed-form solutions for quantifying the ductility of concrete beams with passively restrained concrete. *Journal of Structural Engineering*, *146*(8), 04020154.

Hao, X., Visintin, P. and Oehlers, D.J., 2022a. Simulating the passive confinement of rectangular concrete prisms allowing for size effect. *Proceedings of the Institution of Civil Engineers-Structures and Buildings*, *175*(7), 514–536.

Hao, X., Visintin, P. and Oehlers, D.J., 2022b. Simulating the passive confinement of circular concrete cylinders allowing for size effect. *Australian Journal of Civil Engineering*, *20*(1), 80–114.

Oehlers, D.J., Visintin, P., Chen, J.F., Seracino, R., Wu, Y.F., Lucas, W. 2017. Reinforced concrete behaviour, research, development and design through partial interaction mechanics. *ASCE Journal of Structural Engineering*, *143*(7), 02517002.

Visintin, P., Chen, Y. and Oehlers, D.J., 2015a. Size dependent axial and lateral stress strain relationships for actively confined concrete. *Advances in Structural Engineering*, *18*(1), 1–20.

Visintin, P., Chen, Y. and Oehlers, D.J., 2015b. Simulating the behaviour of FRP-confined cylinders using shear-friction mechanism. *Journal of Composites for Construction*, *19*(6), 04015014.

Part 4

Partial-interaction segmental mechanics

Having visually explained and quantified the mechanical behaviours in the elements of the reinforced concrete members, these elemental behaviours are combined to analyse a small length or segment of the member. To understand the overall mechanism, we start with the simplest case in Chapter 6: that of a segment of the member subjected to flexure and axial loads only. The direct link between the displacement-based moment–rotation approach and the strain-based moment–curvature approach is clearly shown. It is also demonstrated how the moment–rotation displacement-based approach can be used to quantify the flexural rigidity when flexural cracking occurs, as well as the moment–rotation when concrete softening occurs at a hinge. This relationship is not an approximation but, in mechanics terms, an exact solution. The results can be used in a frame-based analysis to quantify the behaviour of a frame from serviceability behaviour to ultimate behaviour and through to collapse.

Chapter 7 expands the segmental approach to members also subjected to transverse shear by using inclined cracks. It shows that the flexural behaviour described in Chapter 6 is just the simplest case of the overall shear behaviour; that is, the same behavioural mechanism applies to flexural failure as to shear behaviour. This chapter quantifies the transverse shear capacity of reinforced concrete members with any type of material and any geometry.

DOI: 10.1201/9781003711407-9

Chapter 6

Flexural segmental analysis

Consider the continuous beam shown in Figure 6.1, with span lengths L, which has been subjected to increasing loads at the mid-points of each of the spans until sufficient plastic hinges have formed to result in a collapse mechanism. The strain-based and slip-based material properties of this beam have been given in Chapters 2 and 3. Three zones of cracking can be seen: two in the sagging (+ve) regions and one in the hogging region (−ve). Within each zone, the elemental mechanisms of tension stiffening and concrete softening, which have been dealt with in Chapters 4 and 5, can be observed.

We will now bring all of the above together and examine a small length of the beam in Figure 6.1, that is, a segment of the beam of length $2L_{seg}$ as shown in Figure 6.2. Bear in mind that each centre of rotation or zone of cracking in Figure 6.1 and each segment of length $2L_{seg}$ in Figure 6.2 is considered to have two hinges, each of length L_{seg}; for this reason, we describe here the collapse mechanism in Figure 6.1 that requires six hinges to form. For all intents and purposes, this is the same as the three hinges that

Figure 6.1 Continuous RC beam at failure. (Reproduced and Modified from Oehlers et al. 2017.)

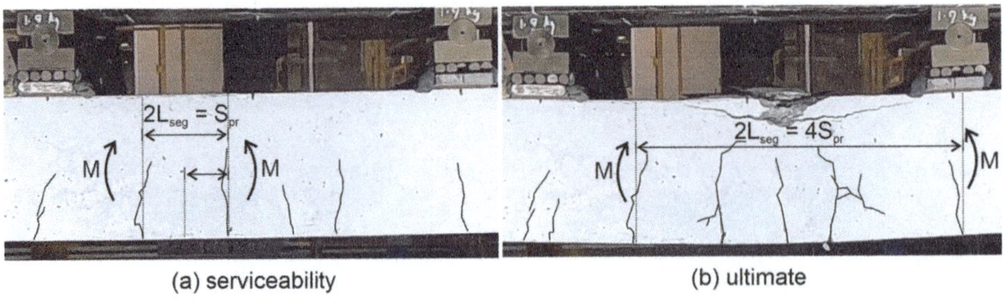

Figure 6.2 Segment of sagging region of beam (a) serviceability limit state; (b) ultimate limit state. (Photos courtesy of Lloyd Dadd.)

DOI: 10.1201/9781003711407-10

are required for the collapse mechanism from conventional structural mechanics, where the hinge length is traditionally defined as $2L_{seg}$ in length. However, the concept of the hinge length being L_{seg} is important, as the ensuing segmental analysis will be shown to determine the rotation of a hinge of length L_{seg}. To further illustrate this new concept, should the centre support in Figure 6.1 be fixed, as in a built-in beam, as well as that on the *RHS*, then for this new concept, there would be two hinges between these fixed supports, each of length L_{def}. Note that the formation and rotation of a hinge in a beam are described in detail in Chapter 8, where moment redistribution is dealt with comprehensively.

Partial-interaction segmental mechanics has already been introduced in Section 1.5. It was shown that there is a fundamental free body with a sloping side at an angle β as illustrated in Figure 1.28d, which can accommodate the external actions of flexure (M_{seg}), transverse shear (V_{seg}) and axial loads (N_{seg}). A special case of this free body occurs when V_{seg} is zero, in which case the segment has vertical sides, meaning β is 90°, as shown in Figure 1.28b. This is the segment we will be dealing with in this chapter, as we are only considering members subjected to axial loads N_{seg} and flexure M_{seg}. Free bodies that are subjected to transverse shear are addressed in Chapter 7. The segmental analysis has already been introduced with the help of Figure 1.29.

To assist the reader, this chapter is divided into four distinct sections:

1. In Section 6.1, titled 'Fundamental concepts of the flexural segmental approach', all the mechanisms that control the flexural segmental approach are described in detail, along with the outcomes. To illustrate the flexural segmental approach, the segment is sliced into small elements to demonstrate that it can be applied to any non-linear material property, such as stress–strain, flexural cracking, fibre stress–crack width and any non-linear time effects such as creep and shrinkage. Importantly, it is shown that it is not directly dependent on the depth of the neutral axis d_{NA} and can easily cope with the complex problem of fibre stress reducing with crack widths.
2. Having described the fundamental principles in Section 6.1, we move on to numerical approaches to finding a solution in Section 6.2, titled 'Numerical flexural segmental analysis'. This approach can be used to determine the applied moment for a given rotation, and consequently curvature, and generate the moment/curvature response by finding individual points along the curve. To find a solution, an iterative approach is used, and the analysis requires the iterative determination of d_{NA} and can handle non-linearity when necessary.
3. Having described the most versatile numerical approach in Section 6.2, Section 6.3 presents a 'Numerical/closed-form solution'. In this approach, the material properties must be linearised, such as the stress–strain distribution of shrinkage, P/Δ, and fibre stress–crack width. For a given rotation, in general, the depth of the neutral axis is derived directly, and the moment at this rotation is also derived directly, meaning iteration is not required. This approach can be used to generate the moment/curvature curve, as in the numerical approach, by generating points along the curve, but generally does not require iteration. This approach also implicitly introduces the concept of transformed sections, although it does not use them explicitly.

4. Having introduced closed-form solutions indirectly in Section 6.3, the remaining sections deal with closed-form solutions directly, specifically the generation of the moment/curvature response, which is very important in design and analysis. Invariably, the properties have to be simplified or linearised. Section 6.4 deals with 'serviceability instantaneous load solutions', Section 6.5 with 'serviceability time-dependent solutions' and Section 6.6 with 'closed-form solutions at ultimate limit state'. Much of the analysis is based on transformed sections, and all of the analyses are grounded in the fundamental principles described in Section 6.1.

6.1 FUNDAMENTAL CONCEPTS OF THE FLEXURAL SEGMENTAL APPROACH

6.1.1 Segmental deformations

The segmental analysis is based on Euler–Bernoulli deformations, as first developed in Visintin et al. (2012) and Visintin et al. (2013a) under instantaneous loads, Visintin et al. (2013b) under time-dependent loads, and Visintin and Oehlers (2018) for fibre-reinforced concrete (FRC), as explained in Section 1.2 with the help of a worked example.

It can be seen in Figure 6.2 that the length of the segment $2L_{seg}$ can be any length that is a multiple of the crack spacing S_p. It can also be seen in Figure 6.1 that the crack spacing S_p is at least an order of magnitude smaller than the span of the beam L. In the flexural segmental analysis of a beam with a moment gradient, it is necessary that the segment is subjected to a constant moment, rather than the moment gradient imposed on the beam. This error is minuscule, as L_{seg} is generally much less than L, and to ensure this occurs, we always make sure L_{seg} is as small as possible. In general, we make $2L_{seg}$ equal to S_p, but sometimes, particularly when there is concrete softening, as in Figure 6.2b, it is necessary to make L_{seg} larger to encompass the concrete softening zone. Note that the moment/curvature (M/χ) results from a Euler–Bernoulli segmental analysis are independent of L_{seg}, meaning one gets the same result regardless of the chosen length.

The Euler–Bernoulli segments in Figure 6.2 are symmetrical about the centre line of the segment, and hence it is only necessary to consider half a segment of length L_{seg} in the analysis, as follows.

6.1.1.1 Euler–Bernoulli deformations

6.1.1.1.1 Uncracked segment

Let us first consider the deformation of an uncracked half-segment, as shown in Figure 6.3, of length L_{seg} that is subjected to pure flexure M_{seg}, where the mid-segment is at 1–1 and the end of the segment is at 3–3. The end of the segment is subjected to a Euler–Bernoulli displacement *A–A*, which causes a rotation θ. The line 3–3 is referred to as the base-line (BL), as it is the position of the end when the member is unstressed. Any deformation from this BL induces a strain and consequently a stress. For example, the contraction Δ_3^* will induce a strain ε_3 of Δ_3^*/L_3, where $L_3=L_{seg}$. From this strain profile, and using the modulus of the concrete, the stress profile can be derived, and

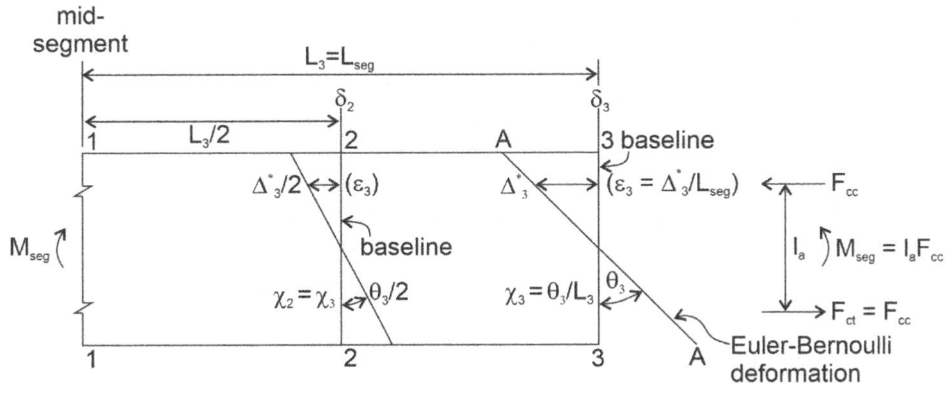

Figure 6.3 Uncracked segment.

consequently, the resultant force in the concrete in compression, F_{cc}. Applying the same analysis to the tension region will yield the resulting tensile force F_{ct}, which, from equilibrium, is equal to F_{cc}, and if the lever arm between the axial forces is l_a, then M_{seg} is equal to $F_{cc}l_a$. Furthermore, the curvature χ_3 is equal to θ_3/L_3. It can be seen that this approach, based on M/θ, can be used to derive the full moment–curvature (M/χ) relationship and is independent of the length of the segment. Furthermore, if the concrete material properties are linear–elastic, it will yield exactly the same results as the classic M/EI approach.

To confirm the above findings, let us now consider the section halfway along at Section 2–2 in Figure 6.3, where L_{seg} is equal to $L_3/2$. To obtain the same strain distribution, we will apply a Euler–Bernoulli rotation of $\theta_3/2$. This will give the same curvature χ_3, axial forces F_{cc} and M_{seg}, and consequently the same M/χ.

6.1.1.1.2 Cracked segment

The behaviour of a cracked segment is shown in Figure 6.4. In Figure 6.4a, there is only one half-crack at the end of the half-segment, such that the segment length L_{seg} is equal to the crack spacing $S_p/2$. Hence, the analysis is that of a segment that encompasses two flexural cracks, as shown in the configuration in the upper part of Figure 6.2a, between adjacent cracks. The Euler–Bernoulli displacement *A–A* is applied in Figure 6.5a with a rotation θ_3, which, as in Figure 6.3, can be used to derive the strain profile in the compression zone and consequently F_{cc} in Figure 6.4a. In the tension zone, the force in the reinforcement F_{rt} is derived from the tension stiffening P/Δ properties from Section 4.3, where Δ is the half-crack width $w/2$, as shown. It can be seen that the rotation θ_3 is the rotation of the half-crack, that is, the rotation of one crack face to the centre line of the crack, which is the Euler–Bernoulli line *A–A*. To the right of the Euler–Bernoulli line, but not shown, is the behaviour of the adjacent half-crack, which is a mirror image of that shown on the left.

Now consider the case where the segment encompasses three cracks, as shown in Figure 6.2b, half of which is depicted in Figure 6.4b, where there are now two half-cracks. The rotation θ_3 at the end of the beam in Figure 6.4c, that is the displacement *A–A*, must occur in both the compression and tension regions. The compression region behaves as illustrated in Figure 6.4a. However, in the tension region, the rotation

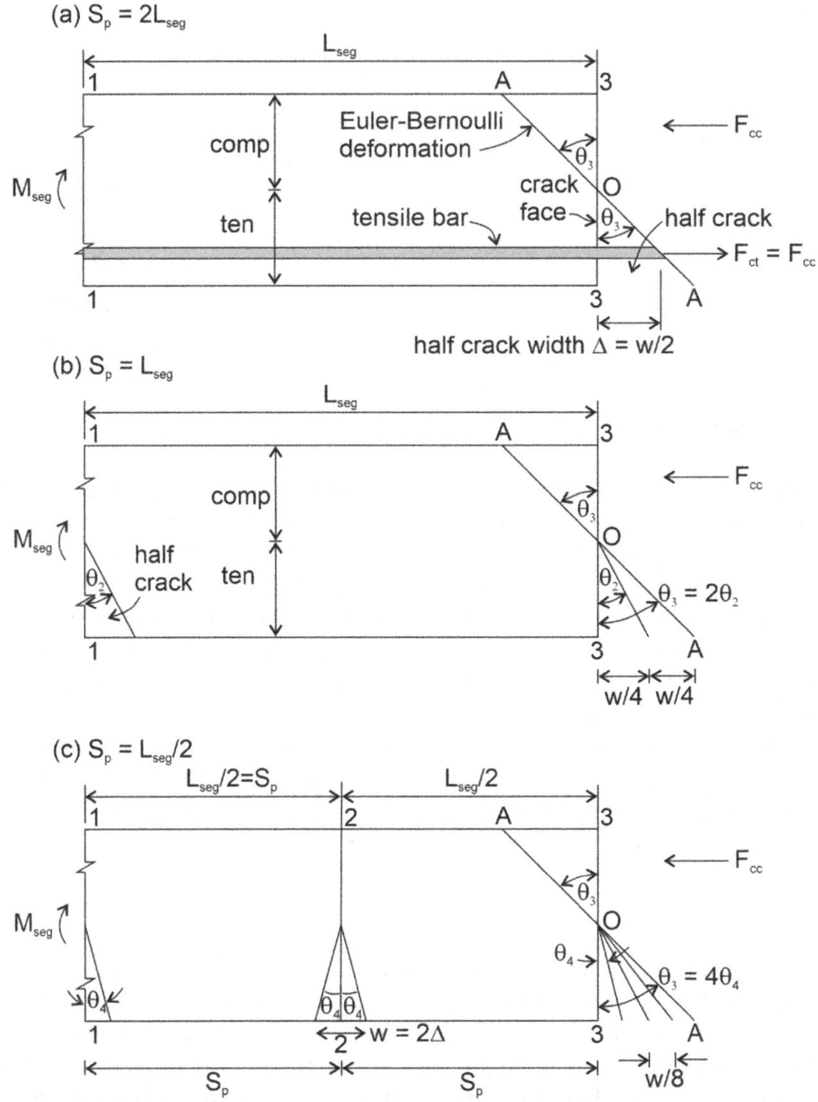

Figure 6.4 Euler–Bernoulli deformation of a cracked segment (a) segment with one crack at the RHS so the crack spacing is $2L_{seg}$; (b) segment with an additional crack at the LHS so the crack spacing is L_{seg}; (c) segment with a further crack at the centre so crack spacing is $L_{seg}/2$.

θ_3 in Figure 6.4b is the accumulation of the half-rotations θ_2 at Section 1–1 and that at Section 3–3. Hence, $\theta_3=2\theta_2$.

Going one step further, consider a segment that encompasses five cracks, as shown in the lower portion of Figure 6.2a, but ignoring the secondary crack that has just started to form on the left-hand side (*LHS*). This is illustrated in Figure 6.4c, where the crack spacing S_p is now equal to $L_{seg}/2$. In the tension region, the rotation θ_3 is now the accumulation of the rotations of four half-crack rotations θ_4; that is, it is not just the rotation at the end θ_4 but the accumulation of all the half-crack rotations, $4\theta_4$.

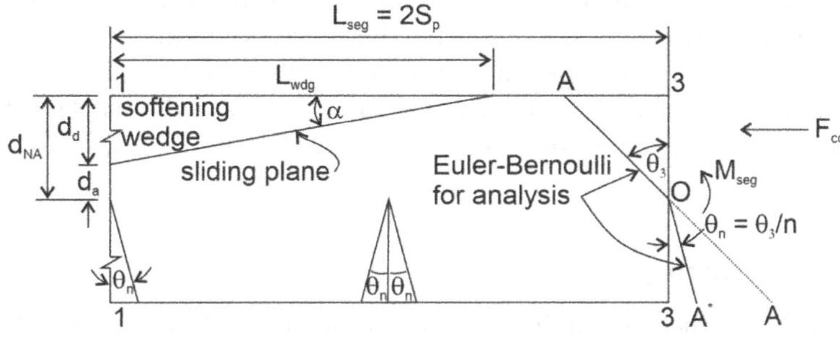

Figure 6.5 Multi-crack segmental analysis.

The analysis of the Euler–Bernoulli deformations of the segments in Figure 6.4a–c is illustrated in Figure 6.5. The half-segment has been sized to encompass a softening wedge of length L_{wdg}, as shown, and consequently requires n half-crack widths, which in this case is $n=4$. A linear Euler–Bernoulli displacement A–O–A at rotation θ_3 has to be applied at the end, as shown. In the tension zone, this comprises the rotation $\theta_n=\theta_3/n$ of n half-crack widths, such that in the segmental analysis in the tension region, only the displacement O–A^* is applied to ensure the linear Euler–Bernoulli displacement A–O–A is achieved. This may, at first glance, appear to be incorrect, but remember that the portion of the Euler–Bernoulli displacement O–A^* is the displacement at the end, whereas O–A is the accumulation of all the displacements, that is $n\theta_n$, which gives a linear Euler–Bernoulli displacement A–O–A.

6.1.1.1.3 Total crack widths

Let us look at the deformations in Figure 6.4 from a different viewpoint. Let us assume that Figure 6.4a represents a member with reinforcement that has a weak bond with the concrete. Hence, the crack spacing $S_p=2L_{seg}$ is large; this often occurs in profiled slabs, where the bond between the concrete and the profiled sheet ribs is very weak compared to that in steel reinforcing bars. We will assume that the concrete deformation A–O at rotation θ_3 remains the same, so that the force F_{cc} remains the same for that deformation. The half-crack width is $w/2$. We are making approximations as we are only trying to visualise what is happening.

Let us now increase the bond strength of the reinforcement and let a crack form at Section 1.1, as in Figure 6.4b. The deformation in the compression region remains as in Figure 6.4a, with a single half-crack; that is, the rotation θ_3 and the depth of the neutral axis remain unchanged. In the tension region, the rotation θ_3 is now shared between the two half-crack rotations θ_2, and hence, the half-crack width has reduced from $w/2$ to $w/4$, such that the total crack width over the segment remains at $w/2$. A further increase in bond produces another crack, as in Figure 6.4c, further reducing the half-crack width to $w/8$, but the total remains at $w/2$.

Broadly speaking, it can be seen that the total crack width, that is, the sum of all the crack widths, is constant at a given moment, no matter how many cracks there are. It has been shown in Chapter 4 that the tension stiffening mechanics for quantifying crack spacing and crack widths are available. However, this mechanics has also shown that there is a large scatter due to the overlap of many competing mechanisms. Perhaps the answer is to first deal with the total crack widths in a zone, as in Figure 6.1, and then determine the scatter through statistics. Furthermore, knowing that the total crack width remains constant and is independent of bond, whilst the crack widths depend on the bond strength, gives the engineer the tool to design for reducing the crack widths until they are narrow enough to allow for autogenous healing or self-healing, to design for longevity.

6.1.1.2 Tension stiffening

The tension stiffening mechanism has been explained in depth in Chapter 4 and is now briefly summarised. Let us consider a reinforcing bar encased in *FRC*. The tension stiffening analysis can quantify the crack spacing S_p in Figure 6.6a; for a specific slip Δ between the crack face and the bar, the force in the bar F_{ts}; and the force in the fibres F_{fib} spanning the crack of width 2Δ. Hence, the total force at the end of the prism P_{pr} is equal to F_{ts} plus F_{fib}. This relationship between the prism force and the reinforcement slip relative to the crack face is referred to as the P/Δ property and can be derived for any situation, as described in Chapter 4, which includes time-dependent variables such as shrinkage.

It is worth noting that there is a longitudinal tensile concrete stress σ_c within the prism, as shown in Figure 6.6b. When the concrete does not contain any fibres, that is, for plain concrete, this stress is zero at the crack faces and a maximum at the centre. When the concrete does contain fibres, that is, for *FRC*, the stress at the crack face starts at σ_{fib}, which is the stress due to the force F_{fib} in Figure 6.6a, and increases as shown. This will cause the prism to extend δ_c, as in Figure 6.6c, such that the total extension is now $\Delta+\delta_c$, as shown. In general, to simplify the problem so that closed-form solutions can be derived, δ_c is ignored as it is generally an order of magnitude smaller than Δ;

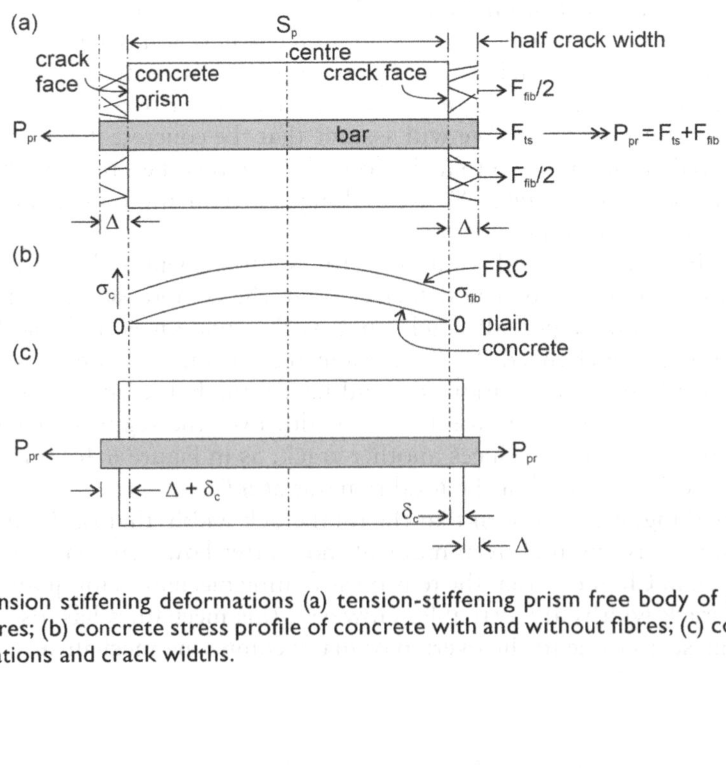

Figure 6.6 Tension stiffening deformations (a) tension-stiffening prism free body of concrete with fibres; (b) concrete stress profile of concrete with and without fibres; (c) concrete deformations and crack widths.

however, it could be included for a more accurate analysis as δ_c is a product of a numerical tension stiffening analysis.

6.1.1.3 Concrete softening

The segmental analysis requires the stress–strain properties of the concrete in compression. Chapter 5 addresses the derivation of these stress–strain properties for all circumstances, such as plain concrete and FRC, as well as for any confinement, such as unconfined, actively confined, and passively confined. The stress–strain response has an ascending portion that requires strain-based properties, as in Section 2.3, and is, therefore, not size-dependent. In contrast, the stress–strain response of the descending portion is size-dependent; that is, it depends on the dimension of the length of the segment L_{seg}, as shown in Figure 6.2, and can be derived theoretically from shear friction analyses, as explained in Chapter 5, or from experimental test results, as described in Chapter 3.

When concrete softening occurs, as illustrated in Figure 6.5, the length of the half-segment L_{seg} needs to encompass the length of the softening wedge L_{wdg}, as shown, to provide the smallest constant moment region for the analysis. As explained in Section 3.1.3, the angle of the softening wedge a in Figure 6.5 can be taken as about 26°. To quantify the length L_{wdg}, the depth of the softening wedge d_d is also required, which is a portion of the depth to the neutral axis d_{NA}. The compressive concrete stress–strain relationship, from Chapters 2 and 3, has an ascending branch from the strain-based properties and a descending branch from the slip-based properties. Hence, the segmental analysis itself will determine the depth of the concrete within the ascending branch d_a and that within the concrete softening descending branch d_d. As d_d is now known, we have L_{wdg} and need to ensure that it is less than L_{seg}. A conservative approach would be to assume that d_d equals d_{NA}, and d_{NA} can be estimated, for example, from linear elastic theory. It must be stressed that overestimating L_{seg} provides only a very minor error in the analysis of plastic hinges, where the rotation capacity is required.

6.1.2 Flexural segmental forces

6.1.2.1 Equilibrium

6.1.2.1.1 Segment properties

Consider an *FRC* segment between two adjacent cracks, as in the upper part of Figure 6.2a. The free-body half-segment is shown in Figure 6.7, where the length of the half-segment L_{seg} is equal to half the crack spacing, which is $S_p/2$ as shown; the width has some arbitrary value b_{seg} and the depth an arbitrary value d_{seg}; and the Euler–Bernoulli displacement is applied to the right-hand side (*RHS*), which is, therefore, where the flexural crack will occur. The vertical line *D–D* is at the mid-length of the whole segment and, because of symmetry with the LHS of the segment, not shown, it is fully fixed horizontally, regardless of what happens. To aid in understanding the behaviour, the segment is sliced into very narrow elements of length $L_e = L_{seg}$ and depth d_{sl}, where $d_{sl} = d_{seg}$.

Prior to any time effects such as creep or shrinkage, the *RHS* of the half-segment in Figure 6.7 is at the vertical line C_1–C_2. This is the base-line (*BL*) for an initially

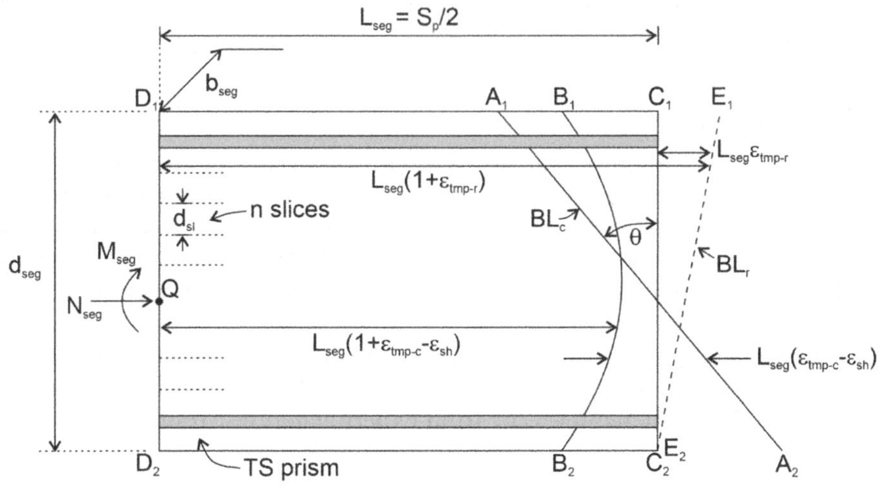

Figure 6.7 Segmental shape.

unstressed segment, such that any imposed horizontal deformations from this vertical line will induce strains. Let us now assume that the concrete component of the segment is subjected to a variable shrinkage strain ε_{sh} and a temperature variation strain $\varepsilon_{tmp\text{-}c}$, where $\varepsilon_{tmp\text{-}c}$ is the coefficient of thermal expansion of the concrete α times the temperature change ΔT, and both vary along the depth d_{seg}. Hence, on the basis that the concrete is totally unrestrained except at D_1–D_2, where it is fixed horizontally, the surface C_1–C_2 will move to B_1–B_2, where the length of an element L_e from D_1–D_2 is given by

$$L_e = L_{seg}\left(1+\varepsilon_{tmp\text{-}c}-\varepsilon_{sh}\right) \tag{6.1}$$

where $\varepsilon_{tmp\text{-}c}$ and ε_{sh} are the strains, that is, for temperature change and shrinkage strain, at a specific level of the segment (Knight et al. 2013). Hence, B_1–B_2 can be plotted and it is the BL for the concrete (BL_c), that is, the position where the concrete is unstressed. Note that variations along the breadth b_{seg} can be accommodated as explained in Section 1.5.2, with the help of Figure 1.29i. We could have derived the BL B_1–B_2 in Figure 6.7 just as easily for expanding or degrading concrete; consequently, it can be any shape.

Let us assume that the reinforcement in Figure 6.7 is only subjected to temperature effects that are at a maximum at the top surface and zero at the soffit. Then the BL for the reinforcement (BL_r) is E_1–E_2, where the expansion is $\varepsilon_{tmp\text{-}r}L_{seg}$ as shown, such that the length of an element L_e is given by Eq. 6.1 without the shrinkage component and using the temperature strain in the reinforcement $\varepsilon_{tmp\text{-}r}$. Time effects can be included in the tension stiffening prism as explained in Chapter 4. As shown on the *LHS* of Figure 6.7, the stress resultants acting on the half-segment are an axial load N_{seg} acting at Level *Q* and a moment M_{seg}, which can also be assumed to act at Level *Q*.

One way of visualising what is happening in Figure 6.7 is that the shape of the segment prior to stressing is D_1–C_1–C_2–D_2; the shape of the concrete, if it were fully unrestrained, that is, unstressed, is D_1–B_1–B_2–D_2; similarly, the shape of the unrestrained reinforcement is D_1–E_1–E_2–D_2; and upon applying the Euler–Bernoulli deformation, the stressed shape changes to D_1–A_1–A_2D_2, where the stresses must be in equilibrium with the externally applied stress resultants.

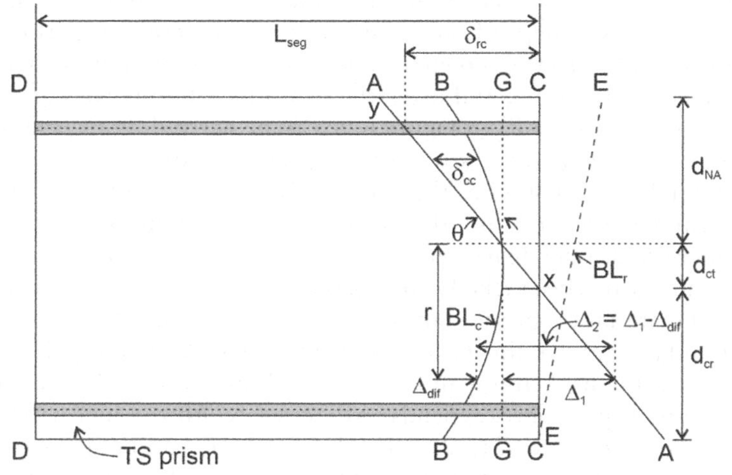

Figure 6.8 Segmental deformations.

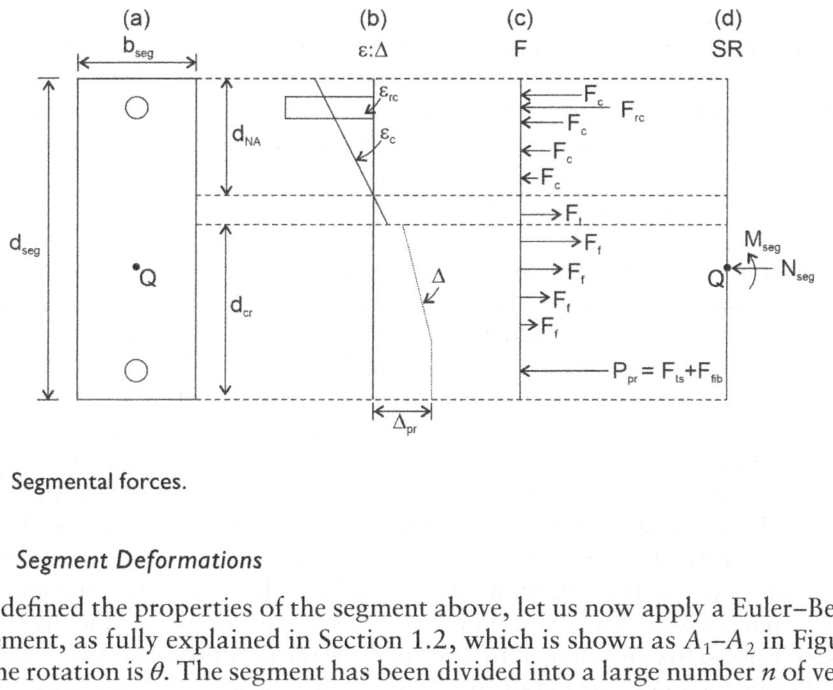

Figure 6.9 Segmental forces.

6.1.2.1.2 Segment Deformations

Having defined the properties of the segment above, let us now apply a Euler–Bernoulli displacement, as fully explained in Section 1.2, which is shown as A_1–A_2 in Figure 6.7, where the rotation is θ. The segment has been divided into a large number n of very thin slices, in which it is assumed that the strains over the depth of each very thin slice are constant. The segment is shown again in Figure 6.8.

As the reinforcement base line (BL_r) in Figure 6.8 is E–E, the contraction of the compressive reinforcement δ_{rc} is the distance between A–A and E–E. This contraction, divided by L_{seg}, is the strain in the compression reinforcement ε_{rc} in Figure 6.9. In the uncracked concrete, the Euler–Bernoulli displacement in Figure 6.8 causes the concrete to contract δ_{cc} between the deformation A–A and the concrete BL (BL_c) B–B. These contractions, divided by L_{seg}, cause the strains ε_c in Figure 6.9b, shown as the unbroken line. It can be seen that within the uncracked concrete region, the strains, and consequently stresses, depend on the strain-based material properties, which can be adjusted for the time effects of creep in concrete and relaxation in the reinforcement. From the strains in Figure 6.9b, and using the strain-based properties of the reinforcement, the reinforcement force F_{rc} in Figure 6.9c can be derived. Using the strains in the concrete ε_c in Figure 6.9b and the stress–strain properties of the concrete in compression from Section 2.3 or Chapter 5, the compressive forces F_c in the concrete elements in Figure 6.9c can be derived. Similarly, for the concrete in tension, F_t.

Below the uncracked concrete in Figure 6.8, specifically below the uncracked concrete in tension d_{ct}, and within the cracked zone d_{cr}, the stresses or forces no longer depend on the strain-based material properties but on the slip-based properties. The displacement between A–A and the vertical line G–G, which intercepts A–A at d_{NA}, is the half-crack width Δ, plotted as the broken line in Figure 6.9b. From Section 2.4, the tensile stress across a crack exerted by the fibres depends on the crack width w, which happens to be 2Δ. Bearing in mind that w is equal to 2Δ, the fibre forces F_f can be derived as shown in Figure 6.9c. It is interesting to note that the fibre force F_f reduces with crack widening, which is easy to simulate in this approach but very difficult to simulate in closed-form solutions. Finally, the force in the tension stiffening prism P_{pr}, as explained in Chapter 4 and Figure 6.6, depends on the half-crack width Δ_{pr}.

Let us take a closer look at the half-crack widths Δ in Figure 6.8. Let r be the distance from the level at which Δ is measured to the neutral axis at the depth d_{NA}, as shown. The half-crack width is shown as Δ_1 between the vertical line G–G and the displacement A–A, which corresponds to the Euler–Bernoulli displacement such that Δ is equal to $r\theta$. This approach is mechanically correct for uncracked segments and will be shown to be correct for cracked segments with a uniform shrinkage strain over the segment depth. However, there is a slight error Δ_{dif} in a cracked segment when the shrinkage strain is not uniform, in which case Δ could have been measured between B–B and A–A to give Δ_2, as shown. This would slightly overestimate Δ, as B–B only occurs when the concrete is totally unrestrained, which it is not. However, the error Δ_{dif} is, for all intents and purposes, very small when compared with the crack widths associated with those required to cause the reinforcement to yield or the crack widths over which fibres are active. Furthermore, this error only occurs when the shrinkage is non-uniform. In the study of very fine cracks, it may be worth using the conservative value of Δ_2.

To further help the reader visualise the segmental analysis, the segments on either side of a crack are shown in Figure 6.10. The Euler–Bernoulli deformation A–A forms the line of symmetry from which deformations are measured, using what are in fact the vertical axes G–G. The deformations Δ on either side of A–A are, of course, for all intents and purposes, horizontal and provide the *FI* strains to determine the stresses in uncracked regions. In cracked regions, they provide the reinforcement slips from which, through *PI*, the forces can be obtained.

6.1.2.1.3 Equilibrium

All the internal forces within the segment in Figure 6.9c that are required to balance the external stress resultants in Figure 6.9d have now been derived. A simple iterative solution described below is to plot the M/θ curve from which the rotation and consequently the curvature at M_{seg} can be derived.

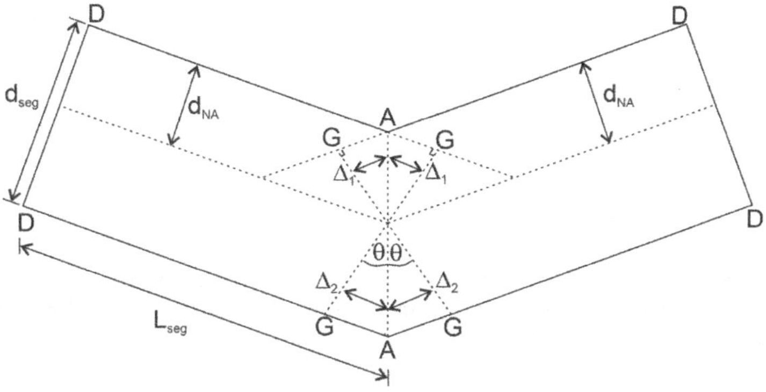

Figure 6.10 Euler–Bernoulli deformation either side of a crack.

The Euler–Bernoulli deformation *A–A* in Figure 6.8 is randomly fixed; that is, both a rotation *q* is chosen and the level at which it intercepts *C–C* is fixed at the level labelled *x*. For this configuration of the Euler–Bernoulli displacement, the strains and crack widths are determined as in Figure 6.9b. From these, the forces in Figure 6.9c can be derived. We need these internal forces to balance the external stress resultants in Figure 6.9d. If the sum of the internal forces in Figure 6.9c does not equal the external force N_{seg}, then we move the level *x* up or down, maintaining the chosen rotation, until it does; that is, we iterate towards a solution. Once horizontal, that is translational, equilibrium has been found, meaning that the sum of the horizontal forces in Figure 6.9c is equal to the external force N_{seg} in Figure 6.9d, we then take moments of the internal forces in Figure 6.9c at the level *Q* where the external axial load N_{seg} is being applied to determine the moment. This is one point on the M/θ variation required. We can then change θ and repeat the iterative procedure to obtain another point and so on. As explained in Section 1.5.2 with the help of Figure 1.29(*l*), moments must be taken at the level of the applied load *Q*. If there are no externally applied loads, meaning N_{seg} is zero, then moments can be taken at any level to obtain the same moment. Note at this stage that the special case of prestressed beams, where an external axial load is applied, is dealt with in detail in Chapter 8. Additionally, the Euler–Bernoulli deformation *A–A* in Figure 6.8 could just as easily be fixed at position *y*; that is for a given rotation, *A–A* could have been moved by varying the horizontal position at *y* to the left or to the right. At this stage, it is not necessary to say that we vary d_{NA} as it is not always clear where d_{NA} is located when dealing with non-linearity, as will be explained later with the help of Figure 6.15. However, when deriving closed-form solutions or when using simplified properties, d_{NA} is a very useful and almost essential parameter.

It can be seen that this numerical approach is incredibly versatile as it can cope with any non-linearity in material properties. Note that in mechanical terms this procedure is not an approximation but an exact solution. Because of this, it can be used to check the accuracy of closed-form solutions, where the mathematics can become very complex, even after making simplifications to the analysis, as will be illustrated in Sections 6.3 onwards.

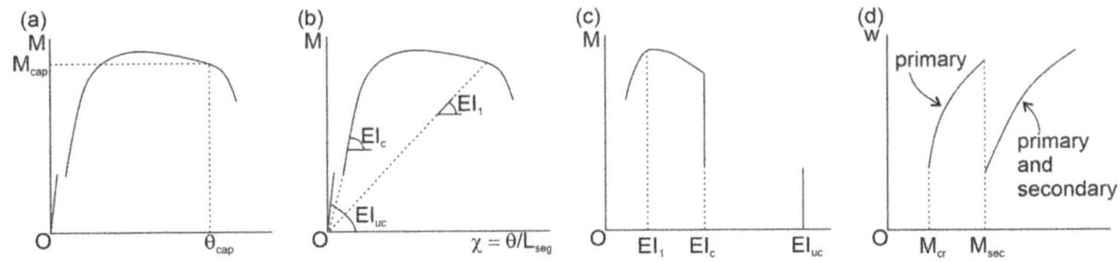

Figure 6.11 Outcomes from segmental analyses (a) moment rotation relationship M/θ; (b) moment curvature relationship M/χ; (c) moment flexural rigidity profile M/EI; (d) crack width moment relationship w/M.

6.1.2.2 Outcomes from segmental approach

The M/θ results from the above approach can be used in many ways. The M/θ results in Figure 6.11a represent the properties of a potential hinge region from serviceability to ultimate failure. Directly from these M/θ results, as shown in Figure 6.11a, the moment capacity M_{cap}, rotation capacity θ_{cap}, the ductile plateau leading up to failure and the softening branch during collapse can be derived. As explained in Chapter 8, this behaviour is essential in controlling moment redistribution within a member and the collapse of a member. Dividing the rotation θ in Figure 6.11a by L_{seg} gives the moment/curvature M/χ in Figure 6.11b, from which the variations in the flexural rigidities EI can be derived, as illustrated in Figure 6.11c. These are not approximations but mechanically exact values that can be used directly in the analysis or design of a member, particularly at serviceability. Finally, from the segmental approach, the variation in crack width with applied moment (w/M) can be derived, as shown in Figure 6.11d, which, for example, could be used in assessing and monitoring any deterioration in a bridge.

Having established the fundamental concepts that underpin the segmental analysis for flexure in this section, these concepts will be applied in the next section. This will ultimately lead to the presentation of a flow chart of the analysis procedure, which can be implemented to write a numerical model.

6.2 NUMERICAL FLEXURAL SEGMENTAL ANALYSIS

6.2.1 Segmental analysis for instantaneous loads

6.2.1.1 Segment dimensions

Following the approach of Visintin et al. (2012) and Visintin et al. (2013a), consider a segment that has the cross-section shown in Figure 6.12a. The segment of length L_{seg} is subjected to a constant moment M_{seg} and an axial load N_{seg}, as shown in Figure 6.12b. It is, therefore, a matter of determining the corresponding Euler–Bernoulli deformation of the segment ends from *A–A* to *C–C* such that equilibrium is achieved.

To perform this analysis under instantaneous loads, consider the segment in Figure 6.12b of length L_{seg}. For ease of analysis, the total segment length $2L_{seg}$ in Figure 6.12a is taken as equal to the crack spacing S_p and, hence, the half-segment length L_{seg} in Figure 6.12b is $S_p/2$.

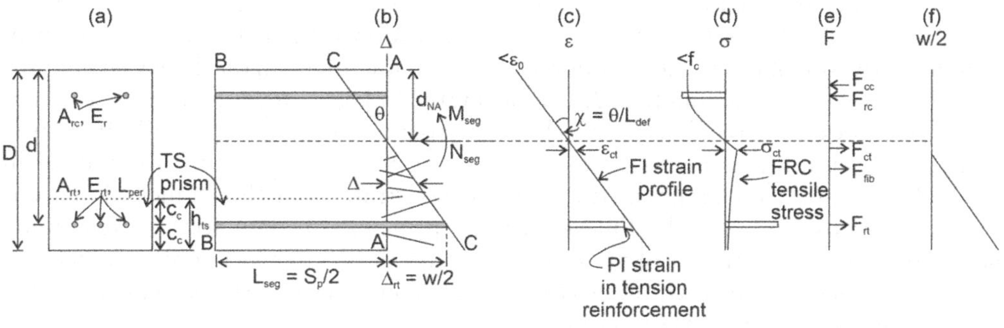

Figure 6.12 Segment under instantaneous loads prior to concrete softening (a) cross section; (b) segment elevation and displacement profile; (c) strain profile; (d) stress profile; (e) force profile; (f) crack width profile.

To determine the crack spacing S_p, the tension stiffening region of the cross-section in Figure 6.12a must be identified, and this is achieved by setting the height h_{ts} and width b_{ts} such that the resultant of the reinforcement force is at the centroid of the prism, as explained in Section 4.1 with Figure 4.1. In the case of the cross-section shown, h_{ts} is simply twice the cover to the centroid of the reinforcement c_{rt} and the width b_{ts} is the full width of the beam. The crack spacing can then be determined by applying partial-interaction theory to the infinite prism analysis presented in Chapter 4.

6.2.1.2 Segment analysis prior to softening

Let us begin by considering a region of the beam in Figure 6.1 that does not contain any plastic hinges but is cracked, as shown in Figure 6.2a; that is, it is in the non-softening region of the beam in Figure 6.1. Having defined the segment dimensions, a rotation θ is imposed on the segment end, and by estimating a neutral axis depth d_{NA}, the Euler–Bernoulli deformation from A–A to C–C, shown as Δ in Figure 6.12b, is established. To determine the internal forces in Figure 6.12e that are induced by the deformation Δ, it is necessary to separately consider those which are strain-based and those which are slip-based, that is, based on partial interaction.

First, consider the forces in Figure 6.12e that occur in the uncracked concrete, that is, the force in the uncracked concrete in tension F_{ct}; that is, the force not softening in compression F_{cc}; and in the reinforcement embedded in uncracked concrete, such as in the compression region, F_{rc}. To apply the strain-based material properties, the distribution of strain in Figure 6.12c, which is induced by the deformation profile in Figure 6.12b, must be determined. From elementary mechanics, the strain in the full-interaction region of the cross-section is simply the deformation Δ divided by the segment length $S_p/2$, such that the full-interaction strain profile in Figure 6.1c is known. The same reasoning also implies that the cross-section curvature χ in Figure 6.12c, which is the increment of end rotation that occurs at each cross-section, is simply the rotation θ divided by the segment length $S_p/2$. Using the strain-based material properties for concrete and reinforcement outlined in Chapter 2, the distribution of stress in the full-interaction region of the cross-section is obtained in Figure 6.12d, from which the forces F_{rc}, F_{cc} and F_{ct} can be determined by integrating the stresses over the cross-sectional area on which they act.

Now consider the partial-interaction force in the tensile reinforcement crossing the crack, P_{rt} in Figure 6.12e. In Section 4.3.5.4, the crack opening stiffness k_{pi} was defined, such that if the slip of the tension reinforcement Δ_{rt} is known from the displacement profile in Figure 6.12b, P_{rt} is simply $k_{pi}\Delta_{rt}$. Importantly, in Figure 6.12c, it can be seen that the strain in the tensile reinforcement is less than that obtained from the full-interaction strain profile, and this is because tension stiffening has been explicitly allowed for in the analysis. This step change in strain also demonstrates that the segmental analysis procedure, whilst applying the Euler–Bernoulli principle of plain sections remaining plain, does not impose the corollary of linear strain profiles. This is important because the analysis, therefore, directly allows for the inclusion of tension stiffening without the need for empirically derived correction factors.

Having quantified all the axial forces in Figure 6.12e, if internal equilibrium with the externally applied stress resultants M_{seg} and N_{seg} is not obtained, it is simply a matter of shifting the neutral axis d_{NA} in Figure 6.12b up or down until equilibrium is achieved. Once equilibrium is obtained with N_{seg}, moments can be taken about the level of N_{seg} to obtain M_{seg} such that the relationship between M_{seg}, N_{seg} and θ is known for a single point on the moment–rotation relationship in Figure 6.11a. The analysis can then be repeated for increasing rotations until the maximum stress in the concrete reaches the concrete compressive strength f_c.

6.2.1.3 Segment analysis after softening

Now let us consider an analysis of the beam in Figure 6.1 in a region where the concrete softening wedge has formed, as shown in Figure 6.2b and Figure 6.13b. Having covered the partial-interaction mechanics of concrete softening in the formation of a size-dependent stress–strain relationship in Chapter 5, the accommodation of concrete softening into the segmental analysis procedure is straightforward. That is, the choice of segment length L_{seg} in Figure 6.13b must be made to accommodate the length of the softening wedge L_{wdg}. The length of the softening wedge can be determined geometrically based on the angle α of the softening wedge, which, as shown in Chapter 3, α varies with confinement, but for which a lower bound approximation is 26° according to Eq. 3.7 (Chen et al. 2015).

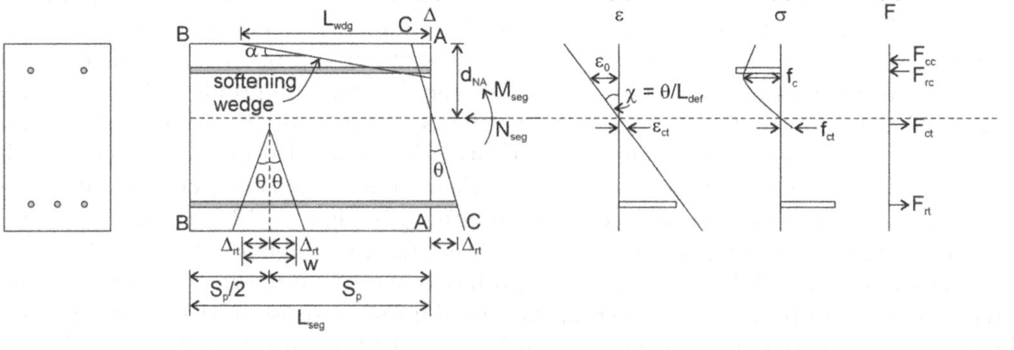

Figure 6.13 Analysis under instantaneous loads after softening. (a) cross-section; (b) segment elevation and displacement profile; (c) strain profile; (d) stress profile; (e) force profile.

As an example, consider the new segment in Figure 6.13b, which now spans a complete crack such that the segment length is taken as $3(S_p/2)$, giving a total rotation of 3θ. The analysis procedure to determine the M/θ relationship is then identical to that described above for the non-softening element, with the exception that the concrete stress–strain relationship is replaced by a size-dependent concrete stress–strain relationship, as described in Section 5.1.2, where the size of the concrete prism is the segment length L_{seg}. Of critical importance in this analysis is to note that the force in the reinforcement is still determined based on the reinforcement deformation of Δ_{rt}, not the accumulated crack opening $3\Delta_{rt}$, because the slip of the reinforcement is not accumulated within the segment. The analysis can then be repeated for all rotations until either the tensile reinforcement ruptures or the concrete can no longer take sufficient force to achieve equilibrium. A flow chart summarising this basic procedure is given in the appendix.

6.2.1.4 Analysis of results

The displacement-based segmental analysis is a powerful tool for analysis at the segment level. However, given that most mechanics techniques to quantify member deflection are framed in the context of an M/χ or moment/flexural rigidity (M/EI) relationship, let us now consider how to obtain these. In Figure 6.12c, it was explained that the curvature is simply the slope of the strain profile and, hence, dividing all rotations in Figure 6.11a by the segment length L_{seg} yields the M/χ relationship in Figure 6.11b. Similarly, taking EI as the slope of the M/χ relationship, the M/EI relationship in Figure 6.11c can be quantified.

Of great importance for serviceability and durability design is the determination of crack widths. Using standard full-interaction strain-based analysis approaches, it is impossible to directly determine crack widths, as the full-interaction approach, by definition, assumes cracks cannot open because the reinforcement and concrete are subjected to the same strain. National codes of practice, therefore, generally use approximate and simplified methods to determine crack widths by either placing a limiting stress on tensile reinforcement or by implementing simplified partial-interaction approaches using reinforcement strains determined from full-interaction analyses (Branson 1963; Bischoff2003). Using the segmental approach, this inconsistency is removed because the relationship between the moment M_{seg} and the crack width w in Figure 6.12b can be obtained by taking the total crack width at the level of the reinforcement to be equal to the slip of the reinforcement from each crack face minus the elastic extension of the concrete shown from the tension stiffening analysis as δ_c in Figure 6.6c, with the resultant shown in Figure 6.12f. Given that the elastic extension of the reinforcement δ_c is at least an order of magnitude less than the crack width, as a conservative assumption, the impact of the extension of the concrete can be ignored. The crack width at any depth can also be extracted from Figure 6.12f because the crack width varies linearly as a consequence of the Euler–Bernoulli deformation Δ in Figure 6.12b, which causes the end rotation θ.

The segmental approach is also extremely powerful in that the analysis can be applied equally to any type of concrete and to any type of internal reinforcement without the need for any empirically derived correction or 'fudge' factor. For example, when considering FRC in Figure 6.12b, the force in the uncracked concrete is determined in an identical way but using a stress–strain relationship for *FRC* instead of that for plain concrete. In the cracked tension region, from the displacement profile in Figure 6.12b, the crack width can be determined as shown in Figure 6.12f and the stress–crack width relationships defined in Section 3.3.1 can be directly applied to give the fibre tensile force F_{fib}. This differs from traditional strain-based analyses, where an equivalent stress–strain relationship must be calibrated instead of simply using the directly measured material property. Similarly, the tension stiffening analysis in Chapter 4 was shown to be generic, allowing it to be applied using any material modulus and any bond–stress–slip relationship. Hence, analysis with *FRP* reinforcement follows the same approach as for steel reinforcement, except for a simple change of material properties.

6.2.2 Segmental analysis of sustained loads

Now let us consider the behaviour of a member that is subjected to sustained applied loads, such that the impact of concrete shrinkage and creep must be considered. To do this, consider the half-segment with L_{seg} equal to $S_p/2$ in Figure 6.14b, where the crack spacing had been determined ignoring any influence of concrete time effects. That is, as an approximation, it is assumed that the cracks form at an early age such that the influence of concrete time effects on crack spacing is negligible. In this example, the segment is only subjected to an applied moment M_{seg} and no external axial load N_{seg} is applied (Visintin et al. 2013b; Hasan et al. 2019).

6.2.2.1 *Base-lines*

Prior to shrinkage taking place or any deformation being applied, the end of the segment is located at *C–C* in Figure 6.14b. Over time, as explained in Section 2.5, the consumption of water for cement hydration and the exchange of moisture with the environment creates a non-linear distribution of shrinkage strain. If all sides of the beam are free to exchange moisture with the environment, and if the concrete were free to deform without any restraint from the reinforcement, the shrinkage deformation would cause the end of the segment *C–C* to contract to *B–B* in Figure 6.14b, where the contraction from *C–C* to *B–B* is equal to $\varepsilon_{sh}L_{seg}$, with ε_{sh} varying with depth. Hence, line *B–B* in Figure 6.14b becomes the base line (BL) for concrete (BL_c) deformations, such that any

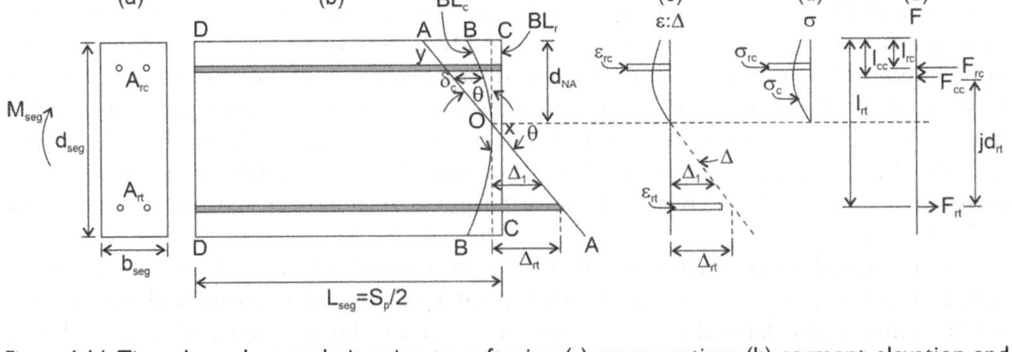

Figure 6.14 Time dependent analysis prior to softening (a) cross section; (b) segment elevation and displacement profile; (c) strain and displacement profile; (d) stress profile; (e) force profile.

movement away from B–B results in the generation of strain and consequently stress in the concrete. Similarly, since the reinforcement does not shrink with time, C–C acts as the BL for the reinforcement (BL_r), such that any movement of the reinforcement away from C–C results in the generation of stress in the reinforcement (Hasan et al. 2019).

6.2.2.2 Analyses

6.2.2.2.1 Deformation due to shrinkage alone

Having defined the BLs, let us first consider the deformation at the end of the beam due to shrinkage alone, that is, there is no applied moment such that $M_{seg}=0$. This is accomplished by determining the Euler–Bernoulli deformation for which, for equilibrium, $N_{seg}=M_{seg}=0$. This will provide the rotation about a vertical line due to shrinkage θ_{sh} and, consequently, the curvature due to shrinkage alone χ_{sh}, which can be used to determine the deflection due to shrinkage.

To do this, a similar approach to that under instantaneous loading is followed, except it is now necessary to consider deformations away from non-concurrent BLs, specifically the separate concrete BL B–B and reinforcement BL C–C in Figure 6.15b (Visintin et al. 2013b). For example, dividing the displacements between the concrete BL B–B and the segment deformation line A_1–A_1 in Figure 6.15b by the segment length $L_{seg}=S_p/2$ yields the distribution of concrete strain ε_c in Figure 6.15c. Similarly, for the displacement A_2–A_2, which induces the strains in Figure 6.15d. It can be observed that the Euler–Bernoulli displacement A_1–A_1 intercepts C–C below the soffit, and the same applies for A_2–A_2. Hence, it could be stated that there are two neutral axis depths in both Figure 6.15c and d. Similarly, dividing the deformation between the reinforcement BL C–C and the segment deformation lines A_1–A_1 and A_2–A_2 yields the reinforcement full-interaction strains ε_r.

Having defined the strains in Figure 6.15c and d, concrete and reinforcement strain-based material properties, including concrete creep, can be applied to determine the distribution of concrete stress and the stress in any reinforcement subjected to a contracting strain. Based on the distribution of stress, the forces in the reinforcement and concrete can be determined, as shown in Figure 6.15e, and the displacement profile A–A can be adjusted up or down until longitudinal force equilibrium is achieved, which in this case is N_{seg} is zero. After this, moments can be taken, and should M_{seg} not equal

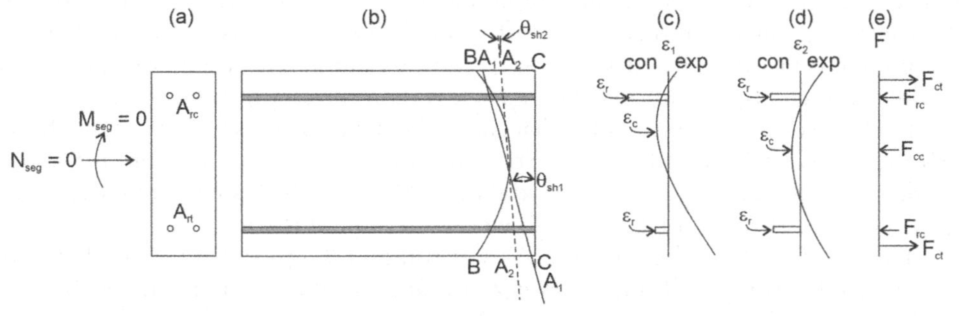

Figure 6.15 Shrinkage deformations in uncracked concrete. (a) cross-section; (b)segment elevation and displacement profile; (c) strain profile for displacement A_1-A_1; (d) strain profile for displacement A_2-A_2; (e) force profile.

zero, the process is repeated until it does, yielding θ_{sh} and consequently χ_{sh}, which is the curvature purely due to shrinkage.

6.2.2.2.2 Deformation due to shrinkage and moment

The above analysis to specifically determine θ_{sh} is now applied with an externally applied moment M_{seg} as illustrated in Figure 6.14. The reinforcement BL remains at *C–C* as it is assumed that the reinforcement is not subjected to time variations. The concrete is subjected to a variation of shrinkage strain ε_{sh} such that the movement from *C–C* to *B–B* is equal to $L_{seg}\varepsilon_{sh}$ where ε_{sh} is the shrinkage at the level being considered. As we now have both BLs, we can apply a Euler–Bernoulli displacement *A–A* where the rotation θ is measured from the vertical, as shown. We need to fix the vertical position of the sloping displacement so we know where it is. It can be fixed at x, where the displacement *A–A* intercepts the vertical line *C–C*, or we could just as easily fix it at y, where *A–A* intercepts the horizontal top fibre. Note that the depth to the neutral axis d_{NA} is of little importance in a numerical analysis. This can also be seen in Figure 6.9c, where there is no need to consider where d_{NA} might be, and also in Figure 6.15, where it can be seen that, in theory, there can be more than one neutral axis, should a neutral axis be defined as the transition from compression to tension.

From the displacement *A–A* in Figure 6.14b, the strain profile in Figure 6.14c can be derived. For example, the concrete contraction δ_c in Figure 6.14b divided by L_{seg} will yield the concrete strain ε_c at that level in Figure 6.14c. Similarly, the reinforcement contraction at the level of the reinforcement between *A–A* and *C–C*, divided by L_{seg}, will give the compressive strain ε_{rc} in Figure 6.14c. In this example, we have assumed that the concrete cracks in tension. The half-crack width Δ is measured from the vertical line through the neutral axis because the applied rotation θ is measured about this vertical axis. This results in the variation of Δ in Figure 6.14c, shown as the broken line, and Δ_{rt} at the level of the tension reinforcement. From the strains in Figure 6.14c, and using the strain–based properties that can account for creep, the strain-based stresses in Figure 6.14d can be determined. From the distribution of stress in the concrete σ_c and from the cross-sectional properties of the member, as shown in Figure 6.14a, the resultant concrete compressive force F_{cc} and its position l_{cc} can be determined, which in this example is shown from the top fibre. Similarly, for the reinforcement in compression, F_{rc} and l_{rc} can be determined. From the variation in Δ in Figure 6.14c, the half-crack width Δ_{rt} at the level of the tension reinforcement can be determined, and from the P/Δ properties, which can include the effect of shrinkage, as discussed in Chapter 4, the reinforcement force (F_{rt}) at the level l_{rt} can be established. It is worth noting that the strain in the tension reinforcement ε_{rt} is less than the strain from a full-interaction analysis because reinforcement slip adjacent to the flexural crack reduces its stiffness.

All the resultant forces and their vertical positions, as shown in Figure 6.14e, have been determined. In this example, no axial force N_{seg} is applied, so it is a matter of finding the position of *A–A* in Figure 6.14b where the sum of the forces in Figure 6.14e is zero. This can be achieved by moving *A–A* at rotation θ_1 up or down to find the position x where the sum of all the forces is zero. Once found, the moments of the forces need to be taken. As there is no axial force N_{seg}, moments can be calculated at any level, such as from the top fibre in Figure 6.14e, to find the applied moment $M_{seg\text{-}1}$ at θ_1. This is one point on the M/θ curve in Figure 6.11a, which can be constructed by varying θ.

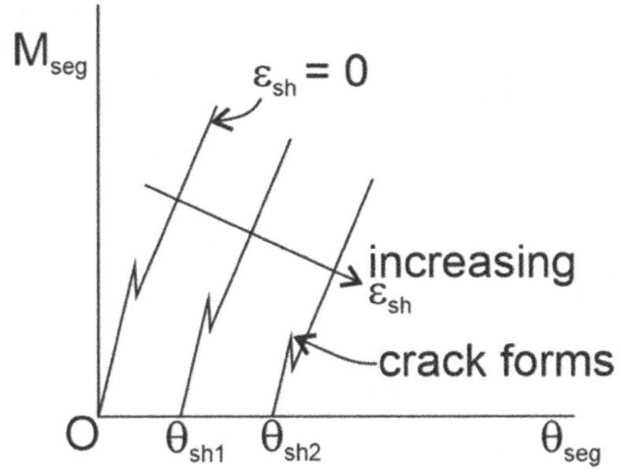

Figure 6.16 M/θ with shrinkage.

Typical results are shown in Figure 6.16. When there is no shrinkage, there is no initial rotation due to shrinkage θ_{sh}, meaning the curve starts at the origin and the step change indicates where cracking first occurs. As shrinkage is applied and increased, the curve is offset as shown by θ_{sh}, with reduced stiffness and reduced cracking moment. It is worth noting that θ_{sh} can be negative if the area of compression reinforcement is much greater than that of the tension reinforcement.

A flow chart of the analysis of a segment allowing for concrete time effects is provided in the appendix at a single point in time. The analysis can be applied for increasing rotations to construct the M/θ, M/χ relationships or the M/EI relationship in an identical manner to that done under instantaneous loads, noting that the relationships are only shown under serviceability loads, as this is the typical application of time-dependent analyses. Where analysis of the time-dependent increase in deflections or crack widths is required, the analysis will need to be repeated to produce a family of M/θ, M/χ relationships or M/EI relationships, with each curve constructed using shrinkage strains and creep coefficients appropriate to the time and environmental conditions.

6.3 NUMERICAL/CLOSED-FORM SOLUTIONS

Instead of visualising the analysis for instantaneous loads and time-dependent loads separately, let us consider the instantaneous load analysis as a simple case of the time-dependent load analysis, allowing both approaches to merge into one. We will start by looking at a basic principle that governs the mechanics, then move on to the segmental analyses.

6.3.1 Symmetry

Consider the uncracked segment in Figure 6.17 whose cross-section is symmetrically reinforced. Let us consider a uniform shrinkage over the depth ε_{sh} that will cause the concrete BL (BL_c) to move to the left from C–C to B–B, a distance δ_{sh} equal to $\varepsilon_{sh}L_{seg}$. The BL_r remains at C–C. Any forces generated within the section due to shrinkage must

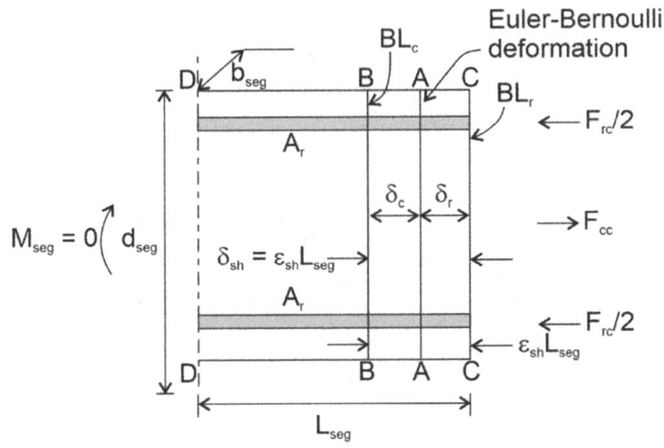

Figure 6.17 Shrinkage in a symmetrical section.

have a translational resultant that is zero, as N_{seg}=0; hence, F_{rt} is equal in magnitude to F_{rc}, and due to symmetry, this will give M_{seg} equal to zero. The only solution is to have a vertical Euler–Bernoulli displacement *A–A* in which the reinforcement contracts δ_r and the concrete expands δ_c. As with the tension stiffening analyses in Chapter 4, the position of *A–A* is determined by ensuring the tensile force in the concrete F_{ct} is equal to the compressive force in the reinforcement F_{rc}. A solution can be easily found as δ_{sh} is equal to $\delta_c+\delta_r$, meaning that the sum of the concrete and reinforcement strains equals the shrinkage strain ε_{sh}.

If one defines the neutral axis depth as the point where there is a transition from compression to tension, then it can be seen that there is no neutral axis depth d_{NA}. Furthermore, the rotation of the Euler–Bernoulli deformation θ is zero, and consequently, the curvature χ is zero; that is, there is no deflection due to shrinkage because the section is symmetrical. We need a non-symmetrical section or a non-symmetrical shrinkage distribution to get deflection due to shrinkage.

6.3.2 Segmental analysis without shrinkage

6.3.2.1 *Full-interaction analyses*

Consider the non-symmetrical segment in Figure 6.18a, which is not subjected to shrinkage; in which the concrete has not cracked; and the materials are linear–elastic. A Euler–Bernoulli deformation *A–A* is imposed, with depth d_{NA} and rotation θ. As the deformation *A–A* is linear, dividing this deformation by L_{seg} means that this deformation will impose a linear strain distribution in the concrete and consequently a linear stress distribution in the concrete. This strain distribution will also give the strains in the reinforcement and consequently the stress in the reinforcement. These stress distributions, along with the dimensions of the cross-section, will give the compressive force in the concrete F_{cc}, the tensile force in the concrete F_{ct}, the tensile force in the reinforcement F_{rt} and the compressive force in the reinforcement F_{rc} in Figure 6.18b. Moments can then be taken at any level, in this case about the top fibre, for which

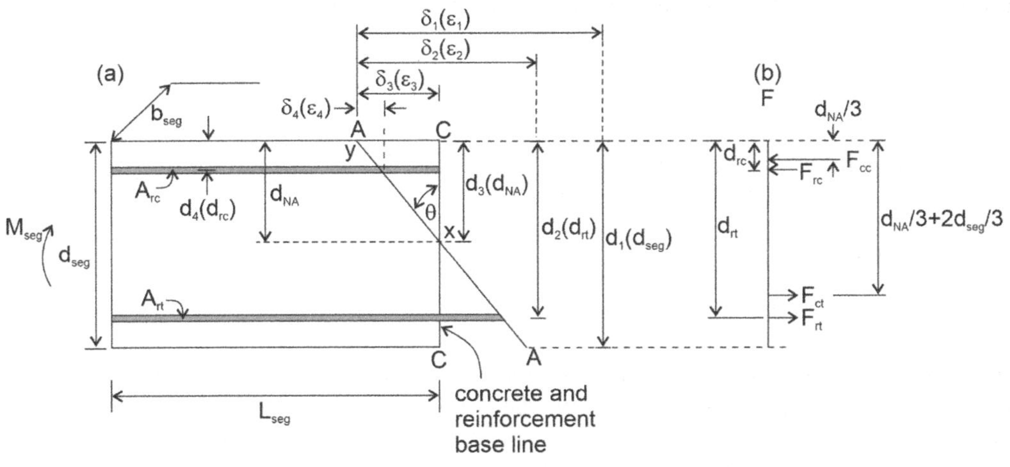

Figure 6.18 Euler–Bernoulli deformation without shrinkage (a) segment elevation and displacement profile; (b) force profile.

the lever arms are given. To find a solution requires the depth of the neutral axis d_{NA}. A convenient way of illustrating the analysis is to consider the deformations of the segment relative to the top of A–A at y, which are the deformations δ_1–δ_4, which, when divided by L_{seg}, will give the strains ε_1–ε_4.

Let us first derive the compressive force in the concrete F_{cc} in Figure 6.18a. The strain at the top fibre ε_3 is equal to $d_{NA}\theta/L_{seg}$; multiplying this by the concrete modulus E_c gives the concrete stress at the top fibre. Bearing in mind that the distribution of concrete stress is triangular and that the cross-section on which it acts is rectangular with width b_{seg}, the concrete compressive force is given by

$$F_{cc} = \frac{E_c d_{NA}^2 b_{seg} \theta}{2L_{seg}} \tag{6.2}$$

The strain in the tension reinforcement ε_{rt} is equal to $\varepsilon_2 - \varepsilon_3$ in Figure 6.18a, which equals $d_{rt}\theta/L_{seg} - d_{NA}\theta/L_{seg}$. Multiplying this strain by the reinforcement modulus E_r and the cross-sectional area A_{rt} gives the tensile force in the reinforcement:

$$F_{rt} = \frac{E_r A_{rt} \theta (d_{rt} - d_{NA})}{L_{seg}} \tag{6.3}$$

Applying the same procedure to the concrete in tension gives

$$F_{ct} = \frac{E_c (d_{seg} - d_{NA})^2 b_{seg} \theta}{2L_{seg}} \tag{6.4}$$

and the reinforcement in compression

$$F_{rc} = \frac{E_r A_{rc} \theta (d_{NA} - d_{rc})}{L_{seg}} \tag{6.5}$$

From Figure 6.18b, for translational equilibrium, the sum of the forces must be zero; that is

$$F_{cc} + F_{rc} = F_{ct} + F_{rt} \tag{6.6}$$

Inserting the above values for these forces from Eqs. 6.2 to 6.5 into Eq. 6.6 gives

$$d_{NA} = \frac{n\left(A_{rt}d_{rt} + A_{rc}d_{rc}\right) + \dfrac{b_{seg}d_{seg}^2}{2}}{n\left(A_{rt} + A_{rc}\right) + b_{seg}d_{seg}} \tag{6.7}$$

where n is the modular ratio E_r/E_c. Having determined d_{NA}, it can be used to quantify the forces in Eqs. 6.2–6.5 for a specific rotation θ. Then, by taking moments, the applied moment M_{seg} for that rotation, in this case about the top fibre, is determined, as no axial forces are being applied. By varying θ, the M/θ variation can be derived, and consequently, M/χ and M/EI.

This approach, using the Euler–Bernoulli displacement, gives exactly the same depth of neutral axis d_{NA} in Eq. 6.7, as one would obtain using transformed sections. Hence, this approach, using the Euler–Bernoulli displacement, although it does not explicitly use transformed sections, does so implicitly; that is, it is based on the same principles as those used to derive transformed sections. It may be stating the obvious, but from Eq. 6.7, it can be seen that d_{NA} is independent of θ, which is not the case when shrinkage or fibres are included in the following section.

6.3.2.2 *Partial-interaction analyses*

6.3.2.2.1 *Tension stiffening*

The above analyses, in which there is no cracking, are *FI* analyses. Let us now allow cracking to occur, which requires a *PI* analysis. From the P/Δ tension stiffening behaviour in Chapter 4, which is summarised in Figure 6.19, the force in the tension

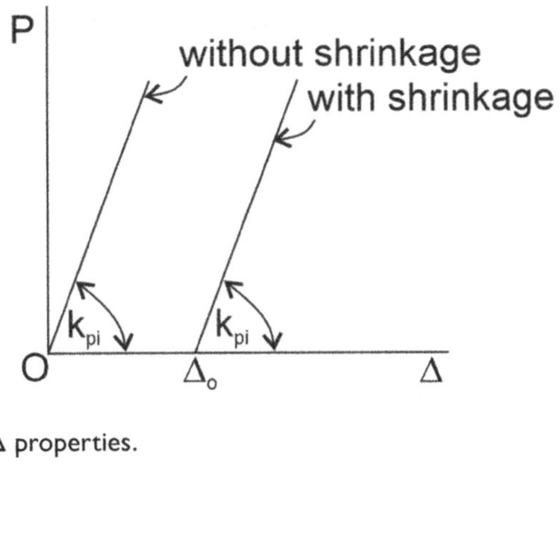

Figure 6.19 Linearised P/Δ properties.

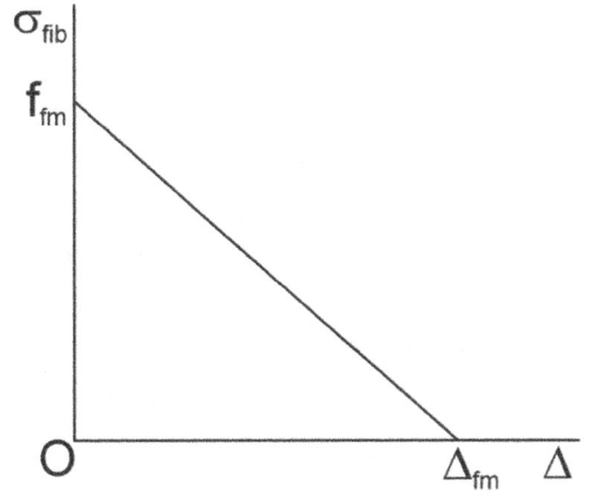

Figure 6.20 Fibre concrete stress–strain only.

reinforcement when there is no shrinkage $F_{rt\text{-}pi}$ is equal to $k_{pi}\Delta$, where k_{pi} is the P/Δ stiffness and consequently, in the metric system, has units of N/mm. From Figure 6.18,

$$F_{rt\text{-}pi} = k_{pi}(d_{rt} - d_{NA})\theta \tag{6.8}$$

Now, placing $F_{rt\text{-}pi}$ from Eq. 6.8 and F_{cc} from Eq. 6.2 into the equilibrium Eq. 6.6 with F_{ct} set to zero, as the concrete has cracked, gives

$$d_{NA}^2 \frac{b_{seg}}{2} + d_{NA}\left(nA_{rc} + \frac{k_{pi}L_{seg}}{E_c}\right) - \left(nA_{rc}d_{rc} + \frac{d_{rt}k_{pi}L_{seg}}{E_c}\right) = 0 \tag{6.9}$$

That is, in this case, a quadratic must be solved to determine d_{NA}, and once solved, all the forces in Figure 6.18b are known from Eqs. 6.2 to 6.6, allowing moments to be taken for any specific curvature. It can be seen that d_{NA} is independent of θ but dependent on k_{pi}. However, the forces in Eqs. 6.2 and 6.8 are dependent on θ.

6.3.2.2.2 Fibres

Another partial-interaction situation involves fibres crossing the crack. A linear idealisation of the material properties of fibre concrete in Section 3.3 is shown in Figure 6.20. When the crack has just formed, the fibre concrete strength is at its maximum f_{fm}. As the crack opens, there is a linear reduction in strength, which reaches zero at the half-crack width Δ_{fm}. It can be seen that the fibre stress depends purely on the half-crack width Δ.

A segment made from *FRC* is shown in Figure 6.21. A Euler–Bernoulli deformation *A–A* has been applied in Figure 6.21a. When the flexural crack has just formed over the height d_{cr} and there is no rotation θ, meaning Δ tends to zero over the full depth of the crack, the fibre stresses remain at their maximum value f_{fm} as shown in Case (1) in Figure 6.21b. As a rotation θ is applied, the widening of the crack causes a linear reduction in strength, as seen in Case (2), where the stress in the bottom fibre is xf_{fm}. A further increase in rotation will lead to Case (3), defined as the limit at which the fibres contribute over the full crack height; that is, the crack width at the bottom fibre Δ_{cm} is equal to Δ_{fm}, such that the fibre stress is zero at this level. An additional increase

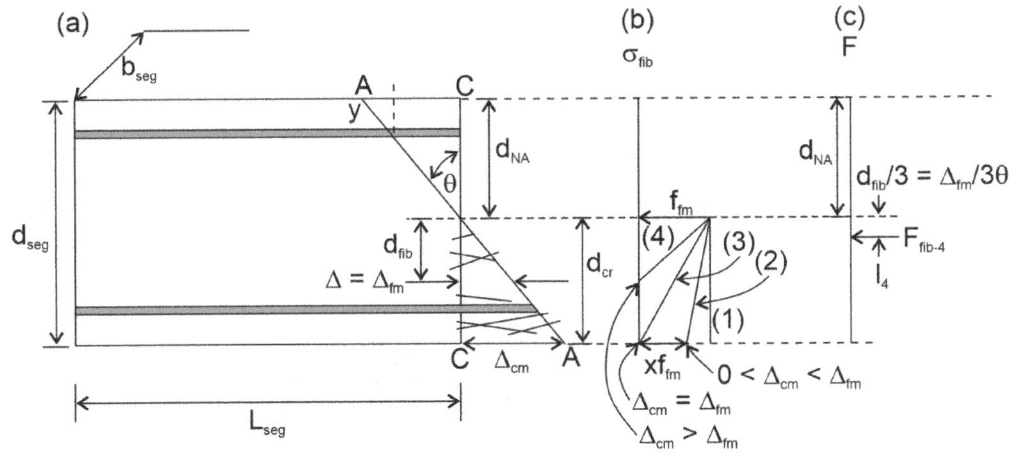

Figure 6.21 Fibre concrete forces (a) segment elevation and deformation profile; (b) fibre stresses in cracked concrete region; (c) resultant concrete fibre force in tension region.

in rotation will result in Case (4), where Δ_{cm} exceeds Δ_{fm}, such that the depth over which the fibre stress is active d_{fib} is governed by Δ_{fm} and θ.

Let us first consider Case (2) in Figure 6.21b, where the half-crack width at the soffit Δ_{cm} is less than Δ_{fm}. In this case, d_{fib} is the depth of the crack, which is $d_{seg}-d_{NA}$. From the geometry,

$$F_{fib\text{-}2} = f_{fm} b_{seg}\left(d_{seg} - d_{NA}\right)\left(1 - \frac{\Delta_{cm}}{2\Delta_{fm}}\right) \tag{6.10}$$

the position of this force from the level of the neutral axis is

$$l_2 = \left(d_{seg} - d_{NA}\right)\left[\frac{\frac{1}{2} - \frac{\Delta_{cm}}{3\Delta_{fm}}}{1 + \frac{\Delta_{cm}}{2\Delta_{fm}}}\right] \tag{6.11}$$

Eventually, the maximum half-crack width Δ_{cm} in Figure 6.21a exceeds Δ_{fm}, beyond which the fibre stress σ_{fib} is zero, as in Case (4). For this illustration of the mechanics, we will only develop a solution for Case (4).

The depth of the segment over which fibre stresses act d_{fib} is shown in Figure 6.21a, where it can be seen that it is limited by Δ_{fm}. The average stress in this region is $f_{fm}/2$, and the depth of this region is Δ_{fm}/θ; hence, the force in this region is

$$F_{fib\text{-}4} = \frac{b_{seg}\Delta_{fm} f_{fm}}{2} \tag{6.12}$$

and the lever arm from the neutral axis depth is given by

$$l_4 = \frac{\Delta_{fm}\left(d_{seg} - d_{NA}\right)}{3\Delta_{cm}} = \frac{\Delta_{fm}}{3\theta} \tag{6.13}$$

whichever is more convenient for use.

Adding this fibre force to the previous analysis that allowed for tension stiffening gives the following equation, which is Eq. 6.9 with the additional fibre term shown last:

$$d_{NA}^2 \frac{b_{seg}}{2} + d_{NA}\left(nA_{rc} + \frac{k_{pi}L_{seg}}{E_c}\right) - \left(nA_{rc}d_{rc} + \frac{d_{rt}k_{pi}L_{seg}}{E_c} + \frac{b_{seg}\Delta_{fm}f_{fm}L_{seg}}{2E_c\theta^2}\right) = 0 \qquad (6.14)$$

This applies, and very importantly, for Case (4) in Figure 6.21b, when

$$\Delta_{cm} > \Delta_{fm} \qquad (6.15)$$

Because of this proviso, an iterative solution would be required to conform to this boundary condition. It can be seen that the quadratic can only be solved for specific rotations θ, that is, curvature χ, but it can be solved with some difficulty to derive points on the M/θ curve. Equation 6.14 shows the parameters that control the derivation of d_{NA}. This example is only meant to demonstrate how it is relatively easy to include the fibre stresses if required, but difficult to find a solution. It is also worth noting that a more accurate solution would be to ensure d_{fib} in Figure 6.21a is above the tension stiffening zone and also to include fibres in the tension stiffening P/Δ component, which can be done as discussed in Section 4.3.3.

6.3.3 Segmental analyses with shrinkage

6.3.3.1 Full-interaction analyses

Now let us move on to members with segments and shrinkage, as shown in Figure 6.22. To find a solution, it has been assumed that the shrinkage strain ε_{sh} is uniform along both the depth d_{seg} and width b_{seg}. The section remains uncracked, although the effect of cracking will be dealt with later. The only difference between the segment without shrinkage in Figure 6.18 and this segment with shrinkage in Figure 6.22 is that the concrete BL has moved by $\varepsilon_{sh}L_{seg}$, creating two BLs. The neutral axis, where the concrete zone transitions into the tensile zone, therefore moves to the left, as shown.

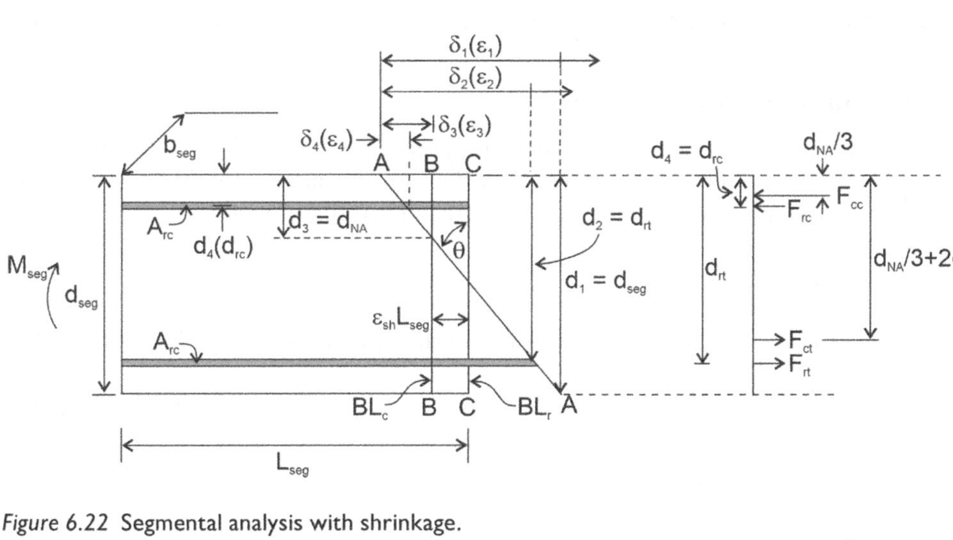

Figure 6.22 Segmental analysis with shrinkage.

The analysis applied to the segment without shrinkage in Figure 6.18 is now applied to the segment with shrinkage in Figure 6.22 to directly compare the effects of shrinkage. The analysis for the compressive force in the concrete is the same as in the non-shrinkage analysis that gave Eq. 6.2; hence

$$F_{cc\text{-}sh} = \frac{E_c d_{NA}^2 b_{seg} \theta}{2L_{seg}} \tag{6.16}$$

The force in the compression reinforcement becomes

$$F_{rc\text{-}sh} = E_r A_{rc} \left[\frac{\theta d_{NA}}{L_{seg}} - \frac{\theta d_{rc}}{L_{seg}} + \varepsilon_{sh} \right] \tag{6.17}$$

that in the tension reinforcement

$$F_{rt\text{-}sh} = E_r A_{rt} \left[\frac{\theta d_{rt}}{L_{seg}} - \frac{\theta d_{NA}}{L_{seg}} - \varepsilon_{sh} \right] \tag{6.18}$$

and that in the concrete in tension

$$F_{ct\text{-}sh} = \frac{E_c \left(d_{seg} - d_{NA} \right)^2 b_{seg} \theta}{2L_{seg}} \tag{6.19}$$

Applying these forces to the equilibrium Eq. 6.6 gives

$$d_{NA\text{-}sh} = \frac{n\left(A_{rt} d_{rt} + A_{rc} d_{rc}\right) + \dfrac{b_{seg} d_{seg}^2}{2} - \dfrac{n \varepsilon_{sh} L_{seg} \left(A_{rc} + A_{rt}\right)}{\theta}}{n\left(A_{rt} + A_{rc}\right) + b_{seg} d_{seg}} \tag{6.20}$$

which is applicable to one specific rotation and one specific shrinkage strain but can be used to construct a family of M/θ variations to assist in analyses of uncracked segments.

6.3.3.2 Partial-interaction analyses

To allow for shrinkage in the tension stiffening prism in Chapter 4, the P/Δ properties are offset, as shown in Figure 6.19, where the offset Δ_0 depends on the shrinkage strain. Using this P/Δ property gives the force in the tension reinforcement as

$$F_{rt\text{-}pi\text{-}sh} = k_{pi} \left(d_{rt} - d_{NA} \right) \theta - k_{pi} \Delta_0 \tag{6.21}$$

The force in the fibres remains as in Eq. 6.12, as this only depends on the rotation θ

$$F_{fib\text{-}pi\text{-}sh} = \frac{b_{seg} \Delta_{fm} f_{fm}}{2} \tag{6.22}$$

Applying Eq. 6.6 with these tensile components produces the following quadratic that can be solved for d_{NA} at specific rotation θ and shrinkage strain $\varepsilon_{sh.}$ This can then be used to find the forces at the specific θ and ε_{sh}, and by taking moments about the lever arms, such as those in Figure 6.22, the moment for that specific rotation can be obtained, giving one point on the M/θ curve. Repeating the procedure will yield the full curve from which the variation in curvature χ can be derived, and from which the variation in the flexural rigidities EI can be determined.

$$d_{NA}^2 \frac{b_{seg}}{2} + d_{NA}\left(nA_{rc} + \frac{k_{pi}L_{seg}}{E_c}\right) - nA_{rc}\left(d_{rc} - \frac{\varepsilon_{sh}L_{seg}}{\theta}\right) + \frac{k_{pi}L_{seg}}{E_c}\left(\frac{\Delta_0}{\theta} - d_{rt}\right) - \frac{b_{seg}\Delta_{fm}f_{fm}L_{seg}}{2E_c\theta^2} = 0 \tag{6.23}$$

Hence, we now have a fundamental equation that allows for shrinkage and fibre stresses in flexurally cracked concrete, which can be solved – albeit with difficulty when the fibre component is included – to construct a family of M/χ for analysis. At the very least, it demonstrates the parameters that govern d_{NA}.

6.4 SERVICEABILITY INSTANTANEOUS LOAD SOLUTIONS

In Sections 6.2 and 6.3, the moment/curvature response was derived through the analysis of individual points on the curve. The aim of the following closed-form solutions is to derive the continuous moment/curvature response. Given that the solutions need to be developed using assumed material properties, their derivations must be split between those considering plain concrete and those considering FRC. Either analysis can be applied with steel or *FRP* reinforcement by taking appropriate material properties for the *FRP* in Section 2.6 and bond properties for the interface between the reinforcement and concrete in Section 3.2.

6.4.1 Plain concrete

Let us first consider plain concrete, which is concrete without fibres, that is subjected to instantaneous loads as developed in Visintin et al. (2018). Importantly, this derivation under instantaneous loads will provide the basis for later derivations with the inclusion of concrete shrinkage and creep. Given that the analysis is being performed at the serviceability limit state, both the concrete and the reinforcement are considered to behave in a linear elastic manner.

6.4.1.1 Instantaneous loads

In routine design, it is most common to perform serviceability analysis by using a transformed section analysis. In this method of analysis, the reinforced concrete section, such as that in Figure 6.23a, is transformed into an equivalent plain concrete section. This is done by replacing the reinforcement (which has a higher elastic modulus) with an increased area of concrete (which has a lower elastic modulus) whilst maintaining the same lever arm, as shown in Figure 6.23b, using the following full-interaction modular ratio:

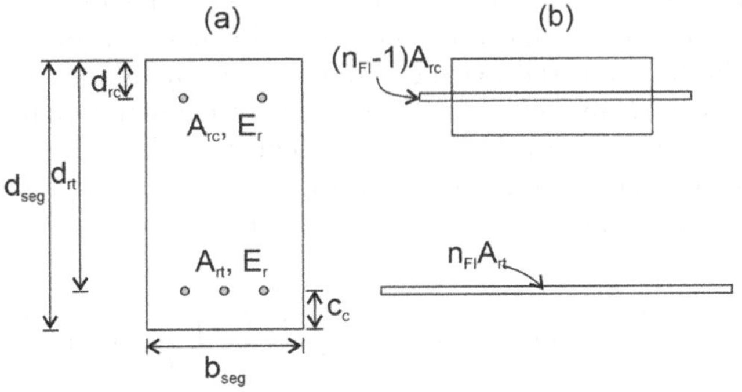

Figure 6.23 Full-interaction transformed section (a) cross section; (b) transformed section using full interaction analysis.

$$n_{FI} = \frac{E_r}{E_c} \tag{6.24}$$

In this way, the resultant forces and moments are identical, but the removal of terms relating to the axial rigidity of the reinforcement allows for simple closed-form solutions. By taking the first moment of area of the transformed section, the following equation can be solved to obtain the neutral axis depth in the cracked concrete.

$$\frac{b_{seg}d_{NA\text{-}cr}^2}{2} + \left(n_{fi} - 1\right)A_{rc}\left(d_{NA\text{-}cr} - d_{rc}\right) = n_{fi}A_{rc}\left(d_{rt} - d_{NA\text{-}cr}\right) \tag{6.25}$$

Similarly, taking the second moment of area gives the moment of inertia of the cracked section I_{cr} as

$$I_{cr} = \frac{b_{seg}d_{NA\text{-}cr}^3}{12} + \left(n_{fi} - 1\right)A_{rc}\left(d_{NA\text{-}cr} - d_{rc}\right)^2 + n_{fi}A_{rc}\left(d_{rt} - d_{NA\text{-}cr}\right)^2 \tag{6.26}$$

Importantly, although this form of transformed section analysis is commonly applied in design, it does not account for the benefits of tension stiffening. As a result, it is necessary to apply correction factors (Branson 1963; Bischoff 2003) if the advantages of tension stiffening are to be considered.

In Section 4.3.5.3, it is shown that it is possible to define a partial-interaction modular ratio n_{PI} by normalising the axial rigidity of the partial-interaction tension stiffening prism in Figure 6.12a with that of transformed reinforcement under full-interaction conditions to yield

$$n_{PI} = \frac{k_{pi}S_p}{E_cA_r} = \left(\frac{\lambda_1 S_p}{\tanh\left(\lambda_1 S_p\right)}\right)n_{FI} \tag{6.27}$$

where the coefficient λ_1 is given by

$$\lambda_1 = \sqrt{k_e\beta} \tag{6.28}$$

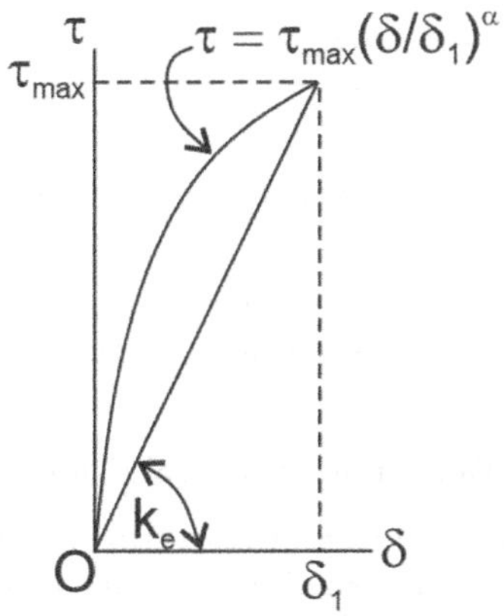

Figure 6.24 τ/δ for serviceability analysis for closed form solutions.

in which

$$\beta = L_{per}\left(\frac{1}{E_r A_r} + \frac{1}{E_c A_c}\right) \tag{6.29}$$

and in which k_e is the secant stiffness of the non-linear τ/δ bond properties in Figure 6.24, which have a rising branch described by

$$\tau = \tau_{max}\left(\frac{\delta}{\delta_1}\right)^{\alpha} \tag{6.30}$$

where δ_1 is the slip at τ_{max} and α quantifies the shape. A detailed explanation is provided in Section 4.3.5.

The crack spacing for non-linear bond properties S_p, required in Eq. 6.27, was derived in Section 4.3.5.2 and is given by

$$S_p = \left[\frac{2^{\alpha}(1+\alpha)}{\lambda_2(1-\alpha)^{1+\alpha}}\right]^{\frac{1}{1+\alpha}}\left[\frac{f_{ct}-f_{pc}}{E_c}\left(\frac{E_c A_c}{E_r A_r}+1\right)\right]^{\frac{1-\alpha}{1+\alpha}} \tag{6.31}$$

where A_r and A_c are the cross-sectional areas of the reinforcement and concrete, respectively, in the tension stiffening prism, where f_{pc} is the tensile strength of FRC after cracking, which is taken as zero for plain concrete, and where, from Eq. 4.81 of Section 4.3.5.2,

$$\lambda_2 = \frac{\tau_{max}}{\delta_1^{\alpha}}\beta \tag{6.32}$$

Hence, in Eqs. 6.25 and 6.26, replacing n_{fi} with n_{PI} for the tension reinforcement yields the following instantaneous partial-interaction neutral axis depth and moment of inertia based on a segmental analysis:

$$\frac{b_{seg} d_{NA\text{-}pi}^2}{2} + (n_{fi} - 1) A_{rc} (d_{NA\text{-}pi} - d_{rc}) = n_{pi} A_{rc} (d_{rt} - d_{NA\text{-}pi}) \tag{6.33}$$

$$I_{cr} = \frac{b_{seg} d_{NA\text{-}pi}^3}{12} + (n_{fi} - 1) A_{rc} (d_{NA\text{-}pi} - d_{rc})^2 + n_{pi} A_{rt} (d_{rt} - d_{NA\text{-}pi})^2 \tag{6.34}$$

where d_{rt} is the depth to the tension reinforcement, that is, the effective depth of the beam, and d_{rc} is the cover to the compression reinforcement.

Worked Example WE-6.1: Derivation of flexural rigidities

Let us consider the reinforced concrete cross-section in Figure E6.1a and determine the difference between the flexural rigidity of the cracked section using a full-interaction analysis compared to a partial-interaction analysis. Take the cross-section to be constructed from concrete with an elastic modulus E_c=30 GPa, compressive strength of 30 MPa, concrete tensile strength of 3 MPa and steel reinforcement with an elastic modulus of E_r=200 GPa.

FULL-INTERACTION ANALYSIS

The full-interaction modular ratio from Eq. 6.24 is

$$n_{FI} = \frac{200{,}000}{30{,}000} = 6.67 \tag{E6.1.1}$$

Applying the full-interaction modular ratio to transform the section in Figure E6.1a yields the section in Figure E6.1b.

From Eq. 6.25,

$$\frac{500 \times d_{NA\text{-}cr}^2}{2} + (6.67 - 1) \times 628 \times (d_{NA\text{-}cr} - 50) = 6.67 \times 1570 \times (700 - d_{NA\text{-}cr}) \tag{E6.1.2}$$

hence

$$d_{NA\text{-}cr} = 147.5 \text{ mm} \tag{E6.1.3}$$

and from Eq. 6.26,

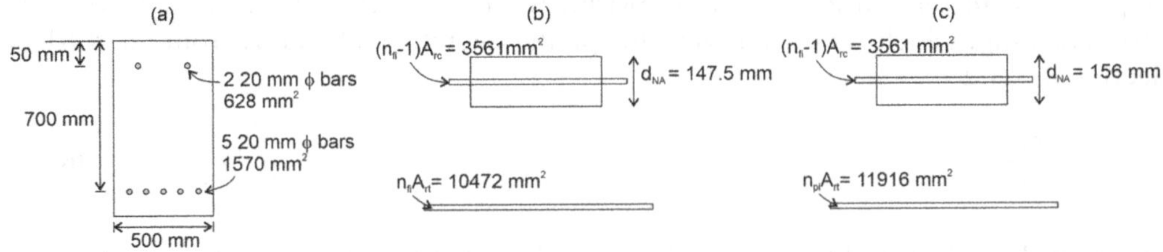

Figure E6.1 Cross-section for analysis under instantaneous loads (a) cross section; (b) transformed cross section full interaction analysis; (c) transformed section using partial interaction analysis.

$$I_{cr}=\frac{500\times147.5^3}{3}+(6.67-1)\times628\times(147.5-50)^2+6.67\times1570\times(d-147.5)^2$$
$$=3763.7\times10^6\ \text{mm}^4 \tag{E6.1.4}$$

which is the *FI* moment of inertia. Now consider the partial-interaction solution.

PARTIAL-INTERACTION ANALYSIS

Applying the *FIB* bond model in Section 3.2.4.1, assuming the concrete to be well confined, the bond properties in Figure 6.24 are $\tau_{max}=13.7$ MPa, $\delta_1=1$ mm and $\alpha=0.4$. Defining the linear bond stiffness as the secant modulus gives k_e=13.7 MPa/mm.

The crack spacing from non-linear bond properties is given by Eq. 6.31. Taking the depth of the tension stiffening prism as twice the cover to the centre of the reinforcement, from Eq. 6.29,

$$\beta=5\times\pi\times20\left(\frac{1}{200{,}000\times1570}+\frac{1}{30{,}000\times2\times50\times500}\right)=1.21\times10^{-6} \tag{E6.1.5}$$

and from Eq. 6.32,

$$\lambda_2=\frac{13.7}{1^{0.4}}\times1.21\times10^{-6}=1.66\times10^{-5} \tag{E6.1.6}$$

Now substituting into Eq. 6.31, the crack spacing is

$$S_p=\left[\frac{2^{0.4}\times(1+0.4)}{1.66\times10^{-5}\times(1-0.4)^{1+0.4}}\right]^{\frac{1}{1+0.4}}\left[\frac{3}{30{,}000}\left(\frac{30{,}000\times2\times50\times500}{200{,}000\times1570}+1\right)\right]^{\frac{1-0.4}{1+0.4}}$$
$$=160.4\ \text{mm} \tag{E6.1.7}$$

Now finding the partial-interaction modular ratio given by Eq. 6.27:

$$n_{PI}=\left(\frac{\lambda_1 S_p}{\tanh(\lambda_1 S_p)}\right)n_{FI} \tag{E6.1.8}$$

which $\lambda_1$is given by Eq. 6.28:

$$\lambda_1=\sqrt{13.7\times5\times\pi\times20\times\left(\frac{1}{200{,}000\times1570}+\frac{1}{30{,}000\times2\times50\times500}\right)}=0.00407 \tag{E6.1.9}$$

Now substituting λ_1 S_p and n_{fi} into Eq. E6.1.8 gives

$$n_{PI}=\left(\frac{0.00407\times160.4}{\tanh(0.00407\times160.4)}\right)6.67=7.59 \tag{E6.1.10}$$

Transforming the section in Figure E6.1a using the full-interaction modular ratio for the compression reinforcement and the partial-interaction modular ratio for the tension reinforcement yields the cross-section in Figure E6.1c, which has an area of tension reinforcement that is 13.8% larger than that of the full-interaction section.

Taking the first moment of area of the cracked section in Figure E1c:

$$\frac{500\times d_{NA\text{-}cr}^2}{2}+(6.67-1)\times 628\times(d_{NA\text{-}cr}-50)=7.59\times 1570\times(700-d_{NA\text{-}cr}) \tag{E6.1.11}$$

$$d_{NA\text{-}cr}=156.2\text{ mm} \tag{E6.1.12}$$

and taking the second moment of area

$$I_{cr}=\frac{500\times 156.2^{\,3}}{3}+(6.67-1)\times 628\times(156.2\ -50)^2+7.59\times 1570\times(d-156.2\ )^2$$
$$=4199.2\times 10^6\text{ mm}^4 \tag{E6.1.13}$$

Comparing the second moment of area obtained using full-interaction assumptions to that obtained with partial-interaction assumptions, it is evident that the inclusion of tension stiffening in this example increases the cracked moment of inertia by approximately 40%.

6.5 SERVICEABILITY TIME-DEPENDENT SOLUTION

6.5.1 Plain concrete

Whilst a straightforward replacement of n_{PI} for n_{FI} is possible under instantaneous loads, when considering time-dependent deformations, it is necessary to consider the impact of residual deformations due to shrinkage alone, that is, θ_{sh} in Figures 6.15 and 6.16. This represents a significant change to the analysis, as it is now necessary to quantify the deformations that occur due to shrinkage alone (line *B–B* in Figure 6.14b), and to do this, the contributions of concrete in tension must be considered.

To simplify the derivation, as is common in practice, let us first make the assumption that the shrinkage strain is uniform throughout the cross-section such that the deformation due to unrestrained shrinkage $\varepsilon_{sh}L_{seg}$ is constant in Figure 6.25b; models to define the magnitude of the shrinkage strain were presented in Section 2.5.2. Furthermore, let us assume that at the serviceability limit, all materials remain in their linear elastic range. This enables us to establish closed-form expressions as in Visintin et al. (2018).

6.5.1.1 *Axial forces*

Applying the basic linear elastic analysis in Figure 6.25b, where there are now two separate BLs, gives the parallel strain profiles in Figure 6.25c that are separated by ε_{sh} and from which the stress profile in Figure 6.25d can be determined, in which the time-dependent elastic modulus, $E_c(t,t_0)$ as defined in Section 2.5.3, is used.

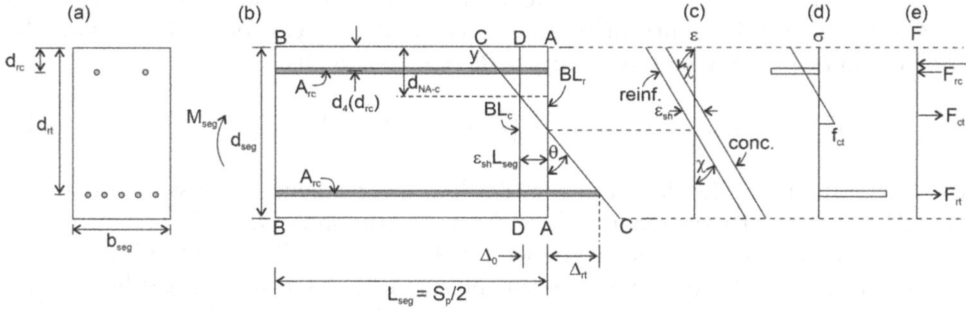

Figure 6.25 Segmental analysis for closed form solution under sustained loads (a) cross-section; (b) segment elevation and deformation profile; (c) strain profile; (d) stress profile; (e) force profile.

Hence, by integrating stresses, the force developed in the concrete in compression in Figure 6.25e is

$$F_{cc} = \frac{1}{2} b_{seg} E_c(t, t_0) d_{NA\text{-}c}^2 \chi \tag{6.35}$$

Similarly, integrating the tensile stress in the concrete in Figure 6.25d, the tensile force developed in the concrete is

$$F_{ct} = \frac{1}{2} b_{seg} E_c(t, t_0) d_{ct}^2 \chi \tag{6.36}$$

where the depth of the tension region is limited by the tensile cracking stress f_{ct} such that

$$d_{ct} = \frac{f_{ct}}{\chi E_c(t, t_0)} \tag{6.37}$$

By substituting Eq. 6.37 into Eq. 6.36, the force in the concrete tension region is

$$F_{ct} = \frac{1}{2} b_{seg} E_c(t, t_0) \left(\frac{f_{ct}}{E_c} \right)^2 \frac{1}{\chi} \tag{6.38}$$

In Figure 6.25c, the strain in the compression reinforcement can be expressed as a function of the curvature and the concrete neutral axis depth as

$$\varepsilon_{rc} = \chi (d_{NA\text{-}c} - d_{rc}) + \varepsilon_{sh} \tag{6.39}$$

where d_{rc} is the cover from the extreme compression face to the centroid of the compression reinforcement.

Assuming the reinforcement behaves elastically, the corresponding force in the compression reinforcement can be determined by multiplying the strain by the axial rigidity of the reinforcement $E_r A_{rc}$ to give

$$F_{rc} = E_r A_{rc} \chi (d_{NA\text{-}c} - d_{rc}) + E_r A_{rc} \varepsilon_{sh} \tag{6.40}$$

To express the force in the reinforcement in terms of the transformed section in Figure 6.23b, the axial rigidity of the reinforcement $E_r A_{rc}$ can be replaced with the full-interaction transformed axial rigidity $n_{FI} E_c(t,t_0) A_{rc}$ to give

$$F_{rc} = (n_{fi} - 1) E_c(t,t_0) A_{rc} \chi (d_{NA\text{-}c} - d_{rc}) + (n_{fi} - 1) E_c(t,t_0) A_{rc} \varepsilon_{sh} \tag{6.41}$$

Now consider the force developed in the tensile reinforcement F_{rt} crossing a crack in Figure 6.25e. To quantify the magnitude of this force, the slip of the reinforcement relative to the crack face must be known. In Section 4.3.5.4, and illustrated in Figure 6.19, it was shown that the slip of the reinforcement could be described using a linear relationship with a constant crack opening stiffness k_{pi}; however, the P/Δ relationship is displaced from the origin by a slip Δ_0, which is the additional slip required to overcome the initial contraction of the reinforcement due to the restraint of shrinkage Δ_0 in Figure 6.19. In Figure 6.19, the force in the reinforcement is given by

$$F_{rt} = k_{pi}(\Delta - \Delta_0) \tag{6.42}$$

where for concrete without fibres, the crack opening stiffness is

$$k_{pi} = E_r A_{rt} \left[\frac{\lambda_1}{\tanh\left(\dfrac{\lambda_1 S_p}{2}\right)} \right] \tag{6.43}$$

and

$$\Delta_0 = \frac{\varepsilon_{sh}}{\tanh\left(\dfrac{\lambda_1 S_p}{2}\right)} \tag{6.44}$$

in which λ_1 has already been defined in Eq. 6.28.

From the displacement profile in Figure 6.25b, the total slip of the tension reinforcement relative to the concrete BL Δ_{rt} is

$$\Delta_{rt} = \theta(d_{rt} - d_{NA\text{-}c}) \tag{6.45}$$

Therefore, substituting Eqs. 6.43–6.45 into Eq. 6.42 and simplifying gives

$$F_{rt} = k_{pi}\, \theta(d_{rt} - d_{NA\text{-}c}) - E_r A_{rt} \varepsilon_{sh} \tag{6.46}$$

Replacing the rotation term in Eq. 6.46 with the equivalent curvature yields

$$F_{rt} = \frac{k_{pi}\, \chi S_p}{2}(d_{rt} - d_{NA\text{-}c}) - E_r A_{rt} \varepsilon_{sh} \tag{6.47}$$

Further replacing the crack opening stiffness k_{pi} with the partial-interaction modular ratio in Eq. 6.27 and simplifying yields

$$F_{rt-pi} = n_{PI} E_c(t,t_0) A_{rt} \chi (d_{rt} - d_{NA\text{-}c}) + (n_{fi} - 1) E_c(t,t_0) A_{rt} \varepsilon_{sh} \tag{6.48}$$

Interestingly, it can be observed that the first term on the RHS is the same as the instantaneous crack opening behaviour, and the second term on the RHS is an offset to allow for the difference in BLs between the reinforcement and concrete in Figure 6.25b.

6.5.1.2 Equilibrium

Summing the forces in the concrete in compression and tension, and the reinforcement in compression and tension given by Eqs. 6.35, 6.38, 6.41 and 6.48 respectively, gives the following expression for longitudinal force equilibrium in Figure 6.25e:

$$\chi^2\left[n_{PI}A_{rt}\left(d_{rt}-d_{NA\text{-}c}\right)-\left(n_{fi}-1\right)A_{rc}\left(d_{NA\text{-}c}-d_{rc}\right)-\frac{1}{2}b_{seg}d_{NA\text{-}c}^2\right]$$
$$-\chi n_{FI}\varepsilon_{sh}\left(A_{rt}+A_{rc}\right)+\frac{1}{2}b_{seg}\left(\frac{f_{ct}}{E_c(t,t_0)}\right)^2=0 \tag{6.49}$$

Taking moments about the top fibre in Figure 6.25e yields

$$M_{seg}=F_{rt\text{-}pi}d_{rt}-F_{rc}d_{rc}-F_{cc}\left(\frac{d_{NA\text{-}c}}{3}\right)+F_{ct}\left(d_{NA\text{-}c}+\frac{2}{3}d_{ct}\right) \tag{6.50}$$

Substituting the magnitudes of the longitudinal forces given by Eqs. 6.35, 6.38, 6.41 and 6.48 into Eq. 6.50 gives

$$M_{seg}=n_{PI}E_c(t,t_0)A_{rt}d_{rt}\chi\left(d_{rt}-d_{NA\text{-}c}\right)-n_{FI}E_c(t,t_0)A_{rt}d_{rt}\varepsilon_{sh}$$
$$-\left(n_{FI}-1\right)E_c(t,t_0)A_{rc}d_{rc}\chi\left(d_{NA\text{-}c}-d_{rc}\right)-\left(n_{fi}-1\right)$$
$$E_c(t,t_0)A_{rc}d_{rc}\varepsilon_{sh}-\frac{1}{6}b_{seg}d_{NA\text{-}c}^3\ \chi$$
$$+\frac{1}{2}b_{seg}\left(\frac{f_{ct}}{E_c(t,t_0)}\right)^2\frac{1}{\chi}\left(d_{NA\text{-}c}+\frac{2f_{ct}}{3\frac{f_{ct}}{E_c(t,t_0)}E_c(t,t_0)}\right) \tag{6.51}$$

which can be further rearranged in terms of the curvature to yield

$$\chi^3\left[n_{PI}A_{rt}d_{rt}\left(d_{rt}-d_{NA\text{-}c}\right)-\left(n_{fi}-1\right)A_{rc}d_{rc}\left(d_{NA\text{-}c}-d_{rc}\right)-\frac{1}{6}b_{seg}d_{NA\text{-}c}^3\right]$$
$$-\chi^2\left[\frac{M_{seg}}{E_c(t,t_0)}+n_{fi}\varepsilon_{sh}\left(A_{rt}d_{rt}+A_{rc}d_{rc}\right)\right]+\chi\left[\frac{1}{2}b_{seg}d_{NA\text{-}c}\left(\frac{f_{ct}}{E_c(t,t_0)}\right)^2\right]+\frac{1}{3}b_{seg}\left(\frac{f_{ct}}{E_c(t,t_0)}\right)^3 \tag{6.52}$$

For a given curvature, Eqs. 6.49 and 6.52 can be solved simultaneously to give $d_{NA\text{-}c}$ and M_{seg}, providing a complete analytical solution to describe force and moment equilibrium in the segment shown in Figure 6.25b. Whilst being complete, they cannot be solved in closed form and hence do not provide a simple path to an analytical solution.

6.5.1.3 Simplified approach

Recognising this limitation, Visintin et al. (2018) attempted to simplify the solution by ignoring the contribution of concrete in tension from Eqs. 6.50 and 6.51, on the basis that shrinkage strains generally cause concrete to crack at very early ages. Hence, ignoring the contribution of the tensile capacity of concrete results in minimal error for plain concrete. This simplification yields longitudinal force and moment equilibrium expressions in Eqs. 6.53 and 6.54.

$$
\begin{aligned}
&n_{PI}E_c(t,t_0)A_{rt}\chi(d_{rt}-d_{\text{NA-}c})-n_{fi}E_c(t,t_0)A_{rt}\varepsilon_{sh}-(n_{fi}-1)E_c(t,t_0)A_{rc}\chi(d_{\text{NA-}c}-d_{rc})\\
&-(n_{fi}-1)E_c(t,t_0)A_{rc}\varepsilon_{sh}-\frac{1}{2}b_{seg}d_{\text{NA-}c}^2\ \chi=0
\end{aligned}
\tag{6.53}
$$

$$
\begin{aligned}
M_{seg}&=n_{PI}E_c(t,t_0)A_{rt}d_{rt}\chi(d_{rt}-d_{\text{NA-}c})-n_{FI}E_c(t,t_0)A_{rt}d_{rt}\varepsilon_{sh}-(n_{PI}-1)E_c(t,t_0)A_{rc}d_{rc}\chi\\
&(d_{\text{NA-}c}-d_{rc})-(n_{FI}-1)E_c(t,t_0)A_{rc}d_{rc}\varepsilon_{sh}-\frac{1}{6}b_{seg}d_{\text{NA-}c}^3\ \chi
\end{aligned}
\tag{6.54}
$$

Equation 6.54 can be rearranged to express curvature as a function of the applied moment:

$$
\chi=\frac{M_{seg}+n_{fi}E_c(t,t_0)\varepsilon_{sh}(A_{rt}d_{rt}+A_{rc}d_{rc})}{E_c(t,t_0)\left[n_{PI}A_{rt}d_{rt}(d_{rt}-d_{\text{NA-}c})-n_{FI}A_{rc}d_{rc}(d_{\text{NA-}c}-d_{rc})-\frac{1}{6}b_{seg}d_{\text{NA-}c}^3\right]}
\tag{6.55}
$$

By recognising that the denominator of Eq. 6.55 is equal to the cross-section flexural rigidity because $\chi=M/EI$, Eq. 6.55 can be written more simply as

$$
\chi=\frac{M_{seg}}{E_c(t,t_0)I_{cr\text{-}pi}}+\frac{n_{FI}E_c(t,t_0)\varepsilon_{sh}(A_{rt}d_{rt}+A_{rc}d_{rc})}{E_c(t,t_0)I_{cr\text{-}pi}}
\tag{6.56}
$$

This simplification is important as it shows that the first term on the RHS of Eq. 6.56 represents the increase in curvature due to an applied moment, and the second term accounts for the offset due to shrinkage alone.

To solve Eq. 6.56, it again involves determining the unknown neutral axis depth for any given moment, which can be done by equating Eqs. 6.53 and 6.54 to yield the following equilibrium expression:

$$d_{NA\text{-}c}^3 - 3e_{eff}d_{NA\text{-}c}^3 + d_{NA\text{-}c}\left[\frac{6}{b_{seg}}\left(A_1 - e_{eff}A_0\right)\right] - \frac{6}{b_{seg}}\left(A_2 - e_{eff}A_1\right) = 0 \tag{6.57}$$

in which

$$e_{eff} = \frac{M_{seg} + E_r\varepsilon_{sh}\left(A_{rt}d_{rt} + A_{rc}d_{rc}\right)}{E_r\varepsilon_{sh}\left(A_{rt} + A_{rc}\right)} \tag{6.58}$$

$$A_0 = n_{PI}A_{rt} + \left(n_{FI} - 1\right)A_{rc} \tag{6.59}$$

$$A_1 = n_{PI}A_{rt}d_{rt} + \left(n_{FI} - 1\right)A_{rc}d_{rc} \tag{6.60}$$

$$A_2 = n_{PI}A_{rt}d_{rt}^2 + \left(n_{FI} - 1\right)A_{rc}d_{ct}^2 \tag{6.61}$$

The following closed-form solution for Eq. 6.57 may be considered impractical for hand-based analysis of simple elements, but it can be easily solved using a spreadsheet.

$$d_{NA\text{-}c} = 2\sqrt{e_{ff}^2 + 2\frac{A_0}{b_{seg}} - 2\frac{A_1}{b_{seg}}}\cos\left\{\frac{1}{3}\text{acos}\left[\frac{e_{eff}^3 - 3\frac{A_0}{b_{seg}}e_{eff}^2 - 6\frac{A_1}{b_{seg}}e_{eff} + 3\frac{A_2}{b_{seg}}}{\left(\sqrt{e_{eff}^2 + 2\frac{A_0}{b_{seg}}e_{eff} - 2\frac{A_1}{b_{seg}}}\right)^3}\right] - \frac{2\pi}{3}\right\} + e_{eff} \tag{6.61a}$$

6.5.2 Fibre concrete

The development of a rational design solution for *FRC* and *UHPFRC* using the segmental approach was proposed by Sturm et al. (2019), using the same method as outlined above numerically for concrete in tension, but allowing for the additional contribution of tension in fibres crossing a crack, as in Visintin and Oehlers (2018). Unfortunately, following this approach, there is no simple relationship connecting moment and curvature; instead, the individual forces in the concrete and reinforcement must be determined separately, after which moments can be calculated. In an attempt to develop a straightforward design approach, a bilinear approximation to the true non-linear moment–curvature relationship was devised. Unfortunately, this approach still necessitates solving a significant number of expressions, and hence, it is not presented here; the reader is instead referred to the original work. It is clear that much further work is required to develop simple design solutions for *FRC*, particularly when time effects are included, and an approach worth further exploration is the development of an equivalent stress block in tension that scales with the maximum crack width.

6.6 CLOSED-FORM SOLUTION AT ULTIMATE LIMIT STATE

In Section 6.1, the basic mechanics of the segmental approach were defined, including the formation of the concrete softening wedge in Figure 6.13b, where it was observed that the length of the segment for analysis depends on the length of the concrete softening wedge. When considering the development of a closed-form analytical solution, this dependency creates a challenge in that the length of the segment is not known and must be assumed. As an alternative for ultimate limit state analysis, Visintin and Oehlers (2016) suggested an approximation whereby the length of the segment in Figure 6.26b could be taken as equal to the length of the wedge, assuming that the wedge extends the full depth of the neutral axis. That is

$$L_{seg} = \frac{d_{NA}}{\tan\alpha} \tag{6.62}$$

In Figure 6.26b, the total rotation from all cracks within the segment θ_T is $2n\theta$, where

$$n = \frac{L_{seg}}{S_p} \tag{6.63}$$

hence substituting Eq. 6.62 into Eq. 6.63 gives

$$n = \frac{d_{NA}}{S_p \tan\alpha} \tag{6.64}$$

In Figure 6.26b, the total slip of the reinforcement from each of the crack faces can be determined geometrically. However, given that it is the slip from each crack face that determines the force developed, this total slip $\Delta_{rt\text{-}t}$ must be divided by $2n$ to give the slip at each crack face Δ_{rt}. That is,

$$\Delta_{rt} = \frac{\theta_T\left(d_{rt} - d_{NA}\right)}{2n} \tag{6.65}$$

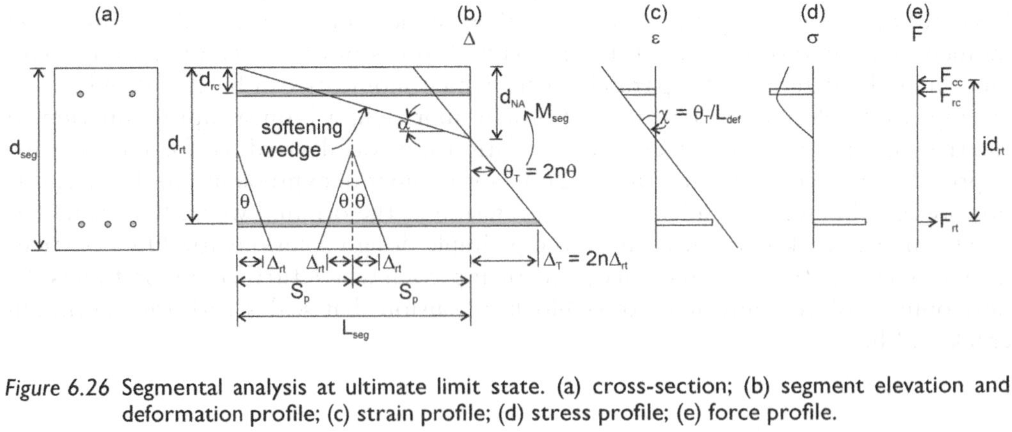

Figure 6.26 Segmental analysis at ultimate limit state. (a) cross-section; (b) segment elevation and deformation profile; (c) strain profile; (d) stress profile; (e) force profile.

and substituting Eq. 6.64 into Eq. 6.65 gives

$$\Delta_{rt} = \frac{S_p\theta_T \tan\alpha\left(d_{rt} - d_{NA}\right)}{2n} \tag{6.66}$$

Then, applying the crack opening stiffness given by Eq. 6.43 gives

$$F_{rt} = k_{pi}\frac{S_p\theta_T \tan\alpha\left(d_{rt} - d_{NA}\right)}{2n} \tag{6.67}$$

Taking the compression reinforcement to be located in the uncracked concrete, the strain in the reinforcement in Figure 6.26c can be determined by dividing the deformation at the level of the compression reinforcement in Figure 6.26b by the segment length. That is

$$F_{rc} = \frac{E_r A_{rc}\theta_T \tan\alpha\left(d_{NA} - d_{rc}\right)}{d_{NA}} \tag{6.68}$$

and in which $E_{rc}A_{rc}$ is the axial rigidity of the compression reinforcement.

If we now consider equilibrium of longitudinal forces in Figure 6.26e, the unknown force in the compression reinforcement is

$$F_{cc} = F_{rt} - F_{rc} \tag{6.69}$$

which, by substituting Eqs. 6.67 and 6.68, yields

$$F_{cc} = \frac{\theta_T \tan\alpha\left[E_r A_{rc}\left(d_{rc} - d_{NA}\right) + S_p k_{pi}\left(d_{rt} - d_{NA}\right)\right]}{2d_{NA}} \tag{6.70}$$

Similarly, taking moments about the level of the tensile reinforcement in Figure 6.26e gives

$$M_{seg} = \left[jd_{rt} + \left(d_{rc} - d_{rt}\right)\right]\frac{E_r A_{rc}\theta_T \tan\alpha\left(d_{rc} - d_{NA}\right)}{d_{NA}} + jd_{rt}\frac{S_p k_{pi}\theta_T \tan\alpha\left(d_{rt} - d_{NA}\right)}{2d_{NA}} \tag{6.71}$$

Equation 6.71, therefore, provides the generic form of the moment–rotation relationship at all load levels following concrete softening, in terms of the unknown concrete compressive lever arm jd_{rt} in Figure 6.14e and the neutral axis depth d_{NA}.

To develop a specific solution to Eq. 6.71, the shape of the concrete stress–strain relationship must be defined. In Section 2.3.1, Popovics' (1973) continuous stress–strain relationship was presented as it provides a continuous stress–strain relationship for concrete and can be easily adapted to allow for the impact of concrete confinement. The form of the model does not, however, lend itself to integration, and hence it cannot be used to develop a closed-form solution to Eq. 6.71. Hence, as an alternative to Popovics' (1973) non-linear concrete stress–strain relationship, let us consider the relationship developed by Hognestad et al. (1955), which can easily be integrated and is shown in Figure 6.27 and given by

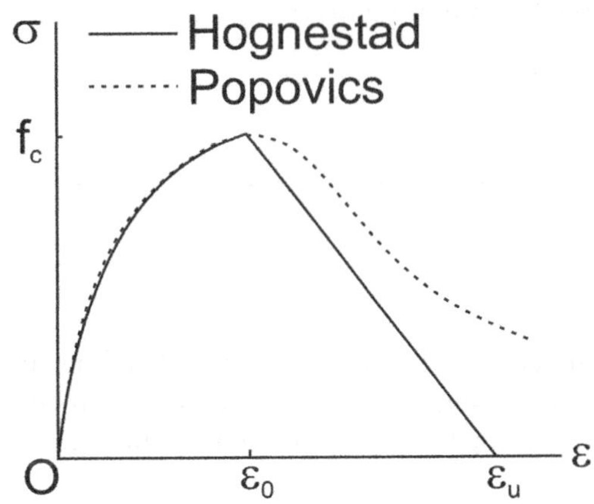

Figure 6.27 Stress–strain relationships.

$$\sigma_c = \begin{cases} f_c\left[\left(\dfrac{2\varepsilon}{\varepsilon_0}\right)-\left(\dfrac{\varepsilon}{\varepsilon_0}\right)^2\right]; & \varepsilon \le \varepsilon_0 \\ f_c - \dfrac{f_c}{\varepsilon_u - \varepsilon_0}(\varepsilon - \varepsilon_0); & \varepsilon > \varepsilon_0 \end{cases} \tag{6.72}$$

The upper term is the ascending branch in Figure 6.27 and the lower term the linear descending branch that connects the peak stress f_c at the strain ε_0 to some strain ε_u at zero stress.

Replacing F_{cc} in Eq. 6.70 with the integral of the concrete compressive force based on the stress in Eq. 6.72 yields

$$\frac{\theta_T \tan\alpha\left[E_r A_{rc}(d_{rc} - d_{NA}) + S_p k_{pi}(d_{rt} - d_{NA})\right]}{2d_{NA}} = bd_{NA}\int_0^{\varepsilon_{tp}} \sigma(\varepsilon_c) \tag{6.73}$$

where the strain at the top fibre of the concrete is

$$\varepsilon_{tp} = \frac{\theta_T d_{NA}}{L_{seg}} \tag{6.74}$$

which, upon substitution of Eq. 6.62, yields

$$\varepsilon_{tp} = \theta_T \tan\alpha \tag{6.75}$$

Solving the RHS of Eq. 6.73 when the strain is less than ε_0 gives

$$F_{cc} = bf_c d_{NA}\left(\frac{\varepsilon_{tp}}{\varepsilon_0}\right)\left(1 - \frac{\varepsilon_{tp}}{3\varepsilon_0}\right) \tag{6.76}$$

and substituting Eq. 6.76 into Eq. 6.73 then allows for the solution of the unknown neutral axis depth

$$d_{NA} = \frac{-B + 3\varepsilon_0^2\left(A_{rc}E + S_p k_{pi}\right)}{4bf_c\left(3\varepsilon_0 - \theta_T \tan\alpha\right)} \tag{6.77}$$

where

$$B = \sqrt{3}\varepsilon_0\sqrt{B_1 + B_2 + B_3 + B_4} \tag{6.78}$$

in which

$$B_1 = 2A_{rc}E_r\varepsilon_0^2\left(A_{rc}E + S_p k_{pi}\right) \tag{6.79}$$

$$B_2 = 16bd_cA_{rc}E_r\left(3\varepsilon_0 - \theta_T \tan\alpha\right) \tag{6.80}$$

$$B_3 = 8bdf_cS_pk_{pi}\left(3\varepsilon_0 - \theta_T \tan\alpha\right) \tag{6.81}$$

$$B_4 = 3S_p^2k_{pi}^2\varepsilon_0^2 \tag{6.82}$$

The centroid of the rising branch of Hognestad's relationship, measured from the top fibre, is given by

$$l_{cc} = d_{NA}\left(1 - \frac{\dfrac{\varepsilon_{tp}}{4\varepsilon_0} - \dfrac{2}{3}}{\dfrac{\varepsilon_{tp}}{3\varepsilon_0} - 1}\right) \tag{6.83}$$

Substituting the top fibre strain given by Eq. 6.75, which must be less than ε_0 into Eq. 6.72 and Figure 6.27, and defining $jd_{rt}=d_{rt}-l_{cc}$ in Figure 6.14e gives

$$jd_{rt} = d_{rt} - \frac{d_{NA}}{4} + \frac{\theta_T d_{NA}}{12\left(\theta_T - 3\varepsilon_0 \cot\alpha\right)} \tag{6.84}$$

Equation 6.71 can now be solved to give the following specific solution to the moment-rotation relationship before concrete softening and whilst the reinforcement is elastic, in which F_{cc} is taken from Eq. 6.70:

$$M_{seg} = \left[d_{rt} + d_{NA}\left(\frac{\dfrac{\varepsilon_{tp}}{4\varepsilon_0} - \dfrac{2}{3}}{\dfrac{\varepsilon_{tp}}{3\varepsilon_0} - 1} - 1\right)\right]F_{cc} + \frac{\theta_T \tan\alpha A_{rc}E\left(d_{rt} - d_{NA}\right)\left(d_{rc} - d_{NA}\right)}{d_{NA}} \tag{6.85}$$

By limiting the force in the tension reinforcement to the yield strength $F_{rt\text{-}y}$, that is, applying the assumption that the reinforcement is elastic perfectly plastic, Visintin and Oehlers (2016) applied the same procedure to develop expressions for the moment–rotation relationship when the tension reinforcement had yielded, prior to concrete softening, and where the axial rigidity of the compression reinforcement is EA_{rc}:

$$M_{seg} = F_{rt\text{-}y}d_{rt} + d_{NA}\left(F_{rt\text{-}y} - F_{rc}\right)\left(\frac{\dfrac{\theta_T \tan\alpha}{4\varepsilon_0} - \dfrac{2}{3}}{\dfrac{\theta_T \tan\alpha}{3\varepsilon_0} - 1} - 1\right) - F_{rc}d_{rc} \tag{6.86}$$

where

$$d_{NA} = -\frac{3F_{rt\text{-}y}\varepsilon_0^2 + E - 3E_rA_{rc}\varepsilon_0^2\theta_T \tan\alpha}{2b_{seg}f_c\theta_T \tan\alpha\left(\theta_T \tan\alpha - 3\varepsilon_0\right)} \tag{6.87}$$

in which

$$E = \sqrt{E_1 + E_2 + E_3} \tag{6.88}$$

and in which

$$E_1 = 9E_rA_{rc}\varepsilon_0^2\theta_T \tan\alpha\left(E_rA_{rc}\theta_T \tan\alpha - 2F_{rt\text{-}y}\right) \tag{6.89}$$

$$E_2 = 12b_{seg}d_{rc}f_cE_rA_{rc}\theta_T^2 \tan\alpha^2 \tag{6.90}$$

$$E_3 = 3F_{rt\text{-}y}^2\varepsilon_0^2 \tag{6.91}$$

When both the compression and tension reinforcement have yielded, the elastic contribution of the compression reinforcement in Eq. 6.85 can be replaced by the compression reinforcement yield force $A_{rc}f_y$. This substitution gives the segment moment:

$$M_{seg} = F_{rt\text{-}y}d_{rt} + F_{rc\text{-}y}d_{rc} + \frac{b_{seg}d_{NA}^2f_c\left(5\varepsilon_0 - 8\theta_T \tan\alpha\right)}{12\theta_T \tan\alpha} - \frac{\left[d_{NA}^2f_c\left(\varepsilon_0 - \theta_T \tan\alpha\right)\right]^2\left(2\varepsilon_0 - 3\varepsilon_u + \theta_T \tan\alpha\right)}{6\theta_T^2 \tan\alpha^2\left(\varepsilon_0 - \varepsilon_u\right)} \tag{6.92}$$

where the corresponding neutral axis depth is given by

$$d_{NA} = -\frac{F_{rt\text{-}y} + F_{rc\text{-}y}}{\dfrac{2b_{seg}f_c}{3} - \dfrac{b_{seg}\left(\dfrac{\varepsilon_0}{\theta_T \tan\alpha} - 1\right)\left(2f_c - \dfrac{f_c\left(\varepsilon_0 - \theta_T \tan\alpha\right)}{\varepsilon_0 - \varepsilon_u}\right)}{2}} \tag{6.93}$$

Once the strain at the top fibre in Figure 6.26c exceeds the strain at the concrete peak stress, the concrete stress block must consider the falling branch in Figure 6.27. In this case, longitudinal force equilibrium in Eq. 6.69 is given by

$$\frac{\theta_T \tan\alpha\left[E_r A_{rc}\left(d_{rc}-d_{NA}\right)+S_p k_{pi}\left(d_{rt}-d_{NA}\right)\right]}{2d_{NA}}=\frac{2b_{seg}d_{NA}f_c\theta_T\tan\alpha}{3\varepsilon_0}-\frac{b_{seg}\left(d_{NA}-\dfrac{d_{NA}\theta_T\tan\alpha}{\varepsilon_0}\right)\left(\theta_T\tan\alpha-2f_c+\dfrac{f_c\varepsilon_0}{\varepsilon_0-\varepsilon_u}\right)}{2} \tag{6.94}$$

which can be solved to give the depth of the neutral axis, including concrete softening, as

$$d_{NA}=-\frac{3\varepsilon_0^2\left(S_p k_{pi}+2E_r A_{rc}\right)-H}{4b_{seg}f_c\left(3\varepsilon_0+\theta_T\tan\alpha\right)} \tag{6.95}$$

in which

$$H=\sqrt{3}\varepsilon_0\sqrt{H_1+H_2+H_3} \tag{6.96}$$

$$H_1=12\varepsilon_0^2 E_r A_{rc}\left(S_p k_{pi}+2E_r A_{rc}\right) \tag{6.97}$$

$$H_2=8b_{seg}d_{rt}f_c S_p k_{pi}\left(3\varepsilon_0+8\theta_T\tan\alpha\right) \tag{6.98}$$

$$H_3=3S_p^2 k_{pi}^2\varepsilon_0^2 \tag{6.99}$$

Now, following the same approach as above, Visintin and Oehlers (2016) derived expressions to describe the moment–rotation relationship after the commencement of softening, where the reinforcement is taken as being elastic–plastic. That is, when the tension reinforcement has yielded and the compression reinforcement is elastic, the moment resisted by the segment is

$$M_{seg}=F_{rt\text{-}y}d_{rt}+\frac{b_{seg}d_{NA}^2 f_c\left(5\varepsilon_0+8\theta_T\tan\alpha\right)}{12\theta_T\tan\alpha}-\frac{\left[b_{seg}d_{NA}^2 f_c\left(\varepsilon_0-\theta_T\tan\alpha\right)\right]^2\left(2\varepsilon_0-3\varepsilon_u+\theta_T\tan\alpha\right)}{6\theta_T^2\tan\alpha^2\left(\varepsilon_0-\varepsilon_u\right)} \tag{6.100}$$

and the corresponding neutral axis depth is

$$d_{NA}=-\frac{3\varepsilon_u\theta_T\tan\alpha\left(F_{rt\text{-}y}-F-E_r A_{rc}\theta_T\tan\alpha\right)-3\varepsilon_0\theta_T\tan\alpha\left(F_{rt\text{-}y}-E_r A_{rc}\theta_T\tan\alpha\right)}{b_{seg}f_c\left[\theta_T\tan\alpha\left(4\varepsilon_0-10\varepsilon_u+3\theta_T\tan\alpha\right)+3\varepsilon_0\left(\varepsilon_0-\varepsilon_u\right)\right]} \tag{6.101}$$

in which the coefficient F is given by

$$F = \sqrt{3(\varepsilon_0 - \varepsilon_u)(F_1 + F_2 + F_3 + F_4)} \tag{6.102}$$

in which

$$F_1 = 3E_r A_{rc} \theta_T^2 \tan\alpha^2 (\varepsilon_0 - \varepsilon_u) \tag{6.103}$$

$$F_2 = 6F_{rt\text{-}y} E_r A_{rc} \theta_T \tan\alpha (\varepsilon_0 - \varepsilon_u) \tag{6.104}$$

$$F_3 = b_{seg} d_{rc} f_c E_r A_{rc} \left(-6\varepsilon_0^2 + 8\varepsilon_0 \theta_T \tan\alpha + 12\varepsilon_0 \varepsilon_u + 6\theta_T^2 \tan\alpha^2 - 20\theta_T \tan\alpha \varepsilon_u\right) \tag{6.105}$$

$$F_4 = 3F_{rt\text{-}y}^2 (\varepsilon_0 - \varepsilon_u) \tag{6.106}$$

and when both the compression and tension reinforcement have yielded

$$M_{seg} = F_{rt\text{-}y} d_{rt} - F_{rc\text{-}y} d_{rc} - \frac{2b_{seg} d_{NA} f_c \left(d_{NA} - \dfrac{5d_{NA}\varepsilon_0}{8\theta_T \tan\alpha}\right)}{3}$$

$$b_{seg}\left[2f_c \frac{f_c(\varepsilon_0 - \theta_T \tan\alpha)}{\varepsilon_0 - \varepsilon_u}\right]\left(d_{NA} - \frac{d_{NA}\varepsilon_0}{\theta_T \tan\alpha}\right) \tag{6.107}$$

$$-\frac{\left\{\dfrac{d_{NA}}{2} - \dfrac{d_{NA}\varepsilon_0}{\theta_T \tan\alpha} + \dfrac{f_c(\varepsilon_0 - \theta_T \tan\alpha)\left(d_{NA} - \dfrac{d_{NA}\varepsilon_0}{\theta_T \tan\alpha}\right)}{\left[12f_c - \dfrac{6f_c(\varepsilon_0 - \theta_T \tan\alpha)}{\varepsilon_0 - \varepsilon_u}\right](\varepsilon_0 - \varepsilon_u)}\right\}}{2}$$

where

$$d_{NA} = -\frac{F_{rt\text{-}y} - F_{rc\text{-}y}}{\dfrac{2b_{seg} f_c}{3} - \dfrac{b_{seg}\left(\dfrac{\varepsilon_0}{\theta_T \tan\alpha} - 1\right)\left[2f_c \dfrac{f_c(\varepsilon_0 - \theta_T \tan\alpha)}{\varepsilon_0 - \varepsilon_u}\right]}{2}} \tag{6.108}$$

It is important to note that the same approach can be applied to sections where the reinforcement yields after the commencement of concrete softening. However, these over-reinforced sections are generally not used in practice, and hence the reader is referred to Visintin and Oehlers (2016) for the complete solution.

Whilst the aforementioned approach can be applied relatively easily to plain concrete, the complexity of the solution for FRC prevents the development of straightforward analytical solutions. Although the concrete compressive stress–strain relationship for FRC is similar, the need to introduce both a strain hardening and strain softening branch to account for the contribution of fibres in tension complicates the derivation of simple analytical solutions. In this scenario, it is recommended that the numerical solution presented in Figures 6.7–6.9 serves as a more powerful analytical tool.

6.7 APPENDIX: FLOW CHART FOR SEGMENTAL ANALYSIS

Figure 6.28 provides a flow chart of the segmental analysis, which can be used in conjunction with the text throughout this chapter as the basis for writing a numerical model.

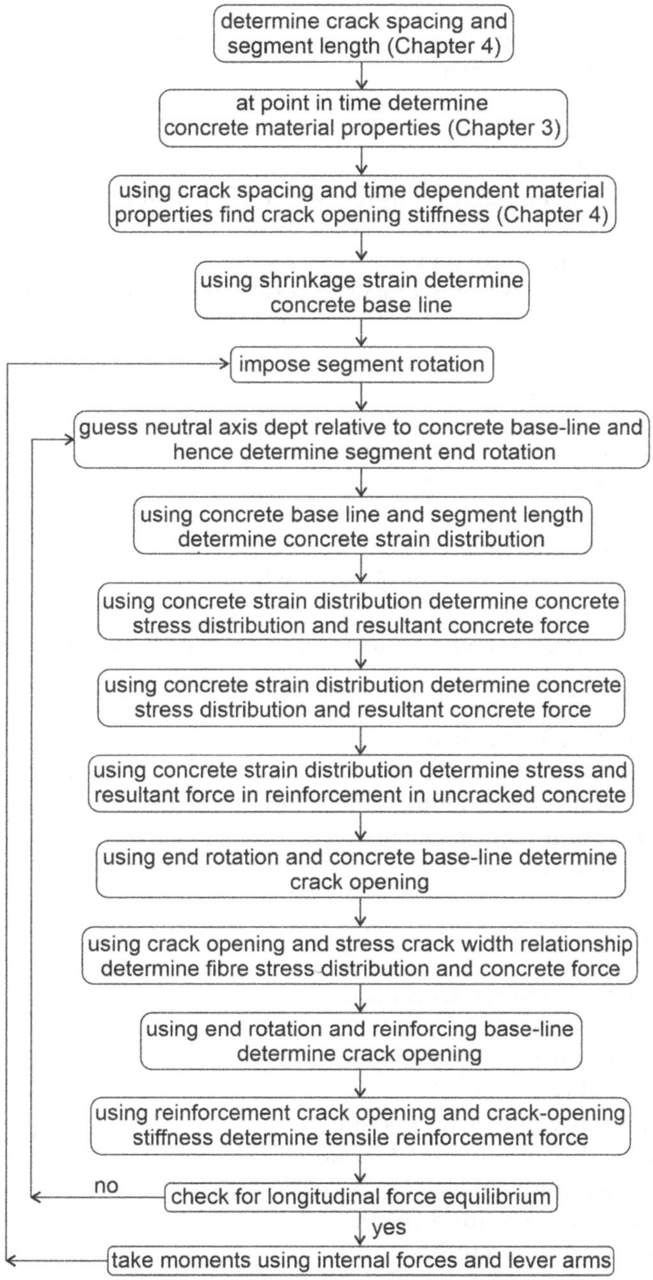

Figure 6.28 Flow chart for segmental analysis.

6.8 NOTATION AND ABBREVIATIONS

A_c cross-sectional area of concrete in *TS* prism
A_r cross-sectional area of reinforcement in *TS* prism
A_{rt} cross-sectional area of tension reinforcement
A_{rc} cross-sectional area of compression reinforcement
BL base-line
BL_c base-line of concrete
BL_r base-line of reinforcement
b width of beam
b_{seg} width of segment
b_{ts} width of *TS* prism
c_c distance of compression reinforcement to upper fibre; compression reinforcement cover
c_{rt} cover to tension reinforcement
c_t distance of tension reinforcement to soffit; tension reinforcement cover
comp compression
con contraction
D total depth of beam
d depth of tension reinforcement to upper compressive fibre; effective depth
d_a d_{NA}-d_d depth of neutral axis not softening
d_{cr} depth of crack
d_{ct} depth of concrete in tension
d_d depth of softening wedge
d_{fib} depth of fibre concrete in tension; d_{fib} where $\Delta<\Delta_{fm}$
d_{NA} depth to neutral axis
$d_{NA\text{-}c}$ depth of concrete in compression; depth of neutral axis based on concrete
$d_{NA\text{-}cr}$ *FI* d_{NA} of cracked section
$d_{NA\text{-}pi}$ d_{NA} from *PI* analysis
$d_{NA\text{-}sh}$ d_{NA} when there is shrinkage
$d_{NA\text{-}ts}$ d_{NA} derived from *TS* analysis
d_{rc} depth of compression reinforcement from top fibre
d_{rt} depth of tension reinforcement
d_{seg} depth of segment
d_{sl} depth of slice
E modulus
EA axial rigidity
EA_{rc} *EA* of compression reinforcement
E_c modulus of concrete
$E_c(t, t_0)$ time-dependent E_c
EI flexural rigidity
EI_c flexural rigidity of cracked section
EI_{uc} flexural rigidity of uncracked section
E_r modulus of reinforcement
E_{rc} E_r of compression reinforcement
E_{rt} modulus of tension reinforcement
exp expansion

F	force; force profile
F_c	compressive force in concrete element
F_{cc}	resultant force in concrete in compression
$F_{cc\text{-}sh}$	F_{cc} with shrinkage
F_{ct}	resultant force in concrete in tension
$F_{ct\text{-}sh}$	F_{ct} with shrinkage
F_f	fibre force in element
F_{fib}	axial force in tension stiffening prism due to fibres
F_r	force in reinforcement
F_{rc}	resultant force in reinforcement in compression
$F_{rc\text{-}sh}$	F_{rc} with shrinkage
$F_{rc\text{-}y}$	F_{rc} at yield
F_{rt}	resultant tensile force in reinforcement
$F_{rt\text{-}c}$	force in tension reinforcement bridging a crack
$F_{rt\text{-}sh}$	F_{rt} with shrinkage
$F_{rt\text{-}pi}$	F_{rt} through *PI* analysis
$F_{rt\text{-}pi\text{-}sh}$	F_{rt} in *TS* prism with shrinkage
$F_{rt\text{-}y}$	F_{rt} at yield
FI	full interaction
FRC	fibre-reinforced concrete
F_t	tensile force in uncracked concrete in a slice
F_{ts}	axial force in tension stiffening prism through reinforcing bar
f_c	concrete compressive strength; stress in concrete in compression
f_{ct}	concrete tensile strength; stress in concrete in tension
f_{fm}	strength of fibre concrete
f_{pc}	tensile strength of *FRC* after cracking
h_{ts}	height of tension stiffening prism
I	second moment of area
I_{cr}	I of cracked section
jd_{rt}	lever arm between F_{cc} and F_{rt}
k_e	τ/δ secant stiffness; bond-slip stiffness
k_{pi}	P/Δ stiffness
L	length; length between supports
L_e	length of element slice
LHS	left-hand side
L_{per}	circumference of reinforcing bar or group of bars
L_{seg}	length of half-segment
L_{wdg}	length of wedge in half-segment
l	lever arm
l_{cc}	lever arm of compression concrete from top fibre
l_{rc}	lever arm of compression reinforcement from top fibre
l_{rt}	lever arm of tension reinforcement from top fibre
M	moment
M_{cr}	moment at which flexural cracking occurs
M_{sec}	moment at which secondary cracks occur
M_{seg}	externally applied moment
N_{seg}	externally applied axial force

n modular ratio E_r/E_c; number of half-crack widths in L_{seg}; equivalent number of cracks in a segment
n_{FI} full-interaction modular ratio
n_{PI} partial-interaction modular ratio
PI partial interaction
P_{pr} resultant axial force in tension stiffening prism
P/Δ tension stiffening parameter that relates P_{pr} to half-crack width Δ
Q level at which N_{seg} is applied
RHS right-hand side
r distance from neutral axis
S_p crack spacing
SR stress resultants
TS tension stiffening
ten tension
V_{seg} externally applied transverse shear
w crack width
x arbitrary level to fix Euler–Bernoulli deformation; proportion of
y arbitrary horizontal position to fix Euler–Bernoulli deformation
α angle of softening wedge; bond-slip parameter to define shape; coefficient of thermal expansion
β angle of crack measured from horizontal; TS parameter
χ curvature
χ_{cr} curvature at first cracking
χ_{sh} curvature due to shrinkage alone
Δ half-crack width in Euler–Bernoulli displacement
Δ^* Euler–Bernoulli displacement in compression region
Δ_{cm} Δ at the soffit of the segment; Δ adjacent to tension stiffening prism
Δ_{dif} error in Δ
Δ_{fm} Δ when f_{fm} is zero
Δ_{pr} half-crack width at level of tension stiffening prism
Δ_{rt} Δ of tension reinforcement measured from a vertical line through centre of rotation at d_{NA}; Δ at level of reinforcement in tension stiffening prism
ΔT change in temperature
Δ_T accumulated Δ in a segment
Δ_0 offset of k_{pi} to allow for shrinkage
δ displacement; displacement profile; bond slip
δ_c displacement of prism face due to internal tensile stresses; concrete displacement that induces compressive stresses
δ_{cc} displacement of concrete in compression
δ_{rc} displacement of reinforcement in compression
δ_{sh} unrestrained concrete displacement due to shrinkage alone
δ_t reinforcement displacement that induces tensile stresses
δ_1 δ at τ_{max}
ε strain; strain profile
ε_c strain in uncracked concrete
ε_{ct} strain in concrete in tension; tensile strain in concrete to cause cracking
ε_r strain in reinforcement

ε_{rc} strain in compression reinforcement
ε_{rt} strain in tension reinforcement
ε_{sh} shrinkage strain
ε_{tmp} strain induced by temperature change
$\varepsilon_{tmp\text{-}c}$ ε_{tmp} for concrete
$\varepsilon_{tmp\text{-}r}$ ε_{tmp} for reinforcement
ε_{tp} ε at top fibre
ε_u maximum strain beyond which stress is zero
ε_0 ε at f_c
θ rotation
θ_{cap} rotation capacity of a hinge
θ_{seg} rotation of segment
θ_{sh} Euler–Bernoulli rotation from shrinkage alone
θ_T total rotation from all cracks in a half-segment
λ *TS* parameter
σ stress
σ_c stress in concrete
σ_{fib} fibre stress; equivalent tensile stress in cracked concrete due to fibres
σ_{rc} stress in compression reinforcement
τ shear stress between reinforcement and adjacent concrete
τ_{max} maximum value of τ
τ/δ bond/slip characteristics of reinforcement

REFERENCES AND DEVELOPMENT

Bischoff, P.H., 2003. Tension stiffening and cracking of steel fiber-reinforced concrete. *Journal of Materials in Civil Engineering*, *15*(2), 174–182.

Branson, D.E., 1963. *Instantaneous and time-dependent deflections of simple and continuous reinforced concrete beams* (No. HPR Report No. 7). Alabama. State Highway Department.

Chen, Y., Visintin, P. and Oehlers, D.J., 2015. Concrete shear-friction material properties: derivation from actively confined compression cylinder tests. *Advances in Structural Engineering*, *18*(8), 1173–1185.

Hasan, N.M.S., Visintin, P., Oehlers, D.J., Bennett, T. and Sobuz, M.H.R., 2019. Time dependent deflection of RC beams allowing for partial interaction and non-linear shrinkage. *Materials and Structures*, *52*, 1–15.

Hognestad, E., Hanson, N.W. and McHenry, D., 1955. Concrete stress distribution in ultimate strength design. *ACI Journal Proceedings*, *52*(12), 455–480.

Knight, D., Visintin, P., Oehlers, D.J. and Jumaat, M.Z., 2013. Incorporating residual strains in the flexural rigidity of RC members with varying degrees of prestress and cracking. *Advances in Structural Engineering*, *16*(10), 1701–1718.

Oehlers, D.J., Visintin, P., Chen, J.F., Seracino, R., Wu, Y. and Lucas, W., 2017. Reinforced concrete behavior, research, development, and design through partial-interaction mechanics. *Journal of Structural Engineering*, *143*(7), 02517002.

Popovics, S., 1973. A numerical approach to the complete stress-strain curve of concrete. *Cement and Concrete Research*, *3*(5), 583–599.

Sturm, A.B., Visintin, P. and Oehlers, D.J., 2019. Rational design approach for the instantaneous and time-dependent serviceability deflections and crack widths of FRC and UHPFRC continuous and simply supported beams. *Journal of Structural Engineering*, *145*(11), 04019138.

Visintin, P. and Oehlers, D.J., 2016. Mechanics-based closed-form solutions for moment redistribution in RC beams. *Structural Concrete*, *17*(3), 377–389.

Visintin, P. and Oehlers, D.J., 2018. Fundamental mechanics that govern the flexural behaviour of reinforced concrete beams with fibre-reinforced concrete. *Advances in Structural Engineering*, *21*(7), 1088–1102.

Visintin, P., Oehlers, D.J. and Haskett, M., 2013a. Partial-interaction time dependent behaviour of reinforced concrete beams. *Engineering Structures*, *49*, 408–420.

Visintin, P., Oehlers, D.J., Haskett, M. and Wu, C., 2013b. Mechanics-based hinge analysis for reinforced concrete columns. *Journal of Structural Engineering*, *139*(11), 1973–1980.

Visintin, P., Oehlers, D.J., Wu, C. and Haskett, M., 2012. A mechanics solution for hinges in RC beams with multiple cracks. *Engineering Structures*, *36*, 61–69.

Visintin, P., Sturm, A.B. and Oehlers, D.J., 2018. Long-and short-term serviceability behavior of reinforced concrete beams: mechanics models for deflections and crack widths. *Structural Concrete*, *19*(2), 489–507.

Chapter 7

Fundamental flexure shear segmental analysis

The segmental flexural analysis of a member under pure flexure, that is, without transverse shear, was covered in Chapter 6. It was shown that both the distribution and opening of flexural cracks can be quantified using the tension stiffening mechanics presented in Chapter 4, and further that concrete crushing can be quantified using the shear friction mechanics presented in Chapter 5. The moment/rotation leading to hinge rotations could then be determined by applying Euler–Bernoulli deformations under pure flexure, the mechanics of which are described in Chapter 6. This flexural analysis was shown, through the use of mechanics, to be able to cope with and quantify concrete confinement, shrinkage or expansion, and the inclusion of fibres in the concrete. As such, it does not require any correction factors; it is, to all intents and purposes, fairly straightforward. The inclusion of transverse shear into the flexural segmental analysis considerably complicates the mechanics (Al-Mhawish et al. 2025b), as will be shown below, but importantly can be incorporated through sound mechanics. The segmental flexural analysis of a member under pure flexure in Chapter 6 is the simplest case of the following fundamental segmental shear flexure analysis in this chapter.

This chapter begins with a visual description of the major forms of cracking that occur in reinforced concrete (*RC*) members, which leads to an idealised segmental free-body diagram for shear flexure mechanical analyses. This is the fundamental free body that governs the behaviour of *RC* members subjected to any combination of stress resultants. This fundamental and generic free body can then be used to determine the strength of the free body after the cracks have occurred. It is shown that the term 'shear failure' is a misnomer, as 'shear failure' in a beam always includes flexure. In fact, a better way of visualising the mechanics is that of flexural failure due to the weakening of the concrete by shear.

7.1 FORMATION OF CRACKS

As flexure is the simplest boundary case of the shear flexure problem, let us first examine the crack formation in a constant moment region before looking at the generic behaviour in a shear flexure region. We will take this opportunity to describe all forms of cracks, not just those directly related to shear and flexure.

DOI: 10.1201/9781003711407-11

7.1.1 Constant moment region

Figure 7.1a shows the flexural failure of a beam in a constant moment region, that is, where there is zero transverse shear. As would be expected, the cracks in the tension region are vertical, β is 90°, and they emanate from the tension face so that their occurrence can be determined from the elementary mechanics in homogeneous regions of $\sigma = My/I$. Furthermore, they are uniformly spaced at a distance S_{pr} that can be determined from the tension stiffening analyses in Chapter 4 of the tension stiffening zone shown. In the compression region, and due to local shear as explained in Chapter 5, the cracks are inclined at an angle α, which is about 26°, when the concrete softens (Chen et al. 2015). It can be seen in Figure 7.1a that concrete crushing caused failure; that is, concrete crushing limited the capacity of the member.

Figure 7.2 also shows flexural failure in the constant moment region where the single crack in the tension region is vertical at β of 90° and, furthermore, emanates from the tension face. Hence, it starts due to flexure; that is, the My/I analysis applies. The wedge in the compression region, even though very small, is inclined at α. It may be worth noting that this beam was designed to debond and hence the single very wide crack. Yet the cause of failure; that is, the reason for the reduction in strength, is still concrete crushing.

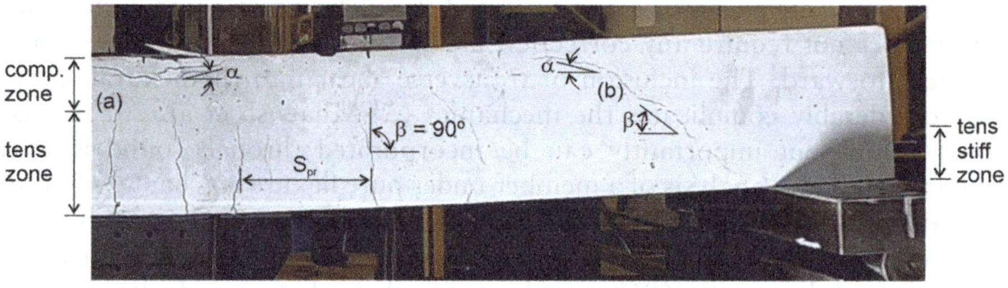

Figure 7.1 Flexural failure and shear flexure crack. (a) Flexural failure through concrete crushing. (b) Shear flexure crack. (Photo courtesy of Lloyd Dadd.)

Figure 7.2 Debonding causing flexural failure through concrete crushing. (Photo courtesy of Lloyd Dadd.)

7.1.2 Constant shear region

Now moving on to a constant shear region. Figure 7.3b shows the shear flexure failure of a beam in a constant shear region. The distribution of cracks just prior to failure is shown in Figure 7.3a, where the two vertical cracks on the left are in the constant moment region and the crack on the right is in the constant shear region. The inclined crack to the right in Figure 7.3a emanates from the tension face, and so it is initiated by flexure; that is, by elastic bending (My/I) analyses. It is near vertical in the tension stiffening region, where the effect of the transverse shear V due to elastic shear ($\tau = Vay^{*}/Ib$) is very small, but then inclines at an angle β in the tension region due to shear.

Shear flexure failure in Figure 7.3b occurred suddenly through the formation of a crack in the compression region at an angle α, which, as explained in Sections 1.4.3 and 5.1.4, is due to the compressive failure of the concrete. The near-horizontal crack parallel to the tension face on the right is due to the dowel action induced by interface slip during shear flexure failure, causing ripping, as explained in Chapter 1. Therefore, this ripping action is a post-sliding action and not a pre-sliding action. Hence in the pre-sliding behaviour, there are two distinct formations of cracks: those in the tension region at an angle β due to initial cracking in a relatively homogeneous material and those in the compression region at an angle α due to concrete softening. It needs to be emphasised that although these cracks connect, they are distinct because they are formed by different mechanisms. A similar example of shear flexure failure in Figure 7.3 is clearly seen in Figure 7.4.

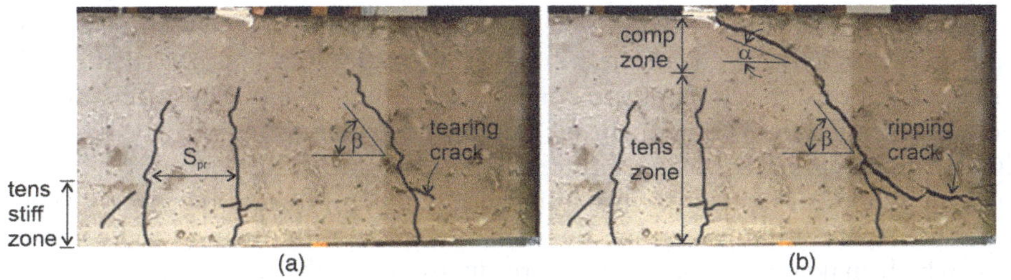

Figure 7.3 Crack development in shear flexure failure. (a) crack pattern prior to failure in compression region; (b) crack pattern after shear failure. (Modified from Oehlers and Visintin 2021.)

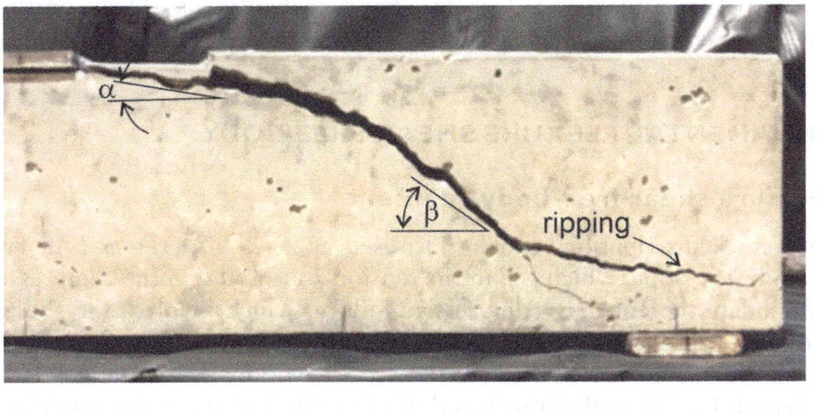

Figure 7.4 Shear flexure failure through concrete crushing. (Modified from Dadd et al. 2025.)

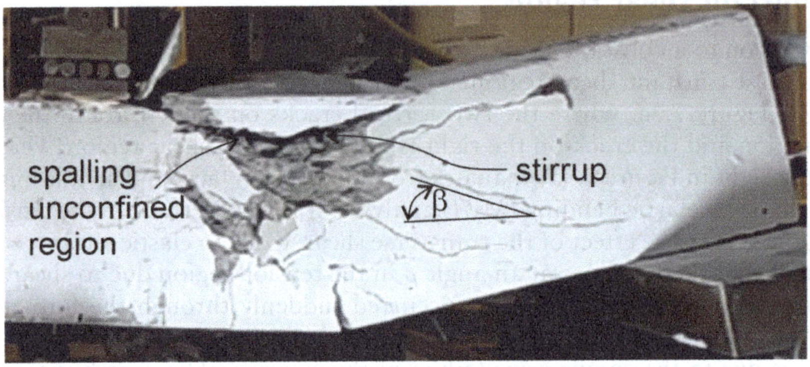

Figure 7.5 Shear flexure failure in member with stirrups. (Photo Courtesy of Lloyd Dadd.)

Figure 7.1b shows an inclined crack in a constant shear region just prior to shear flexure failure. This crack in the constant shear region in Figure 7.1b looks similar to that in the constant shear region in Figure 7.3a. However, the crack in Figure 7.1b does not extend to the tension face and, hence, must have emanated near the mid-depth of the beam, where the elastic shear stress ($\tau = Vay^*/Ib$) is at its maximum. The inclination of the crack in the tension zone is shown as β. Failure of a similar beam in a constant shear region is shown in Figure 7.3b, where the crack in the compression zone is at α and the inclined crack in the tension region at β does extend to the tension face, suggesting that it emanated due to flexure. Hence, whether or not the initial crack emanated due to flexural stresses (My/I) or due to shear stresses (Vay^*/Ib), they have two distinct slopes: at α in the compression region due to concrete softening or about to soften, and at β in the tension region due to cracking in the homogeneous region.

Figure 7.5 shows shear flexure failure in the constant shear region in a beam with stirrups. A single stirrup is just visible in the failed compression region. As explained in Section 5.2, the stirrups confine the concrete encased by them in the compression region, which makes it stronger and more ductile than the unconfined concrete surrounding the encased concrete. The inclined crack at β is visible, but the crack at α in the confined concrete is hidden by the spalling of the weaker unconfined concrete that surrounds the stirrup-encased concrete.

Further examples of flexure failure are shown in Figure 7.6. Those on the left are where the initial crack was instigated by flexure and those on the right by shear.

7.2 FUNDAMENTAL FLEXURE SHEAR FREE BODY

7.2.1 Flexure shear free-body segment

The above crack distributions can be idealised as A–B–C or D–E–F in Figure 7.7. There is a bilinear slope, which is β in the tension zone and α in the compression zone. To accommodate the concrete compressive failure, which involves the sliding planes B–C or E–F at α, the length G–E must be at least twice the depth to the neutral axis, should α be considered to be 26°, as discussed in Sections 1.4.3 and 5.1.4. This then gives the length L_{def}. Note that this length is not critical to the ensuing analyses.

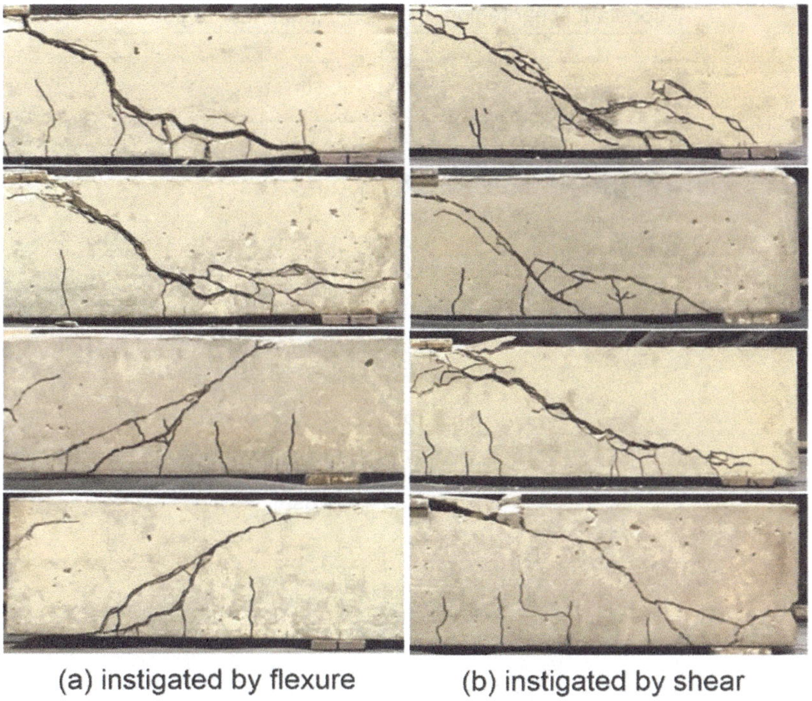

Figure 7.6 Further examples of shear flexure failure. (a) instigated by flexure; (b) instigated by transverse shear (Modified from Dadd et al. 2025.)

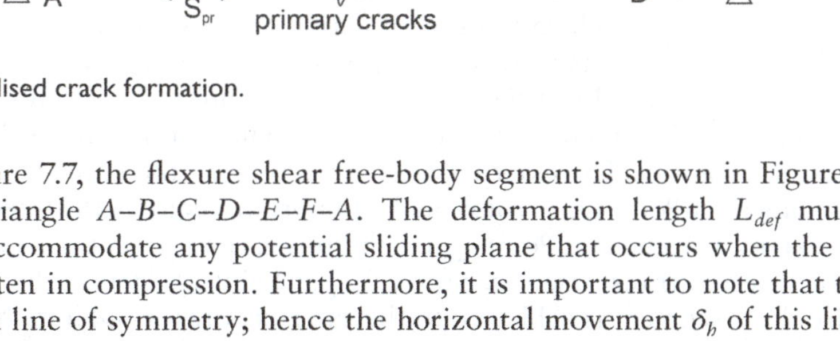

Figure 7.7 Idealised crack formation.

From Figure 7.7, the flexure shear free-body segment is shown in Figure 7.8 as the truncated triangle *A–B–C–D–E–F–A*. The deformation length L_{def} must be long enough to accommodate any potential sliding plane that occurs when the concrete is about to soften in compression. Furthermore, it is important to note that the vertical line *C–F* is a line of symmetry; hence the horizontal movement δ_h of this line must be zero, meaning that *C–F* is fully restrained horizontally. This is in contrast to the surface *A–F–E*, which is vertically unrestrained.

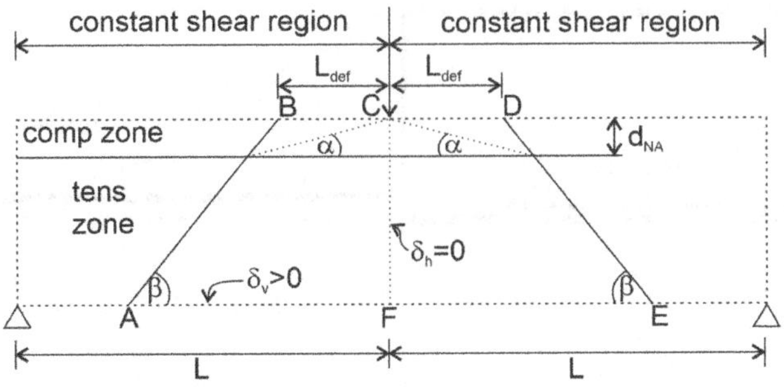

Figure 7.8 Flexure shear segment.

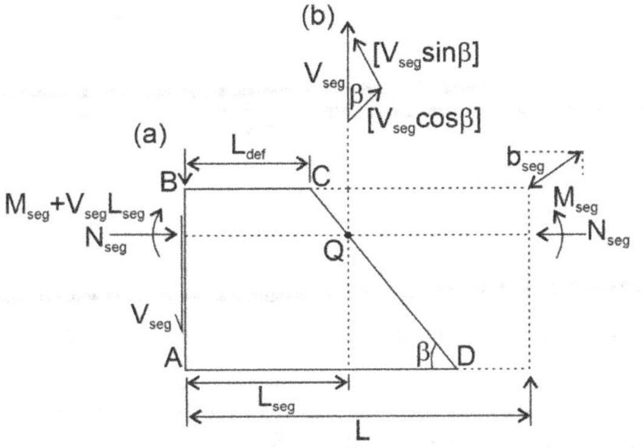

Figure 7.9 Flexure shear applied external stress resultants.

7.2.2 Flexure shear applied stress resultants

As the beam in Figure 7.8 is symmetrical about its centre line *C–F*, it is only necessary to consider one half. The right-hand side of the beam is shown in Figure 7.9. For the flexural segmental analysis in Chapter 6, it was necessary to consider a segment within which the moment was constant. In this flexure shear segment analysis, it is necessary to consider a segment in which the transverse shear is constant.

Along the potential sliding plane *C–D* in Figure 7.9, the segment is subjected to an external axial force N_{seg} in which the line of action crosses the sliding plane *C–D* at Point *Q* at the distance L_{seg}, as shown. This point is important because, as explained in Section 1.5.2 with the help of Figure 1.29 (*l*), any internal moments taken about the sliding plane must be taken about this Point *Q*. The beam is also subjected to an external variable moment. The external moment acting on the segment M_{seg} must be the applied moment at Point *Q*, as rotational equilibrium requires the internal moment to equal the external moment at Point *Q*. Furthermore, the segment is also subjected to an applied or external transverse shear force V_{seg} at Point *Q*. Hence, the sliding plane of the flexure shear segment is subjected to the external stress resultants N_{seg}, M_{seg} and V_{seg} at Point *Q*, which is positioned through the intersection of the line of action of N_{seg} and the sliding plane.

The stress resultants at Point Q in Figure 7.9 apply whether the segment is in a constant shear region or in a variable shear region. However, for analysis, it is assumed that the segment is subjected to constant shear because the variation in shear over the length of the segment is generally small. It is worth bearing in mind in the ensuing analyses that there is always a direct relationship between V_{seg} and M_{seg} based on the distribution of loads and fundamental engineering mechanics. When there is no applied axial force N_{seg}, as in this beam, then Point Q in Figure 7.9 can be positioned anywhere along the sliding plane C–D. The transverse shear force in Figure 7.9b has been resolved along and perpendicular to the sliding plane. The component of the transverse shear along the sliding plane $V_{seg} \sin \beta$ has to be resisted by the sliding plane and is weakened by the tensile force normal to the sliding plane $V_{seg} \cos \beta$.

7.2.3 Flexure shear analysis

7.2.3.1 *Euler–Bernoulli displacements*

To determine the internal forces within the segment in Figure 7.9 that balance the external stress resultants N_{seg}, M_{seg} and V_{seg}, the sliding plane C–D is subjected to a Euler–Bernoulli displacement. Let us first consider the compression zone shown in Figure 7.10, where the inclined face O–F is rotated by θ such that the axis of rotation is at Point O. At a distance r from the centre of rotation, there is a Euler–Bernoulli displacement of Δ across the arc that is

$$\Delta = r\theta \tag{7.1}$$

The concrete deformation is shown in Figure 7.10a. The concrete at Point A of the face is displaced Δ_{cc}, which has a horizontal component $\Delta_{cc} \sin \beta$. This horizontal displacement is resisted by a horizontal element of length $l_{cc\text{-}h}$, which has a horizontally fixed boundary C–D, the centreline of a symmetric free body; hence the horizontal concrete strain is

$$\varepsilon_{cc-h} = \frac{\Delta_{cc} \sin \beta}{l_{cc-h}} \tag{7.2}$$

The concrete displacement Δ_{cc} in Figure 7.10a also has a vertical component Δ_{cc} cos β, which is $l_{cc\text{-}v}$ from the unrestrained horizontal free surface D–O. As the surface is

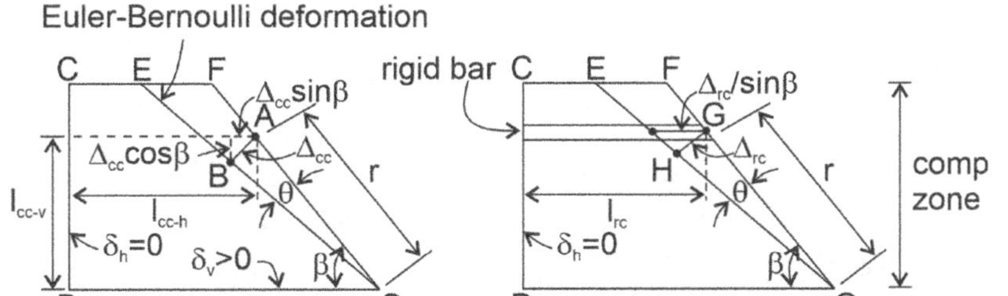

Figure 7.10 Euler–Bernoulli deformation within compression zone. (a) Concrete deformation. (b) Compression reinforcement deformation.

not directly restrained, it will move and reduce the vertical strain and, as such, will be ignored.

The deformation of a reinforcing bar in the compression zone is shown in Figure 7.10b. If the bar is considered to be flexible, then the Euler–Bernoulli deformation Δ_{rc} would have two components, as for Δ_{cc} in Figure 7.10a. However, it is unlikely that the bar could bend to accommodate these deformations. Instead, let us assume that the bar is rigid, meaning it remains horizontal as shown in Figure 7.10b, such that the horizontal deformation is $\Delta_{rc}/\sin\beta$. If the distance to the rigid boundary is l_{rc}, as shown, then the strain in the bar is

$$\varepsilon_{rc} = \frac{\Delta_{rc}}{\sin\beta\; l_{rc}} \tag{7.3}$$

Now consider the cracked tension zone in Figure 7.11. The upper part of Figure 7.11a shows the deformation of a fibre that is so thin it could be considered very flexible and easy to deform into any shape, as in Figure 1.13. Between the crack faces, the fibre spans the crack width w_{fib}, which means the fibre is spanning a total width of $2\Delta_{fib}$. The fibre force P_{fib} is therefore perpendicular to the crack face and is a function of the crack width such that

$$P_{fib} = f\left(w_{fib}\right) = f\left(2\Delta_{fib}\right) \tag{7.4}$$

which has two components, namely a horizontal component P_{fib} sin β and a vertical component P_{fib} cos β.

The lower part of Figure 7.11a shows the deformation of a tension bar where the crack width is $2\Delta_{rt}$. As with the compression bar in Figure 7.10b, if it is assumed to be rigid, meaning it cannot deflect vertically, then the horizontal *E-B* deformation is $\Delta_{rt}/\sin\beta$. Consequently, the P/Δ approach from tension stiffening in Chapter 4 can be used to derive the horizontal tensile force in the bar, where the horizontal component required Δ_{rt}^* is

$$\Delta_{rt}^* = \frac{\Delta_{rt}}{\sin\beta} \tag{7.5}$$

Figure 7.11b shows the deformation of a stirrup. Once again, assuming it is rigid, as in the horizontal bar in Figure 7.11a, the vertical measurement of the half-crack width Δ_{st} is $\Delta_{st}/\cos\beta$ such that required Δ is

$$\Delta_{st}^* = \frac{\Delta_{st}}{\cos\beta} \tag{7.6}$$

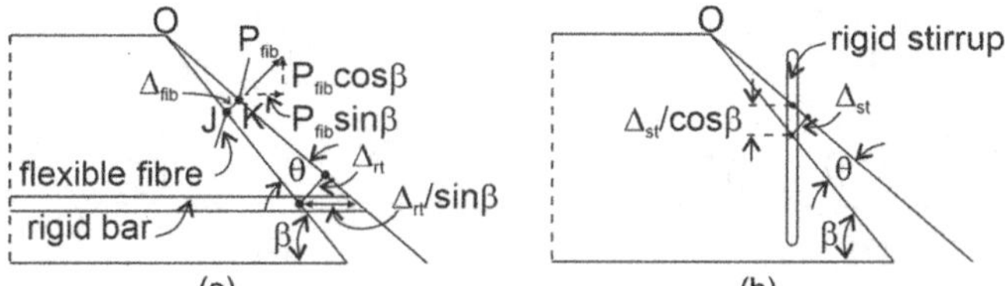

Figure 7.11 Euler–Bernoulli deformation within tension zone. (a) Tension reinforcement deformation; (b) Stirrup deformation.

7.2.3.2 Euler–Bernoulli forces

In a similar way to what has been done in Zhang et al. (2016) and Sturm et al. (2021), a Euler–Bernoulli displacement has been applied to the sliding plane *C–D* in Figure 7.9, as shown in Figure 7.12a. To determine the internal forces, the simplest solution is to slice the segment in Figure 7.12a into small horizontal elements of depth d_e and varying lengths l_e and of width into the page of b, and also vertical elements of width h_e and length l_v as shown.

Let us first consider the concrete horizontal deformation shown in the concrete compression region in Figure 7.10a. The Euler–Bernoulli deformation causes the varying horizontal displacements shown as Δ^* in Figure 7.12a. As shown in Figure 7.10a for the concrete component, Δ^* has a value of $\Delta_{cc} \sin\theta$. This horizontal deformation Δ^* is resisted by the horizontally fixed boundary *C–F* in Figure 7.12. Hence, in the compressive regions of the beam, the local Euler–Bernoulli displacement $\Delta^*=\Delta_{cc}\sin\beta$ causes a strain $\Delta_{cc}\sin\beta/l_e$, which is assumed to be uniform over the depth of the element d_e. The variation in strain is no longer linear because of the varying lengths l_e and takes the form shown in Figure 7.12b. From the strain distribution in Figure 7.12b and from the concrete stress/strain relationship as defined in Section 2.3 and Chapter 5, which can include the effects of passive confinement, the horizontal stress can be determined, and consequently, the resultant compressive force F_{cc} in Figure 7.12c.

Should there be any compressive reinforcement in the segment, then the compressive reinforcement, as shown in Figure 7.10b, has a horizontal contraction $\Delta^*=\Delta_{rc}/\sin\beta$ and, hence, a horizontal strain $(\Delta_{rc}/\sin\beta)/l_{rc}$, where l_{rc} is the length of the bar within the segment, as shown in Figure 7.10b. From the reinforcement material properties in Section 2.6, the compressive stress in the reinforcement can be determined, and consequently, the resultant force in the compression reinforcement F_{rc} in Figure 7.12c.

Consider now the tension zone of the segment, which in this analysis is assumed to be cracked, as in Figure 7.11a. The horizontal reinforcement has a horizontal component

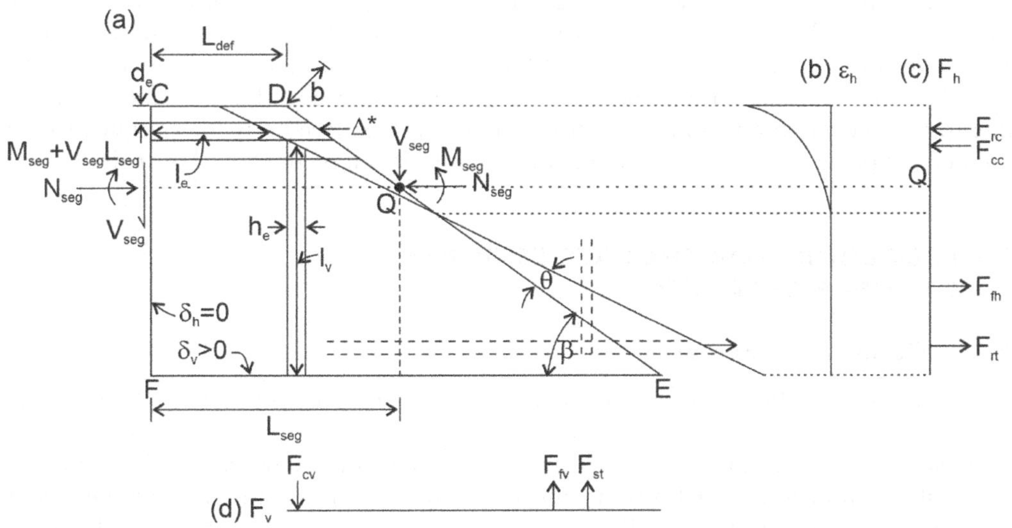

Figure 7.12 Euler–Bernoulli flexure shear analysis (a) segment elevation and displacement profile; (b) strain profile; (c) longitudinal force profile; (d) transverse force profile.

of Δ^* at the level of the tension reinforcement, which is $\Delta_{rt}/\sin\beta$, and this can be used with the P/Δ properties from Chapter 4 to derive the resultant force in the reinforcement F_{rt} in Figure 7.12c. Note that unlike the force in the compression reinforcement F_{rc}, this force does not rely on bearing against the horizontally fixed boundary C–F in Figure 7.12a but simply relies on two crack faces separating, that is, the crack width, or more specifically, the half-crack width. Similarly, the fibres crossing the crack in Figure 7.11a have a force P_{fib}, which depends on the half-crack width Δ_{fib} and can be derived from the fibre properties in Section 3.3, bearing in mind that the fibre properties depend on the crack width $2\Delta_{fib}$. The horizontal component $P_{fib}\sin\beta$ is the horizontal resultant fibre force F_{fh} in Figure 7.12c.

Now let us consider the effects of the vertical components of the forces in both Figures 7.10a and 7.11. Starting with the concrete in compression in Figure 7.10a, the vertical component of Δ^* in Figure 7.12 is $\Delta_{cc}\cos\beta$ from Figure 7.10a. This component acts on the vertical element of width h_e in Figure 7.12a; however, the boundary along F–E at a distance l_v is not directly restrained vertically, subsequently lessening the vertical strain and consequently reducing the vertical compressive force F_{cv} in Figure 7.12d. Quantifying this compressive force would be difficult, and it is suggested that it could be ignored.

In contrast, the vertical E-B half-crack opening at a stirrup in Figure 7.12, shown in Figure 7.11b, is equal to $\Delta_{st}\cos\beta$. This can be used directly with its P/Δ properties in Chapter 4 to achieve the force in the stirrup, as it is not dependent on a rigid boundary reaction but on crack face separation. The resultant stirrup force F_{st} is shown in Figure 7.12d. Similarly, the vertical component of the fibre force $P_{fib}\cos\beta$ in Figure 7.11a is only dependent on crack face separation, and the resultant of this vertical component is shown as F_{fv} in Figure 7.12d.

Note that when F_{cv} in Figure 7.12d is removed or ignored as suggested above, it appears that there is no longer vertical equilibrium. However, this is not the case, as there is a shear force acting along D–E which has a vertical component, as will be explained later. The purpose of this exercise was to show how these forces can be derived, and at this stage, not to consider equilibrium, which will be discussed later. Except for the vertical components of the induced forces in Figure 7.12, it can be seen that the above analysis is the same as for the flexural analysis in Figure 1.29 and in the flexural segmental analysis in Chapter 6. This is the best free body for a segment with an inclined plane we have at the moment; no doubt it can be improved over time.

7.3 PROCEDURE FOR THE SAFE ESTIMATE OF SHEAR CAPACITY

7.3.1 Generic overview

From the review of shear failure in Section 7.1, we can observe the following phenomena.

1. Sometimes shear failure is a gradual failure, as illustrated in Figures 7.1b and 7.3, where an inclined crack first forms at a transverse shear load V_{cr}, extends under increased load and then fails at a transverse shear load V_{sld}, such that there is a warning of shear failure.

2. Alternatively, it can be catastrophic, as shown in Figure 7.4, where the inclined crack forms at V_{cr} and the section immediately fails without any prior warning, such that V_{sld} is less than V_{cr}.
3. Furthermore, the angle of the sliding plane in the tension region of the beam β can vary from being almost vertical, as in Figure 7.3, to steeply inclined, as in Figures 7.4 and 7.5.
4. It can also be observed that the inclined crack can extend to the soffit of the beam, as in Figure 7.6a, which suggests that flexure had a significant role in its initiation. Alternatively, it may not extend to the soffit, as in Figure 7.6b, suggesting that transverse shear had a significant role.

To cope with these behaviours, it has often been suggested that two independent mechanisms control the shear capacity, as illustrated in Figure 1.31. The first mechanism causes the initial crack in a relatively pristine or uncracked beam, where the transverse shear that causes the initial cracks is V_{cr}. After the initial crack has formed, the second mechanism causes sliding along the initial crack at a transverse shear load V_{sld}. When $V_{sld} > V_{cr}$, this explains the above Phenomenon 1, and when $V_{sld} < V_{cr}$, this explains Phenomenon 2. If the variations in V_{sld} and V_{cr} have the shapes shown in Figure 1.31, then V_{cr} is required to determine the minimum strength at β_{cap}. However, if V_{sld} has the shape as for V_{stir}, then there is no need for V_{cr}.

The fundamental flexure shear free-body in Section 7.2 provides us with the mechanics for analysing the behaviour of an inclined plane, whether it is cracked or not. The difficulty lies not with the mechanics of this free body, which is relatively easy to understand, but with the causes of failure, which is more of a material problem. The sliding capacity V_{sld}, which requires non-linear analyses, has been quantified in Section 7.4. Although empirical and semi-empirical solutions are available, there does not appear to be any simple, mechanically sound procedure for predicting V_{cr} or β (Zhang1997; Iskhakov and Ribakov 2014). To help the reader, let us examine two mechanically sound procedures, even though they do not work. Should the segment have stirrups, as shown in Figure 7.5, there is a possibility that the post-sliding shear capacity V_{post} is greater than the pre-sliding capacity V_{sld}. This analysis will also quantify the ductility of the flexure shear failure, which is important in assessing the collapse of a member or structure, and is covered in the post-sliding analysis in Section 7.5.

7.3.2 Initial crack formation

A shear span of a beam that incorporates the free-body diagram in Figure 7.12a is shown in Figure 7.13a. To perform the Euler–Bernoulli analysis in Figure 7.12a, it is necessary to determine the angle β of the potential sliding plane in the tension region, which is the value of β in Figures 7.1–7.6. Central to the analysis in Figure 7.12 is Point *Q*, where the external stress resultants V_{seg}, M_{seg} and N_{seg} must be in equilibrium with the internal forces generated along the sliding plane. Point *Q* therefore signifies both the position and the combination of external stress resultants (V_{seg}, M_{seg} and N_{seg}) that cause cracking in a homogeneous body, as shown in Figure 7.13a.

From Figure 7.13a,

$$M_{seg} = K_{LA} V_{seg} \tag{7.7}$$

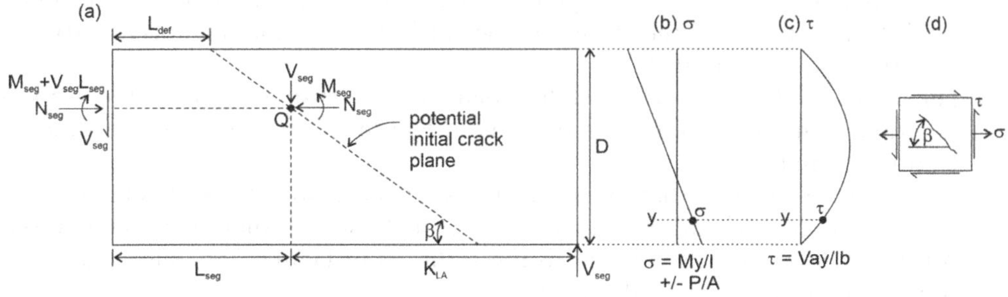

Figure 7.13 Analysis of initial cracking (a) segment elevation; (b) axial stress profile; (c) shear stress profile; (d) combined stresses on element.

where for this case in particular, K_{LA} is the lever arm as shown. However, Eq. 7.7 serves to remind the reader that there is a direct relationship between the transverse shear force V_{seg} and the applied moment M_{seg} for a particular loading arrangement. It is also worth noting in Figure 7.13a that N_{seg} could be the axial force in a column or a prestressing force, which is quantified in Chapter 9.

7.3.2.1 *Elementary analysis of initial cracking*

Importantly, under initial loading, the uncracked member in Figure 7.13a can be considered homogeneous, even though there may be some microcracking. As such, the elementary mechanics in Figure 7.13b to determine the axial stress σ, and in Figure 7.13c to determine the shear stress τ, can be used.

It is a question of determining the level y–y in Figure 7.13b and c at which the maximum principal stress occurs. Hence, apply a combination of V_{seg}, M_{seg} and N_{seg} to the cross-section. For a specific level y–y, determine the axial stress σ in Figure 7.13b and the shear stress τ in Figure 7.13c. The combined stresses are shown in Figure 7.13d, which gives the principal stress at level y–y as

$$\sigma_{pri} = \frac{\sigma}{2} + \sqrt{\frac{\tau^2 + \sigma^2}{2}} \tag{7.8}$$

and the inclination of the principal stress β as

$$\tan 2\beta = 2\tau / \sigma \tag{7.9}$$

One would think that this elementary analysis would provide the inclination of the crack β that is required. Unfortunately, it does not do so. Performing the analysis, it can be shown that flexure dominates; that is, β is predominantly 90°, and it is only when the shear–span K_{LA} is much less than $0.5D$, where D is the depth of the member, that β reduces. This contradicts what is happening in practice, as can be seen in the shear failure photographs. Hence, this approach using elementary linear analyses does not work. Something else is happening.

7.3.2.2 Fundamental free-body analysis

Let us now try an alternative approach using the fundamental free body, shown in Figure 7.12a, on a pristine beam without cracks in Figure 7.13a. The above analysis suggests that the transverse shear stresses have very little effect on the formation of the initial crack, so we will only consider flexure and come back to the transverse shear later. We will also ignore any applied axial loads, that is, N_{seg} in Figure 7.13a.

Let us start with the potential sliding plane that goes through Point Q in Figure 7.13a, but with β equal to 90°; that is, the potential sliding plane is vertical. From the elementary analyses illustrated in Figure 7.13b,

$$V_{seg-90} = \frac{f_{ct} I}{y_{max} K_{LA}} \tag{7.10}$$

where f_{ct} is the concrete tensile strength, I is the second moment of area and y_{max} is the distance from the neutral axis to the tension face.

Now applying this analysis when the slope is β, we consider an element at the soffit of the sliding plane with a width and depth of one unit, as shown in Figure 7.14.

The element of the sliding plane in Figure 7.14 is subjected to an axial force σ_{ap}, as shown, which has a shear component τ of σ_{ap} cos β acting downwards and a tensile component σ_t normal to the inclined plane equal to $\sigma_{ap} \sin^2 \beta$. Dividing this force σ_t by the area of the sliding plane gives the normal stress as $\sigma_{ap} \sin^2 \beta$. When σ_t is equal to f_{ct}, this gives

$$V_{seg-\beta} = V_{seg-90} \frac{1}{\sin^2 \beta} \tag{7.11}$$

That is, the shear at which flexural cracking occurs increases by the inverse of $\sin^2 \beta$. Unfortunately, this also does not provide a useful prediction of V_{cr}, but it is an improvement over that in Section 7.3.2.1. However, it does provide the required shape of V_{cr} as shown in Figure 1.31. It is also worth noting in Figure 7.14 that the shear due to σ_{ap}, that is, σ_{ap} cos β, is in the opposite direction to that induced by the applied shear V_{seg},

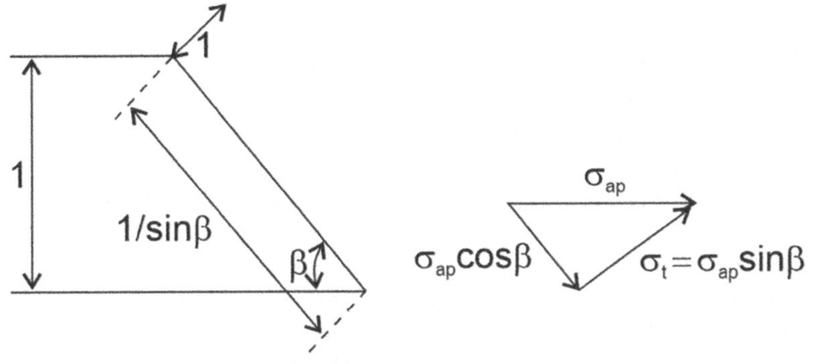

Figure 7.14 Element at base of free body.

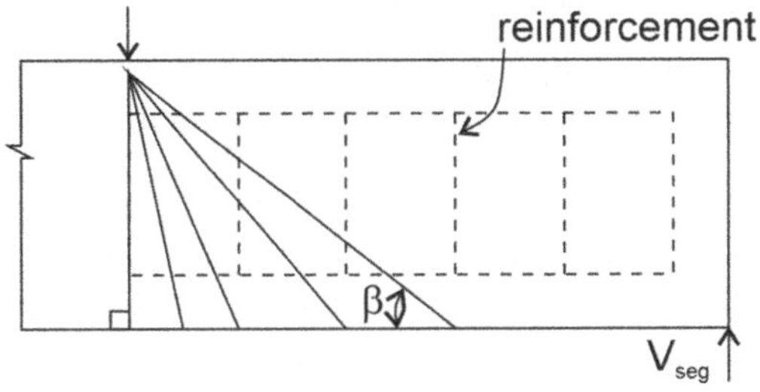

Figure 7.15 Series of analyses to determine shear capacity.

and hence, this is another reason for ignoring this shear component in the analyses. It is clear that much more research is required on the topic of the initial crack.

7.3.2.3 Safe estimate of shear capacity

There is still much research required to predict the shear load that causes the initial crack and the angle at which it occurs, should there be a solution. The question is how do we overcome this lack of information in analysis and design. A safe solution, one that provides the actual value or a lower bound, is to analyse the beam at different values of β to determine the lowest V_{sld} to use in design. This can also be done for a beam with stirrups, as shown in Figure 7.15, just to emphasise to the reader that, even for a beam with stirrups, it is necessary to determine the strength of a sliding plane between stirrups.

7.4 PRE-SLIDING ANALYSIS

Having formed the initial crack, let us now determine the transverse shear V_{sld} at the maximum capacity just prior to the onset of sliding along this initial crack.

7.4.1 Shear transfer components pre-sliding

The components that can transfer shear along the sliding plane D–E in Figure 7.16a are shown in Figure 7.16b, where V_{cc} is the shear transferred by the concrete in compression, V_{ct} is the shear transferred by the uncracked concrete in tension, V_{cr} is the shear transferred across a crack in the concrete and V_d is the shear transferred by dowel action of the reinforcing bars.

The concrete in compression in Figure 7.16b can transfer the shear V_{cc} through shear friction, as there is a normal compressive force σ_n along the potential sliding plane that can be derived from Figure 7.12c. Prior to sliding, this shear is limited by V_u, as explained in Section 3.1.2. The concrete in the uncracked tension zone in Figure 7.16b can also transfer shear V_{ct}, but this is severely limited as σ_n is now tensile. Furthermore, as shear flexure failure is an ultimate limit state, the depth of the uncracked tensile zone is very small. Hence, it may be expected that V_{ct} tends to zero. The shear transferred in the cracked tension zone, V_{cr}, can also be considered negligible, as once cracking occurs, the transfer of shear is through aggregate interlock, which requires interface slip that

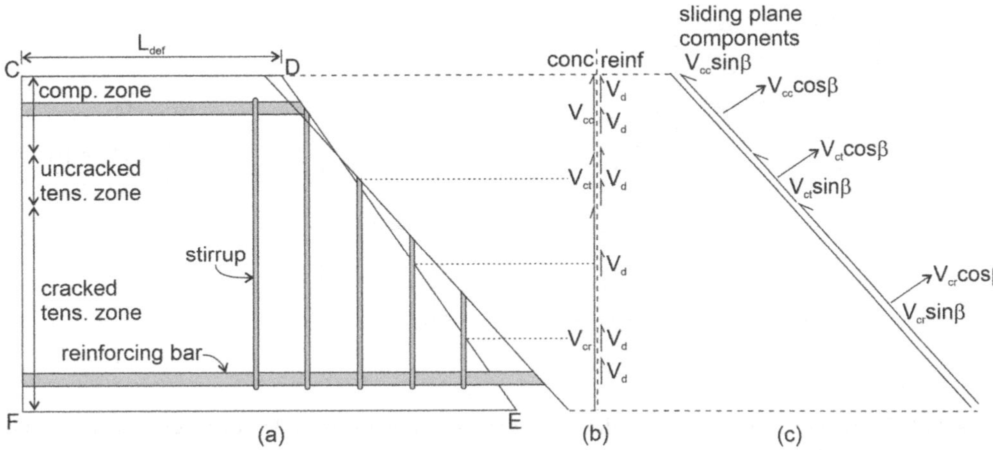

Figure 7.16 Shear transfer components (a) segment elevation and displacement profile; (b) distribution of shear force in reinforcement through dowel action and through concrete; (c) concrete sliding components along inclined plane.

does not occur as we are dealing with the pre-sliding capacity. Hence, V_{cr} is negligible. Finally, any reinforcement crossing the sliding plane – whether it is longitudinal bars or stirrups – can transfer shear by dowel action V_d, but dowel action requires interface slip, which does not occur, as once again we are dealing with the pre-sliding capacity. Hence, V_d is zero. Note that the post-sliding capacity, that is, the capacity when slip occurs and which covers all the components in Figure 7.16b, is dealt with in Section 7.5. From the above discussion, it can be deduced that the major transfer of shear along the sliding plane in Figure 7.16 prior to slip is V_{cc}, which occurs along the compression zone.

The shears transferred by the concrete in Figure 7.16b are resolved into their components along the sliding plane in Figure 7.16c. It can be seen that there is always a normal tensile force, that is, the cos β components, which reduces any applied normal stress σ_n. In the compression zone, there is a large compressive σ_n due to the longitudinal compressive force, such as that shown in Figure 7.12c. Hence, the reduction due to V_{cc} cos β in Figure 7.16c can be withstood. However, in the uncracked tension zone, σ_n is tensile; hence, the V_{ct} cos β component further increases the tensile normal stress, further reducing the effectiveness of this region in resisting shear. The same can be said for the cracked tension zone. Somehow, V_{cr} cos β has to be resisted by the concrete across the crack width, in which there is no confining stress, making it difficult to envisage even when slip occurs. Hence, once again, it is the concrete in compression that resists a substantial amount, if not all, of the transverse shear in deriving the pre-sliding capacity.

7.4.2 General pre-sliding analysis

The shape of the flexure shear segment is shown in Figure 7.17, where it can be seen that it is controlled by the length of the compression zone L_{def} and by the slope of the inclined plane β. These parameters are subject to the normal scatter that is associated with RC. For example, it would be possible to estimate L_{def}, then perform the following

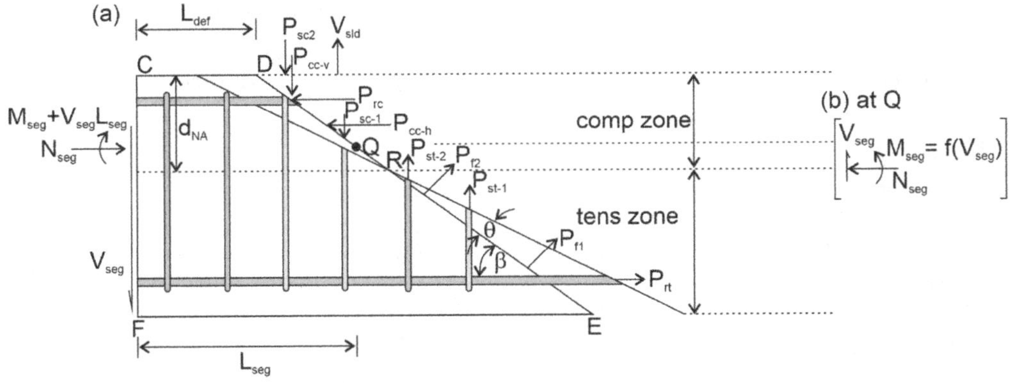

Figure 7.17 Component forces in pre-sliding analyses (a) segment free body; (b) external stress resultants acting on sliding plane.

analyses to determine the depth of the neutral axis d_{NA}, then adjust L_{def} as in Figure 7.8, and if it is not close to the estimated value, repeat the analyses, that is, iterate towards a solution. However, the sliding plane may have emanated from a flexural crack, as in Figure 7.3, which itself would affect L_{def} and lessen the value of iterating. Similarly for the slope β. Alternatively, β could be varied to find the minimum pre-sliding capacity, but this is unlikely to occur where β was formed, as this depends on the homogeneous material and not on the shear sliding capacity. However, the following analyses provide a mechanical tool for assessing any shape of the flexure shear segment. Let us now assume that the shape of the flexure shear segment, that is, L_{def} and β, is fixed and the segment contains both longitudinal tension and compression bars and lateral stirrups, as in Figure 7.17.

The basic principles of a Euler–Bernoulli flexure shear analysis have been described with the help of Figure 7.12 and are similar to those described in Zhang et al. (2016) and Sturm et al. (2021), although this previous analysis was conducted with a different and incomplete resolution of forces on the free body. The external stress resultants on the *LHS* of the free body are shown on the left side of Figure 7.17a. The external stress resultants on the sliding plane are shown in Figure 7.17b. They act at Point *Q* as explained with the help of Figure 7.9. It is a question of determining the internal forces acting on the sliding plane in Figure 7.17a that balance the external stress resultants acting on the sliding plane in Figure 7.18b. In this pre-sliding analysis, the shear force acting on the sliding plane in Figure 7.17a shown as V_{sld}, is V_{cc} in Figure 7.16b and is resisted by the portion of the potential sliding plane in the concrete in compression. Note that V_{sld} acts vertically on the concrete compression zone, similar to *V* in Figure 5.18, and consequently reduces the strength of the concrete, as explained in Chapter 5. All the other shear forces are zero, as explained in Section 7.4.1. A Euler–Bernoulli rotation θ is applied to the sliding plane in Figure 7.17a, with the axis of rotation at Point *R* at an arbitrary depth d_{NA} as shown.

In the compression zone in Figure 7.17, the horizontal component of the contraction resisted by the fixed boundary produces a compressive horizontal force in the reinforcement P_{rc} and in the concrete $P_{cc\text{-}h}$. Furthermore, the vertical component of the contraction produces a compressive force in the stirrups P_{sc} and in the concrete

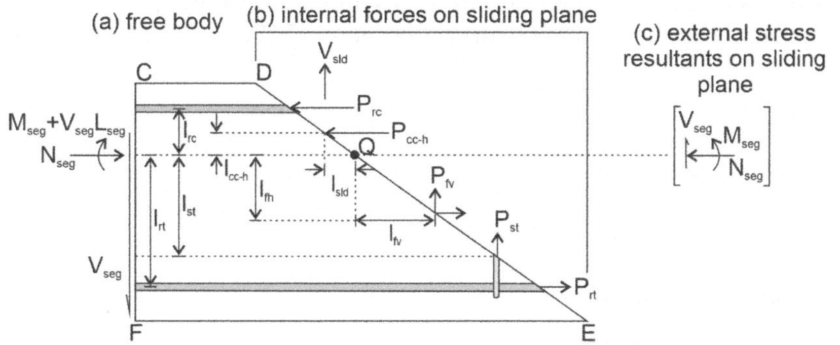

Figure 7.18 Euler–Bernoulli flexure shear force components (a) segment free body; (b) internal forces acting on sliding plane; (c) external stress resultants acting on sliding plane.

$P_{cc\text{-}h}$, but as these forces are indeterminate, being resisted by a free boundary and consequently diminished, they will be ignored. In the tension region, none of the forces depend on the reaction against a fixed boundary, as we are ignoring the concrete in tension. Instead, all of the forces depend on the crack width and can be derived from the *PI* tension stiffening P/Δ relationship from Chapter 4. The force in the stirrups is shown as P_{st} and that in the longitudinal tension reinforcement as P_{rt}. Any fibres spanning the crack are straightened across the crack and exert a tensile force normal to the crack, P_f, that depends on the crack width, as discussed in Section 3.1.5. Being normal to the crack face, they have both a horizontal and a vertical component, as shown in Figure 7.11a.

7.4.3 Equilibrium

From the above section, it can be seen that for a fixed Euler–Bernoulli deformation as in Figure 7.17a, all the internal forces along the sliding plane can be derived. These must be in equilibrium with the external stress resultants in Figure 7.17b. The axial force N_{seg} is an externally applied axial force that is applied at level Q on the sliding plane. It could be an axial force in a column or a post-tension force in a cable. Note that N_{seg} for prestressed beams or segments is dealt with in full in Chapter 9. The applied transverse shear force V_{seg} in Figure 7.17 is the applied transverse shear force at Point Q. Finally, the moment M_{seg} is the applied moment at Point Q, which is directly related to V_{seg} through the loading configuration applied to the beam; for example, in Figure 7.13a, the transverse shear force at Point Q on the sliding plane is V_{seg} and the moment M_{seg} at Point Q is $V_{seg}K_{LA}$.

The resultant internal component forces V and P in Figure 7.17 and their lines of action or lever arms l are shown in Figure 7.18, where P_{rc} is the total force in the compression reinforcement, the resultant of which acts at l_{rc} from Q; $P_{cc\text{-}h}$ is the total force in the concrete in compression, with the resultant acting at $l_{cc\text{-}h}$; P_{st} is the total force in the stirrups in tension acting at l_{st}; P_{fv} is the vertical component of all the fibre forces acting at l_{fv}; P_{fh} is the horizontal component of the fibre forces acting at l_{fh}; P_{rt} is that of the tensile reinforcement at l_{rt}; and finally, the shear force acting on the sliding plane V_{sld} acts at l_{sld}.

Equilibrium is required of the internal forces along the sliding plane in Figure 7.18b with the externally applied stress resultants in Figure 7.18c. For vertical equilibrium,

$$V_{seg} = V_{sld} + P_{st} + P_{fv} \tag{7.12}$$

and for horizontal equilibrium,

$$N_{seg} = P_{rc} + P_{cc-h} - P_{fh} - P_{rt} \tag{7.13}$$

and for rotational equilibrium,

$$M_{seg} = P_{rc}l_{rc} + P_{cc-h}l_{cc-h} + P_{st}l_{st} + P_{fv}l_{fv} + P_{fh}l_{fh} + P_{rt}l_{rt} - V_{sld}l_{sld} \tag{7.14}$$

It can be seen in Eq. 7.12 that the shear to be resisted by the sliding plane V_{sld}, in this pre-sliding analysis, is reduced by the stirrups and fibres in tension. However, in Eq. 7.13, the fibres increase the compressive forces acting on the sliding plane. It may also be worth noting that N_{seg} could be an axial load in a column or a tendon force in a post-tensioned member. In Eq. 7.14, it can be seen that on the *RHS*, the shear force component is negative, whilst all the others are positive. Hence, to achieve the required moment, all the positive components have to increase to cope with transverse shear.

7.4.4 Deriving flexure shear capacity accurately

The above approach is essentially a flexural approach using an inclined plane, in which the transverse shear weakens the concrete stress–strain property.

7.4.4.1 Full moment rotation approach

Let us start with the simplest problem of concrete without fibres and a beam or slab without stirrups, such that from Eq. 7.12 V_{sld} is equal to V_{seg}, and from the loading configuration on the beam, M_{seg} is known as it is a function of V_{seg}. Numerous approaches can be made, but the most straightforward approach is to derive the moment–rotation response. The maximum moment M_{max} will then also give the maximum transverse shear V_{max}. The steps in the analysis are as in the following flow chart:

1. From Eq. 7.12, V_{sld} is equal to V_{seg}.
2. Estimate or guess V_{sld}, which is $V_{sld\text{-}est}$. From $V_{sld\text{-}est}$, derive $M_{seg\text{-}est}$ as there is a direct relationship between V and M depending on the beam configuration.
3. Derive the concrete compressive σ_c/ε_c for concrete subjected to the vertical transverse shear force $V_{sld\text{-}est}$, which can be derived from Section 5.4.
4. Apply a small rotation θ_x, as shown in Figure 7.18, similar to Figure 7.17.
5. Move the Euler–Bernoulli displacement up or down at the fixed θ_x until Eq. 7.13 is satisfied. If there are no applied axial forces N_{seg}, then move the displacement up or down until the axial force is zero.
6. Take moments about Point Q as in Eq. 7.14 to determine the derived moment $M_{seg\text{-}der}$.

7. If the derived moment $M_{seg\text{-}der}$ is not close to the estimated moment $M_{seg\text{-}est}$, then iterate; that is, vary $V_{seg\text{-}est}$ and consequently obtain a new σ_c/ε_c response for the new transverse shear and repeat the above steps until it does.
8. This gives one point on the moment–rotation response.
9. Increase the rotation and repeat the above analyses to obtain further points.
10. Repeat the analysis until a maximum moment M_{max} is achieved, from which V_{max} can be determined. These are the combinations of stress resultants that cause failure.

7.4.4.2 Allowance for stirrups and fibres

Allowing for both stirrups and fibres in the concrete can follow the above approach, except that it requires more iterations. The transverse shear V_{sld} no longer equals V_{seg} but can be derived from Eq. 7.12, which will vary with the rotation in Figure 7.17. Every time the Euler–Bernoulli displacement is moved, the tensile stirrup force and the fibre force change, and consequently, V_{sld} changes, requiring a new concrete stress–strain for the analysis, which from Section 5.2.3 can also allow for passive confinement, which is a great advantage. But it can be done. No doubt simplifications can be found, but at least we have a model that can find an answer for use in developing analysis approaches.

7.4.4.3 Shear resistance at specific moment

An alternative approach to the moment–rotation method above is to determine whether a member can resist a required shear load V_{req}. This simplifies the approach as the concrete stress–strain response is now constant because the shear V_{req} is constant. For V_{req}, the moment M_{req} is known from the loading arrangement. The steps are as follows:

1. From Eq. 7.12, V_{req} is equal to V_{seg} and M_{req} is also known as it is a function of V_{req}.
2. Derive the concrete compressive σ_c/ε_c for concrete subjected to the vertical transverse shear force V_{req}, which can be derived from Section 5.4.
3. Apply a rotation θ_x in Figure 7.18, as shown in Figure 7.17.
4. Move the Euler–Bernoulli displacement up or down until Eq. 7.13 is satisfied.
5. Take moments about Point Q as in Eq. 7.14 to determine the derived moment $M_{req\text{-}der}$.
6. If the derived moment $M_{req\text{-}der}$ is not close to the required moment M_{req}, then vary the rotation until it is.
7. If there is no solution, then the member has failed; otherwise, if there is a solution, the member can resist the required transverse shear.

7.4.5 Approximate shear flexure approach

The moment–rotation approach in Section 7.4.4 is an accurate method, as it simulates the continuously varying concrete σ/ε properties with the continuously varying transverse shear. However, it requires a large number of iterations. An alternative approach, and another way of visualising the problem, is to conduct a moment–rotation analysis, initially ignoring the effect of the transverse shear V_{sld} on the stress–strain concrete properties. Then, check to determine whether the transverse shear on the concrete has weakened it sufficiently to cause failure. It must be emphasised that, as with all mechanical simplifications, this is an approximation.

7.4.5.1 Shear flexure approach

7.4.5.1.1 Mechanics

By ignoring the effect of transverse shear on the concrete compressive σ/ε properties, these properties can be obtained from Chapter 5 and will remain constant. A moment–rotation analysis is then performed with gradually increasing rotation. Let the results at one specific rotation θ_x be those in Figure 7.19; that is, we have determined M_{seg}, and consequently, from the distribution of applied loads V_{seg} and from Eq. 7.12, we have V_{sld} as the other parameters in the equation have been derived from the moment–rotation analysis.

The vertical transverse shear force V_{sld} in Figure 7.19 has been resolved along the sliding plane and normal to the sliding plane in Figure 7.19b. The longitudinal concrete compressive force P_{cc}, which was derived from the moment–rotation analysis, has also been resolved in the same directions in Figure 7.19c. The combinations of these stress resultants are shown in Figure 7.19d, where it can be seen that P_{cc} increases the shear force to be resisted, which is detrimental, and V_{sld} reduces the normal force, which is also detrimental. These stress resultants can be used to derive the shear stress τ_{calc} and the normal stress $\sigma_{N\text{-}calc}$ on the sliding plane in the region where the concrete is in compression. From the shear friction properties, such as those shown in Figure 1.4 or in Section 3.1, a specific capacity V_u at zero slip can be determined for a confinement stress $\sigma_{N\text{-}calc}$; at zero slip, we are conducting a pre-sliding analysis. If $\tau_{calc} < V_u$, then shear sliding failure has not occurred for that M_{seg}, and the applied rotation can be increased. At each rotation, the ability to resist the shear is determined until a rotation is reached at which the shear can just be resisted. This is then the maximum moment M_{max} and consequently from which the maximum shear V_{max} is derived.

No doubt simpler solutions can be found, which is the aim of research. However, the researcher now has a generic free-body model that is based on sound mechanics to try to find a simple analysis procedure.

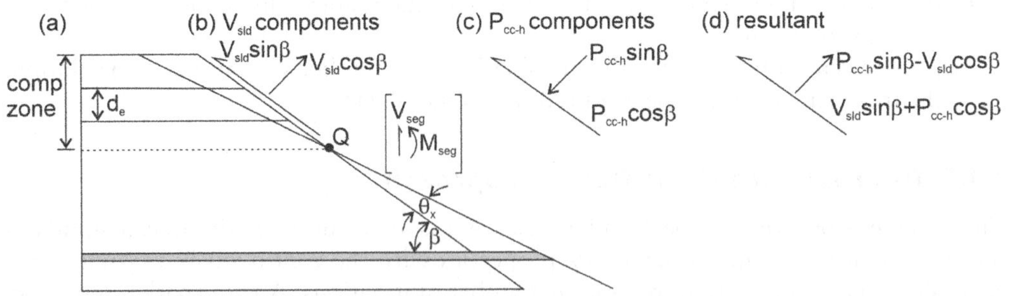

Figure 7.19 Approximate shear flexure approach (a) segment elevation; (b) V_{sld} components; (c) $P_{cc\text{-}h}$ components; (d) resultant forces.

7.4.5.1.2 Further simplification of mechanics

The above approach, with just tension reinforcement in Figure 7.19, can be further visualised and simplified as in Figure 7.20 (Al-Mhawish et al. 2025a). The possible sliding plane *D–E* in Figure 7.20a often emanates from an existing crack at spacing S_{cr} that can be derived from Chapter 4. Although the spacing can be accurately predicted through tension stiffening analyses, the location of a flexural crack is governed by semi-random phenomena, as explained in Chapter 4. It is suggested that L_{def} at the top fibre should be just large enough to accommodate the wedges associated with concrete softening.

As explained previously in this chapter, in the flexural analyses of an inclined plane, as in Figure 7.19, the resultant horizontal compressive force $F_{cc\text{-}R}$ and the level at which it acts, that is, Point *Q* in Figure 7.20a, can be determined. As a simplification to the analyses, let the transverse shear force V_Q be assumed to act also through this Point *Q*. Hence, the theoretical moment at Point *Q* is given by

$$M_Q = F_{cc-R} l_{cc} = V_Q L_{sp} \tag{7.15}$$

where, from Figure 7.20, L_{sp} is the shear span at Point *Q*, such that the theoretical moment at the end of the shear span, M_{theo}, can be determined as shown in Figure 7.20b. For validation, it is the theoretical moment M_{theo} that needs to be compared with the experimental moment at failure M_{exp}.

The Euler–Bernoulli analysis in Section 7.2.3 and in Figure 7.10 and 7.11 is now performed as in Figure 7.21, where the compressive zone has been divided into *n* slices. As in Figure 7.21a, β_x is fixed, then a small rotation θ_x is applied. The *E-B* deformation moves up and down until horizontal equilibrium is achieved, which in this case the resultant axial force N_{seg} is zero.

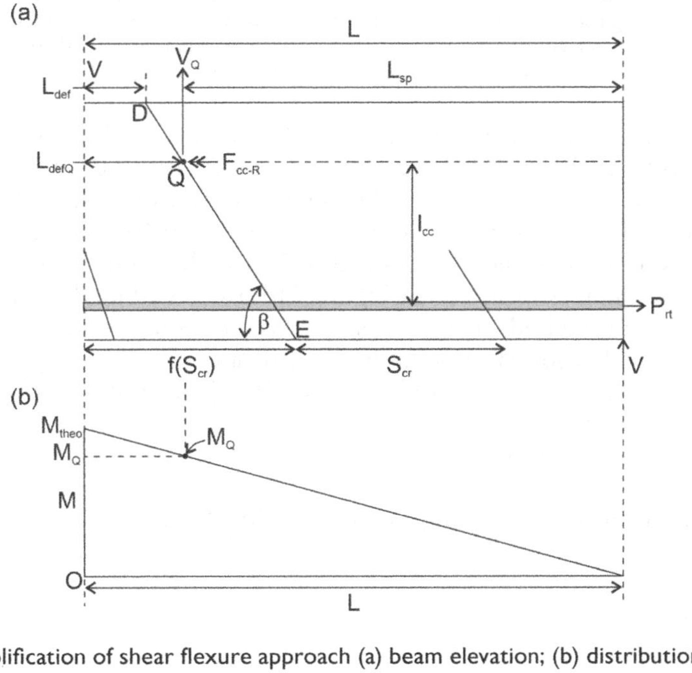

Figure 7.20 Simplification of shear flexure approach (a) beam elevation; (b) distribution of moment.

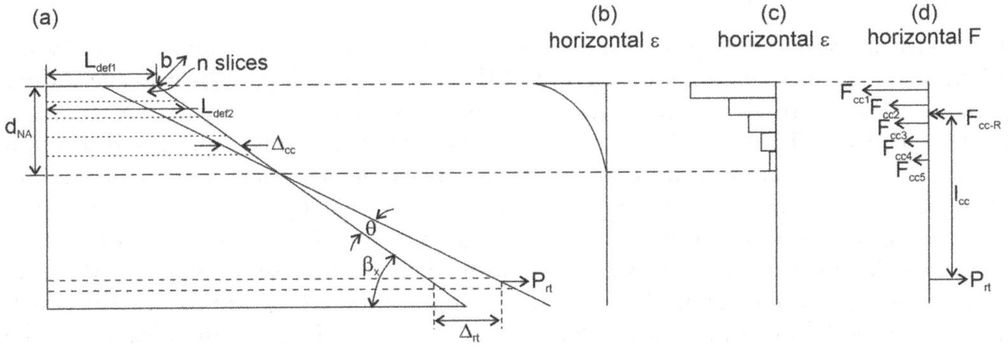

Figure 7.21 E-B analysis of sloping plane (a) segment elevation and Euler Bernoulli displacement profile; (b) distribution of horizontal strain on compression region; (c) distribution of horizontal stress in sliced segment; (d) force profile.

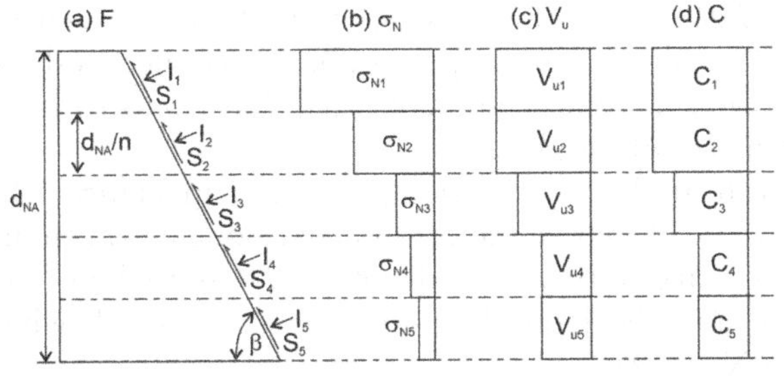

Figure 7.22 Shear capacity of compression region (a) division of compression region into elements and showing the individual normal forces I and shear forces S on the inclined plane of each element; (b) distribution of normal stress to sliding plane σ_N; (c) distribution of material shear capacities V_u; (d) distribution of element shear capacities C.

The moment M_Q is then taken as in Figure 7.21d and in Eq. 7.15, which from Eq. 7.15 also gives V_Q. We now have both M_Q and V_Q for this β_x at this θ_x. It is now a question of determining whether the potential sliding plane can resist this shear force V_Q, which will determine its shear capacity C_Q derived from the shear friction properties.

The compression zone of the beam in Figure 7.21 is shown in Figure 22a. The horizontal concrete compressive forces in Figure 21d are distributed as in Figure 7.21c. At failure, there is a rigid body displacement across the sliding plane, and the transverse shear force V is divided equally among the n segments.

In Figure 7.22a, the inclined shear force S_x over the element of depth d_{NA}/n is given by

$$S_x = F_{ccx} \cos \beta + (V / n) \sin \beta \tag{7.16}$$

such that the total shear force acting along the inclined plane S_{tot} is

$$S_{tot} = \sum_{1}^{n} S_x \tag{7.17}$$

It is now a question of whether that shear force along the inclined plane S_{tot} can be resisted; that is, whether the capacity of the inclined plane is greater or less than the total shear force S_{tot}.

The normal force I_x over the element of depth d_{NA}/n is given by

$$I_x = F_{ccx} \sin\beta - (V / n)\cos\beta \tag{7.18}$$

The cross-sectional area of the sliding plane A_{sldx} over the element of depth d_{NA}/n and width b is given by

$$A_{sldx} = \frac{bd_{NA}}{n\sin\beta} \tag{7.19}$$

Therefore, the normal stress on the sliding plane σ_{Nx} is

$$\sigma_{Nx} = I_x / A_{sldx} \tag{7.20}$$

which is shown in Figure 7.22b.

From the shear friction properties in Chapter 3, the shear stress required to initiate shear sliding V_u depends on σ_N, hence the distribution of V_{ux} can be determined as in Figure 7.22c. Multiplying by A_{sldx} gives the distributions of capacities in Figure 7.22d, in which the total of the capacities is given by

$$C_{tot} = \sum_{1}^{n} C_x \tag{7.21}$$

It is now a question of when shear failure occurs. Does it occur in a brittle fashion when one of the shear forces S_x in Figure 7.22a exceeds one of the capacities C_x in Figure 7.22d, or does it occur in a ductile fashion when the total shear force S_{tot} exceeds the total capacity C_{tot}? For most concretes, as in Figure 3.4, the shear capacity τ_{SF} remains constant at V_u after slip occurs and does not suddenly reduce. Failure along a sliding plane is associated with a rigid body displacement on either side of the sliding plane. Hence, because of the ductility, all the capacities can be achieved simultaneously, particularly when the *SF* properties are ductile.

For the fixed value β_x in Figure 7.21, θ_x is gradually increased until member failure is achieved, resulting in Point *A* in Figure 7.23. The above analysis is then repeated at varying β_x to determine the minimum strength, as shown in Figure 7.23, marked as $V_{min}(M_{min})$, to remind the reader that failure occurs due to both the applied moment and applied shear, and not due to shear alone.

7.4.5.2 Starting with shear capacity

A novel and useful approach that can lead to closed-form solutions is to start with the shear capacity and then, from a moment–rotation analysis, determine when it is reached.

The free-body diagram is shown in Figure 7.24b. The external stress resultants applied to the free body at the sliding plane are shown in Figure 7.24c, which consists of a transverse shear force V_{seg} and a moment M_{seg}. As in this example, no external load

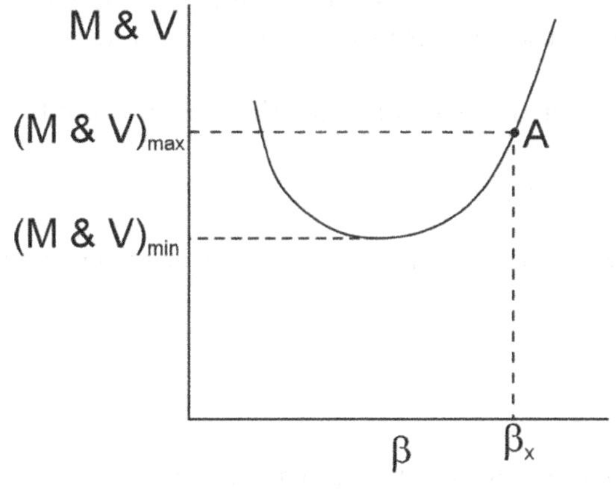

Figure 7.23 Variation of capacities with β.

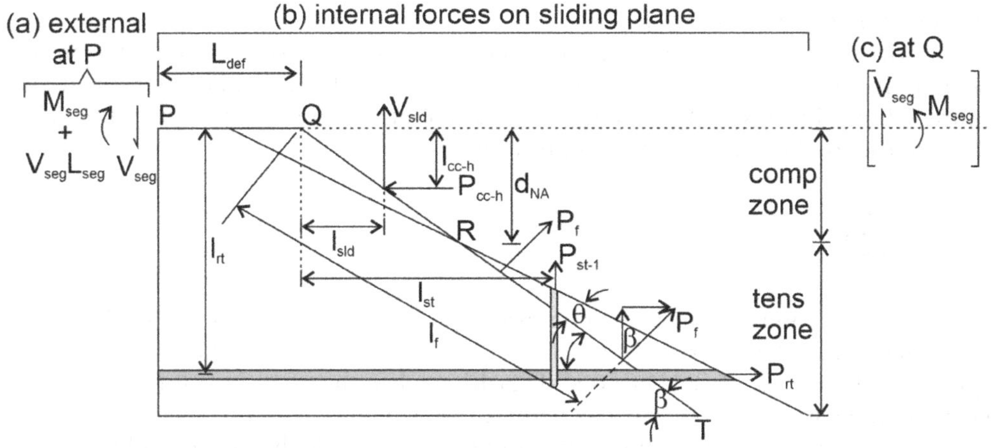

Figure 7.24 Shear flexure approach (a) external forces on free body on LHS at location P; (b) segment free body and the internal forces on sliding plane; (c) external stress resultants applied to free body at location Q.

N_{seg} is being applied to the free body; moments can be taken anywhere along the sliding plane. We have chosen Point *Q* at the top fibre. Hence, the external or applied moment at Point *Q* is what is required for M_{seg}. The transverse shear at Point *Q* is the applied transverse shear V_{seg}, which is the same throughout the free body, as we are dealing with a constant shear region.

A Euler–Bernoulli analysis will give all the internal forces acting on the sliding plane, as shown in Figure 7.24b, and these need to balance the external stress resultants in Figure 7.20d. To simplify the analysis and lead to a closed-form solution, the concrete can be assumed to have zero tensile capacity and to be uniformly stressed in compression. The internal forces in Figure 7.24b consist of the transverse shear forces resisted directly by the sliding plane in compression V_{sld}, which, for ease of analysis, can be considered uniform along the compression zone Q–R due to rigid body displacements when sliding eventually occurs. It can be seen from the triangle of forces that this transverse shear force V_{sld} has a component along the sliding plane, $V_{sld} \sin \beta$, and a component normal to the sliding plane, $V_{sld} \cos \beta$. The lever arm of V_{sld} from Q is shown as l_{sld}. It is worth noting that all of the lever arms of the internal forces are taken about Point Q, as this is where the external moment is taken. The force in the tension reinforcement is P_{rt}, the resultant force in the fibres crossing the crack is P_f, the resultant vertical force in the stirrups is P_{st} and the resultant horizontal compressive force in the concrete is $P_{cc\text{-}h}$.

From vertical translational equilibrium, the vertical external stress resultant in Figure 7.24c must equal the vertical internal stress resultant in Figure 7.24b; that is,

$$V_{seg} = V_{sld} + P_{st} + P_f \cos \beta \tag{7.22}$$

For horizontal translational equilibrium, the external applied horizontal stress resultant N_{seg} is zero in this example, which must equal the internal horizontal stress resultants in Figure 7.24b; that is,

$$N_{seg} = 0 = P_{cc-h} - P_f \sin \beta - P_{rt} \tag{7.23}$$

And finally, for rotational equilibrium, the external applied moment M_{seg} in Figure 7.24c, which is applied at Point Q, must equal the internal moment in Figure 7.24b, also about Point Q. Therefore,

$$M_{seg} = P_{st} l_{st} + P_f l_f + P_{rt} l_{rt} + P_{cc-h} l_{cc-h} - V_{sld} l_{sld} \tag{7.24}$$

The surface area of the potential sliding plane in Figure 7.24b, extending from Point Q to R with a width b, is given by Eq. 7.19, where in this analysis $n=1$ and d_{NA} is the depth to the neutral axis from the compression face.

The shear friction capacity of the sliding plane at zero slip, as we are dealing with the pre-sliding capacity, over this area A_{sld} is derived from the *SF* properties at zero slip, which is V_u in Figure 1.4 and is given in Section 3.1.4.1 in the following form:

$$V_u = a f_c + b \sigma_n \tag{7.25}$$

where it can be seen that V_u depends on both the concrete compressive strength f_c and the normal stress to the sliding plane σ_N. Therefore, the *SF* capacity of the sliding plane Z_{cap}, which is the maximum shear force it can resist, is the product of Eqs. 7.19 and 7.25; that is,

$$Z_{cap} = V_u \frac{b d_{NA}}{\sin \beta} \tag{7.26}$$

where from Eq. 7.25, V_u depends on σ_N. From Figure 7.24b, there are two forces acting on the sliding plane Q–R, whose normal components to the sliding plane are $P_{cc\text{-}h} \sin \beta$, which is compressive, and $V_{sld} \cos \beta$, which is tensile. Dividing these normal forces by A_{sld} in Eq. 7.19 and rearranging gives

$$\sigma_n = \frac{P_{cc-h}\sin^2\beta}{bd_{NA}} - \frac{V_{sld}\sin\beta\cos\beta}{bd_{NA}} \tag{7.27}$$

where it can be seen that the shear component reduces the confinement and, hence, weakens the resistance to shear.

In Figure 7.24b, the shear force resisted by the sliding plane is V_{sld}. Let the maximum shear force that the sliding plane can resist be $(V_{sld})_{max}$. The resolution of this shear force along the sliding plane $(V_{sld})_{max}$ sin β is equal to its capacity Z_{cap} in Eq. 7.24, less the shear along the sliding plane due to $P_{cc\text{-}h}$, which is $P_{cc\text{-}h}$ cos β. Rearranging this equation gives

$$\left(V_{sld}\right)_{max} = \frac{Z_{cap}}{\sin\beta} - \frac{P_{cc-h}}{\tan\beta} \tag{7.28}$$

Substituting $(V_{sld})_{max}$ from Eq. 7.28 into Eq. 7.22 gives the maximum externally applied shear force that can be resisted as

$$\left(V_{seg}\right)_{max} = \frac{Z_{cap}}{\sin\beta} - \frac{P_{cc-h}}{\tan\beta} + P_{st} + P_f\cos\beta \tag{7.29}$$

A Euler–Bernoulli rotation analysis, as in Figure 7.24, can be used to determine the combination of parameters in Eq. 7.29 that are required to cause failure. The procedure can be simplified as much as possible to determine closed-form solutions, such as by assuming that the concrete has zero tensile strength and exhibits a linear–elastic stress–strain response in compression. As in Section 7.4.4, numerous approaches are available. For example,

1. For a specific applied V_{seg}, and consequently, from the loading arrangements, an applied M_{seg}.
2. Apply rotation θ_x to the Euler–Bernoulli analysis in Figure 7.24.
3. Vary d_{NA} until there is horizontal equilibrium, as in Eq. 7.23; this gives $P_{cc\text{-}h}$ and P_f.
4. From vertical equilibrium in Eq. 7.22, we have V_{sld}.
5. From rotational equilibrium in Eq. 7.24, we get a derived value of M_{seg}.
6. If the derived M_{seg} in Step 5 does not equal the applied M_{seg} in Step 1, then vary qx until it does.
7. Once it does, check for shear failure as follows:
8. From Eq. 7.27, obtain σ_n and for this value of σ_n, get V_u from Eq. 7.25.
9. From Eq. 7.26, obtain Z_{ap}.
10. From Eq. 7.29, obtain the maximum shear strength $(V_{seg})_{max.}$ If this is less than or greater than the applied V_{seg} in Step 1, iterate until it is equal; that is, go back to Step 2 and vary θ_x.

7.4.6 Concept of shear failure

It is general practice to assume that a member has a specific shear capacity and a specific moment capacity. From the analyses above, using the shear flexure free body, it can be seen that these two components are directly related as it gives the applied moment acting on the sliding plane M_{sld}=M_{seg} not to cause failure, but when failure occurs;

and also the transverse shear acting on the sliding plane $V_{sld} = V_{seg}$ when failure occurs. Neither of these stress resultants are capacities, as neither would cause failure by themselves; rather, it is their combination, $M_{sld} + V_{sld}$, that causes failure.

7.5 POST-SLIDING ANALYSIS

The post-sliding behaviour is the same as the pre-sliding behaviour in Section 7.4, with an additional passive-confinement mechanism associated with interface slip *S*, which is explained in Section 1.4.4 and Chapter 5 and illustrated in Figure 7.25. The element has a single vertical crack in which transverse reinforcement crosses the crack at right angles, which could be stirrups or longitudinal reinforcement. The concrete elements *on* either side of the crack act as rigid bodies such that the slip *S* imposed is the same throughout the depth. The slip *S* causes a separation h_{cr} which depends on the shear friction properties in Sections 1.3.1 and3.1.2.

The separation in Figure 7.25 is resisted by tension in the reinforcing bars of magnitude P_s, which balances the induced compression in the concrete of magnitude P_s; the latter is referred to as confinement or confinement of the sliding plane. It can be seen that this passive-confinement mechanism is in local equilibrium, as the overall resultant force is zero, such that it does not affect the global equilibrium. The confinement σ_n improves the ability of the crack to resist slip, as can be seen from the shear friction properties in Sections 1.3.1 and 3.1.2. The passive-confinement is balanced by tension in the bars. Hence, the ability of the bar to resist global forces is diminished; that is, a gain in confinement is balanced by a loss in the reinforcement capacity to resist tensile forces. You cannot have something for nothing.

The passive-confinement mechanism in Figure 7.25 is shown in Figure 7.26 for an inclined sliding plane. In this case, the effective crack width is $h_{cr+}S \sin \beta$. Let P_{st} be the tensile force in the reinforcement due to the global behaviour and P_{sl} that due to passive confinement. Once again, as far as passive confinement is concerned, ignoring P_{st}, there is local equilibrium. The passive-confinement forces are resolved into their components in Figure 7.26b. It can be seen that the shear forces are of equal magnitude but opposite

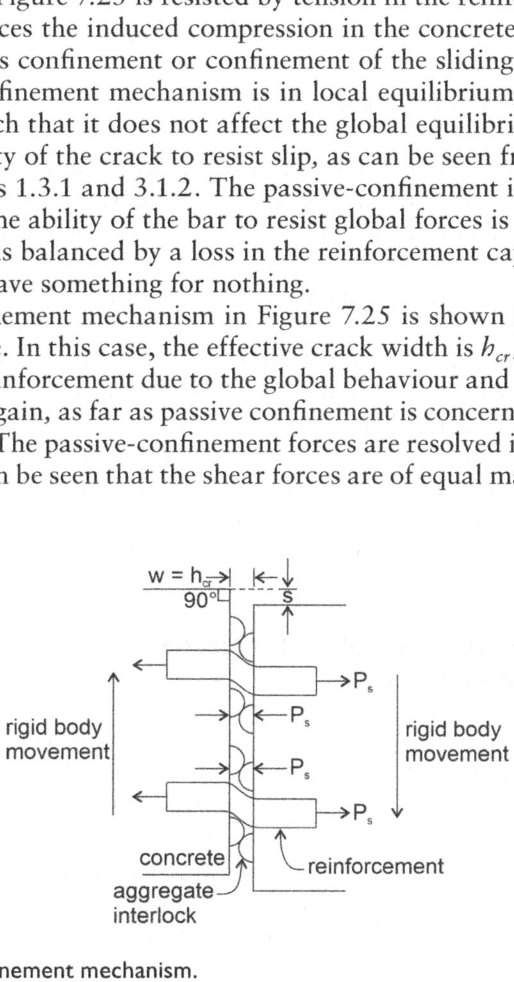

Figure 7.25 Passive-confinement mechanism.

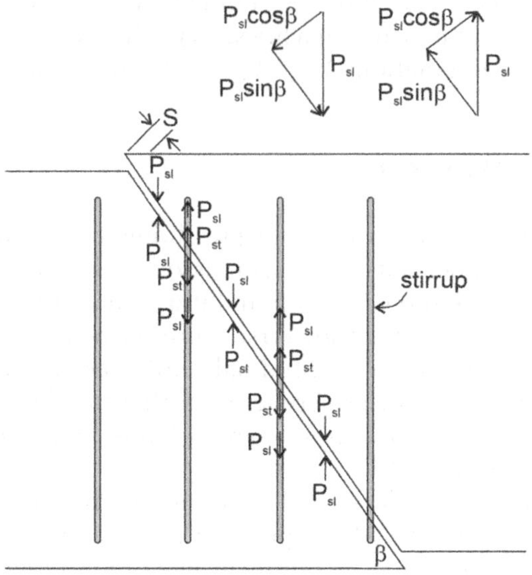

Figure 7.26 Passive confinement across inclined plane.

direction, so they do not change the overall shear distribution. As before, the passive confinement across the concrete-encased sliding plane, which is beneficial, is balanced by tension in the reinforcement, which reduces the ability of the reinforcement to resist global forces.

The great benefit of passive confinement is that it allows the stirrups that cross the sliding plane in the compression zone, such as those shown in Figure 7.17a, to passively confine the concrete once sliding occurs. Prior to sliding, the forces in the stirrups in the compression zone, shown as P_{sc}, tend to zero and have little effect. Once sliding through the compression zone occurs, there is a separation due to aggregate interlock, which induces the passive-confinement mechanism and consequently increases the shear resistance of the sliding plane. This mechanism can be added to the others in the pre-sliding moment–rotation analysis, where, for example, in Section 7.4.4, the shear capacity is no longer at the appropriate V_u at zero slip but at τ a slip S.

7.6 DERIVING CLOSED-FORM SOLUTIONS FOR SHEAR CAPACITY

In the method described in Section 7.4.5.1, an iterative numerical approach is required to find a solution (Al-Mhawish et al. 2025a). This approach is accurate as it can cope with any non-linearity, but each beam has to be analysed individually. In contrast, a closed-form solution is generic (Al-Mhawish et al. 2026). The art of developing closed-form solutions is to simplify the mechanics and the material properties without any significant loss of accuracy. To illustrate the development of closed-form solutions, let us attempt to do this for the shear flexure approach in Section 7.4.5.1 (Al-Mhawish et al. 2026).

In the numerical approach in Section 7.4.5.1, the depth of the neutral axis d_{NA} was derived automatically through the Euler–Bernoulli iterations. To simplify the mechanics, let us assume it can be derived from the elementary analysis of the second moment of area of the transformed cracked section, as shown in Figure 7.27, where Point Q is the centroid of the transformed compression region; A_{sp} is the cross-sectional area of the compression region, which is the area of the potential sliding plane when it is vertical; and d_c is the distance between the centroid of the compression region and the tension reinforcement.

The mechanics already illustrated in Figure 7.19 for the numerical approach is shown again in Figure 7.28 with subtle alternative representations to help in the derivation of a closed-form solution. The forces acting on the inclined plane are shown in Figure 7.28a, where V_{sld} is the vertical portion of the applied shear force V_Q acting on the potential sliding plane of the compression zone; F_{cc} is the horizontal resultant compressive force in the concrete; P_{rt} *is* that in the tension reinforcement; P_{fh} is the resultant horizontal component of the fibre force; P_{st} *is* the resulting vertical force in the stirrups, which could be derived by assuming the stirrups are uniformly smeared over the width w_{st}; and P_{fv} is the resulting vertical component of the fibre force. The lever arms from Point Q are also shown.

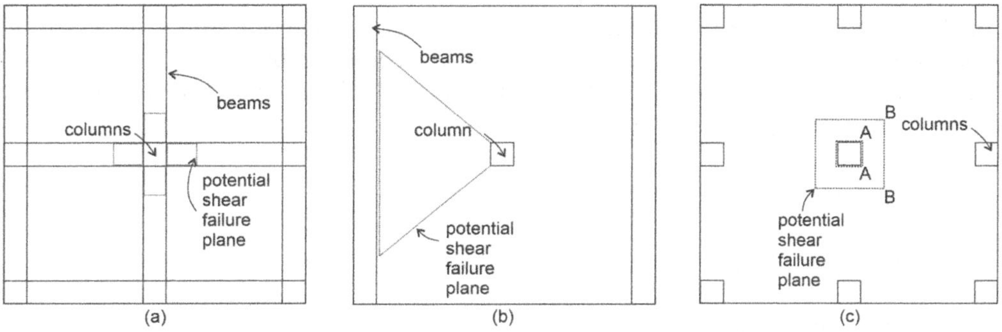

Figure 7.27 Transformed cracked section.

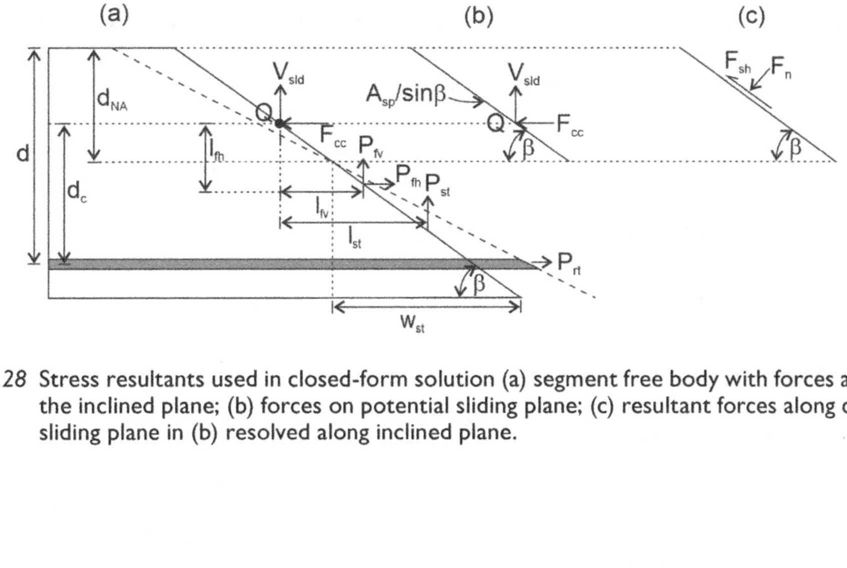

Figure 7.28 Stress resultants used in closed-form solution (a) segment free body with forces acting on the inclined plane; (b) forces on potential sliding plane; (c) resultant forces along concrete sliding plane in (b) resolved along inclined plane.

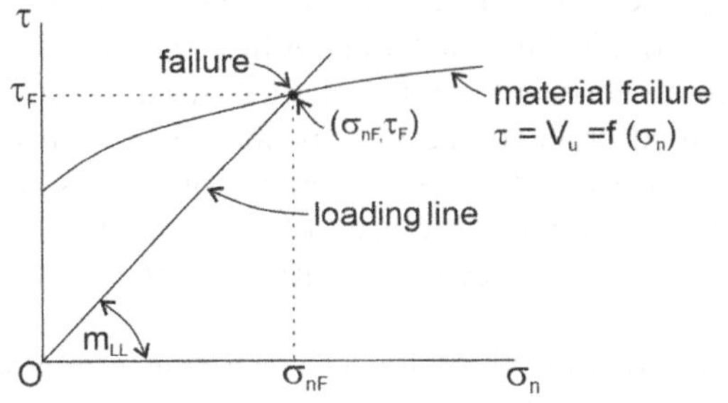

Figure 7.29 Shear friction material properties at failure.

The segment in Figure 7.28a is subjected to an externally applied moment at Point Q of M_Q and also an applied shear force of V_Q, both acting through Point Q; this is yet another simplification to help in the mechanics. The resulting forces on the inclined potential sliding plane of cross-sectional area $A_{sp}/\sin\beta$ are shown in Figure 7.28b. It raises the question of when, as the loads on the beam are gradually increased, these combined forces cause failure through their shear friction strengths being achieved. This is illustrated in Figure 7.29, where we need to derive through mechanics the increase in τ and σ_n, which is the slope m_{LL} as the beam is gradually loaded, as shown by the loading line. Where this loading line intercepts the shear friction material properties $V_u=f(\sigma_n)$, which are given in Section 3.1, failure occurs at the combination (σ_{nF}, τ_F), as shown.

The inclined components of the forces in Figure 7.28b are shown in Figure 7.28c, where the normal force F_n is given by

$$F_n = F_{cc}\sin\beta - V_{sld}\cos\beta \tag{7.30}$$

and the shear force F_{sh} by

$$F_{sh} = F_{cc}\cos\beta + V_{sld}\sin\beta \tag{7.31}$$

which, when divided by the cross-sectional area of the inclined plane $A_{sp}/\sin\beta$, gives the average normal stress σ_n and average shear stress τ. By using the average stress values, it is assumed that these stresses are uniformly distributed over the potential sliding plane, which is yet another simplification.

From Figure 7.28a, for horizontal equilibrium,

$$P_{rt} + P_{fh} = F_{cc} \tag{7.32}$$

and for vertical equilibrium,

$$V_Q = V_{sld} + P_{st} + P_{fv} \tag{7.33}$$

and for rotational equilibrium,

$$M_Q = P_{rt}d_c + P_{fh}l_{fh} + P_{fv}l_{fv} + P_{st}l_{st} \tag{7.34}$$

Substituting P_{rt} from Eq. 7.32 into Eq. 7.34 and rearranging give

$$F_{cc} = \left(M_Q^{+} P_{fh} d_c - P_{fh} l_{fh} - P_{fv} l_{fv} - P_{st} l_{st}\right) / d_c \tag{7.35}$$

Now substituting into Eq. 7.30, F_{cc} from Eq. 7.35 and V_{sld} from Eq. 7.33 and bearing in mind that

$$V_Q = \frac{M_Q}{L_{spQ}} \tag{7.36}$$

where L_{spQ} is the shear span of Point Q in Figure 7.28a from the support, or, as another definition, L_{spQ} is the ratio of the applied M_Q/V_Q at Point Q, gives the following expression for the normal stress σ_n as

$$\begin{aligned} \sigma_n \frac{A_{sp}}{\sin\beta} &= M_Q\left(\frac{\sin\beta}{d_c} - \frac{\cos\beta}{L_{spQ}}\right) - P_{st}\left(\frac{l_{st}\sin\beta}{d_c} - \cos\beta\right) \\ &\quad - P_{fv}\left(\frac{l_{fv}\sin\beta}{d_c} - \cos\beta\right) + P_{fh}\left(\sin\beta - \frac{l_{fh}\sin\beta}{d_c}\right) \end{aligned} \tag{7.37}$$

Similarly, for Eq. 7.31, the shear stress τ is given by

$$\begin{aligned} \tau \frac{A_{sp}}{\sin\beta} &= M_Q\left(\frac{\cos\beta}{d_c} + \frac{\sin\beta}{L_{spQ}}\right) - P_{st}\left(\frac{l_{st}\cos\beta}{d_c} + \sin\beta\right) \\ &\quad - P_{fv}\left(\frac{l_{fv}\cos\beta}{d_c} + \sin\beta\right) + P_{fh}\left(\cos\beta - \frac{l_{fh}\cos\beta}{d_c}\right) \end{aligned} \tag{7.38}$$

Eliminating M_Q from Eqs. 7.37 and 7.38 gives

$$\begin{aligned} \tau = m_{LL}\sigma_n &+ \frac{P_{st}\sin\beta}{A_{sp}}\left[m_{LL}\left(\frac{l_{st}\sin\beta}{d_c} - \cos\beta\right) - \left(\frac{l_{st}\cos\beta}{d_c} + \sin\beta\right)\right] \\ &+ \frac{P_{fv}\sin\beta}{A_{sp}}\left[m_{LL}\left(\frac{l_{fv}\sin\beta}{d_c} - \cos\beta\right) - \left(\frac{l_{fv}\cos\beta}{d_c} + \sin\beta\right)\right] \\ &- \frac{P_{fh}\sin\beta}{A_{sp}}\left[m_{LL}\left(\sin\beta - \frac{l_{fh}\sin\beta}{d_c}\right) + \left(\cos\beta - \frac{l_{fh}\cos\beta}{d_c}\right)\right] \end{aligned} \tag{7.39}$$

in which

$$m_{LL} = \left[\frac{\dfrac{\cos\beta}{d_c} + \dfrac{\sin\beta}{L_{spQ}}}{\dfrac{\sin\beta}{d_c} - \dfrac{\cos\beta}{L_{spQ}}}\right] \tag{7.40}$$

Equation 7.39 is the fundamental equation required to determine the shear capacity of the mechanism in Section 7.4.5.1.

Examining the components of Eq. 7.39, first consider a beam without stirrups and fibres. This leaves only the first term in Eq. 7.39, which is $\tau = m_{LL}\sigma_n$, plotted as the 'loading line' in Figure 7.29. Where this loading line intercepts the material failure envelope $\tau = V_u = f(\sigma_n)$ gives both the normal stress σ_{nF} and shear stress τ_F at failure, which can be substituted into either Eq. 7.37 or 7.38 to give M_{QF}, which is now the moment at failure, and through Eq. 7.36, the shear at failure $V_{QF.}$

The other terms in Eq. 7.39 account for the effects of stirrups and fibres. It can be seen that the second and third terms, which deal with vertical stirrups and the vertical component of the fibre force, have, as would be expected, exactly the same form, as both are addressing vertical forces. The difficulty here is to know how these forces, namely P_{st} and P_v, vary as the load is applied. As shown in Figure 7.28a, it can be seen that P_{rt}, P_{st} and the fibre force P_f are all related by the crack width. Perhaps this will provide a solution, and we leave this to the reader to pursue.

7.7 PUNCHING SHEAR

7.7.1 Behaviour

Up until this point, we have considered shear failure in beams, as illustrated in Figure 7.30a, or in one-way slab systems. In the system shown in Figure 7.30 b, the load is transferred in one direction, causing the shear failure plane to extend over either the full width or a large proportion of the member width. Conversely, in shallow flat-slab elements, particularly where there is two-way bending, such as that shown in Figure 7.30c, it is possible for failure to occur around the perimeter of a column, such that the column effectively punches through the slab.

In the scenario in Figure 7.30c, as explained in Al-Mhawish et al. (2025a), a tension region occurs at the top surface of the slab around the perimeter of the column face, and as the load increases, it extends diagonally towards the bottom face of the slab. For analysis, it is important to remember that the inverse of the free-body in Figure 7.12 can be applied to each face of the column in Figure 7.30c, with the extension that the width

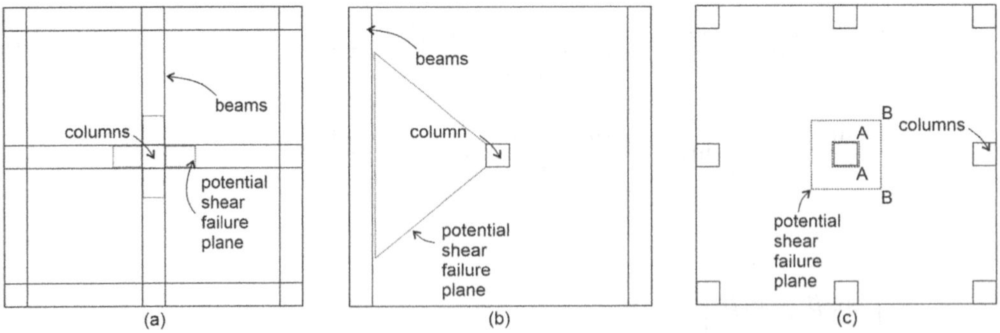

Figure 7.30 Shear failure associated with slabs. (a) Slab and beam floor system; (b) one-way slab; (c) punching two-way slab.

of the slab increases the further the failure surface extends away from the column face. This is shown in Figure 7.30c, where if failure were to occur at the column face *A–A*, the width of the shear failure zone would be equal to the width of the column, and if failure were to occur further away from the column along *B–B*, the area of concrete and reinforcement to resist shear would increase. For analysis, it is therefore necessary to quantify the shear capacity at a range of locations away from the column face, where the applied shear and bending moment also change, in order to identify the weakest potential failure location. Having identified the shear strength of a given failure plane, such as *B–B* in Figure 7.30c, a lower bound estimate of the punching shear strength is obtained by multiplying the shear capacity by 4 to capture the full failure surface around the perimeter of the column. This approach yields a lower bound estimate of the shear capacity, as it assumes that failure occurs simultaneously along the full failure surface, ignoring any post-sliding capacity and any redistribution of loads that may occur during failure.

7.7.2 Mechanics

The mechanics of punching are fairly simple to understand but complex to quantify (Al-Mhawish et al. 2025a). In research, it is always best to start with the simplest problem, which in this case is shown in Figure 7.31. The member consists of a circular slab of diameter D supported all the way round its perimeter with a shear load V; it is loaded at its centre with a punching load P; and is reinforced with a grid mesh where the longitudinal or vertical bars have an area A_{rv} with a spacing S_v, and the transverse or horizontal bars have an area A_{rh} with a spacing S_h. To simplify the problem so that

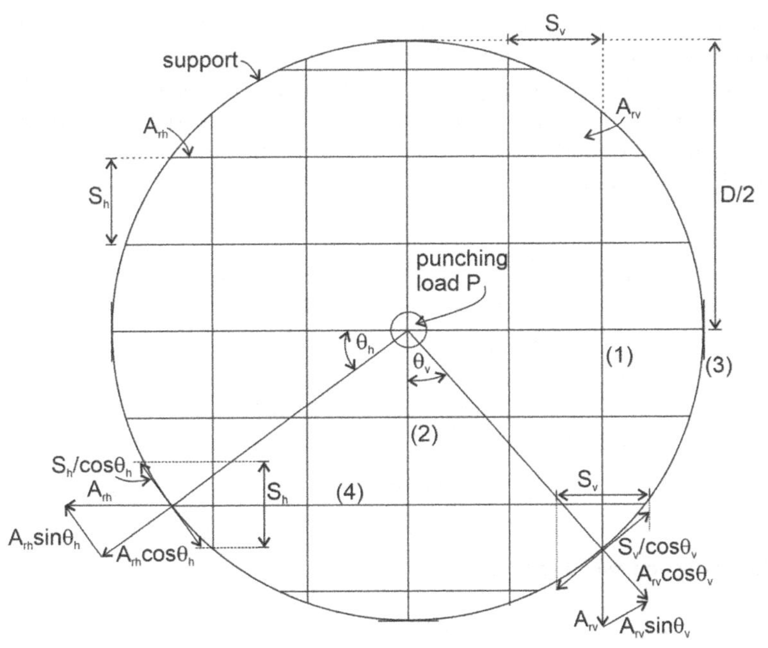

Figure 7.31 Simplest punching shear problem.

it can be analysed using the techniques developed in this chapter, it is first necessary to convert the grid reinforcement to radial reinforcement, which is reinforcement that emanates from the centre of the slab to the perimeter.

Consider the vertical or longitudinal bar labelled (1) in Figure 7.31. The cross-section of the bar is A_{rv}, which can be resolved tangential to the slab as $A_{rv} \sin \theta_v$ and normal to the slab as $A_{rv} \cos \theta_v$. The tangential component does not act on the slab and so can be ignored. The normal component is the equivalent radial component of the longitudinal bar, which acts over a length of the circumference of $S_v/\cos \theta_v$. Hence, the density of the radial reinforcement, that is, the amount of radial reinforcement per unit length of circumference, is

$$\rho_{rv} = \frac{A_{rv} \cos\theta_v}{S_v / \cos\theta_v} = \frac{A_{rv} \cos^2\theta_v}{S_v} \tag{7.41}$$

The density of the radial reinforcement from the vertical bars varies from a maximum at bar (2), which has a value of A_{rv}/S_v, to zero at bar (3), as this does not have a radial component.

The same approach can be used to determine the radial component of the horizontal bar (4), which gives

$$\rho_{rh} = \frac{A_{rh} \cos^2\theta_h}{S_h} \tag{7.42}$$

This also varies from zero to A_{rh}/S_h. Hence, the variation of the radial reinforcement density can be determined around the circumference of the slab by summing both components in Eqs. 7.41 and 7.42.

Having converted the grid reinforcement to radial reinforcement, we now have a slab that is symmetrical about its centre. A cross-section is shown in Figure 7.32. Any of the analyses developed in this chapter can be applied to this slab, such as that shown in Figure 7.19. The only difference, as shown in Figure 7.32, is that the width varies from b being zero where the load is applied to $b = \pi D$ at the supports, in which the failure plane has a width of $b = \pi L_{def}$.

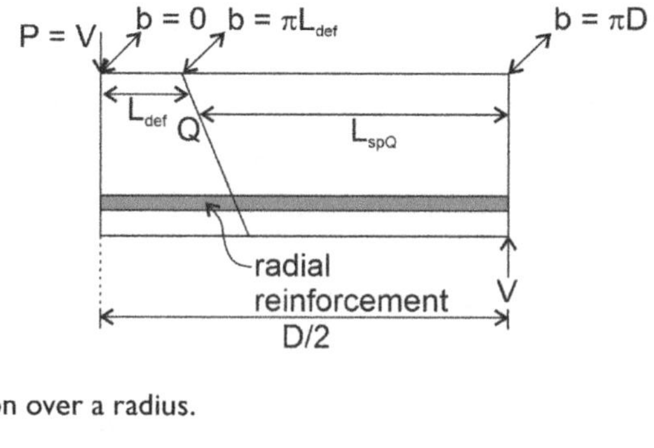

Figure 7.32 Cross-section over a radius.

7.8 NOTATION AND ABBREVIATIONS

A_{rh}	area of transverse reinforcement in grid reinforcement
A_{rv}	area of longitudinal reinforcement in grid reinforcement
A_{sld}	surface area of potential sliding plane
A_{sp}	area of potential sliding plane in transformed section
A_{sc}	area of compression reinforcement
A_{st}	area of tension reinforcement
b	width of member
b_{seg}	width of segment
C	shear capacity
C_{tot}	ΣC_x
$comp$	compression
D	depth of member; effective depth; diameter of circular slab
d	effective depth
d_e	depth of horizontal element
d_c	distance of centroid of A_{sp} to tension reinforcement
d_{NA}	depth to neutral axis; depth of compression zone
E_c	modulus of concrete
E_s	modulus of reinforcement
E-B	Euler–Bernoulli
F	force; force profile
F_c	force in concrete
F_{cc}	force in concrete in compression
$F_{cc\text{-}R}$	resultant force in concrete in compression
F_{fv}	resultant vertical force in fibres
F_h	horizontal force profile
F_{rc}	resultant force in compression reinforcement
F_{rt}	resultant force in tension reinforcement
F_{st}	resultant force in stirrups
F_V	vertical force profile
f_c	compressive strength of concrete
f_{ct}	tensile strength of concrete
h_{cr}	crack widening due to aggregate interlock
h_e	with of vertical element
I	moment of inertia
I_x	normal force to sliding plane
K_{LA}	shear span; $M_{seg}=V_{seg}K_{LA}$
L	length of shear span
LHS	left-hand side
$L_{cc\text{-}h}$	length of compression concrete in segment
L_{def}	length of compression region
$L_{def\text{-}Q}$	L_{def} at Q
L_{seg}	distance from position of applied stress resultants at Point Q to LHS of free body
l	lever arm of resultant force; lever arm of resulting force with same subscript
l_{cc}	lever arm between P_{rt} and $F_{cc\text{-}R}$

l_e	length of element
l_{rc}	length of compression reinforcement in segment; lever arm of compression reinforcement
l_v	length of vertical element
M	moment
M_{exp}	experimental moment at failure
M_{max}	maximum moment for a beam subjected to flexure and transverse shear
M_Q	moment at Q
M_{req}	derived from V_{req} for specific beam and loading
M_{seg}	externally applied moment to segment
M_{theo}	theoretical moment at shear span
My/I	elementary analysis $My/I \pm P/A$
m	modular ration; E_s/E_c
m_{LL}	slope of loading line; τ/σ_n
N_{seg}	externally applied axial force
n	number of slices within d_{NA}
P	internal force acting on sliding plane; resultant force acting on sliding plane; punching load
PI	partial interaction
P/Δ	force in bar at crack face against slip of bar at crack face
$P_{cc\text{-}h}$	horizontal component of force in concrete in compression
$P_{cc\text{-}v}$	vertical component of force in concrete in compression
P_f	fibre force across crack width w; resultant of P_f
P_{fib}	fibre force bridging crack
P_{fh}	horizontal component of fibre force
P_{fv}	vertical component of fibre force
P_{rc}	compressive force in longitudinal compression reinforcement
P_{rt}	tensile force in longitudinal tensile reinforcement
P_s	local interface forces due to passive confinement
P_{sc}	compressive force in stirrups in compression
P_{sl}	local forces due to passive confinement along inclined plane
P_{st}	tensile force in bar; tensile force in stirrups in tension; global stirrup tensile force
Q	point at which the external stress resultants act; in line with N_{seg}; intercept of $F_{cc\text{-}R}$ with sliding plane
R	rotation centre of Euler–Bernoulli displacement
RC	reinforced concrete
r	radius of rotation of $E\text{-}B$ rotation
$reinf$	reinforcement
S	interface slip; shear force along sliding plane
S_{cr}	crack spacing
S_h	spacing of transverse reinforcement
S_{tot}	ΣS_x
S_v	spacing of longitudinal reinforcement
SF	shear friction
S_{pr}	crack spacing; primary crack spacing
$stiff$	stiffening
$tens$	tension

V	resultant transverse shear
V_{cc}	shear resisted by concrete in compression
V_{cr}	transverse shear to cause initial crack; shear resisted by cracked concrete
V_{ct}	shear resisted by concrete in tension
V_{max}	derived from M_{max}
V_Q	transverse shear force through Q
V_{req}	transverse shear that the beam is required to resist
V_{seg}	externally applied transverse shear force
$V_{seg\text{-}\beta}$	shear to cause initial crack in inclined plane
$V_{seg\text{-}90}$	shear to cause initial crack in vertical plane
V_{sld}	transvers shear resisted by concrete in compression; transverse shear at the commencement of failure; transverse shear resisted by sliding plane
V_{stir}	V_{sld} for beam with stirrups
V_u	shear friction material strength at zero slip
w	rotational width of crack; crack width
w_{st}	width of smeared stirrups acting on segment
w_{fib}	rotational width of crack at fibre; crack width at fibre
Vay	elementary analysis VAy^*/Ib
V_{cc}	transverse shear resisted by concrete in compression
V_{cr}	transverse shear resisted across crack; transverse shear force to initiate cracking
V_{ct}	transverse shear resisted by concrete in tension
V_d	transverse shear resisted by reinforcement dowel action
V_{post}	post-sliding shear capacity
V_{seg}	transverse shear force applied to segment
V_u	shear friction shear strength at zero slip
y	distance from neutral axis
$y–y$	horizontal level of member
Z_{cap}	SF capacity of sliding plane
α	angle of crack in compression region
β	angle of crack in tension region; inclination of principal stress
β_{cap}	β that gives the minimum strength
β^*	β for inclined reinforcing bar
Δ	E-B deformation
Δ_{cc}	E-B deformation in concrete in compression
Δ_{fib}	E-B deformation of fibres
Δ_h	horizontal component of Δ^*
Δ_{rc}	E-B deformation of compression reinforcement
Δ_{rt}	E-B deformation of tension reinforcement; half-crack width at tension reinforcement
Δ_{rt*}	component of Δ_{rt} required for P/Δ analysis
Δ_{st}	half-crack width at stirrups; E-B deformation at stirrups
Δ_{st}^*	component of Δ_{st} required for P/Δ analysis
Δ_v	vertical component of Δ^*
Δ^*	varying E-B displacements whose components are Δ_h and Δ_v
δ_h	horizontal movement
$\delta_h=0$	horizontally restrained fully
δ_v	vertical movement

$\delta_v>0$	partially restrained vertically
ε	strain
$\varepsilon_{cc\text{-}h}$	ε_h for compression reinforcement
ε_{rc}	strain in reinforcement in compression
ε_h	horizontal strain profile
θ	rotation; Euler–Bernoulli rotation
σ	stress profile; stress
σ_{app}	applied axial load
σ_n	normal stress to potential sliding plane
σ_{nF}	normal stress at beam failure
σ_{pri}	principal tensile stress
σ_t	tensile force normal to inclined plane
ρ_{rh}	density of radial reinforcement from transverse bars
ρ_{rv}	density of radial reinforcement from longitudinal bars
τ	shear stress profile; shear stress
τ_F	shear stress at beam failure

REFERENCES AND DEVELOPMENT

Al-Mhawish, K.H.A., Sturm, A.B., Oehlers, D.J. and Visintin, P., 2025a. A comprehensive review and analysis of punching shear in reinforced concrete slabs without transverse reinforcement. *Journal of Structural Engineering*, *151*(11), 04025176.

Al-Mhawish, K.H.A, Visintin, V. and Oehlers, D.J., 2025b. Interaction between shear failure and flexural failure of RC beams with only tension reinforcement. Submitted for publication to *ASCE Journal of Structural Engineering.*

Al-Mhawish, K.H.A, Visintin, V. and Oehlers, D.J., 2026. *Mechanics based closed form solution for the shear capacity of RC beams. In preparation.*

Chen, Y., Visintin, P. and Oehlers, D.J., 2015. Concrete shear-friction material properties: Derivation from actively confined compression cylinder tests. *Advances in Structural Engineering*, *18*(8), 1173–1185.

Dadd, L., Bennett, B., Xie, T. and Visintin, P., 2025. Shear behaviour and environmental impact of single and multi-generation recycled aggregate concrete beams. *Engineering Structures*, *322*, 119200.

Iskhakov, I. and Ribakov, Y., 2014. Exact solution of shear problem for inclined cracked bending reinforced concrete elements. *Materials & Design*, *57*, 472–478.

Oehlers, D.J. and Visintin, P., 2021. A displacement based approach for the analysis, development, and design of reinforced concrete. *Structural Concrete*, *22*, 2105–2125.

Sturm, A.B., Visintin, P. and Oehlers, D.J., 2021. Mechanics of shear failure in fiber-reinforced concrete beams. *Journal of Structural Engineering*, *147*(3), 04020344.

Zhang, J.P., 1997. Diagonal cracking and shear strength of reinforced concrete beams. *Magazine of Concrete Research*, *49*(178), 55–65.

Zhang, T., Visintin, P. and Oehlers, D.J., 2016. Shear strength of RC beams with steel stirrups. *Journal of Structural Engineering*, *142*(2), 04015135.

Part 5

Partial-interaction member mechanics

The segmental flexural analysis in Chapter 6 is central to the analysis of members. It provides all the properties, in particular moment–curvature, required for all member analyses. The use of these segmental properties in member analyses varies with the interface bond strength between the components of the member. Various scenarios are used as examples to illustrate the effect of bond strength.

Chapter 8 deals with flexural members with a strong bond, such as the bond between steel reinforcing bars and concrete. It demonstrates how the results of a single segmental analysis can be used to analyse the serviceability behaviour, from moment redistribution to the collapse of RC beams. Chapter 9 takes this scenario one step further to simulate and analyse flexural members with axial loads, which includes both post-tensioned and pre-tensioned members.

Having dealt with members with a strong bond, the difference in the analysis of members with a weak bond is considered next. Chapter 10 addresses profiled slabs and composite steel and concrete beams where, due to the weak bond, the segmental properties now vary along the length of the member. Also considered part of this category are externally FRP or steel-plated RC beams, where the primary objective is often to prevent plate debonding, particularly to ensure that it does not debond catastrophically.

In all of these member analyses, the flexural segmental analysis from Chapter 6 is used directly and, as mentioned previously, it is central to all analyses. Chapter 7 is used to ensure that premature failure does not occur due to transverse shear forces.

DOI: 10.1201/9781003711407-12

Chapter 8

Serviceability to moment redistribution RC member analyses

The material properties of the components of an *RC* member, such as the concrete properties and the interface bond/slip, were described in Chapters 2 and 3. Using these material properties, the mechanics of small elements of the member, such as tension stiffening and concrete softening, were then covered in Chapters 4 and 5. This provided enough information to fully simulate the behaviour of short lengths of the member, that is, segments, in which there was only local debonding within the segment, which was addressed in Chapters 6 and 7. The properties from these segmental analyses can now be used in Chapter 8 to simulate the behaviour of members in which all the properties, such as those in Figure 6.11, required to analyse the whole member can be derived from one segmental analysis. Another way of visualising this is that only local debonding occurs within the segment. The approaches in this chapter cover virtually all *RC* structures.

In Chapter 9, prestressed members, both pre-tensioned and post-tensioned, are discussed. If the prestressing tendon is horizontal, only one segmental analysis, as in this chapter, is required. However, if the tendon is sloped, then several segmental analyses are needed to account for the variation in the tendon position. Chapter 10 deals with profiled slabs and composite steel and concrete beams, in which numerous segmental analyses are necessary to account for the variation in the axial forces due to the accumulation of bond forces along the member. Chapter 11 addresses the retrofitting technique in which *FRP* or steel plates are adhesively bonded to the sides of RC members, where it will be shown that two segmental analyses are required: one for the unbonded region, as in this chapter, and one for the debonded region, which acts as a prestressed member, as in Chapter 9.

This chapter starts with a precis of the M/θ segmental analysis in Chapter 6 to explain the information that is now available for the analysis of a member subjected to flexure. This information is then used in the numerical analysis of *RC* members, particularly moment redistribution (MR), to emphasise the importance of designing members for ductility. This is then followed by closed-form solutions for moment redistribution (MR).

Note that we promised at the start of the book that we would explain all concepts, no matter how trivial they appear. Bear this in mind in the following descriptions of basic mathematics and mechanics.

DOI: 10.1201/9781003711407-13

8.1 SUMMARY OF RESULTS FROM SEGMENTAL ANALYSES

The Euler–Bernoulli deformation of a segment in the hogging (–ve) region of a member is shown in Figure 8.1. The segment is subjected to a constant moment M and is positioned between flexural cracks at a spacing S_{pr}. There are two hinges, and the length of each hinge is L_{def}, which is $S_{pr}/2$, where S_{pr} is the crack spacing. It was shown in Section 1.2 that the rotation θ divided by L_{def} is the curvature χ,which is d^2y/dx^2, and rearranging gives Eq. 8.1. It can be seen that the rotation θ at the hinge face, as shown in Figure 8.1, is χL_{def}, which is the integration of the curvature over L_{def} and represents the slope of the hinge at the hinge face $(dy/dx)_{hng}$, also shown in Figure 8.1. Hence, the rotation θ at the hinge face is the slope of the hinge at that point, which is also the slope of the adjacent beam.

$$\theta = \frac{d^2y}{dx^2} L_{def} = \chi L_{def} = \left(\frac{dy}{dx}\right)_{hng} \tag{8.1}$$

The M/θ results from a segmental analysis are illustrated in Figure 8.2, which can be idealised as an ascending branch with its associated cracked, uncracked, and non-linear variations. This is followed by a plastic plateau, often assumed to be horizontal, at a moment capacity M_{cap} that starts at a rotation θ_{asc} and remains fairly level until there is rapid deterioration at a rotation θ_{cap}, as shown by the descending branch.

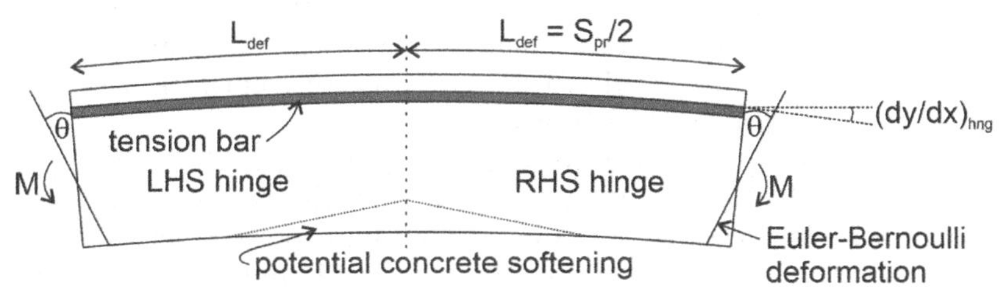

Figure 8.1 Segment in hogging (–ve) region.

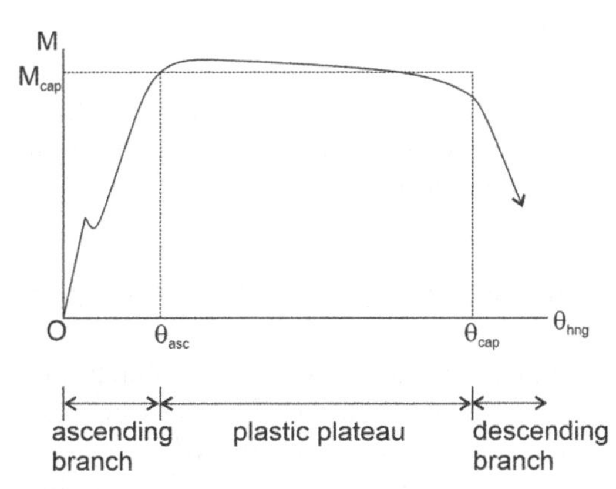

Figure 8.2 Typical M/θ results.

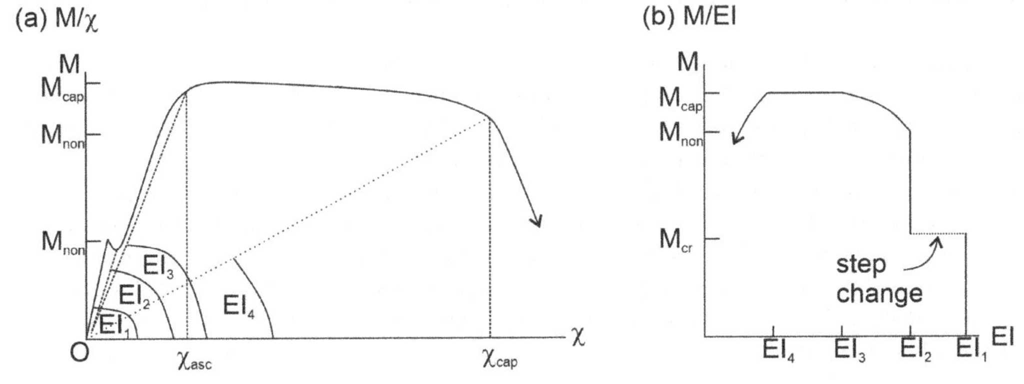

Figure 8.3 Variations in curvature and EI, (a) typical moment curvature M/χ variation; (b)typical moment flexural rigidity variation M/EI extracted from the M/χ variation.

Should $\theta_{cap} - \theta_{asc}$ be large, then the behaviour is considered to be ductile; however, should it be small or non-existent, then the behaviour is considered to be brittle.

It needs to be remembered that it has been shown in previous chapters, in particular Chapter 6, that this segmental M/θ analysis can directly allow for such variables as concrete shrinkage, expansion and creep; concrete tensile cracking and compressive softening; reinforcement relaxation; the environment, which includes bond deterioration; temperature gradients; fibres in concrete in tension and compression regions; and flexure, shear and axial stress resultants. The emphasis here is on directly allowing for these variables, that is, not indirectly through the use of empirical or semi-empirical correction factors as required in M/χ approaches.

Dividing the rotation θ in the M/θ variation in Figure 8.2 by L_{def} gives the moment/curvature (M/χ) variation in Figure 8.3a. From this, the tangent stiffnesses can be derived, such as for the uncracked section EI_1 and that in the cracked region EI_2, as well as the secant stiffnesses at different points such as EI_3 and EI_4. Hence, the abscissa in Figure 8.3a comprises all of the following variables:

$$\chi = \frac{d^2y}{dx^2} = \frac{\theta}{L_{def}} = \frac{M}{EI} \tag{8.2}$$

The flexural rigidities in Figure 8.3a are shown in Figure 8.3b, which can be used in a stiffness analysis. Figure 8.3 provides all the information we require to analyse a member where there is only local debonding.

8.2 NON-LINEAR NUMERICAL ANALYSIS OF BEAMS

When conducting a non-linear analysis, there is absolutely no need to revert to the use of flexural rigidities as in Figure 8.3b, as the variations in curvature can be used directly. Flexural rigidities are, however, useful when simplifications are required to find closed-form solutions. Let us assume that the non-linear M/χ properties in Figure 8.3a were derived from the segmental analysis of a specific beam and are now available to analyse the behaviour of the whole beam under various loading conditions and restraints. No attempt has been made to cover all scenarios; however, the following examples have been chosen to illustrate the numerical non-linear approach, particularly with regard to the formation of hinges leading to the failure of the beam. In the following sections, the classical process of moment redistribution (MR) will be described, but the reader is also referred to the work of Oehlers et al. (2004) and Oehlers (2006) for additional treatment.

8.2.1 Symmetrical simply supported beam

Let us start with the simplest example of a symmetrically loaded, symmetrically restrained, statically determinate, simply supported beam, as in Figure 8.4a. The only possible positions for hinge formation are on either side of the mid-span, as shown. It is common practice to assume that there is only one hinge at mid-span, but in reality, there are two hinges, such that the beam on the *RHS* is a mirror image of the beam on the left-hand side (*LHS*).

The beam is subject to symmetrical loading such that the loading on the RHS is a mirror image of that on the LHS. This symmetrical loading produces the static distribution of moment in Figure 8.4b, where the maximum static moment is M_{st} at mid-span; for example, should the applied load be w kN/m, then the static moment is $wL^2/8$. The deformation of the beam under this static loading is shown in Figure 8.4c, where at the supports, the deflection y is zero and the slope dy/dx is not zero; and at mid-span, the deflection is y_{max} and the slope dy/dx is zero. These are the boundary conditions that must be adhered to.

Because of the symmetry in Figure 8.4, we only need to consider half of the beam, as shown in Figure 8.5a. For this statically determinate beam example, we will apply

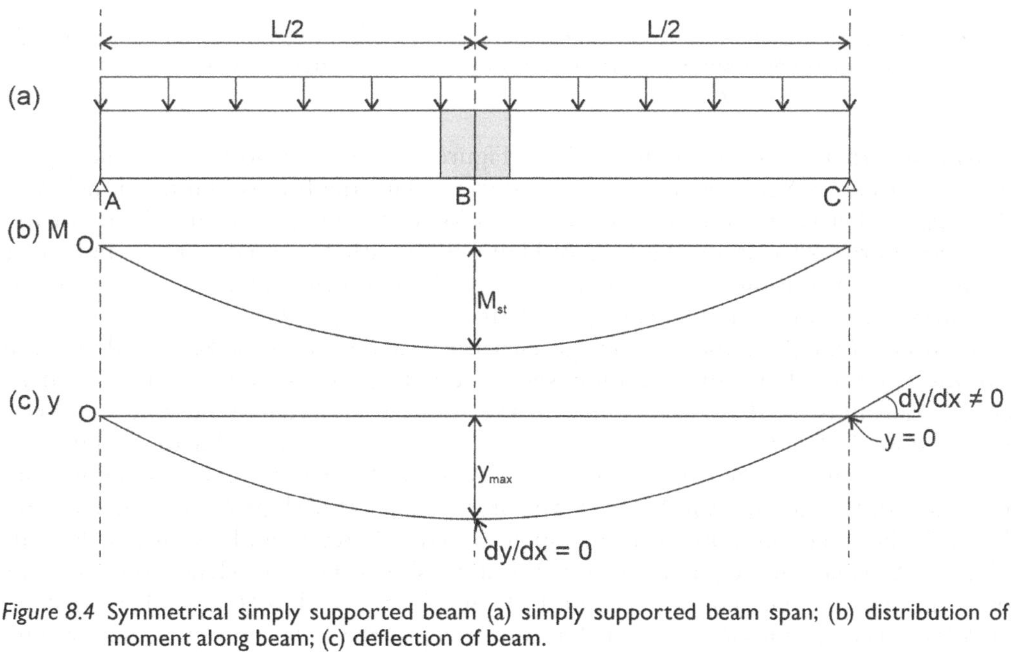

Figure 8.4 Symmetrical simply supported beam (a) simply supported beam span; (b) distribution of moment along beam; (c) deflection of beam.

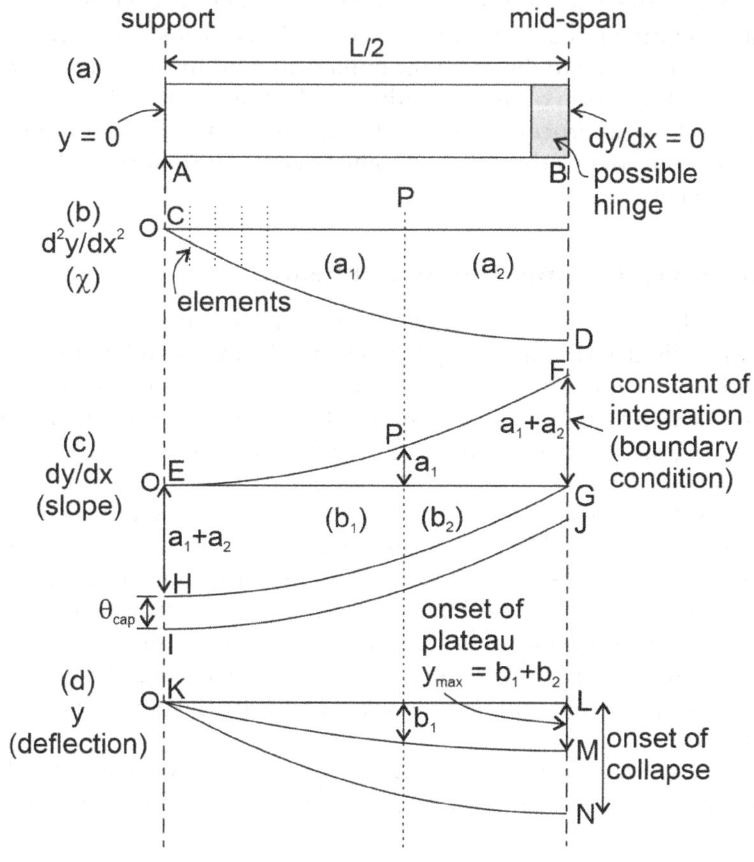

Figure 8.5 Numerical analysis of simply supported beam: (a) beam half-span to be considered for analysis; (b) distribution of curvature d^2y/dx^2 ; (c) distribution of slope dy/dx; (d) distribution of deflection y.

a load such that the static moment M_{st} in Figure 8.4b is equal to the moment capacity M_{cap} in Figure 8.2. We are, therefore, dealing with the full ascending branch and deriving the behaviour just prior to the formation of the hinge in Figure 8.5a. For the variation in the moment on the *LHS* of Figure 8.4b with M_{st} equal to M_{cap}, and using the non-linear variation in curvature in the ascending branch in Figure 8.2, we can plot the variation in curvature *C–D* along the half beam in Figure 8.5b.

We need to integrate the variation in curvature shown in Figure 8.5b to derive the variation in slope. This can be accomplished by dividing the distribution into very small elements, as illustrated, and accumulating the areas of these elements to find the variation of slope *E–F* in Figure 8.5c. For example, the slope at Position *P* with a magnitude of a_1 is represented by the area to the left of Position P in Figure 8.5b, and so on. The total area of the curvature in Figure 8.5b corresponds to the slope *F–G* in Figure 8.5c. From the boundary conditions at mid-span in Figure 8.4c, it can be seen that *F–G* in Figure 8.5c is not the required value as the slope should be zero. Hence, the constant of integration is *F–G*, and the variation *E–F* must be adjusted to *H–G* so that the slope at Point *G* is zero, with the greatest slope occurring at the *LHS* support, as required by the boundary conditions. Numerical integration of the curvature *H–G* in the same way provides the variation of the vertical deflection *K–M* in Figure 8.5d, where the maximum deflection *L–M* is y_{max}. As M_{st} is equal to M_{cap}, we are dealing with the behaviour just prior to a hinge forming.

Should the beam be displaced further down than y_{max} in Figure 8.5d, that is, if it is now under displacement control, the static moment distribution in Figure 8.4b with M_{st} equal to M_{cap} remains unchanged; however, to accommodate this displacement, the hinge has to rotate. The maximum rotation of the hinge without a loss of strength is θ_{cap} in Figure 8.2. Applying this rotation will cause the slope distribution *H–G* in Figure 8.5c to displace downwards as shown to *I–J*; the shape remains unchanged as the static moment distribution remains the same. Integration of this variation will cause the beam to deflect more to *K–N* in Figure 8.5d. Hence, the displacement *L–N* is the deflection at the onset of collapse; collapse is defined as when the strength reduces, that is, the static moment has to be reduced to maintain equilibrium. It can be seen that the maximum deflection has increased from *L–M* to *L–N* due to stable hinge rotation, meaning a ductile hinge will allow more energy to be absorbed. This type of analysis can be used to study the collapse of a member and can even be used to follow the descending branch in Figure 8.2.

8.2.2 Symmetrical encastre or built-in beam

8.2.2.1 Deformation prior to the onset of hinge formation

Now moving onto a statically indeterminate beam such that the ends of the beam in Figure 8.4a are now rotationally restrained, and hinges can form adjacent to these supports. Once again, the beam is symmetrically loaded, so only half needs to be considered. The *LHS* of this encastre, that is built-in beam, is shown in Figure 8.6a. There are two possible hinges in this half beam, meaning there are four in the whole beam. The deformation of this half beam is shown in Figure 8.6b, where at the support the deflection and slope are zero, and at mid-span the slope is also zero but not the deflection, which is at its maximum $y_{max.}$

The variation in moment along the beam is shown in Figure 8.6c, where M_{hog} is the restraining moment at the support, M_{sag} the sagging (+ve) moment at mid-span, and M_{st} the maximum of the static applied moment. If the behaviour of the beam were linear-elastic, then M_{st} would equal $1.5M_{hog}$, but this is not the case as we are dealing with the non-linear curvature properties in the ascending branch in Figure 8.3a. Hence an iterative approach must be used to find the relationship between M_{st} and M_{hog}. To perform the iterative analysis, let us apply a static moment M_{st} that is expected to be less than that required to cause a hinge. Then, guess M_{hog}. Hence, the distribution of moment in Figure 8.6c is now fixed.

We will assume in this analysis that the hinge first forms at the support in Figure 8.6a; however, a similar procedure can be used when a hinge first forms at mid-span. Furthermore, we will assume that the M/χ relationship in Figure 8.3a applies to both the hogging (−ve) and sagging (+ve) regions, although it would be very easy to incorporate an M/χ relationship for each region. For the distribution of moment in Figure 8.6c and using the curvature properties in the ascending branch of Figure 8.3a, the variation in curvature in Figure 8.6d can be derived. Starting from the *LHS* of Figure 8.6d,

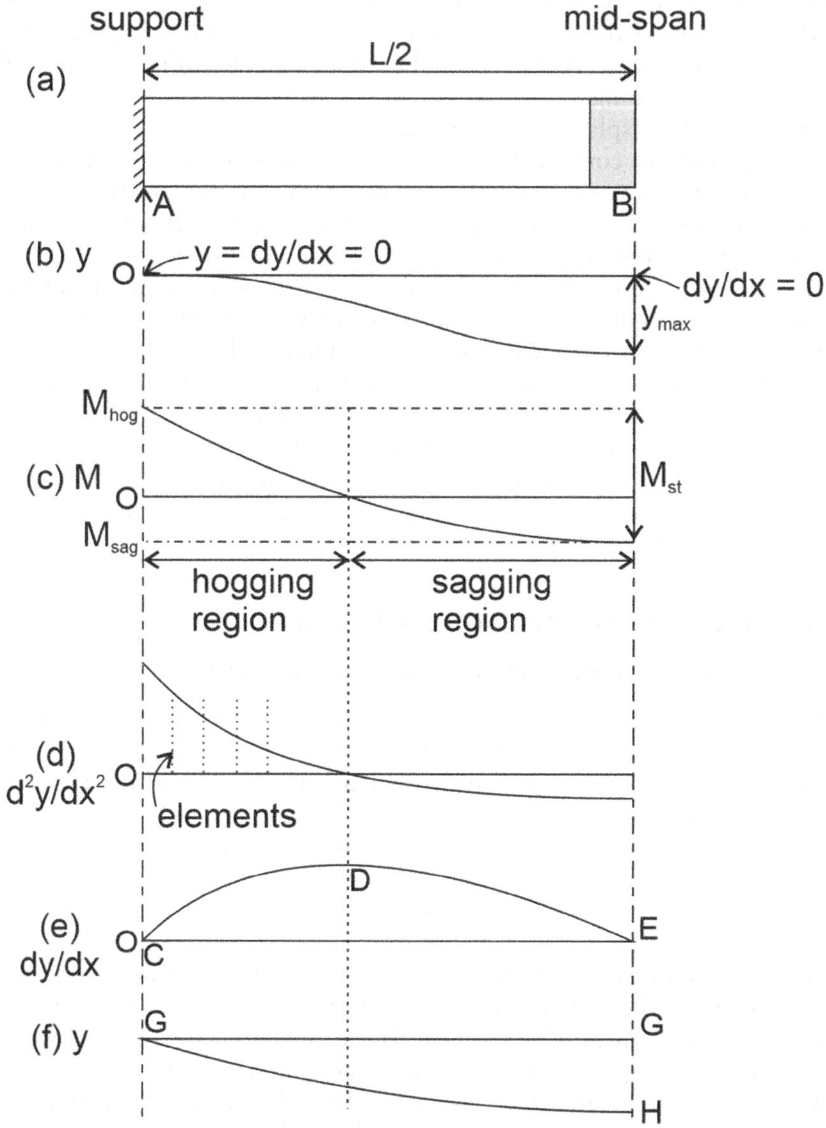

Figure 8.6 Deformations up to the onset of a hinge formation: (a) beam half-span of a statically indeterminate beam; (b) distribution of deflection y; (c) distribution of moment M; (d) distribution of curvature d^2y/dx^2; (e) distribution of slope dy/dx at onset of hinge formation; (f) distribution of deflection y at onset of hinge formation.

where the slope is known to be zero, numerical integration of this curvature will give the variation in slope *C–D–E* in Figure 8.6e. Should the slope at mid-span at Point *E* not be zero, then vary the guessed value of M_{hog} until it is. Once M_{hog} is fixed, the variation in slope *C–D–E* in Figure 8.6e is known, and the integration of this variation will give the variation of the deflection *F–H* in Figure 8.6f. This is the deflected shape for specific M_{hog} and M_{st} in Figure 8.6c.

The above analysis can be repeated with gradually increasing static moments M_{st} in Figure 8.6c until the hogging (–ve) moment M_{hog} is equal to the moment capacity in Figure 8.3a of M_{cap}. As M_{st} is gradually increased, it needs to be checked that the capacity of the sagging (+ve) region is not exceeded. If it were, then the first hinge would occur in the sagging region first, but in this example, we will assume that the hinge forms at the supports. If *C–D–E* in Figure 8.6e now represents the variation in the slope when M_{hog} is M_{cap}, then the integration of this slope will give the deflected shape *F–H* in Figure 8.6f at which the hinge is about to form.

Before we proceed any further, it is worth noting that although we consider MR to occur only at the ultimate limit state, it happens any time there is a variation in *EI* along the span of the beam. This could be in an uncracked beam because of the differences in reinforcement or in a beam with one section being uncracked and one being cracked. The critical point is that although we know that a static distribution of moment, M_{st} in Figure 8.4, exists in an *RC* structure at all but collapse, the distribution of moment is rarely elastic.

8.2.2.2 Deformation after onset of hinge formation

From the results of the above analysis using Figure 8.6, let $M_{st\text{-}1}$ in Figure 8.7a be the static moment at which the support hinge first reaches its capacity, so that the moment at the support is the hogging capacity $M_{cap\text{-}hog}$. This distribution has been labelled M_1

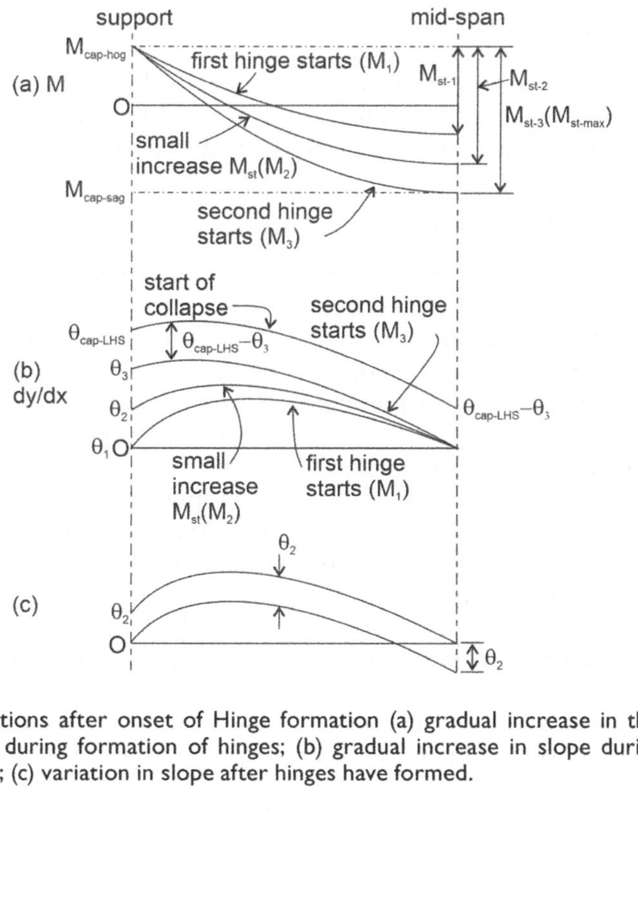

Figure 8.7 Deformations after onset of Hinge formation (a) gradual increase in the distribution of moment during formation of hinges; (b) gradual increase in slope during the formation of hinges; (c) variation in slope after hinges have formed.

and represents the distribution of moment when the hinge capacity is reached and the hinge is about to rotate. Also from the above analysis in Figure 8.6, the variation in slope is determined for M_1 and shown in Figure 8.7b, where the rotation at the support is θ_1 and is zero as the hinge is about to rotate.

Let us now slightly increase the static load to raise the static moment to $M_{st\text{-}2}$ in Figure 8.7a, ensuring that no further hinges form, and the distribution is labelled M_2. For this moment distribution M_2 and consequently using Figure 8.3a for this curvature distribution, integration of the curvature from the left-hand support would yield the lower curve in Figure 8.7c, where the slope at mid-span is θ_2, which is simply the constant of integration. Hence, the curve must be moved upwards as shown to satisfy the boundary condition. Consequently, the support hinge has rotated θ_2, and this variation is shown in Figure 8.7b with the slope of θ_2 on the *LHS*. We will assume that $\theta_2 < \theta_{asc}$ in Figure 8.2, meaning a hinge is not about to form.

Let us further increase the static moment to $M_{st\text{-}3}$, which corresponds to distribution M_3 in Figure 8.7a, where the hinge just starts to form at mid-span at the capacity $M_{cap\text{-}sag}$. Hence, $M_{st\text{-}3}$ is equal to the sum of the hogging (−ve) and sagging (+ve) capacities. Integration of the curvatures for this distribution of moment will give the variation in slope in Figure 8.7b, where the rotation at the support is θ_3.

With further downward displacement of the beam, both hinges rotate, but the moment distribution M_3 in Figure 8.7a remains unchanged, so the shape of the slope distribution labelled θ_3 in Figure 8.7b also remains unchanged. Thus, when the downward displacement is applied, both hinges will rotate the same amount, as shown in the upper curve of Figure 8.7b. If the support hinge first reaches its rotation capacity $\theta_{cap\text{-}LHS}$, then the rotations can increase by $\theta_{cap\text{-}LHS} - \theta_3$ without any loss of strength. Hence, integrating the rotation labelled $\theta_{cap\text{-}LHS}$ will yield the deflection just prior to failure when the strength begins to reduce. Of course, further downward displacements can be applied, which would mean following the descending branch in Figure 8.2a, but the static moment would have to be reduced to find a solution.

8.2.3 Non-symmetrical encastre or built-in beam

Now let us consider a non-symmetrical encastre beam as shown in Figure 8.8a, where the applied loads, distributions of beam stiffnesses, and moment capacities may vary along the beam. Hence, the only known boundary conditions, once hinges start to form, are that the deflections at the supports are zero.

The beam is designed such that the moment capacity of the *LHS* hinge is $M_{cap\text{-}LHS}$ in Figure 8.8b, that of the *RHS* is $M_{cap\text{-}RHS}$, and that of the potential sagging hinge is $M_{cap\text{-}sag}$. Furthermore, it is designed such that the hinges at the supports will form first and then redistribute moment to allow the sagging hinge to form. The static moment distribution is fixed, as it depends on the distribution of the applied loads, which is constant. Therefore, the distribution of moments in Figure 8.8b is fixed. It is a question of determining whether the support hinges have sufficient rotational capacities to allow the sagging hinge to form and, therefore, utilise the full capacity of the beam. It is also a matter of determining the maximum deformation of the beam before collapse due to loss of strength.

From the ascending M/χ properties, such as those in Figure 8.3a, for the different parts of the beam, the variation in curvature for the distribution of moment in Figure 8.8b

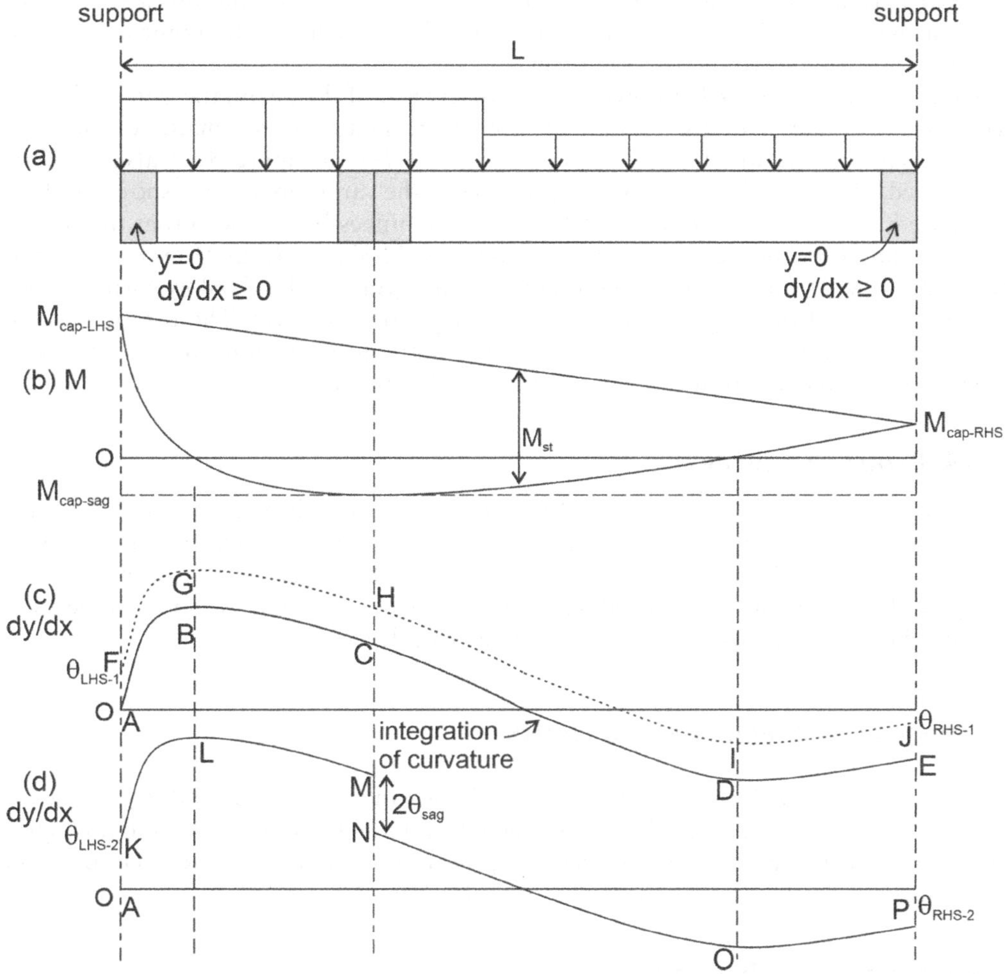

Figure 8.8 Non-symmetrical encastre beam. (a) non- symmetric encastre beam; (b) distribution of moment; (c) variation in slope during the formation of support hinges; (d) variation in slope during the formation of hinge between supports.

can be derived. Numerical integration of this variation in curvature, starting from the *LHS*, gives the variation in slope, represented as *A–B–C–D–E* in Figure 8.8c. The shape of this distribution is fixed, as it is derived from the fixed distribution of moment in Figure 8.8b. However, there are no boundary conditions for the slope to fix it in a vertical direction. We do know that the integration of the slope variation, such as *A–B–C–D–E*, must result in deflections at the supports that are zero. Hence, it is a question of moving *A–B–C–D–E* vertically until the integration yields zero deflection at the supports.

Let us assume that the integration of the variation *F–G–H–I–J* in Figure 8.8c, which has the same shape as *A–B–C–D–E*, results in zero deflection at the supports. Hence, the *LHS* hinge must rotate by $\theta_{LHS\text{-}1}$ and the *RHS* hinge must rotate by $\theta_{RHS\text{-}1}$ to achieve the distribution of moment in Figure 8.8b. Assuming that these rotations are less than their capacities θ_{cap}, then the beam is said to be able to redistribute moment from the hogging (−ve) regions to the sagging (+ve).

Should there be any further downward displacement of the beam, the sagging hinges will start to rotate. The distribution of moment in Figure 8.8b remains unchanged, so the shape of the distribution of slope *F–G–H–I–J* in Figure 8.8c also remains unchanged. However, there is now a step change in the variation of slope shown as $2\theta_{sag}$ in Figure 8.8(d due to the rotation of both sagging hinges by θ_{sag}, such that the shapes *F–G–H* and *H–I–J* in Figure 8.8c are the same as the shapes *K–L–M* and *N–O–P* in Figure 8.8d. As previously, the variation in slope in Figure 8.8d is fixed by ensuring that the integration of the slope gives zero deflection at the supports. The rotation of the sagging hinges, $2\theta_{sag}$, can be gradually increased until one of the four hinges reaches its capacity θ_{cap}, after which the strength of the member reduces.

8.2.4 Propped cantilever

Propped cantilevers are often used in *RC* research to study shear failure and the rotational capacities of beams. A propped cantilever is a non-symmetrical member, as in the non-symmetrical member in Figure 8.8a, but in this case, the *RHS* hinge has no capacity. The same analysis procedure as in the previous section on non-symmetrical encastre beams can be used to analyse propped cantilevers.

8.3 MR CLOSED-FORM SOLUTIONS

The mechanics of *MR* and collapse were explained in Section 8.2, and it was shown how numerical solutions could be found when dealing with non-linearity and non-symmetry. As is normal practice in research, simplifications are made to derive closed-form solutions that clearly show the parameters controlling MR.

8.3.1 Moment redistribution

Consider the symmetrical encastre beam in Figure 8.9a subjected to a uniformly distributed load of w kN/m such that the maximum static moment M_{st} is $wL^2/8$. From elementary linear–elastic mechanics, the hogging (−ve) elastic moment at the supports M_1 in Figure 8.9b is twice the sagging (+ve) elastic moment M_3 at mid-span, such that M_{st} is equal to elastic moments M_1+M_3. To help in the distribution of longitudinal reinforcement, the structural engineer needs to reduce M_1 to M_2 and increase M_3 to M_4, such that M_2+M_4 is still equal to M_{st} as shown. However, these are no longer elastic moments derived from linear–elastic theory but from non-linear theory in which hinges are required. This procedure is referred to as *MR* from the hogging (−ve) region to the sagging (+ve) region. MR can also occur from the sagging (+ve) to hogging (−ve) regions, in which case M_4 and M_2 are the elastic moments, and M_3 and M_1 are the non-elastic moments that require hinges.

Structural engineers use the following moment redistribution factor K_{MR} to define the amount of MR that is required:

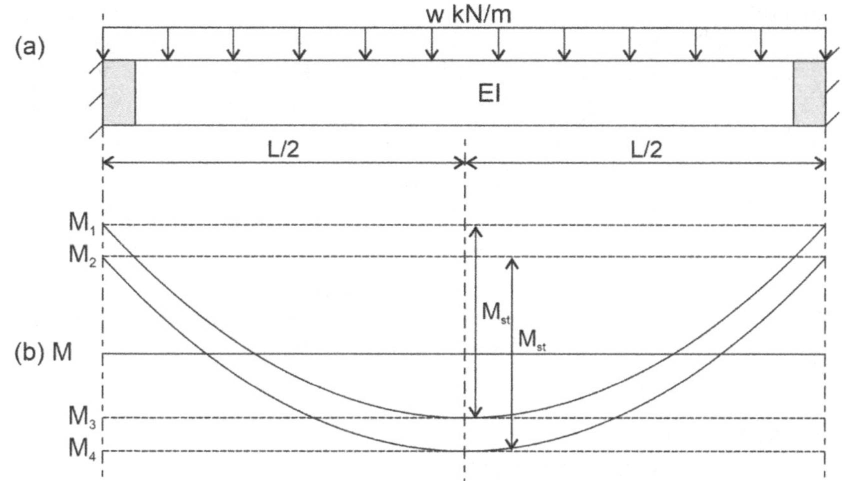

Figure 8.9 Moment redistribution. (a) symmetric encastre beam; (b) redistribution of moment.

$$K_{MR} = \frac{M_{el} - M_{MR}}{M_{el}} \tag{8.3}$$

where M_{el} is the linear–elastic moment for a specific static moment distribution M_{st} and M_{MR} is the non-linear moment required from MR. The factor K_{MR} for hogging to sagging in Figure 8.9b is $(M_1 - M_2)/M_1$ and that for sagging to hogging is $(M_4 - M_3)/M_4$.

8.3.2 MR in symmetrical encastre, built-in Beam

To find a solution through simplification, following the approach of Oehlers et al. (2004) and Haskett et al. (2010), let us assume that the flexural stiffness of the encastre or built-in beam, EI, is constant and that the member is subjected to a uniformly distributed load of w kN/m. Hence, the encastre beam is symmetrical. More complex scenarios are also considered in Sturm et al. (2020).

8.3.2.1 MR from hogging (−ve) to sagging (+ve) regions

Consider first *MR* from the hogging (-–ve) regions to the sagging regions (+ve). The symmetrical encastre beam is shown in Figure 8.10a, where the restraining moments are M_{hog}, and the maximum static moment is M_{st}, which is equal to $wL^2/8$, as shown in Figure 8.10b, and which is also equal to $M_{cap\text{-}hog}$ plus $M_{sag\text{-}hog}$. As we are redistributing from the hogging (−ve) to the sagging (+ve) region, $M_{cap\text{-}hog}$ is less than the elastic value *of* $2M_{st}/3$.

The behaviour of the beam in Figure 8.10 can be visualised as having two components: that due to the static load M_{st}, and that due to the restraining moments at the supports, M_{hog}. The deformation of the beam due to the static loads acting alone is labelled M_{st} in Figure 8.10c, where the deflection is y_{st}, and the rotation or slope at the supports is θ_{st}. From elementary mechanics for linear elastic components, by integrating

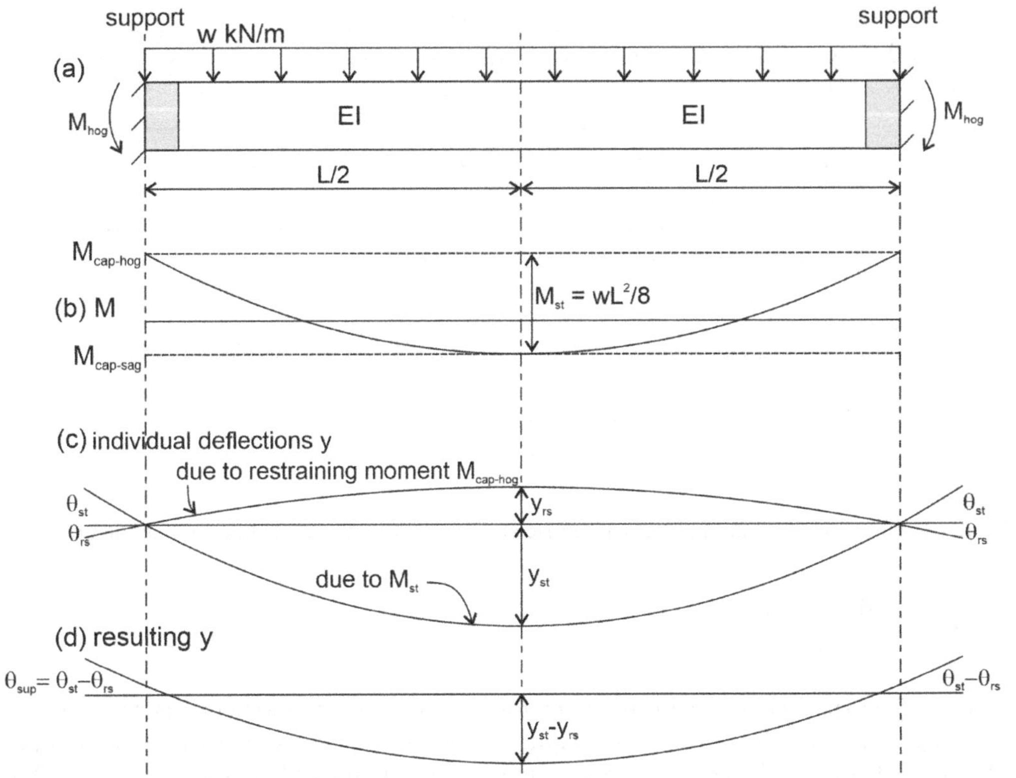

Figure 8.10 MR from hogging (−ve) to sagging (+ve) regions. (a) encastre beam where hinges first form at the supports; (b) distribution of moment; (c) individual deflections due to the static moment and restraining moments; (d) resulting deflection.

the curvature along the beam, the slope at the supports is given by the second component in the following equation:

$$\theta_{st} = \frac{wL^3}{24EI} = \frac{M_{st}L}{3EI} \tag{8.4}$$

The third component is derived by eliminating w from the second component, as M_{st} is equal to $wL^2/8$.

The deformation due to the restraining moments $M_{cap\text{-}hog}$ in Figure 8.10b is labelled as $M_{cap\text{-}hog}$ in Figure 8.10c and produces an upward deflection y_{rs} and end rotations in the opposite direction, θ_{rs}. Once again, from elementary mechanics:

$$\theta_{rs} = \frac{M_{cap\text{-}hog}L}{2EI} \tag{8.5}$$

Hence, the resulting rotation at the supports, θ_{sup}, as shown in Figure 8.10d, is

$$\theta_{sup} = \theta_{st} - \theta_{rs} = \frac{M_{st}L}{3EI} - \frac{M_{cap\text{-}hog}}{2EI} \tag{8.6}$$

Substituting the appropriate values into Eq. 8.3, the MR factor for redistribution from hogging (–ve) to sagging (+ve) is

$$K_{MR\ hog\text{-}sag} = \frac{\frac{2}{3}M_{st} - M_{cap\text{-}hog}}{\frac{2}{3}M_{st}} \tag{8.7}$$

From Eqs. 8.6 and 8.7,

$$K_{MR\ hog\text{-}sag} = \frac{2EI\theta_{sup}}{2EI\theta_{sup} + M_{cap\text{-}hog}L} \tag{8.8}$$

where θ_{sup} can be obtained from Eq. 8.6 and must not exceed the hinge rotation capacity θ_{cap} in Figure 8.2 for redistribution to occur.

It can be seen from Eq. 8.8 the complexity of moment redistribution. To determine the importance of the parameters, consider the following inverse of Eq. 8.8:

$$\left(K_{MR\ hog\text{-}sag}\right)^{-1} = 1 + \frac{M_{cap\text{-}hog}L}{2EI\theta_{sup}} \tag{8.9}$$

where MR increases as the *RHS* increases. Hence, *MR* increases as the capacity of the hinge, M_{cap}, increases, and very importantly, as the span L increases. Furthermore, *MR* increases as the flexural rigidity, *EI*, reduces and the rotation at the support, θ_{sup}, reduces.

8.3.2.2 MR from sagging (+ve) to hogging (–ve) regions

Now consider MR from the sagging (+ve) region to the hogging (–ve) region. The *LHS* of the beam in Figure 8.10a is shown in Figure 8.11a. In this case, the hinge is at mid-span and has a capacity of $M_{cap\text{-}sag}$.

In Figure 8.11, we are dealing with the deflection of a cantilever where the static load causes a downward deflection due to M_{st} in Figure 8.11c. The end rotation θ_{st} and the moment at the hinge $M_{cap\text{-}sag}$ cause an upward deflection labelled $M_{cap\text{-}sag}$ and a rotation θ_{sg}, such that the rotation at the hinge θ_{mid} is equal to θ_{st} minus θ_{sg}. Using the same procedure as that used in Figure 8.10:

$$\theta_{mid} = \theta_{st} - \theta_{sg} = \frac{wL^3}{48EI} - \frac{M_{cap\text{-}sag}L}{2EI} = \frac{M_{st}L}{6EI} - \frac{M_{cap\text{-}sag}L}{2EI} \tag{8.10}$$

In this case,

$$K_{MR\ sag\text{-}hog} = \frac{\frac{1}{3}M_{st} - M_{cap\text{-}sag}}{\frac{1}{3}M_{st}} \tag{8.11}$$

and from the above,

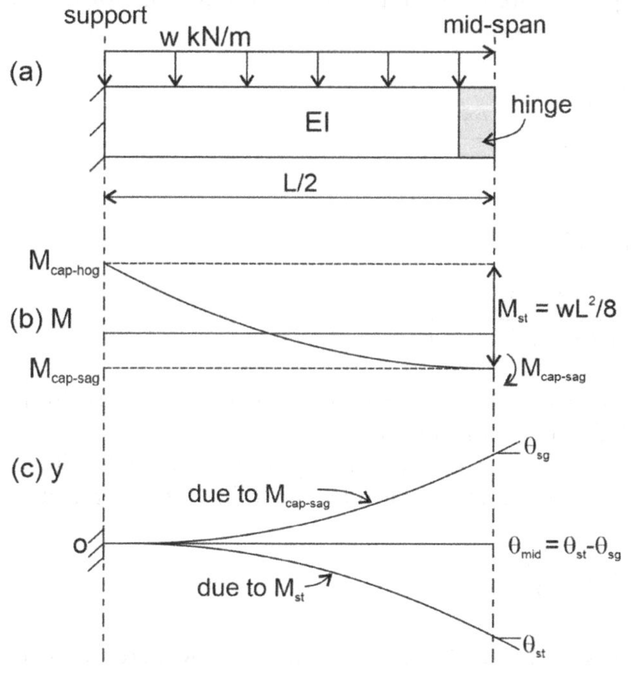

Figure 8.11 MR from sagging (+ve) to hogging (−ve) regions. (a) encastre beam where hinge first forms at mid-span; (b) distribution of moment; (c) distribution of deflection components.

$$K_{MR\ sag\text{-}hog} = \frac{2EI\theta_{mid}}{2EI\theta_{mid} + M_{cap\text{-}sag}L} \tag{8.12}$$

8.4 NOTATION AND ABBREVIATIONS

dy/dx	slope
$(dy/dx)_{hng}$	slope of hinge at segment face; θ_{hng}
d^2y/dx^2	curvature; χ
E	modulus
EI	flexural rigidity
I	moment of inertia
K_{MR}	moment redistribution factor
$K_{MR\ hog\text{-}sag}$	K_{MR} for MR from hogging to sagging region
$K_{MR\ sag\text{-}hog}$	K_{MR} for MR from sagging to hogging region
L	length of beam
LHS	left-hand side
L_{def}	deformation length of half segment
M	moment
M_{cap}	moment capacity
$M_{cap\text{-}hog}$	moment capacity of hogging (−ve) hinge
$M_{cap\text{-}LHS}$	moment capacity of LHS hogging (−ve) hinge

$M_{cap\text{-}RHS}$ moment capacity of RHS hogging (−ve) hinge
$M_{cap\text{-}sag}$ moment capacity of sagging (+ve) hinge; sagging (+ve) moment distribution
M_{cr} moment to cause flexural cracking
M_{el} linear elastic moment for a specific applied static moment
M_{hog} maximum moment in hogging (−ve) region; hogging (−ve) moment distribution
M_{MR} non-linear moment required for MR; change in elastic moment due to MR
M_{non} moment at the commencement of non-linearity
M_{sag} maximum moment in sagging (+ve) region
M_{seg} external moment applied to segment
M_{st} maximum static moment; static moment distribution
$M_{st\text{-}max}$ maximum static load that can be applied
MR moment redistribution
M/EI moment flexural rigidity relationship
M/χ moment curvature relationship
RHS right-hand side
S_{pr} crack spacing
w uniformly distributed load
y deflection
y_{max} maximum deflection of beam
y_{rs} deflection due to the restraining moments
χ curvature
χ_{asc} χ at end of ascending branch of M/χ relationship
χ_{cap} χ at start of descending branch of M/χ relationship
θ rotation; Euler-Bernoulli rotation
θ_{asc} hinge rotation end of ascending branch
θ_{cap} hinge capacity; hinge rotation at commencement of descending weakening branch
$\theta_{cap\text{-}LHS}$ rotation capacity of LHS hinge;
θ_{hng} rotation of hinge
θ_{LHS} rotation of LHS hogging (−ve) hinge
θ_{mid} rotation of hinge at mid-span
θ_{RHS} rotation of RHS hogging (−ve) hinge
θ_{rs} support slope due to restraining moments
θ_{sag} rotation of sagging hinge
θ_{sg} slope due to sagging (+ve) moment distribution
θ_{st} slope due to static moments; slope at support due to static loads
θ_{sup} rotation at support

REFERENCES AND DEVELOPMENT

Haskett, M., Oehlers, D.J., Mohamed Ali, M.S. and Wu, C., 2010. Analysis of moment redistribution in fibre-reinforced polymer plated RC beams. *Journal of Composites for Construction*, *14*(4), 424–443.

Oehlers, D.J., 2006. Ductility of FRP plated flexural members. *Cement and Concrete Composites*, *28*(10), 898–905.

Oehlers, D.J., Ju, G., Liu, I.S.T. and Seracino, R., 2004. Moment redistribution in continuous plated RC flexural members. Part 1: neutral axis depth approach and tests. *Engineering Structures*, 26, 2197–2207.

Sturm, A.B, Visintin, P. and Oehlers, D.J., 2020. Closed-form expressions for predicting moment redistribution in reinforced concrete beams with application to conventional concrete and ultrahigh performance fibre reinforced concrete. *Structural Concrete*, 21, 1577–1596.

Chapter 9

Prestressed members

The previous Chapter 8 dealt with the behaviour of members in which there is only local debonding within a segment. Prestressed members fall into this category, as the prestressing force is an external stress resultant N_{seg} applied to the segment in a segmental analysis. Hence, only one segmental analysis is required to provide all the M/χ properties needed to analyse the whole member. Unless the tendon force is inclined, in which case each segment would have an externally applied load N_{seg} acting at different levels, several segments would need to be analysed to find a reasonable solution.

We will start with the simplest behaviour, that of a post-tensioned unbonded member with a horizontal tendon. The effect of bonding the tendon will then be explained, followed by pre-tensioned members with horizontal tendons. It will then be shown how these analyses can be applied to inclined tendons. Having defined the unique forces in a free-body segment due to post- and pre-tensioning, these forces will then be used in a tension stiffening analysis to account for flexural cracking. Finally, the mechanics of the dispersal of the bearing force in post-tensioned members will be covered.

9.1 SEGMENTAL ANALYSIS OF POST-TENSIONED MEMBER

9.1.1 Application of post-tension axial force

The segment of the beam shown in Figure 9.1a has an area of top reinforcement A_{rt}, a bottom reinforcement area A_{bt}, and the cross-sectional area of the tendon is A_{tnd}, which is at Level *Q*; should there be several tendons, then Level *Q* is at the level of the resultant force of all the tendons. The tendon can be made of any material, such as steel or *FRP*. The prestressed beam is to be post-tensioned with a force $N_{seg\text{-}pos}$ that produces a strain in the tendon ε_{pos} such that

$$N_{seg-post} = f\left(\varepsilon_{pos}\right) \tag{9.1}$$

For example, if the tendon is elastic with a modulus E_{tnd}, then $N_{seg\text{-}pos}$ would be $A_{tnd}E_{tnd}\varepsilon_{\text{pos}}$.

The free-body half-segment first developed in Knight et al. (2013a) is shown in Figure 9.1a, which has a length L_{def} of half the anticipated crack spacing S_{pr}. The *RHS* of the half-segment *C*–*C* is the mid-span of the segment and, hence, by symmetry, it is fixed longitudinally and is the datum line. The *LHS A*–*A* is the unrestrained side of the

DOI: 10.1201/9781003711407-14

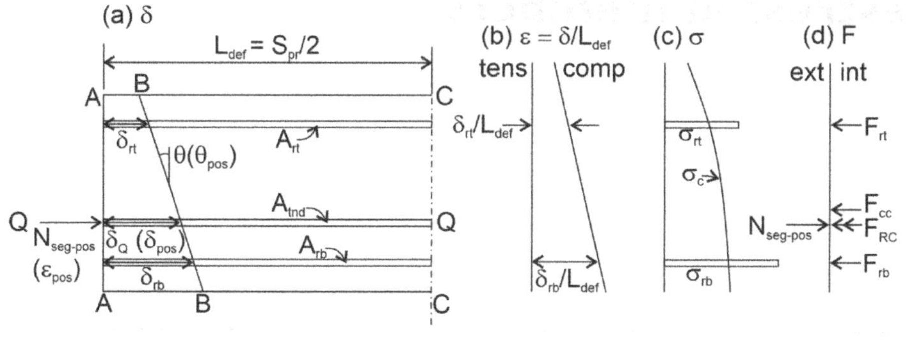

Figure 9.1 Application of post-tensioning (a) segment and deformation profile; (b) strain profile; (c) stress profile; (d) force profile.

segment on which the Euler–Bernoulli displacement is applied. As we are not considering any residual stresses or deformations, such as those due to shrinkage or thermal gradients, *A–A* is the base-line that represents the position of the segment side where the reinforcement and concrete have zero stress; any deformations from this base-line divided by L_{def} produce strains and consequently stresses. Note that the analysis of this free body follows the general procedures outlined in Chapter 6 for an un-prestressed beam, where the iterative procedure is fully described and allowance can be made for any residual stresses or strains.

In Figure 9.1a, the post-tensioning force $N_{seg\text{-}pos}$ is applied on the *LHS* at the level of the tendon, shown as Level *Q*. This post-tensioning force is an externally applied stress resultant that must be balanced by internal forces within the segment. To find the internal forces required for equilibrium, a Euler–Bernoulli deformation *B–B* is applied. The position of this displacement is fixed by both the rotation θ and the deformation δ_Q at Level *Q*. For a specific location of *B–B* in Figure 9.1a, the strain profile is fixed as shown in Figure 9.1b and consequently the stresses in Figure 9.1c, as well as the internal forces displayed on the *RHS* of Figure 9.1d, where F_{rt} is the resultant force in the top reinforcement; F_{cc} is the resultant compressive force in the concrete; F_{rt} is the resultant force in the bottom reinforcement and F_{RC} is the resultant of all the internal forces. It is a matter of varying δ_Q and θ until the resultant internal force F_{RC} aligns with the externally applied axial force $N_{seg\text{-}pos}$ shown on the *LHS* of Figure 9.1d. At equilibrium, let the rotation in Figure 9.1a be θ_{pos} and the deformation at Level *Q*, which is δ_Q, be δ_{pos}; these two parameters fix the Euler–Bernoulli deformation upon the application of the post-tensioning force.

9.1.2 Application of applied moment prior to cracking

We will now assume that the tendon duct is grouted such that there is a bond between the tendon and the adjacent concrete, so the tendon now acts as a reinforcing bar with a residual strain due to the post-tensioning of ε_{pos}. The deformation *B–B* from the application of the post-tensioning force in Figure 9.1a can be considered a residual deformation and applies prior to the application of the moment. This is shown in Figure 9.2a as line *B–B*. A moment M_{seg} is now applied such that at Level *Q*, the stress resultants are M_{seg} and $N_{seg\text{-}pos}$. No matter what deformations occur, the external stress resultant $N_{seg\text{-}post}$

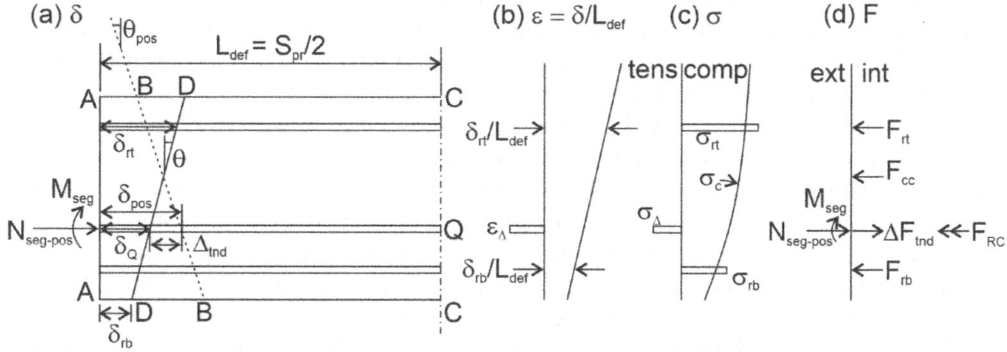

Figure 9.2 Addition of applied moment whilst uncracked (a) segment and deformation profile; (b) strain profile; (c) stress profile; (d) force profile.

remains constant as it is required to balance the residual stresses due to post-tensioning in Figure 9.1. It has been explained in Section 1.5.2 using Figure 1.29(*l*) why M_{seg} must be assumed to act at the level of the resultant tendon force.

The Euler–Bernoulli deformation *D–D* in Figure 9.2a at θ and δ_Q is applied. This causes the grouted tendon, now acting as reinforcement, to extend Δ_{tnd} as shown, such that the increase in strain in the tendon due to flexure ε_Δ is Δ_{tnd}/L_{def} as shown in Figure 9.2b. This increase in strain in the tendon results in an increase in stress σ_Δ in Figure 9.2c and a local increase in force in the tendon of ΔF_{tnd} in Figure 9.2d, which, if the tendon were linear–elastic, would be $A_{tnd}E_{tnd}\varepsilon_\Delta$. This is an internal force in the tendon due to flexure and is not to be confused with or considered part of the external post-tensioning force $N_{seg\text{-}pos}$. The Euler–Bernoulli deformation is varied, as in the previous example, by varying δ_Q and θ, until the resultant of the internal forces F_{RC} shown on the *RHS* of Figure 9.2d is in equilibrium with the external stress resultants shown on the *LHS*, that is, F_{RC} is equal to $N_{seg\text{-}post}$; and taking moments about Level *Q* gives the moment M_{seg}.

From Figure 9.2, it is important to note that the local change in force in the tendon ΔF_{tnd} is not a global increase in the external post-tensioning force but a local internal increase due to flexure. Hence it is shown on the *RHS* of Figure 9.2d and is part of the forces that make up the internal stress resultant F_{RC}. Another way of saying this is that the external post-tensioning force $N_{seg\text{-}pos}$ does not change, once grouting occurs, under flexure, but remains constant to balance the residual forces due to post-tensioning. It is also important to realise that in this analysis of an uncracked member, any length L_{def} of the half-segment can be used; that is, varying the length L_{def} gives exactly the same results.

9.1.3 Application of applied moment after cracking

Let us take the above analysis of post-tensioned members a step further, where the applied external moment causes flexural cracking, using a similar approach to that in Knight et al. (2013a,b, 2014, 2015). The steps in the iterative analysis are exactly the same as illustrated in Figure 9.2. However, in this case, the Euler–Bernoulli deformation causes flexural cracking, as shown in Figure 9.3a. The force in the bottom, now tension reinforcement, F_{rb} can be determined from the P/Δ properties based on the tension stiffening behaviour explained in detail in Chapter 4 for un-prestressed prisms, which depends on Δ_{rb}, which depends on the bond-slip properties at the interface between the reinforcement and the adjacent concrete.

The increase in the force in the grouted tendon ΔF_{tnd} shown in Figure 9.3c also depends on the interface bond and the half-crack width Δ_{tendon} and can be determined from the tension stiffening analyses in Chapter 4, as the residual strain due to post-tensioning does not affect the P/Δ from un-prestressed prisms in Chapter 4. This is because the tendon is grouted after it has been post-tensioned, and upon grouting, the interface bond is under zero stress. The development of the local internal forces ΔF_{tnd} at a crack is illustrated in Figure 9.4, where a tendon is encased in concrete. In Figure 9.4a, a crack has just initiated but not opened, so the crack width w is zero. Because no interface slip has occurred, the bond stress τ between the tendon and the adjacent concrete is zero, so the tendon force $N_{seg\text{-}pos}$ remains unchanged, as shown.

Now let the crack width w widen to $2\Delta_{tnd}$, as shown in Figure 9.4b. From the tension stiffening mechanism in Chapter 4, the interface slip induces a shear stress distribution

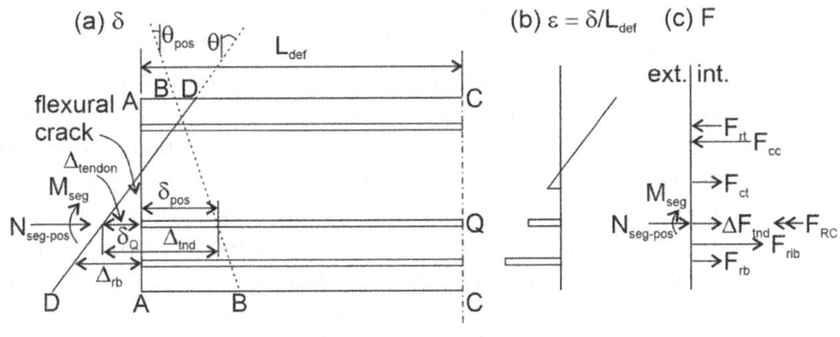

Figure 9.3 Addition of applied moment to cracked member (a) segment and deformation profile; (b) strain profile; (c) force profile.

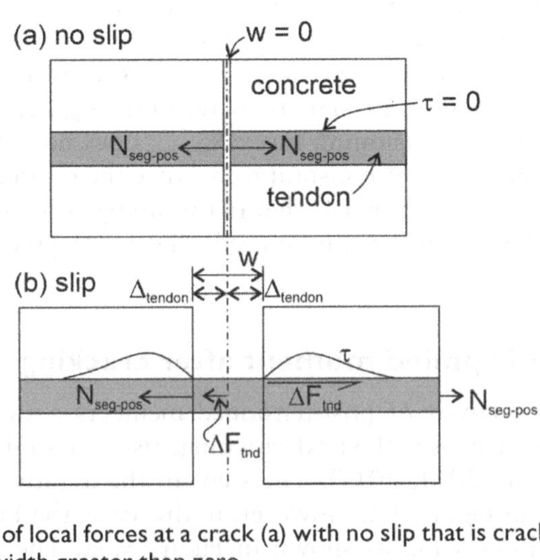

Figure 9.4 Development of local forces at a crack (a) with no slip that is crack width w is zero; (b) with slip so crack width greater than zero.

τ of magnitude ΔF_{tnd}, causing the force in the tendon to increase locally by ΔF_{tnd}, as shown. Hence, any further local increase in the tendon force is due to the local bond developing, which is why ΔF_{tnd} has been shown on the *RHS* of Figure 9.3c alongside the internal forces. As shown in Figure 9.3c, the analysis can include any fibre force F_{fib} and the tensile force in the uncracked concrete F_{ct} if required.

9.2 SEGMENTAL ANALYSIS OF PRE-TENSIONED MEMBER

9.2.1 Application of pre-tension axial force

Now let us move on to a pre-tensioned member. The application of the pre-stressing force is illustrated in Figure 9.5a. The tendon ends are anchored, and a pre-tensioning force N_{seg} is applied to the tendon, which causes a tendon strain ε_{pre}. Concrete is poured around the stressed tendon, and after hardening, the end anchors are released. This causes the segment face *A–A* to contract δ_Q at Level *Q* and rotate θ, as shown, which results in a reduction in strain in the tendon of δ_Q/L_{def}. Hence, the tendon strain has reduced to ε_{pre} - (δ_Q/L_{def}), reducing the tendon force to $N_{seg\text{-}pre}$. As in the previous analyses for post-tensioning, it is a matter of varying θ and δ_Q in Figure 9.5a until there is equilibrium in Figure 9.5b, that is, the internal resultant force F_{RC} is equal to and in line with the external force $N_{seg\text{-}pre}$.

When equilibrium is achieved in Figure 9.5, let δ_Q be δ_{pre} and θ be θ_{pre}, that is, δ_{pre} and θ_{pre} fix the Euler–Bernoulli deformation on the release of the pre-stressing tendon. The tendon force is now given by

$$N_{seg\text{-}pre} = f\left(\varepsilon_{pre} - \frac{\delta_{pre}}{L_{def}}\right) \tag{9.2}$$

Comparing the pre-tensioning Eq. 9.2 with the post-tensioning equivalent of Eq. 9.1, it can be seen that in the segmental analyses for applying the tendon force, the difference is that in the post-tensioning analyses, the tendon force is constant, while in the pre-tensioning analyses, it varies with the position of the Euler–Bernoulli deformation.

It is also important to realise that on the release of the tendon in Figure 9.5, the contraction of the tendon δ_{pre} is also equal to the contraction of the adjacent concrete, as there is full interaction. Because these contractions are the same, no interface bond

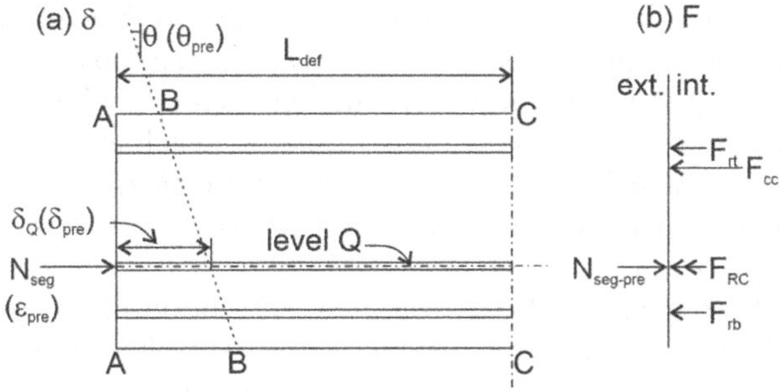

Figure 9.5 Application of pre-tensioning force (a) segment and deformation profile; (b) force profile.

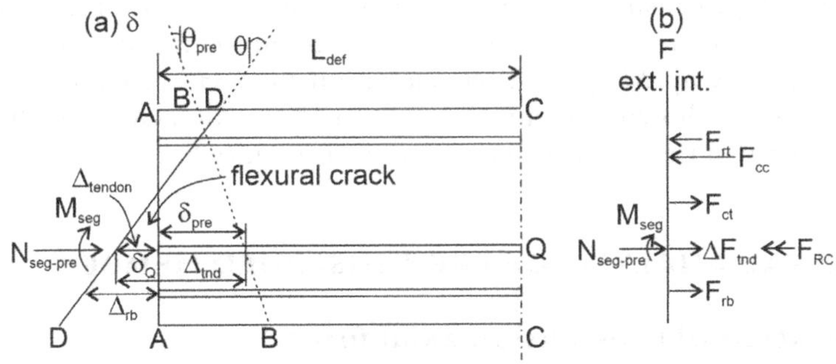

Figure 9.6 Pre-tensioned segmental analyses (a) segment and deformation profile; (b) force profile.

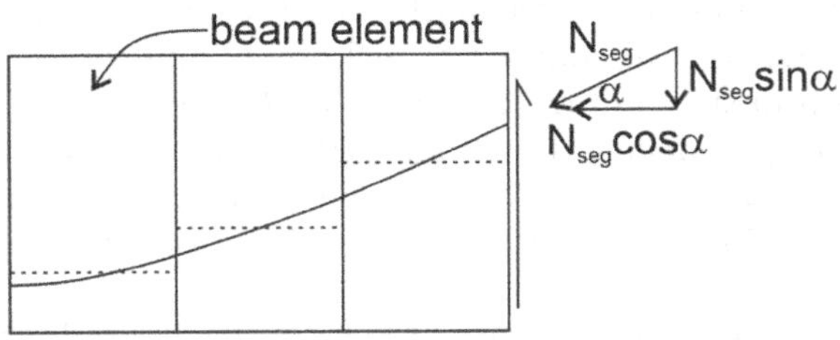

Figure 9.7 Beams with inclined tendons.

develops between the tendon and the adjacent concrete, so the effect of cracking, as demonstrated in Figure 9.4, will be the same as in the post-tensioned analysis.

9.2.2 Application of applied moment

The application of an external applied moment follows exactly the same procedure as for the post-tensioned analysis for both cracked and uncracked sections. The segmental analysis is summarised in Figure 9.6.

9.3 INCLINED TENDONS

The above analyses have been derived for tendons that are horizontal. It is suggested that this procedure can be applied to beams in which the tendons are inclined, as shown in Figure 9.7a, by slicing the beam into very small elements in which the tendons can be assumed to be horizontal. As can be seen in Figure 9.7, each element will have its own tendon force $N_{seg} \cos \alpha$, which will also vary with the vertical position or height of the tendon. Hence, each element will have unique M/θ properties that can be used in the analyses of the whole beam.

It also needs to be remembered that the inclined tendon exerts a downward shear force $N_{seg} \sin \alpha$ in Figure 9.7, which is beneficial as it is in the opposite direction to the applied shear force $V_{seg.}$ Hence, V_{seg}-$N_{seg} \sin \alpha$ can be used in the flexural shear analyses in Chapter 7.

9.4 TENSION STIFFENING ANALYSES

In the above moment–rotation segmental analyses, any value can be used for the segment length L_{def} when cracking or concrete softening has not occurred; all values of L_{def} will give exactly the same results and will correspond with moment–curvature analyses. The following tension stiffening analysis techniques, which allow for the compressive force in the concrete due to prestressing, can be used to determine the crack spacing for prestressed members; however, it is suggested that any reasonable value of crack spacing can be chosen.

In the partial-interaction tension stiffening analysis technique that is fully described in Chapter 4, the essence of the procedure is to determine the slip–strain across an interface, which is the difference between the strains on either side of the interface, as this gives the change in slip across the interface that is the subject of the following procedures.

9.4.1 Post-tensioning

When a post-tensioning force is being applied, there is no bond between the tendon and the adjacent concrete, so we are dealing with a no-interaction analysis. To allow for the compressive stress in the concrete due to post-tensioning, it is first necessary to determine the distribution of stresses and strains when the post-tensioning force is first applied prior to grouting. The cross-section of the tension stiffening prism is shown in Figure 9.8b. The choice of size is fully explained with the help of Figure 4.1 in Chapter

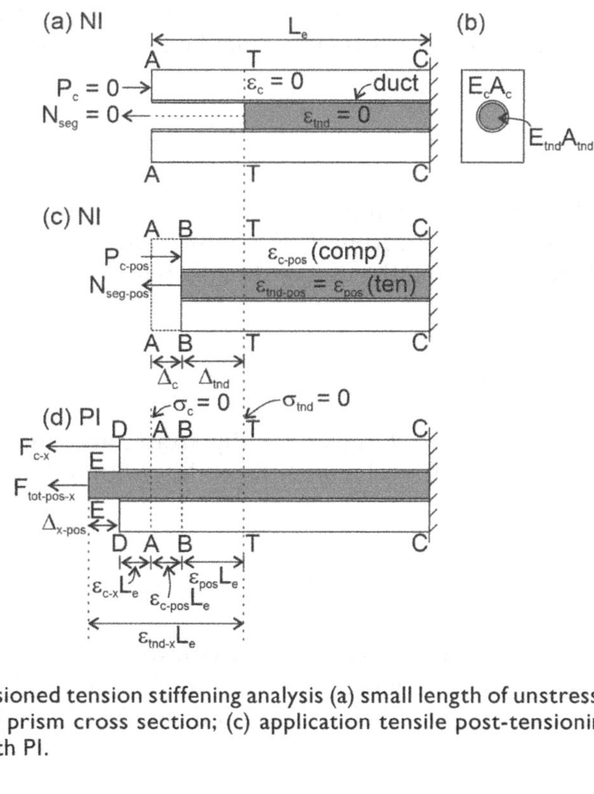

Figure 9.8 Post-tensioned tension stiffening analysis (a) small length of unstressed prism; (b) tension stiffening prism cross section; (c) application tensile post-tensioning force; (d) grouted prism with PI.

4. A small length of the unstressed prism is shown in Figure 9.8a. The length of the concrete component is L_e, whereas the length of the tendon is shown shorter because the post-tensioning procedure changes the length of the tendon to accommodate the stresses and strains required for both equilibrium and compatibility. Position *T–T* is the base-line for the tendon, which is the position at the end of the tendon when it is not strained. Similarly, *A–A* is the base-line for the concrete, which is when it is unstrained.

The tensile post-tensioning force $N_{seg\text{-}pos}$ is now applied in Figure 9.8c, such that for equilibrium, the compressive force in the concrete $P_{c\text{-}pos}$ is equal to $N_{seg\text{-}pos}$. This causes the concrete face to contract Δ_c to position *B–B*, such that the strain in the concrete $\varepsilon_{c\text{-}pos}$ is Δ_c/L_{def}. The tendon needs to extend Δ_{tnd} to achieve compatibility; this is done by increasing the strain to ε_{pos} and adjusting the length of the tendon until compatibility and equilibrium are achieved. If, for example, the materials were linear–elastic, then the compressive force in the concrete $P_{c\text{-}pos}$ would equal $E_cA_c\varepsilon_{c\text{-}pos}$ and the tensile force in the tendon N_{seg} would be $E_{tnd}A_{tnd}\varepsilon_{\text{pos}}$. Equating these forces gives

$$\varepsilon_{c\text{-}pos} = \frac{E_{tnd}A_{tnd}}{E_cA_c}\varepsilon_{pos} \tag{9.3}$$

The tendon is now grouted so that we move from a no-interaction analysis to a partial-interaction (*PI*) analysis, which is shown in Figure 9.8d. For convenience, let us determine the displacements relative to *T–T*, which is the end of the tendon when it is unstressed. From the strains in Figure 9.8c, the deformation from *T–T* to *B–B* is $\varepsilon_{pos}L_e$, and the deformation from *B–B* to *A–A* is $\varepsilon_{c\text{-}pos}L_e$, which is the position of the unstressed concrete. Any increase in concrete strain $\varepsilon_{c\text{-}x}$ due to a load being applied to the tendon and through the interface bond would cause the concrete to extend $\varepsilon_{c\text{-}x}L_e$ to position *D–D*. Note that the subscript x simply denotes the element in the tension stiffening analysis in Figure 4.4 in Chapter 4. From Figures 9.2 and 9.3, the total force in the tendon $F_{tot\text{-}pos}$ is

$$F_{tot\text{-}pos} = N_{seg\text{-}post} + \Delta F_{tnd} \tag{9.4}$$

The total force in the tendon, shown as $F_{tot\text{-}pos\text{-}x}$ in Figure 9.8d, causes a strain $\varepsilon_{tnd\text{-}x}$ and extends the tendon by $\varepsilon_{tnd\text{-}x}L_e$ from *T–T* to *E–E*. This leaves an incompatibility between the face of the tendon and the face of the concrete of $\Delta_{x\text{-}pos}$. The displacements in Figure 9.8c give

$$\Delta_{x\text{-}pos} = \left[\left(\varepsilon_{tnd\text{-}x} - \varepsilon_{c\text{-}x}\right) - \left(\varepsilon_{c\text{-}pos} + \varepsilon_{pos}\right)\right]L_e \tag{9.5}$$

Dividing both sides of Eq. 9.5 by L_e gives the slip-strain, and adding the effect of a shrinkage strain ε_{sh}, as derived from Figure 4.6, as this will no doubt occur, gives the slip–strain as

$$\frac{\Delta_{x\text{-}pos}}{L_e} = \left(\varepsilon_{tnd\text{-}x} - \varepsilon_{c\text{-}x}\right) + \varepsilon_{sh} - \left(\varepsilon_{c\text{-}pos} + \varepsilon_{pos}\right) \tag{9.6}$$

which can be used in the numerical analysis of tension stiffening in Figure 4.4to derive crack spacing and is also an alternative approach to that in Section 9.1.3 for deriving *P*/Δ. The first term in Eq. 9.6 is the slip–strain in an element without any residual stresses for an analysis as in Figure 4.4. The second term is the detrimental increase in the slip–strain due to shrinkage, and the third term is the beneficial reduction due to post-tensioning.

9.4.2 Pre-tensioning

The pre-tensioning analysis is shown in Figure 9.9. In this case, we are dealing with full interaction throughout. The element prior to the release of the tendon is shown in Figure 9.9a. On the release of the tendon at the ends of the prism, both the concrete and tendon contract Δ_{prism} to position B–B in Figure 9.9b, where the tensile force in the tendon $N_{seg\text{-}pre}$ is equal to the compressive force in the concrete $P_{c\text{-}pre}$, meaning they are equal and opposite. It can also be seen that the compressive strain in the concrete $\varepsilon_{c\text{-}pre}$, which is equal to Δ_{prism}/L_e, is also the reduction in strain in the tendon, such that the tendon strain $\varepsilon_{tnd\text{-}pre}$ is now $\varepsilon_{pre}-\varepsilon_{c\text{-}pre}$. Should the materials remain linear–elastic, then $P_{c\text{-}pre}$ is equal to $E_cA_c\varepsilon_{c\text{-}pre}$, and $N_{seg\text{-}pre}$ is equal to $E_{tnd}A_{tnd}$ $(\varepsilon_{pre}-\varepsilon_{c\text{-}pre})$. Equating these forces gives the concrete strain upon releasing the tendon as

$$\varepsilon_{c\text{-}pre} = \frac{E_{tnd}A_{tnd}}{E_C A_c + E_{tnd}A_{tnd}}\varepsilon_{pre} \tag{9.7}$$

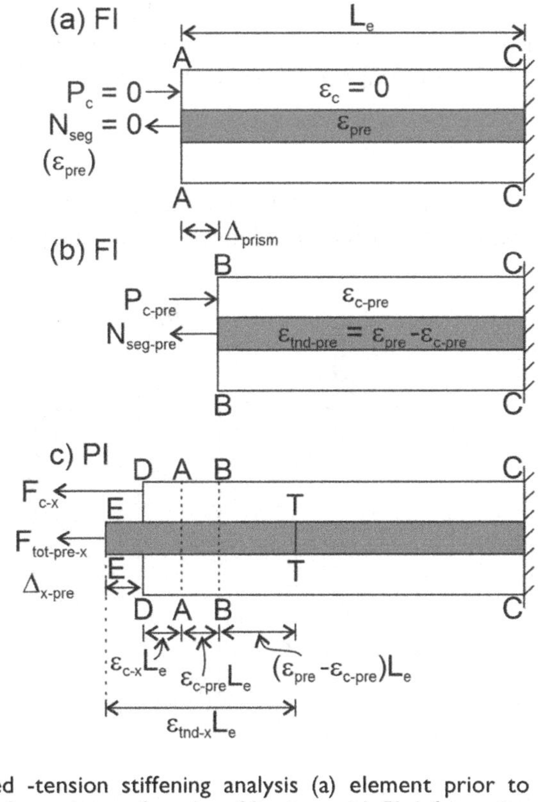

Figure 9.9 Pre-tensioned -tension stiffening analysis (a) element prior to release of the tendon; (b) element after release of tendon; (c) prism with PI deformations.

The deformations are shown in Figure 9.9c, where $F_{tot\text{-}pre\text{-}x}$ from Figure 9.6 is

$$F_{tot\text{-}pre} = N_{seg\text{-}pre} + \Delta F_{tnd} \tag{9.8}$$

and, as in the previous example, end slip Δ_x can be derived and arranged as the slip–strain:

$$\frac{\Delta_{x\text{-}pre}}{L_e} = \left(\varepsilon_{tnd\text{-}x} - \varepsilon_{c\varepsilon\text{-}x}\right) - \varepsilon_{pre} \tag{9.9}$$

Hence, in this pre-tensioning example, pre-tensioning is beneficial in reducing the slip–strain by ε_{pre}, and the slip-strains can be used in the tension stiffening analysis in Figure 4.4 of Chapter 4.

9.5 POST-TENSIONING BEARING PLATE ANALYSIS

The application of the post-tensioning force in Section 9.1 requires a bearing plate at either end of the tendon, such that this bearing plate imposes a concentrated force on the cross-section at the ends of the beam. The dispersal of this concentrated force can cause the member to split, as already described in Section 1.4.5 when considering the dowel action of reinforcing bars. Let us now look at this behaviour more closely for patch loads by first considering a strip load for which there is a closed-form solution (Oehlers and Johnson 1981; Oehlers 1989;Oehlers and Bradford 1995) and then a patch load where there is a semi-empirical approach (Oehlers and Johnson 1981).

9.5.1 Strip load

A symmetrically loaded prism subjected to a concentrated strip load is shown in Figure 9.10. The strip load is of width b_a and length h_a, and the prism is of width b_c and

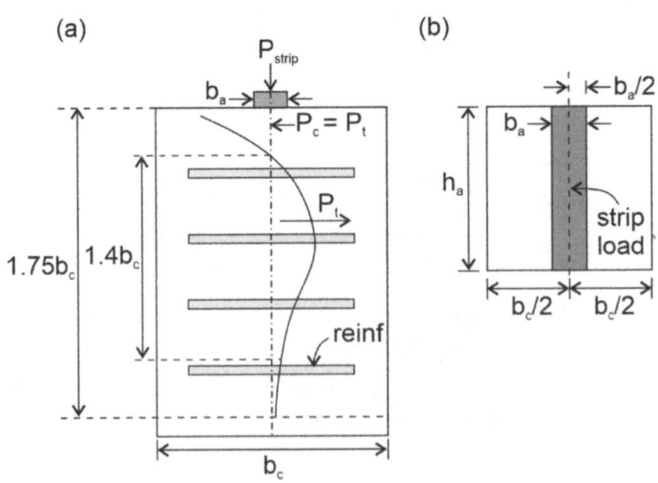

Figure 9.10 Prism subjected to symmetrical strip load. (a) elevation; (b) plan.

length h_a, such that the strip load extends the full length of the prism, as can be seen in Figure 9.10a. The bearing plate on which the strip load acts should be stiff enough to uniformly spread the load over the width b_a.

A strip load P_{strip} is applied as shown in Figure 9.10a. The dispersal of this concentrated line load causes lateral stresses within the prism over about the length $1.75b_c$. Adjacent to the applied load, the lateral stresses are compressive; that is, they provide very useful passive confinement to the concrete in axial compression, and, as explained in Section 5.2.3, substantially enhance the compressive strength of the concrete. Further from the applied load in Figure 9.10a are lateral tensile stresses, which can cause splitting. For equilibrium, the resulting lateral tensile force P_t is equal to the lateral compressive force P_c, and the region over which tension occurs is roughly $1.4b_c$.

If splitting should occur in an unreinforced prism, then P_t tends to zero, in which case the confinement P_c also reduces to zero. The concrete in axial compression is no longer laterally restrained and, in all probability, fails in compression through the formation of a wedge, as described in Chapter 5. If the prism is laterally reinforced, as in Figure 9.10a, then this can take up the lateral tensile forces required to balance the dispersal, which in turn would provide some lateral compressive confinement for equilibrium. It is suggested that the amount of lateral confinement maintained has more to do with the stiffness of the transverse reinforcement than its strength. Alternatively, should sufficient fibres be present in the concrete, then the tensile strength of the concrete does not reduce to zero but, at virtually zero crack width of the splitting crack, can substantially increase the tensile strength, as explained in Section 2.4.1.2 and Section 3.3.1. It appears that this would be a very good application of fibre concrete, not just for its strength but also for its ability to fail in tension in a ductile fashion. It can be seen that there are many ways to prevent collapse due to splitting, but surely the simplest approach would be to first try to design against splitting.

From closed-form solutions (Oehlers and Johnson 1981; Oehlers 1989; Oehlers and Bradford 1995), the lateral force P_t in Figure 9.10a is given by

$$P_t = \frac{1}{\pi}\left(1 - \frac{b_a}{b_c}\right)^2 P_{strip} \tag{9.10}$$

When b_a is of zero width, that is, when there is a line load of zero width, then there is maximum dispersal, and the lateral force P_t is at its maximum of

$$P_t = \frac{1}{\pi} P_{strip} \tag{9.11}$$

In contrast, when $b_a=b_c$ in Eq. 9.10, then P_t is zero, as there is no lateral dispersal. Bear in mind that the lateral force P_t is both tensile and detrimental, and compressive and beneficial.

Furthermore, from the closed-form analysis, P_{strip} to cause splitting, P_{split}, is given by

$$P_{split} = \frac{0.6 b_c h_a f_{ct} \pi}{\left(1 - \dfrac{b_a}{b_c}\right)^2} \tag{9.12}$$

Inserting Eq. 9.12 into Eq. 9.10 gives the lateral force at the commencement of splitting as

$$P_{t\text{-}split} = 0.6 b_c h_a f_{ct} \tag{9.13}$$

which is independent of the lateral dispersal parameter $(1-b_a/b_c)^2$ and is shown to be governed by the cross-sectional area of the prism $b_c \times h_a$.

Hence, we now have the load to cause splitting for a strip load, P_{split}, and the lateral force when this occurs, $P_{t\text{-}split}$. The latter would be of importance should one try to design for strength after splitting, such as using transverse reinforcement or, in particular, fibre concrete.

9.5.2 Patch load

Rarely are bearing pads strips; they are more likely to be patches. The end of a beam of width b_{seg} and depth d_{seg}, where the bearing plate, or patch load, of width D and depth E is applied eccentrically, is illustrated in Figure 9.11. As with the tension stiffening analysis in Section 4.1, the analysis procedure used here requires a symmetrical section, such as that with width b and depth d, as shown. The symmetrical section bxh should be as large as possible to fit within $b_{seg} \times d_{seg}$. If the rectangular patch load is extended to form the strip 5–6–7–8 in Figure 9.11, then the splitting load, from Eq. 9.12, would give

$$P_{split\text{-}b} = \frac{0.6 b h f_{ct} \pi}{\left(1 - \dfrac{D}{b}\right)^2} \tag{9.14}$$

where the subscript b in $P_{split\text{-}b}$ reflects lateral dispersal over the width b in Figure 9.11.

Similarly, if the patch load covered 1–2–3–4, then the splitting strength for dispersal in the h direction is

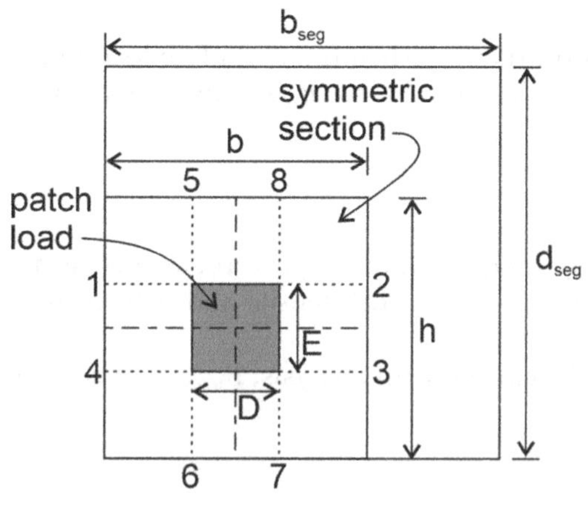

Figure 9.11 Prism subjected to non-symmetrical patch load.

$$P_{split\text{-}h} = \frac{0.6bhf_{ct}\pi}{\left(1-\frac{E}{h}\right)^2} \qquad (9.15)$$

To determine the splitting load for the patch load in Figure 9.11, the following reciprocal relationship (the form of which is used in the design of columns with bi-axial bending) is employed (Oehlers and Johnson 1981).

$$\frac{1}{P_{split\text{-}patch}} = \frac{1}{P_{split\text{-}b}} + \frac{1}{P_{split\text{-}h}} \qquad (9.16)$$

where $P_{split\text{-}patch}$ is the load required to cause the patch load to split. This approach is mechanically correct at its extremities. For example, when the length $E = h$ in Figure 9.11, then from Eq. 9.15, $P_{split\text{-}h} \rightarrow \infty$, so Eq. 9.16 gives the correct answer.

It needs to be remembered that for the patch loads, the transverse force at splitting, $P_{t\text{-}split}$, is the same in both directions; that is

$$P_{t\text{-}split} = 0.6bhf_{ct} \qquad (9.17)$$

In conclusion, Eq. 9.16 can be used to determine when splitting occurs when the bearing plate acts as a patch load. Should splitting occur, then at least Eq. 9.17 could be used to determine the minimum amount of transverse reinforcement required to maintain equilibrium. Furthermore, it should be a great help in designing for post-splitting using fibre reinforcement.

9.6 NOTATION AND ABBREVIATIONS

A_c	cross-sectional are of concrete tension stiffening prism
A_{rb}	cross-sectional area of bottom reinforcement
A_{rt}	cross-sectional area of top reinforcement
A_{tnd}	cross-sectional area of tendon
b	width of effective prism
b_a	width of strip load or patch load
b_c	width of prism
b_{seg}	width of beam or segment
$comp$	compression
D	width of patch load
d_{seg}	depth of beam or segment
E	modulus; depth of patch load
E_c	concrete modulus
E_{tnd}	tendon modulus
ext	external stress resultants
F	force; stress resultants
F_c	force in concrete in tension stiffening prism
F_{cc}	resultant concrete compressive force
F_{ct}	resultant concrete tensile force

F_{fib}	resultant fibre force
F_{RC}	resultant of all the internal forces
F_{rb}	resultant force in bottom reinforcement
F_{rt}	resultant force in top reinforcement
$F_{tot\text{-}pos}$	total force in post-tensioned tendon after cracking
$F_{tot\text{-}pre}$	total force in pre-tensioned tendon after cracking
f_{ct}	tensile strength of concrete
h	height of effective prism
h_a	depth of strip load; depth of prism
int	internal forces
L_e	element length
N_{seg}	external pre-tensioning force prior to release; axial force in tendon
$N_{seg\text{-}pos}$	external post-tensioning force
$N_{seg\text{-}pre}$	external pre-tensioning force after release
P_c	axial force on concrete; lateral confining compressive force
$P_{c\text{-}pos}$	P_c after post-tensioning
$P_{c\text{-}pre}$	P_c after release of pre-tensioned prism
P_{split}	load to cause splitting in prism with a strip load
$P_{split\text{-}b}$	P_{split} when there is transverse dispersal over the with b
$P_{split\text{-}h}$	P_{split} when there is transverse dispersal over the with h
$P_{split\text{-}patch}$	load on patch to cause splitting
P_{strip}	strip load; line load
P_t	lateral tensile force due to load dispersal
$P_{t\text{-}split}$	P_t when splitting occurs
$Q\text{-}Q$	level of resultant post-tensioning force
S_{pr}	crack spacing
ten	tension
w	crack width
α	tendon inclination
Δ	increase; half-crack width
Δ_c	contraction of concrete prism after post-tensioning
ΔF_{tnd}	local increase in tendon force due to M_{seg}
Δ_{prism}	contraction of pre-tensioned prism after release
Δ_{rb}	Δ at level of bottom reinforcement
Δ_{tendon}	half-crack width at tendon
Δ_{tnd}	increase in tendon length due to applied moment M_{seg}; extension of tendon after post-tensioning
$\Delta_{x\text{-}post}$	interface slip at crack in post-tensioned prism
$\Delta_{x\text{-}pre}$	interface slip at crack in pre-tensioned prism
δ	deformation
δ_{pre}	displacement of tendon after pre-tensioning
δ_{pos}	displacement of tendon after post-tensioning
δ_Q	Euler–Bernoulli deformation at Level Q
δ_{rb}	deformation of bottom reinforcement
δ_{rt}	deformation of top reinforcement
ε	strain
ε_c	concrete strain

$\varepsilon_{c\text{-}pos}$	strain in concrete after post-tensioning
$\varepsilon_{c\text{-}pre}$	strain in concrete after pre-tensioning
ε_{pos}	post-tensioning strain in tendon
ε_{pre}	pre-tensioning strain in tendon prior to release
ε_{tnd}	tendon strain
$\varepsilon_{tnd\text{-}pos}$	ε_{tnd} after post-tensioning
$\varepsilon_{tnd\text{-}pre}$	ε_{tnd} after pre-tensioned prim released
ε_{Δ}	increase in tendon strain due to Δ_{tnd}
θ_{pos}	Euler–Bernoulli rotation after post-tensioning
θ_{pre}	Euler–Bernoulli rotation after pre-tensioning
σ	stress
σ_c	stress in concrete
σ_{rb}	stress in bottom reinforcement
σ_{rt}	stress in top reinforcement
σ_{tnd}	stress in tendon
σ_{Δ}	increase in tendon stress due to Δ_{tnd}
τ	interface bond shear stress

REFERENCES AND DEVELOPMENT

Knight, D., Oehlers, D.J. and Mohamed Ali, M.S., 2013a. Short term partial interaction behaviour of RC beams with prestressed FRP and steel. *Journal of Composites for Construction*, *18*(1). https://doi.org/10.1061/(ASCE)CC.1943-5614.0000408.

Knight, D., Visintin, P. and Oehlers, D.J., 2015. Displacement-based simulation of time-dependent behaviour of RC beams with prestressed FRP and steel tendons. *Structural Concrete*, *2015*(3), 406–417.

Knight, D., Visintin, P., Oehlers, D.J. and Jumaat, M.Z., 2013b. Incorporating residual strains in the flexural rigidity of RC members with varying degrees of prestress and cracking. *Advances in Structural Engineering*, *16*(10). https://doi.org/10.1260/1369-4332.16.10.1701.

Knight, D, Visintin, P., Oehlers, D.J. and Mohamed Ali, M.S., 2014. Simulating RC beams with unbonded FRP and steel prestressing tendons. *Composites: Part B*, *60*, 392–399.

Oehlers, D.J., 1989. Splitting induced by shear connectors in composite beams. *Journal of Structural Division of the American Society of Engineers*, *115*(2), 341–362.

Oehlers, D.J. and Bradford, M.A., 1995. *Composite Steel and Concrete Structural Members: Fundamental Behaviour.* Oxford: Pergamon, Elsevier Science Ltd.

Oehlers, D.J. and Johnson, R.P., 1981. The splitting strength of concrete prisms subjected to surface strip or patch loads. *Magazine of Concrete Research*, *33*(116), 171–179.

Chapter 10

Profiled slabs and composite steel and concrete members

Chapter 8 dealt with *RC* beams in which only local debonding occurred, such that the segmental analysis of one segment, as illustrated in Chapter 6, could be used to derive the mechanical properties, as in Figure 6.11, along the whole length of the member. This relied upon intermediate crack (*IC*) debonding not occurring, as explained in Section 1.4.1 with the help of Figure 1.12 and in Chapter 4. Should *IC* debonding occur, then the reinforcement debonds along the length of the member, resulting in global debonding, which is covered in this Chapter 10 and in the ensuing Chapter 11.

The numerical flexural shooting method of analysis used in global debonding is first explained using a composite steel and concrete beam as an example. This is also used to explain how the properties derived from a local segmental analysis are used in a global analysis. Hence, the analysis can still incorporate time-dependent properties and the inclusion of fibres. The procedure is then applied to profiled slabs.

Importantly, it is shown that the local M/χ properties derived from the M/θ segmental analyses in Chapter 6 and used in Chapter 8 are still required for the global analysis, but these properties vary along the length of the member due to the variation in the internal axial load, which is the interface bond force. This demonstrates a link between the analysis for local debonding and global debonding.

10.1 FLEXURAL SHOOTING METHOD OF ANALYSIS

10.1.1 Composite steel and concrete beam

The shooting method for axially loaded members has already been explained in Chapter 4 for a tension stiffening analysis. Taking one step further, the shooting method is now explained for a member subjected to both flexure and axial loads based on the approach of Oehlers and Bradford (1995) and Chen et al. (2024). The half-span of a simply supported composite steel and concrete beam in Figure 10.1 is used to illustrate the procedure, as the two components are visually separate. The variation of the externally applied moment to the beam M_{ext} is shown in Figure 10.1a, where the half-span length is $L/2$. The deformed shape of the composite beam is shown in Figure 10.1b,b, where the concrete slab has a depth h_s and the steel beam a depth h_b.

Slip Δ at the interface between the concrete slab and the steel beam in Figure 10.1b can be resisted by friction, chemical bond and mechanical anchors such as stud shear connectors. This shear force over the half-span has a total value of P and causes at

DOI: 10.1201/9781003711407-15

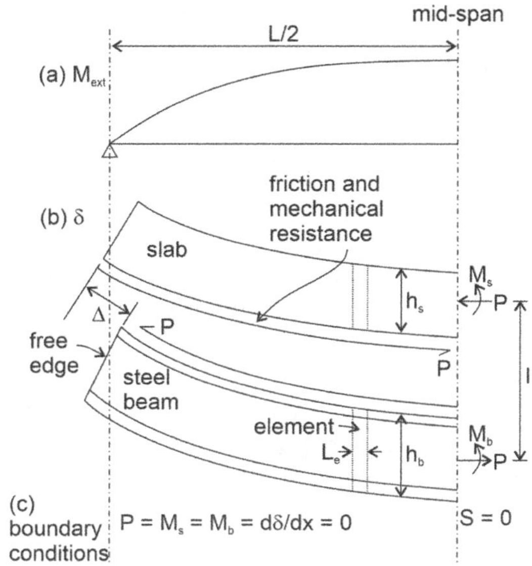

Figure 10.1 Simply supported composite steel and concrete beam (a) distribution of externally applied moment; (b) reinforced concrete and steel section deformations; (c) boundary conditions for analysis.

mid-span a compressive force P in the slab and a tensile force P in the steel beam. The flexural deformation of the composite beam causes at mid-span a moment M_s in the slab and a moment M_b in the beam, as shown. Hence, if the externally applied moment at mid-span is M_{ext}, then from equilibrium,

$$M_{ext} = M_s + M_b + Pl \tag{10.1}$$

where l is the lever arm between the axial forces P. This equation applies at any section along the composite beam.

Known boundary conditions are required in the ensuing shooting method of analysis. It is, therefore, worth noting that on the *LHS* of the beam in Figure 10.1, the internal axial force P and moments M_s and M_b are zero; as the moments are zero, so are the flexural strains, and consequently the slip–strain ds/dx as shown in Figure 10.1c. Should the beam be symmetrical and symmetrically loaded about the mid-span, the slip s at mid-span is zero, which is another boundary condition. Should the beam not be symmetric, then it should be noted that the *RHS* of the beam has the same boundary conditions as the *LHS* of the beam.

10.1.2 Composite beam shooting analysis

As in Oehlers and Bradford (1995), for analysis, the composite beam is sliced into very narrow elements of length L_e, an example of which is shown in Figure 10.1b, where $L_e = L$. As with the segmental analysis in Section 6.1.2.1, because the elements are very small, it will be assumed that the element is in a constant moment region. Starting from the *LHS* of the composite beam, the first two elements are shown in Figure 10.2. Take,

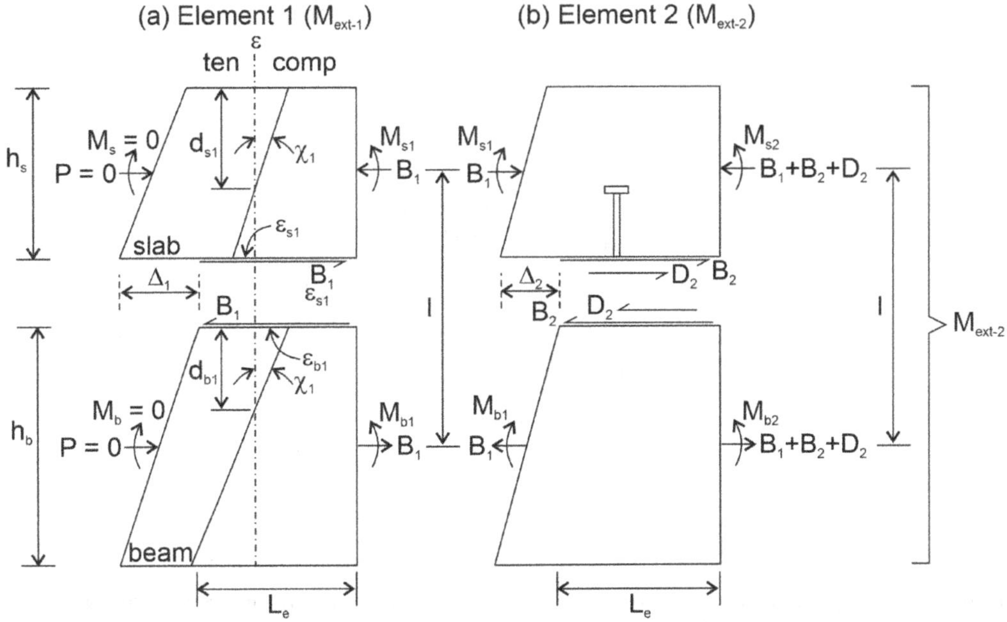

Figure 10.2 Flexural shooting analysis for composite beam (a) element 1 adjacent to support of simply supported composite beam; (b) element 2 adjacent to element 1.

for example, Element 2 in Figure 10.2b. The mean of the stress resultants should be used in the constant moment analysis, but for ease of explanation, we will assume that the stress resultants on the *RHS* act within the whole element. There is virtually no error, as the effective length L_e is probably two orders of magnitude smaller than the span length L.

The first element is shown in Figure 10.2a, where the stress resultants on the free edge are zero. The *LHS* is the free edge in Figure 10.1b, where the deformation of the composite beam induces an interface slip of Δ, which is Δ_1 in Figure 10.2a. The direction of this interface slip induces an interface bond force B_1, which could be due to friction and therefore independent of slip. However, if the interface were adhesively bonded, it would depend on the bond-slip properties, as discussed in Section 3.2, and would be dependent on the magnitude of the slip. The bond force B_1 acts to the right on the slab element and to the left on the beam element, as shown. Hence, for equilibrium on the *RHS*, there is a compressive force B_1 acting on the slab and a tensile force B_1 acting on the beam. From the applied moments in Figure 10.1a, the moment on the first element is $M_{ext\text{-}1}$, as shown in Figure 10.2a, which induces curvature. It is assumed that there is no vertical separation at the interface, so the curvature χ_1 is the same in both components. It is a question of finding the distribution of the curvature, that is, χ_1, d_{s1} and d_{b1} (where d_s and d_b are the depths to the neutral axes, as shown) in the strain profile in Figure 10.2a when Eq. 10.1 is satisfied. To do this requires the $M_{s1}/\chi_1/d_1$ relationship for both the concrete and steel elements when there is an axial force of B_1.

The M/χ relationships required for the analysis of the concrete element in Figure 10.2a can be derived from the Euler–Bernoulli M/θ segmental analysis already described in Chapters 6 and 8. Hence, time effects and fibres can be easily incorporated.

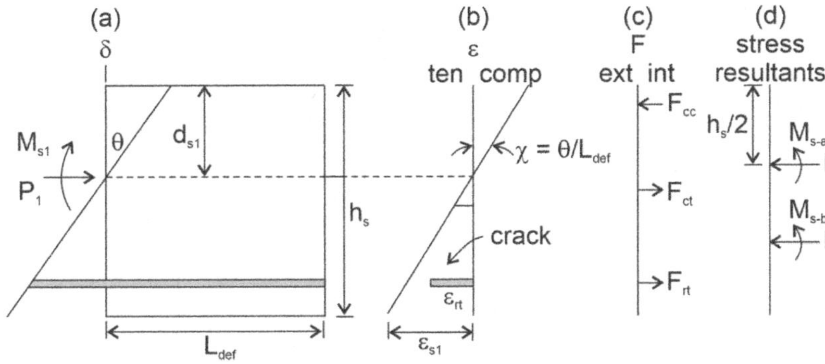

Figure 10.3 Segmental-analysis of concrete component of composite beam (a) segment and displacement profile; (b) strain profile; (c) force profile; (d) stress resultants.

The segmental analysis of any slab component of the composite beam is illustrated in Figure 10.3. The segment is subjected to an axial compressive load P_1 in Figure 10.3a, which for the first element in Figure 10.2a is equal to the bond force B_1. For a fixed value of θ_1 and d_{s1} in Figure 10.3a, the strain profile in Figure 10.3b, the stress distribution, and consequently the force profile in Figure 10.3c can be derived. It is a question of varying d_{s1} in Figure 10.3a until the resultant force in Figure 10.3c is equal to a compressive force of P_1. Moments can now be taken at any level, such as at mid-depth, to find the moment $M_{s\text{-}a}$ in Figure 10.3d. Please note that as no external forces are being applied, the moment can be taken at any level, such as below the mid-depth, where the moment is $M_{s\text{-}b}$. Both combinations of stress resultants will give the same force distribution in Figure 10.3c. It is also worth noting that whether the slab is uncracked or cracked, the effective strain in the concrete slab at the interface with the steel beam is ε_{s1} in Figure 10.3b, which is also shown in the concrete element in Figure 10.2a.

The M/χ for the steel beam in Figure 10.2a can be obtained in the same way or using more conventional techniques, as cracking does not occur. For the same value of θ_1 as in the concrete segment above, d_{b1} can be varied until there is a resultant tensile force B_1; then M_{b1} can be determined. If Eq. 10.1 is not satisfied, then θ_1 has to be varied until equilibrium is achieved. Additionally, from the segmental analysis of the slab in Figure 10.2a, ε_{s1} for $\chi_1$can be obtained, which is the slab strain adjacent to the interface and is also shown in Figure 10.3b. Similarly, for the steel beam in Figure 10.2a, ε_{b1}can be obtained, which is the strain in the beam's top fibre adjacent to the interface. The sum of ε_{s1} and ε_{b1} gives us the interface slip–strain.

The deformations adjacent to the interface in Figure 10.2a are shown exaggerated in Figure 10.4. The interface slip on the *LHS* is Δ_1. It is a question of determining the change in this slip over the element length. From Figure 10.2a, there is a tensile strain in the slab at the interface of ε_{s1}. This causes the bottom of the slab to expand by $\varepsilon_{s1}L_e$ as shown in Figure 10.4a. Similarly, at the top of the beam in Figure 10.2(a, there is a compressive strain ε_{b1}, which causes a contraction of $\varepsilon_{b1}L_e$, as shown in Figure 10.4a. Hence, the slip has reduced to

$$\Delta_2 = \Delta_1 - (\varepsilon_{s1} + \varepsilon_{b1})L_e \qquad (10.2)$$

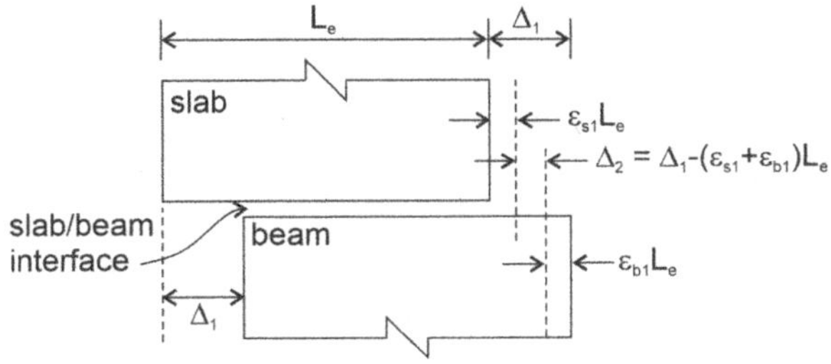

Figure 10.4 Slip-strain in composite steel and concrete beam.

where the bracketed term is the slip–strain *ds/dx*. Note that should the expansion of the slab $\varepsilon_{s1}L_e$ in Figure 10.4a be greater than $\Delta_1 - \varepsilon_{b1}$, then the slip will increase.

The slip in the second element Δ_2 in Figure 10.2b is now known and consequently the interface force B_2, and should there be mechanical connectors such as stud shear connectors, the dowel force D_2, which depends on the slip. Hence, the axial force on the *RHS* has increased to $B_1+B_2+D_2$. The process for Element 1 can now be repeated for Element 2. It can be seen in this example that the accumulated bond force increases and the slip reduces. The analysis is continued until a known boundary condition is reached. If that known boundary condition is not satisfied, then the assumed value of Δ_1 in Figure 10.2a is changed until the boundary condition is satisfied. It is worth emphasising that as you move along the beam, the axial force in this example increases; hence, a new M/θ analysis is required for each element.

To repeat, if the composite beam is symmetric about mid-span, the boundary conditions are shown in Figure 10.1c. Hence, starting on the *LHS*, as we have done, we can choose or guess a value for Δ_1 and work towards mid-span. If the slip at mid-span is not zero, then we change Δ_1 and repeat the analyses until the boundary condition is satisfied. If the beam is not symmetric, then the slip at mid-span would not be zero, but the slip–strain at the right-hand support would be zero, so this can be used as a boundary condition to be achieved. Great care must be taken in deriving the slip from the slip–strain, as explained using Figure 10.4.

10.2 PROFILED SLABS

The shear slip bond in profiled slabs is very weak compared with that associated with reinforcing bars and is transmitted through ribs of virtually any cross-section, in which the bond is enhanced through protrusions and indentations. These ribs do not have a defined perimeter length as occurs with circular reinforcing bars, and hence, it is difficult to use the concept of a bond-slip/shear stress (τ/δ); instead, we use the concept of shear flow q.

It has been explained in Chapter 1, with the help of Figure 1.5, that one way of determining the bond-slip properties is to test bars with short embedment lengths, as illustrated in Figure 10.5a. These embedment lengths are short enough that the interface slip δ can be assumed to be uniform. Stating the obvious, the shear stress τ for the slip δ is given by the axial force P divided by the perimeter length of the bar πD and divided

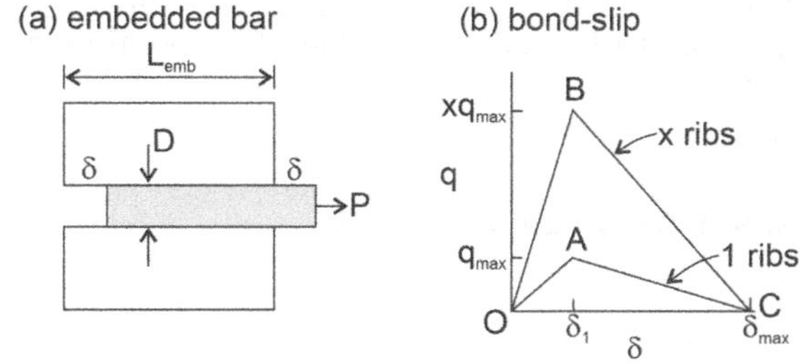

Figure 10.5 Shear flow (a) test specimen with short embedded length; (b) shear flow bond slip properties.

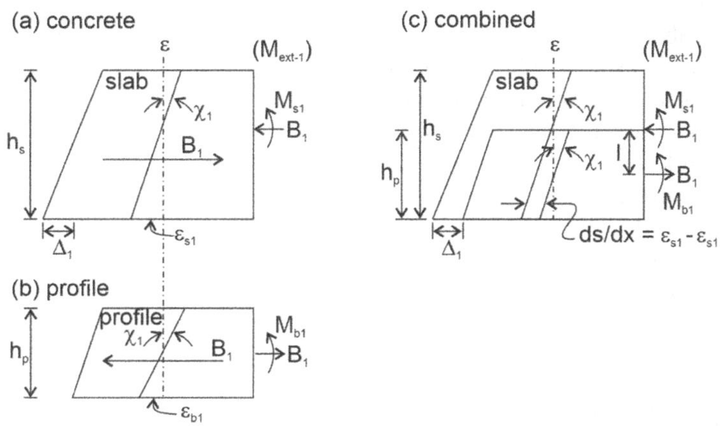

Figure 10.6 Profiled slab analyses (a) reinforced concrete segment; (b) profile sheet segment; (c) combined segment.

by the embedment length $L_{emb.}$ However, profiled sheet ribs do not have an obvious perimeter length, so it is much more convenient to deal with shear flows q, that is

$$q = \frac{P}{L_{emb}} \tag{10.3}$$

which is the force per unit length of the rib and is specific to that particular rib. Hence, the bond-slip properties are now in terms of shear flow, as in Figure 10.5b, where the shear flow variation for a single rib is $O\text{–}A\text{–}C$, and that for x ribs is $O\text{–}B\text{–}C$.

Profiled slabs behave in exactly the same way as composite beams, as shown in the mechanics in Figure 10.2 and illustrated in Figure 10.6 (Chen et al. 2024). The elements of the profiled slab are shown separated in Figure 10.6a and b. Of note is that the bond forces B derived from qL_e can be shown to act anywhere, as they do not affect the analyses. Both elements are combined in Figure 10.6c, where it can be seen that the strain profiles are parallel, as they have the same curvature, so that the slip-strain can be measured at any level.

Global debonding can also occur, and it is quite likely to happen as the interface bond strengths are relatively weak compared with those in reinforcing bars. Weak bond has been shown in Chapter 4 to cause large crack spacings, which can lead to very wide cracks that can, in turn, lead to *IC* debonding. The *IC* debonding analysis is covered in the following Chapter 11.

10.3 NOTATION AND ABBREVIATIONS

B	bond force
$comp$	compression
D	dowel force
d_b	depth to neutral axis in beam
d_s	depth to neutral axis in slab
ds/dx	slip–strain
F	force profile
F_{cc}	resultant compressive force in concrete
F_{ct}	resultant tensile force in concrete
F_{rt}	resultant force in tension reinforcement
h_b	height of beam
h_p	height of profile
h_s	height of slab
IC	intermediate crack
L	span of beam
LHS	left-hand side
L_{def}	deformation length; half-length of segment
L_e	element length
L_{emb}	embedded length
l	lever arm between internal axial forces
M	moment
M_{ext}	externally applied moment
M_b	internal moment in beam
M_s	internal moment in slab
M/χ	moment/curvature variation
M/θ	moment/rotation variation
P	axial force
q	shear flow
q_{max}	maximum shear flow capacity
RC	reinforced concrete
RHS	right-hand side
S	slip
ten	tension
x	number
χ	curvature; curvature profile
Δ	interface slip at end
δ	deformation; interface slip
δ_{max}	slip at zero shear
δ_1	slip at τ_{max}
ε	strain; strain profile

ε_b strain in beam at interface
ε_s strain in slab at interface
θ Euler–Bernoulli rotation
τ shear stress
τ_{max} maximum shear stress
τ/δ bond-slip variation

REFERENCES AND DEVELOPMENT

Chen, S., Visintin, P., Sturm, A.B. and Oehlers, D.J., 2024. Quantifying the serviceability flexural benefits of using UHPFRC in profiled slabs. *Structural Concrete*, *2024*, 1–16.

Oehlers, D.J. and Bradford, M.A., 1995. *Composite Steel and Concrete Structural Members: Fundamental Behaviour*. Oxford: Pergamon.

Chapter 11

FRP retrofitted reinforced concrete members

This chapter deals directly with *IC* debonding, as illustrated in Figure 1.12c and explained in detail in Chapter 4; that is, it addresses global debonding. Should *IC* debonding occur, the reinforcement debonds along the length of the member. This is unlikely to happen with reinforcing bars, as generally the bond is very strong, but can occur in profiled slabs and in externally *FRP* plated *RC* members, where the bond is usually very weak. The mechanics will be illustrated using a retrofitted beam that is plated on its soffit, where the plate could be steel but is more commonly *FRP*.

It will be shown how the properties derived from a local segmental analysis in Chapter 6 are used directly in this global analysis of plate debonding. Hence, the analysis can still incorporate time-dependent properties and the inclusion of fibres. The procedure is applied to an externally plated *RC* beam. First, the transition from local debonding to global *IC* debonding is explained, which also elucidates the mechanics of crack closure. This is followed by the analysis of an *FRP* retrofitted *RC* beam where debonding occurs over the length of the beam, demonstrating that sometimes global debonding is gradual and stable, while at other times it is unstable, leading to catastrophic failure. It is also shown that as global debonding develops, the *RC* member behaviour transitions from that of a standard *RC* beam to a type of post-tensioned member, requiring a change in analysis techniques.

The following analyses may initially appear complex, but they are used to illustrate the mechanisms required to cause *IC* debonding. However, the actual debonding analyses required for design to predict when *IC* debonding occurs and its development are straightforward, and the steps are summarised in the last Section 11.4.

11.1 SHOOTING ANALYSIS OF PLATED *RC* BEAMS

The shooting method of analysis, already explained in great detail in Section 10.1 with the help of Figure 10.2 for composite beams, is utilised in this analysis of plated beams; it is suggested that this section be reread before continuing. In Figure 10.2a for the composite beam, it was convenient to start the analysis at the end of the beam where the applied moment is zero. For plated beams, *IC* debonding starts at the position of maximum moment where the cracks are widest, as shown in Figure 6.1. Hence, the first element of the shooting method will start at this position.

DOI: 10.1201/9781003711407-16

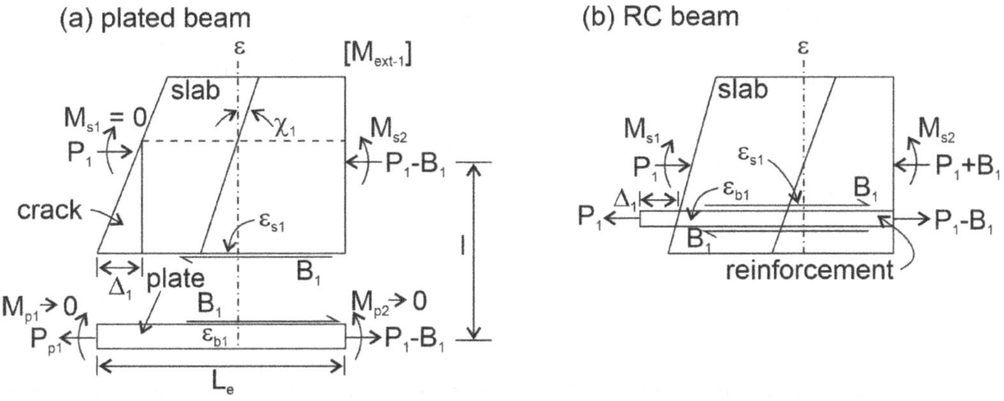

Figure 11.1 Shooting analysis for *RC* beams (a) slab and plate element of a plated beam; (b) slab and reinforcement element of RC beam.

Following the approach of Oehlers et al. (2016a,b), Element 1 of the shooting method of analysis for a plated beam is shown in Figure 11.1a. At the crack on the left-hand side *(LHS)*, the plate extends Δ_1 from the crack face and the tensile force in the plate is P_1. The plate depth is likely two orders of magnitude smaller than the depth of the slab, so any flexural component within the plate, M_p, can be ignored. However, it could be included in exactly the same way as the steel beam in the composite steel and concrete beam in Figure 10.2 in Section 10.1.2. In this analysis, we will ignore flexure in the plate.

For equilibrium on the *LHS* of Figure 11.1a, the axial force on the slab is a compressive force P_1 and there is also a moment M_{s1}. We will assume that when we conduct the segmental analysis, the element is subjected to the stress resultants on the *RHS* of Figure 11.1a.

$$M_{ext} = M_s + Pl \tag{11.1}$$

where M_{est} is the external moment at this section. For additional help in understanding this procedure, compare Figure 10.2 with Figure 11.1. They are almost identical, except that the steel beam component in Figure 10.2 is replaced by the plate in Figure 11.1.

Slip Δ_1 across the interface of length L_e in Figure 11.1a induces a bond force on the plate B_1, which depends on the interface slip as it depends on the bond-slip τ/δ properties. Hence, the tensile force on the *RHS* of the plate has reduced to $P_1–B_1$, from which the plate strain ε_{b1} can be determined. From equilibrium, the shear on the slab is of the same magnitude but in the opposite direction, reducing the compressive force in the slab to $P_1–B_1$ that is by the same magnitude. The concrete component can now be analysed using the standard segmental approach, as shown in Figure 11.2, to obtain the M/χ relationship for an axial load P_1. It is a matter of determining in Figure 11.1a the curvature χ_1 and the strain at the interface ε_{s1} when the internal moment $(P_1–B_1)$ plus M_{s2} is equal to the external moment $M_{ext\text{-}1}$, that is, when Eq. 11.1 is satisfied. This will provide the slab strain at the interface ε_{s1}. Hence, we now have the interface slip–strain $\varepsilon_{b1}–\varepsilon_{s1}$ that changes the interface slip. It may be worth noting that the shear forces B_1 at the interface are not included directly in the overall analysis for equilibrium to satisfy Eq. 11.1, as they are equal and opposite and act along the same plane; thus, they cancel out.

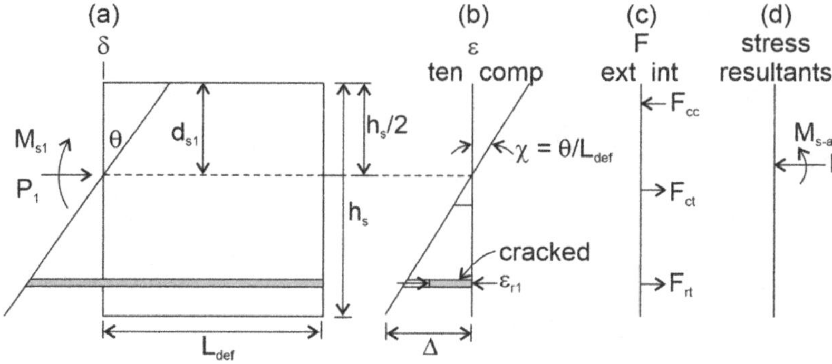

Figure 11.2 Segmental analysis of concrete component (a) concrete segment and displacement profile; (b) strain profile; (c) force profile; (d) stress resultants.

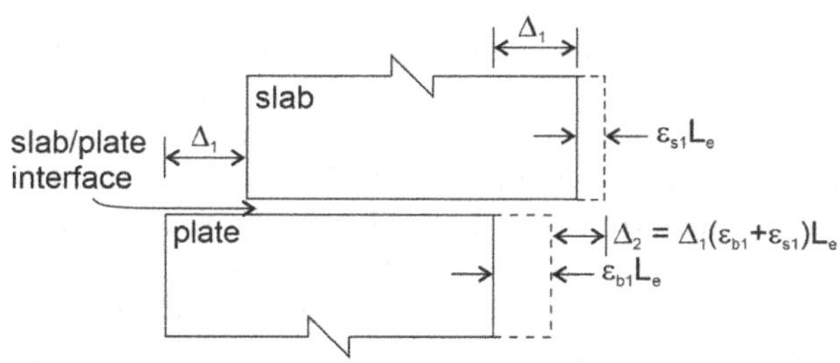

Figure 11.3 Slip–strain in plated beam.

Great care must be taken in deriving the slip from the slip–strain, as explained using Figure 11.3 for the case when the tensile plate strain ε_{b1} is greater than the tensile slab strain ε_{s1}. In this case, both components expand as shown, but because the plate expands more than the slab, the slip reduces to Δ_2, which is Δ_1 less $(\varepsilon_{b1}-\varepsilon_{s1})L_e$. This can be compared to Δ_2 from the analysis of the composite beam in Section 10.1.2 using Figure 10.4, where one component strain is compressive and the other tensile. Returning to the plated beam in Figure 11.3, should $\varepsilon_b<\varepsilon_s$, then the slip would increase.

The analysis of an *RC* beam in which the reinforcement is now within the slab is shown in Figure 11.1b. It is exactly the same as for the plated beam; however, bear in mind that the slab interface strain ε_{s1} is now at the level of the reinforcement, as shown. Hence, if it is necessary to check for debonding in encased reinforcement with weak bond, which might occur with *FRP* reinforcing bars, then this technique can be used.

11.2 TRANSITION FROM LOCAL TO GLOBAL DEBONDING

It was shown in Chapter 4, in a tension stiffening analysis with the help of Figure 4.4, and in Chapter 1, with the help of Figure 1.12c, how the force in the bar can have a limit that depends on the interface bond-slip properties. This is referred to as the intermediate crack (*IC*) debonding resistance P_{IC}, which can be quantified both numerically and in closed-form solutions. This tension stiffening mechanism depends on an uncracked prism. It will be demonstrated how this mechanism can occur in beams through the closure of flexural cracks. To illustrate the transition from local to global debonding, we will first consider debonding in a constant moment segment with two cracks, as covered in Chapter 6, in which only local debonding occurs. We will then move on to debonding in a constant moment segment with a single crack, which uses the same analysis techniques as that with two cracks, but in this case, only global debonding occurs. Finally, we will examine a beam with a moment gradient to show the mechanism of crack closure, which allows for the transition from local to global debonding.

11.2.1 Local debonding in a segment

The reader is now familiar with the mechanics of a two-crack segment, which has been explained in detail in Chapter 6. In this approach, we did not follow the slip along the reinforcement to determine the force in the reinforcement; instead, we used the P/Δ results from tension stiffening analyses, which did follow the bond-slip, to determine the reinforcement force across a crack. Obviously, the P/Δ approach is much more convenient in analysis. To explain the debonding mechanism, we will now return to first principles and follow the slip along the reinforcement in a flexural member, as was done in Oehlers et al. (2005) and Liu et al. (2007). The shooting method of analysis (Oehlers and Bradford 1995) has already been explained with the help of Figure 11.1.

Let us start with the analysis of a segment with two cracks, as shown in Figure 11.4b, which is the standard model used in Chapter 6; as usual, we will only consider a half-segment. For ease of explanation, it is assumed that the bond-slip properties are bilinear, as shown in Figure 11.4a, such that the maximum shear stress τ_{max} occurs at a slip δ_1, and beyond a slip of δ_{max}, there is zero shear stress, indicating that debonding has occurred.

The shooting method of analysis, as explained with the help of Figure 11.1b, will start on the *RHS* of the half-segment in Figure 11.4b and move to the left to reach the known boundary condition of zero slip s at the mid-segment. Let O–D–C_2 in Figure 11.4c represent the slip distribution when the slip at the crack face is δ_{max}, meaning that Δ_{rt} in Figure 11.4b is equal to δ_{max}. For this distribution of slip, and using the bilinear bond-slip properties in Figure 11.4a, the variation in shear stress O–D–C in Figure 11.4d can be constructed. Be reminded that the area under O–D–C times the bond interface area is the force P in the reinforcement. Hence, the area under O–D–C is proportional to P. This force is invariably limited by the short half-crack spacing $S_{cr}/2$. Let us now increase Δ_{rt} in Figure 11.4b to obtain the slip distribution O–A–B–C_1 in Figure 11.4c. As the slip between B and C_1 is greater than δ_{max}, local debonding has occurred, so there is only shear across O–A–B. This results in the shear stress distribution O–A–B–C in Figure 11.4d, where in B–C the shear stress is zero; that is, it is unbonded over the length shown as $L_{db\text{-}l}$. It can also be seen that the area under O–A–B is less than that under O–D–C, indicating that the bond force B has reduced.

The reinforcement forces across a two-crack segment are shown in Figure 11.4b, where the force at the reinforcement ends is P and the bond force is B. Hence, at the mid-segment, the reinforcement force is $P-B$, and the force in the concrete is B; the force in the concrete B is a measure of the benefits of tension stiffening, which reduces the deflection of the beam. The reduction in the bond force has no effect on the

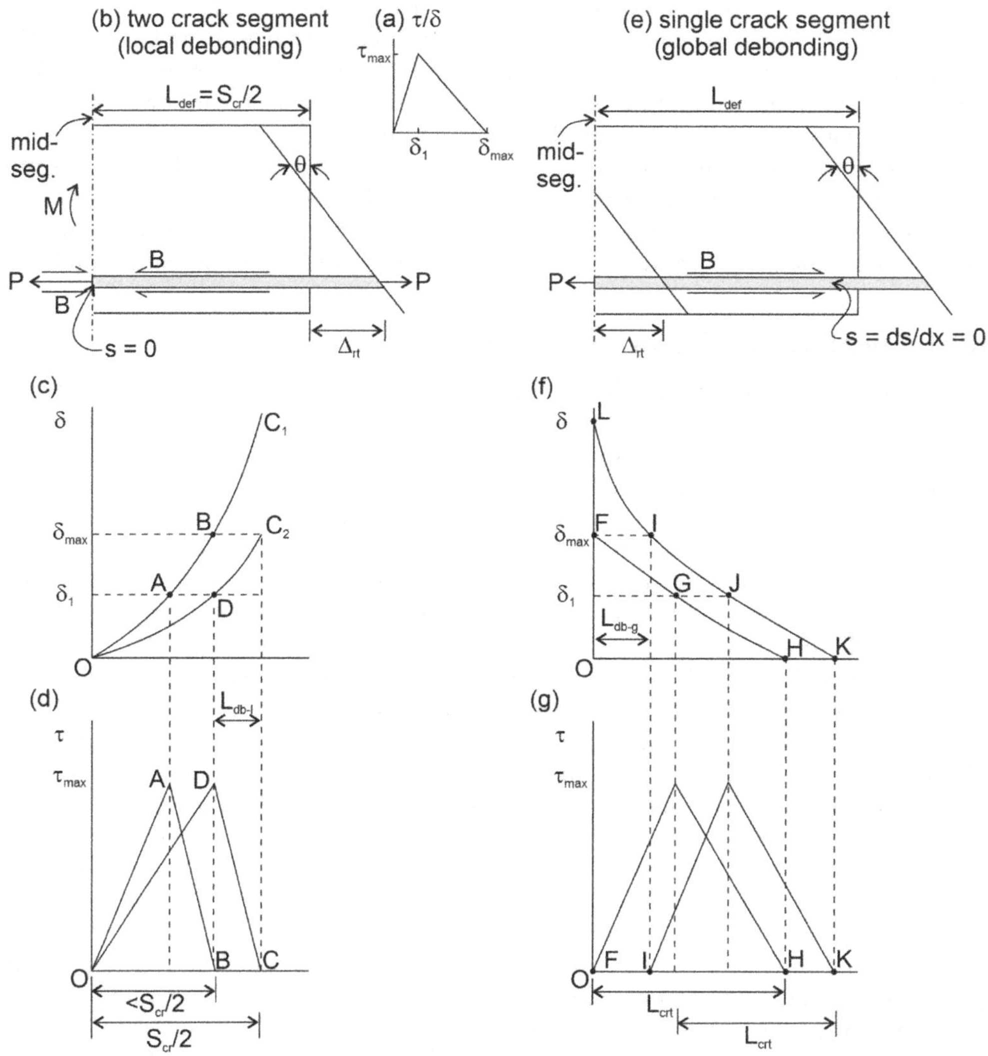

Figure 11.4 Local and global debonding in a segment (a) bilinear bond properties; (b) local debonding in a two crack segment; (c) slip distribution for (b); (d) bond stress distribution for (b); (e) global debonding for single crack segment; (f) slip distribution for (e); (g) bond stress distribution for (e).

reinforcement force P but reduces the benefits of tension stiffening. Very importantly, local debonding does not reduce the strength of the reinforcement, making it conceivable that $P>P_{IC}$.

11.2.2 Global debonding in a segment

Now consider debonding in a segment with a single crack as shown in Figure 11.4e, where L_{def} is long enough to allow the full bond strength to be developed. The shooting analysis in Figure 11.1b is now applied from the *LHS* of Figure 11.4e, and the boundary condition used on the *RHS* is where there is full interaction; that is, where both the slip *s* and the slip–strain *ds/dx* are zero at the same point. Let the slip distribution *F–G–H* in Figure 11.4f occur when the shear stress distribution is fully developed; that is, when the bond-slip τ/δ in Figure 11.4a is fully developed without any restriction along the length of the reinforcement. This length is referred to as the critical length L_{crt}, and the reinforcement force at which this occurs is referred to as the IC debonding resistance P_{IC}, which is equal to the maximum value of *B* and occurs at a slip at the crack face of Δ_{IC}; let the reinforcement strain at which P_{IC} occurs be ε_{IC}. Should Δ_{rt} be increased above Δ_{IC}, then this simply shifts the bond distribution from *F–G–H* to *I–J–L* in Figure 11.4g, the shape remaining the same. This transition from *F* to *I* is the global debonded length $L_{db\text{-}g}$, where $\Delta_{rt} - \Delta_{IC}$ is equal to $L_{db\text{-}g}\varepsilon_{IC}$. In Chapter 4, P_{IC} and L_{crt} were derived from an axial tension stiffening analysis. This is an alternative approach using a flexural analysis.

In global debonding, as in Figure 11.4e, the force in the bar *P* depends solely on the bond force *B* and is limited by the bond strength, which is in contrast to that in Figure 11.4b, where the force in the bar is not limited by the bond. This is the major conceptual difference between local debonding and global debonding. There is a maximum value P_{IC} that depends on the bond-slip τ/δ characteristics, which requires a minimum length of reinforcement L_{crt} to achieve. Once reached at Δ_{IC}, any further increase in the crack width is accommodated by global debonding, where the bond distribution moves as a rigid body.

Note that we are using the bond-slip characteristics shown in Figure 11.4a, where the shear stress $\tau$is zero when the slip is greater than δ_{max}. Often, after the slip δ_{max}, there is assumed to be a friction component of τ that is constant. Hence, should there be a frictional component – which is unlikely to occur in externally plated beams but does occur in internally encased reinforcement – then this analysis is a lower bound; however, friction can easily be included in a numerical analysis.

11.2.3 Transition from local to global debonding in a beam

11.2.3.1 Crack closure numerical model

The propagation of global debonding, as illustrated in Figure 11.4e–g, depends on an uncracked segment on both sides of the single crack, which rarely occurs naturally in beams where there are usually multiple cracks. For global debonding to occur, crack closure is required. Consider a segment of a beam between two flexural cracks at a spacing S_{cr} as shown in Figure 11.5b. The force in the reinforcement on the *LHS* is P_{LHS}, and the reinforcement has slipped relative to the crack face Δ_{LHS}. For the *RHS*, we have P_{RHS} and Δ_{RHS}. Flexural crack closure depends on a moment gradient, as in Figure 11.5a. The segment in Figure 11.5b can be considered to consist of disturbed regions *D* adjacent to the cracks, where linear strain profiles cannot be used by themselves, and an undisturbed region *UD* between these cracks, where linear strain profiles can be used. To find a solution to this shooting problem, we need to define sufficient boundary conditions at the ends of the segment that are in the disturbed region.

In these analyses, we are dealing with a moment gradient, as in Figure 11.5a. Even though the segmental analysis in Chapter 6 is based on a constant moment region, it is suggested that it would provide a very good approximation of the force in the reinforcing bar at a crack that is both P_{LHS} and P_{RHS} which can then be used as boundary conditions.

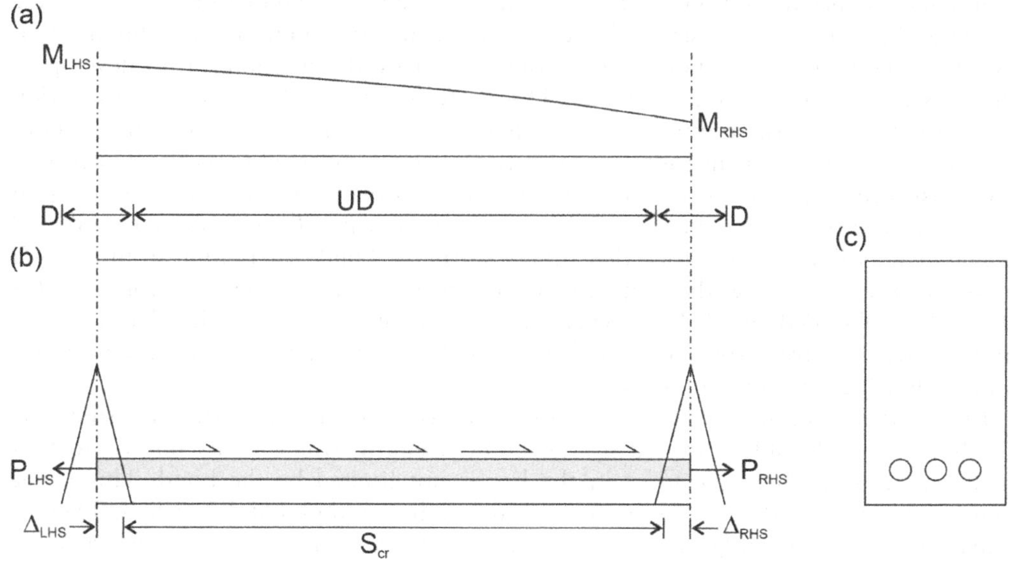

Figure 11.5 Segmental analysis with varying moment (a) moment gradient along segment; (b) segment elevation; (c) segment cross section.

At the time this research was conducted, the segmental analysis was not available, so the classic rectangular stress blocks were used. This was more than sufficient to illustrate the mechanism of crack closure but probably not sufficient to quantify it (Oehlers et al. 2005; Liu et al. 2007).

In contrast to the disturbed regions in Figure 11.5b, the undisturbed region analysis follows the standard analysis as shown in Figure 11.1b, where P_1 is equal to P_{LHS} in Figure 11.5b, and where the boundary condition to aim for is the force in the reinforcement on the right-hand side $P_{RHS.}$ Since P_1 in Figure 11.1b is now known as P_{RHS}, the slip Δ_1 is estimated and the analysis proceeds to the *RHS*, where the force is P_{RHS}. If the first estimate of Δ_1 does not achieve P_{RHS}, then it is varied until it does.

11.2.3.2 Transition from local to global debonding

To illustrate the crack closure mechanism (Liu et al. 2007; Oehlers et al. 2005), a segment of a member with four cracks at a spacing S_{cr} was analysed, although only the slips between Cracks 1 and 3 are shown in Figure 11.6. The member is subjected to a moment gradient, where the maximum moment occurs at Crack 1.

Applying the load to the member in Figure 11.6 causes crack widening. The initial distributions of slips, *A–B–C* and *D–E–F*, are close to what would be predicted from the two crack segment analysis in Figure 11.4e; the crack width at Crack 2 is *C–D*. With further increases in load and the further opening of Crack 1, the slip distribution *A–B–C* becomes non-symmetrical as in *G–H*, tending towards that for the single crack analysis in Figure 11.4f; the width of Crack 2 has now reduced to *H–I*. A further increase in load, which causes an increase in the width of Crack 1, results in the slip distribution moving to *K–L–M*, where Crack 2 has now closed. An additional increase in the applied load and in the width of Crack 1 leads to the slip distribution *N–P* when Crack 3 closes.

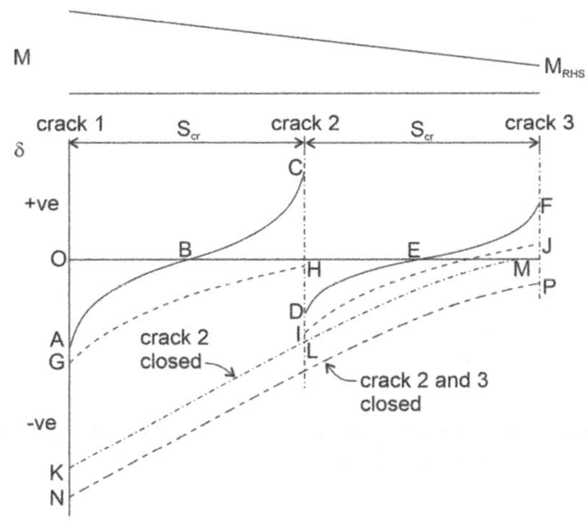

Figure 11.6 Crack closure.

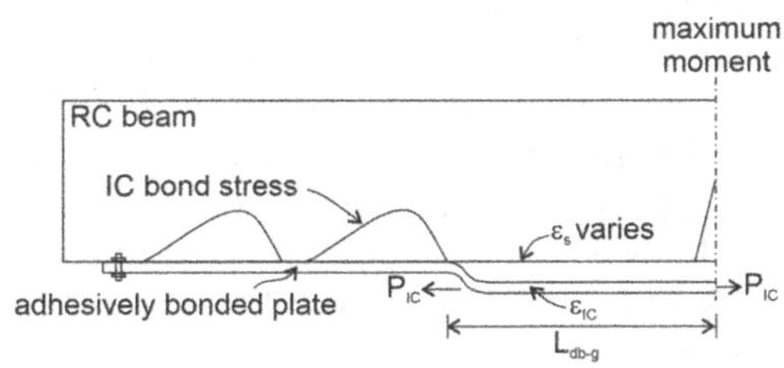

Figure 11.7 Global debonding of plated beam.

Thus, there is a mechanism by which cracks can close and *IC* debonding can occur along the beam. These numerical analyses showed that the force in the reinforcement can initially exceed P_{IC} because of the two crack mechanism in Figure 11.4b, which allows, in theory, unlimited forces in the bar. However, there is then a rapid, unstable reduction in the bar force due to the single crack debonding mechanism in Figure 11.4e and the crack closure mechanism in Figure 11.6, leading to a stable *IC* debonding force P_{IC}. Hence, P_{IC} should be used in analyses, as it would ensure ductile rather than brittle behaviour.

11.3 GLOBAL DEBONDING OF PLATED MEMBER

11.3.1 Plated member

It is now common practice to adhesively bond plates, such as *FRP* or steel plates, to the soffits of *RC* members, as shown in Figure 11.7. The beam is subjected to a moment gradient, being at a maximum on the *RHS*. The plate can debond prematurely from its *LHS* end, and this can be prevented by terminating the plate in a low moment region or simply bolting the plate end, as shown. The plate can also debond from the position of maximum moment on the *RHS* when the force in the plate reaches the IC

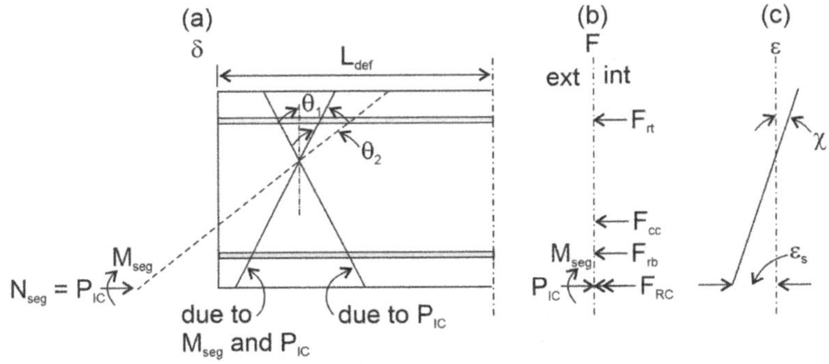

Figure 11.8 *IC* debonded post-tensioned segmental analysis (a) segment elevation and displacement profile; (b) force profile; (c) strain profile.

debonding resistance P_{IC} at a strain ε_{IC}. Let us assume that the plate has debonded the distance $L_{db\text{-}g}$. No matter what this length is, through the mechanics of *IC* debonding, as explained using Figures 11.4e–g, the force in the plate remains at P_{IC} and the strain at ε_{IC}. Therefore, in Figure 11.7, the extension of the debonded plate Δ_p is $\varepsilon_{IC}L_{db\text{-}g}$.

Consider the slip distribution *L–I–J–K* in Figure 11.4f in which global debonding has occurred, as shown in the shear stress distribution *I–J–K* in Figure 11.4g, where *O–I* is the global debonded length $L_{db\text{-}g}$. At the crack face at Point *L* in Figure 11.4f, the reinforcement slip is δ_{max} plus $\varepsilon_{IC}L_{db\text{-}g}$, whereas at Point *I* at the end of the debonded region, the reinforcement slip is δ_{max}. Hence, the increase in length of the reinforcement over the debonded region is $\varepsilon_{IC}L_{db\text{-}g}$. For compatibility, the increase in length of the slab adjacent to the reinforcement must be the same.

Following the approach in Oehlers et al. (2016a,b) over the debonded length $L_{db\text{-}g}$ in Figure 11.7, the debonded plate acts as a post-tensioned tendon. However, unlike the post-tensioned tendon in Chapter 9, where the force in the tendon varies with flexure, the axial force in the plate remains at P_{IC}, which is characteristic of the *IC* debonding mechanism. Hence, the segmental analysis is that shown in Figure 11.8a, where N_{seg} remains constant at P_{IC}, in contrast to the post-tensioning mechanism in Section 9.1, where it varies with flexure. Thus, the standard M/θ segmental analysis can be applied to derive the M/χ response. Note that after global debonding has occurred, which requires crack closure, cracks can then open up again. The analysis in Figure 11.8 also derives the strain in the *RC* beam adjacent to the plate, ε_S. The integration of this variation of strain over the length $L_{db\text{-}g}$ in Figure 11.7 gives the extension of the bottom fibre of the beam Δ_b adjacent to the plate over the debonded length $L_{db\text{-}g.}$

Let us now fix the debonded length $L_{db\text{-}g}$ in Figure 11.7 to determine what is required to achieve this debonded length. For a given fixed moment gradient, we now apply the segmental analysis in Figure 11.8 to determine the variation in ε_s over the debonded length. Integration of this strain variation over the debonded length gives the extension of the beam soffit Δ_b. The extension of the plate Δ_p over the debonded length has already been shown to be $\varepsilon_{IC}L_{db\text{-}g}$.

Bearing in mind that the strain in the plate ε_{IC} is independent of the applied load, that is, the applied moment distribution. If the extension of the beam Δ_b is less than the extension of the plate Δ_p over the debonded length $L_{db\text{-}g}$ in Figure 11.7, then the applied load on the beam would have to be increased for the strain in the beam to reach that in the plate. Hence, debonding up to the distance $L_{db\text{-}g}$ is stable; that is, it would gradually progress with increasing applied load. Alternatively, if the extension of the beam Δ_b is greater than that in the plate Δ_p, then debonding is unstable as it would require a reduction in the applied load to achieve compatibility. In this case, at the given applied load, there would be a step change in the debonded length. This is not unusual and is akin to loading an initially uncracked *RC* beam, where on cracking there is a step change in the deflection to accommodate the change in flexural stiffness due to cracking. It is a question of increasing the debonded length until there is compatibility; that is, Δ_p is equal to Δ_b. This procedure can continue with increased debonded length to simulate debonding along the length of the beam.

11.3.2 Prestressed plate beam

The debonding mechanism of the prestressed plate in Figure 11.9 is identical to that in an un-prestressed plate as shown in Figure 11.7. However, allowance has to be made for the prestressing force, as illustrated in Figure 11.9 from Oehlers et al. (2016a,b). The prestressing force after the release of the prestressing anchors $N_{seg\text{-}pre}$ has been quantified in Chapter 9 using Figures 9.5 and 9.6. The *IC* debonding resistance P_{IC} depends mainly on the bond-slip characteristic and, as such, is independent of any residual stresses due to prestress; that is, it is the same as that for the unbonded plate. Hence, the force in the plate to develop the *IC* shear stress distribution τ_{IC} over the length L_{crit} when the crack width is Δ_{IC} is now $N_{seg\text{-}pre}$ plus P_{IC}, as shown in Figure 11.9a.

Increasing the crack width in Figure 11.9b will cause debonding over the length $L_{db\text{-}g}$ such that the increase in the crack width is $\varepsilon_{IC}L_{db\text{-}g}$. The properties of the bonded pre-tensioned region can be obtained from Chapter 9 using Figures 9.5 and 9.6. The debonded region now acts as a post-tensioned member where the properties can be derived from Figure 11.8, but with P_{IC} replaced with $P_{IC}+N_{seg\text{-}pre}$. As with the un-prestressed beam in Figure 11.7, the debonding path in Figure 11.9b can be followed by

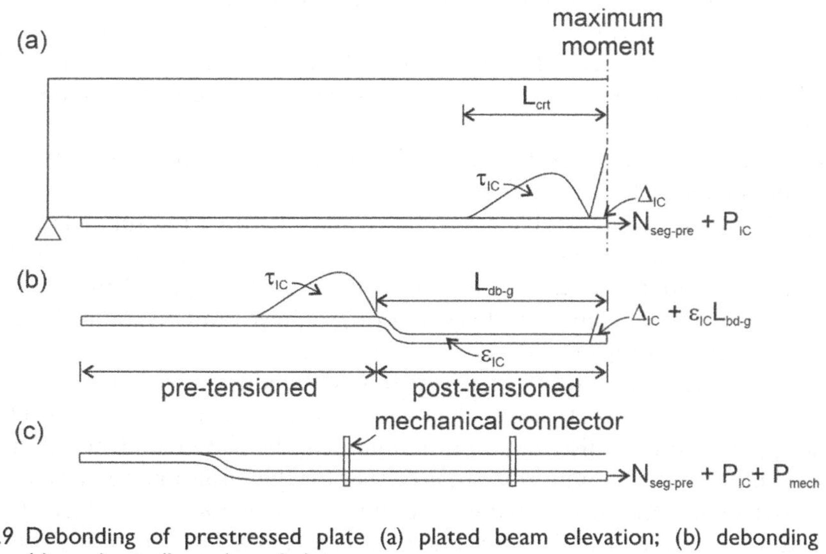

Figure 11.9 Debonding of prestressed plate (a) plated beam elevation; (b) debonding of plate; (c) mechanically anchored plate.

ensuring compatibility between the length of the debonded plate and the length of the soffit of the beam adjacent to the debonded plate.

It is also worth noting that if mechanical connectors are used, as shown in Figure 11.9c, then the force required to continue IC debonding would have to be increased by the resistance of all the mechanical connectors in the debonded region, denoted as P_{mech}. Bear in mind that the force on the connector depends on the interface slip, which can be extracted from the PI analyses of global debonding (Oehlers et al. 2016b). An example of slip during the global debonding of a segment is shown in Figure 11.4f, and it is this slip that induces forces in the mechanical shear connectors.

11.4 SUMMARY OF SIMPLE DEBONDING ANALYSIS

The analyses required to quantify IC debonding are summarised as follows:

1. From a tension stiffening analysis in Chapter 4, derive P_{IC}, which will give the IC debonding strain ε_{IC}. This can be derived either numerically or using closed-form solutions.
2. From the standard RC segmental analysis in Chapter 6, derive the $(M/\chi)_{unb}$ curve for the unbonded segment so that we have the properties of the unbonded region when required, such as in determining the deflection of the beam. Also, from this analysis, derive the external moment M_{ext} when the force in the plate is P_{IC}, which denotes the onset of debonding. This will give the distribution of the moment along the beam at the onset of debonding.
3. From Chapter 9, derive $(M/\chi)_{deb}$ for the post-tensioned analysis where the tendon force is constant at P_{IC}. This analysis will give the variation of soffit strain ε_s with moment.
4. Now assume the plate has debonded over a short length $L_{db\text{-}g}$.
5. The extension of the plate in the debonded region Δ_{pl} is $\varepsilon_{IC} L_{db\text{-}g}$, which remains constant regardless of the applied moment distribution.
6. The distribution of the strain in the soffit of the beam at the interface over the debonded region ε_s was determined from the analysis for determining $(M/\chi)_{deb}$. Integration of this strain over the debonded region will give the extension of the soffit of the beam Δ_{RC}.
7. It is now necessary to check for compatibility between the beam and the plate over the debonded length $L_{db\text{-}g}$.
8. If $\Delta_{RC} < \Delta_{pl}$, then we have stable debonding. This is because the applied moment distribution would have to be increased to achieve the boundary condition $\Delta_{RC} = \Delta_{pl}$. It should be noted that Δ_{RC} increases with increasing applied moment, but Δ_{pl} remains unchanged. Hence, in this situation, we have gradual stable debonding as the applied moment is increased.
9. In contrast, if $\Delta_{RC} > \Delta_{pl}$, then we have unstable debonding, as the only way of finding a solution for compatibility is to reduce the applied moment distribution. Therefore, if the applied moment distribution is maintained, the plate will debond until stable debonding occurs. It is important to note that for this situation, there is no solution while the applied moment is maintained until stable debonding is reached.
10. The above analyses can be continued by increasing $L_{db\text{-}g}$ to follow debonding along the beam.

11.5 NOTATION AND ABBREVIATIONS

B	bond force
$comp$	compression
D	deformed region
d_s	depth to neutral axis in slab
ds/dx	slip–strain
ext	external stress resultants
F	force profile
F_{cc}	resultant compressive force in concrete
F_{ct}	resultant tensile force in concrete
F_{RC}	resultant of the internal forces
F_{rb}	resultant force in bottom reinforcement
F_{rt}	resultant force in tension reinforcement
h_s	depth of concrete component
IC	intermediate crack
int	internal forces
L	span of beam
LHS	left-hand side
L_{crt}	minimum length to develop P_{IC}
$L_{db\text{-}g}$	global debonded length
$L_{db\text{-}l}$	local debonded length
L_{def}	deformation length; half-length of segment
L_e	element length
M	moment
M_{ext}	externally applied moment
M_{LHS}	moment on the LHS
M_p	moment in plate
M_{RHS}	moment on the RHS
M_s	internal moment in slab
M_{seg}	moment applied to segment
N_{seg}	axial force acting on segment
$N_{seg\text{-}pre}$	axial force in prestressed tendon after release
P	axial force
P_{IC}	IC debonding resistance
P_{LHS}	axial force on the LHS
P_{mech}	shear force on mechanical shear connectors in debonded region
P_{RHS}	axial force on the RHS
RC	reinforced concrete
RHS	right hand side
S_{cr}	crack spacing
s	slip
ten	tension
UD	undeformed region
χ	curvature; curvature profile
Δ	interface slip at end; interface slip at crack face; element slip
Δ_{IC}	slip at crack face to first develop P_{IC}

Δ_{LHS}	reinforcement slip at LHS
Δ_{pl}	extension of plate over debonded region
Δ_{RC}	extension of soffit of RC member over debonded region
Δ_{RHS}	reinforcement slip at RHS
Δ_{rt}	tension reinforcement slip at crack face
δ	deformation; interface slip
δ_{max}	slip at zero shear
δ_1	slip at τ_{max}
ε	strain; strain profile
ε_b	strain in plate
ε_{IC}	strain in reinforcement when subjected to P_{IC}
ε_s	strain in slab at interface
θ	Euler–Bernoulli rotation
τ	shear stress
τ_{IC}	IC debonding shear stress distribution
τ_{max}	maximum shear stress

REFERENCES AND DEVELOPMENT

Liu, I.S.T., Oehlers, D.J. and Seracino, R., 2007. Study of intermediate crack debonding in adhesively plated beams. *Journal of Composites for Construction*, *11*(2), 175–183.

Oehlers, D.J. and Bradford, M.A., 1995. *Composite Steel and Concrete Structural Members: Fundamental Behaviour*. Oxford: Pergamon.

Oehlers, D.J., Liu, I.S.T. and Seracino, R., 2005. The gradual formation of hinges throughout reinforced concrete beams. *Mechanics Based Design of Structures and Machines*, *33*, 373–398.

Oehlers, D.J., Visintin, P. and Lucas, W., 2016a. Flexural strength and ductility of FRP-plated RC beams: fundamental mechanics of incorporating local and global IC debonding. *Journal of Composites for Construction*, *20*(2), 04015046.

Oehlers, D.J., Visintin, P. and Lucas, W., 2016b. Fundamental mechanics governing FRP-plated RC beams with anchored and prestressed FRP plates. *Journal of Composites for Construction*, *20*(6), 04016047.

Part 6

Long-term and sustainable design

Having now explained all facets of RC member behaviour, it is an appropriate time to determine the effects of material deterioration over time and the closely related topic of design for environmental sustainability. It will be shown that this is not a daunting task, as once the deterioration of the particle quantities in Part 2 has been quantified, they can be easily input into the elemental and segmental behaviours to quantify their effect on member behaviour. Corrosion is first dealt with in Chapter 12, where it is shown that the application of the segmental mechanics can simulate the progressive damage of a structure due to reinforcement corrosion, as well as how this understanding can be used to estimate the remaining life of a structure.

Finally, in Chapter 13, we examine environmentally sustainable design from a structural engineering perspective. Importantly, in this chapter, we show how a fundamental understanding of the mechanics of reinforced concrete greatly assists structural engineers in designing sustainable solutions. That is, the mechanics developed in the preceding chapters provide engineers with the tools required to assess existing structures and design approaches to repair and strengthen them. Alternatively, if it is necessary to construct a new structure, we have shown how it is possible to utilise new types of concrete or reinforcement that can either directly or indirectly minimise the whole-of-life environmental impact. As part of this chapter, we also explain the basic principles of life cycle assessment and the major factors that the structural engineer may need to consider to ensure that different design alternatives can be reliably compared.

DOI: 10.1201/9781003711407-17

Chapter 12

Impact of corrosion on RC member behaviour

Up until this point, this book has largely focused on quantifying the mechanical properties of reinforcement and concrete, as well as the behaviour of reinforced concrete elements when subjected to instantaneous or sustained loads. In this discussion, we have assumed that the structure is designed to have sufficient strength and stiffness so that it will remain serviceable over some period of time termed the 'service life'.

In the absence of an ultimate limit state event such as an earthquake that makes the structure uninhabitable or a development decision to demolish a structure, service life is typically defined by damage to the reinforcement caused by corrosion. To explain why, consider the reinforcing bar embedded in concrete with cover *c* shown in Figure 12.1a. Concrete is a naturally alkaline material (pH 12–13) and hence when steel is placed within the concrete, a reaction takes place that forms a passive corrosion-resistant layer

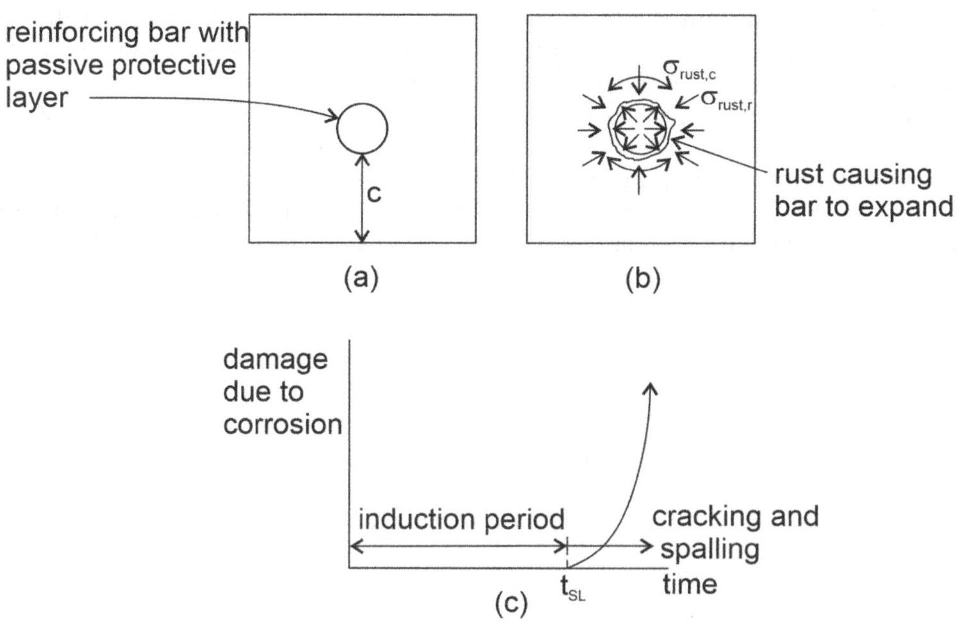

Figure 12.1 Link between reinforcement and reinforced concrete service life (a) tension-stiffening cross section; (b) impact of reinforcement corrosion; (c) link between corrosion damage and time.

DOI: 10.1201/9781003711407-18

on the surface of the bar. Thus, reinforcement embedded within concrete has some natural resistance to corrosion.

Importantly, this passive layer is not permanent, and it exists only while the concrete surrounding the reinforcement remains alkaline. Hence, the induction period in Figure 12.1c is defined as the time taken for the passive layer to be destroyed, which is commonly done by the penetration of carbon dioxide or chloride ions through the concrete to the reinforcement, as will be described in Sections 12.1.1 and 12.1.2. Once the passive layer surrounding the reinforcement is lost, damage due to corrosion in Figure 12.1c is rapid. It is, therefore, common to take the end of the induction period as a conservative estimate of the service life t_{SL}, although it is also possible to be less conservative and define the service life using another indicator, such as the initiation of corrosion cracking or spalling.

The impact of corrosion of reinforced concrete is significant. The corrosion of reinforcement converts steel to rust, and because rust has very little strength or stiffness, corrosion can be thought of in terms of a change in the bar diameter. The loss of steel, therefore, has a significant impact on the strength and ductility of the reinforcement. The conversion of steel to rust also affects the surface of the reinforcing bar, as shown in Figure 12.2, and as a consequence, a significant change in the bond between the reinforcement and concrete may occur. Furthermore, the products of corrosion are expansive; that is, they have a higher volume than that of the original steel, and as a result, corrosion causes the reinforcing bar to expand, creating radial compression $\sigma_{rust.r}$ and circumferential tensile stresses $\sigma_{rust.c}$, as illustrated in Figure 1.5 in Section 1.3.2. Hence, a small amount of corrosion can be considered beneficial as it creates a confining effect. However, as corrosion continues, the circumferential stresses exceed the concrete cracking stress, leading to splitting cracks or spalling along the length of the bar, as shown in Figure 12.1b.

In this chapter, the role of carbonation and chloride ion ingress in defining the induction period in Figure 12.1c will first be discussed. The impact of corrosion on the strain and slip-based material properties will be described in more detail in Sections 12.2.1 and 12.2.2 before incorporating them into tension stiffening and segmental analyses to show the impact at the member level in Sections 12.3 and 12.4, respectively.

Figure 12.2 Example of loss of surface roughness as bar corrodes. (Photo courtesy of Mark Peacock.)

12.1 IMPACT OF CONCRETE ON SERVICE LIFE

12.1.1 Chloride penetration

Chlorides present in seawater and de-icing salts are a common cause of durability issues. Although the chlorides have little impact on the concrete itself, they reduce the alkalinity of the concrete, resulting in the destruction of the passive corrosion-resistant layer on the steel surface. Hence, in Figure 12.1c, as a lower bound, the service life t_{SL} could be taken as the induction period required for chlorides to travel through the concrete cover and reach a critical concentration at the surface of the reinforcement. It is for this reason that in near-coastal environments or where chloride ions are present, national codes of practice require an increased concrete cover to be specified.

The transport of chlorides through concrete occurs via a combination of capillary suction, which transports ions through the pore water, and ion diffusion, which causes the difference in ion concentration within the pore water solution (FIB 2013; Li et al. 2019). Chloride penetration is generally modelled based on the application of Fick's second law of diffusion.

$$\frac{\partial C}{\partial t}=\frac{\partial}{\partial x}\left(D\frac{\partial c}{\partial x}\right)-s \tag{12.1}$$

which describes the distribution of the concentration of chloride ions C in space x and time t as a function of the diffusion coefficient D. The sink term s in Eq. 12.1 accounts for the fact that chlorides may react with the cement and become chemically bound, thereby removing them from the diffusion process. Conversely, in the presence of other reactants, it is possible that chlorides previously bound to the cement may be released; in this case, the sink term is taken as a negative number, as it acts to enhance the rate of diffusion.

Given the challenge in modelling the coupled transport and chemical reaction process, in Eq. 12.1 it is common to ignore the sink term s and instead replace the diffusion coefficient D with an effective diffusion coefficient D_{eff} which changes in space and time and includes any changes in diffusion rate due to chemical reactions (FIB 2013; Shafikhani and Chidiac 2019). Equation 12.1 can then be solved using the effective diffusion coefficient and the mathematical error function *erf* to give the spatial distribution of chlorides at any given time t as follows:

$$c(x,t)=c_s\left[1-erf\left(\frac{x}{2\sqrt{D_{eff}t}}\right)\right] \tag{12.2}$$

where

$$erf(z)=\frac{2}{\sqrt{\pi}}\int_0^z e^{-t^2}dt \tag{12.3}$$

When applying Eq. 12.2, it is important to remember that the diffusion process depends on concrete humidity, temperature, and the concentration of chloride ions, and these are all assumed to be constant. Hence, the time period over which Eq. 12.2 yields accurate predictions must be considered. Cyclic wetting and drying of concrete can also significantly increase the rate of damage, since during periods of wetting, the ion concentration near the surface of the concrete is elevated, creating a significant variation in ion concentration and pore humidity. This causes ions to be drawn into the concrete, where ion concentration is lower, via diffusion and pore suction. It is for this reason that structures located in tidal zones, e.g. jetties, are particularly susceptible to corrosion. In these scenarios, it is therefore common to have large concrete covers or to use more corrosion-resistant reinforcement, such as galvanised or epoxy-coated steel or glass fibre reinforced polymers.

According to Khan et al. (2017), Eq. 12.2 can be further simplified by replacing the error function with a simple parabolic distribution to yield

$$c(x,t) = c_s\left[1 - \frac{x}{2\sqrt{3D_{eff}t}}\right]^2 \tag{12.4}$$

Equation 12.4 can then be solved by setting x to the reinforcement cover c and solving for the time t_{SL} to reach a critical chloride concentration c_{crit} at the level of the reinforcement if the surface chloride concentration is c_s.

$$t_{SL} = \frac{1}{12D_{eff}}\left[\frac{c}{1 - \sqrt{\dfrac{c_{crit}}{c_s}}}\right]^2 \tag{12.5}$$

According to (FIB 2013), the critical chloride concentration can be taken as between 0.4% and 0.6% of the weight of cement in the concrete or 0.05% of the weight of the concrete. The effective diffusion coefficient in units of m^2/s has also been estimated by (FIB 2013) to vary with the mean concrete compressive strength f_{cm} and is different for concretes with and without supplementary cementitious materials. The reader is referred to the source material (FIB 2013) for a specific material model.

Worked Example WE-12.1: Chloride ingress

As an example, let us use the simplified method in Eq. 12.5 to determine the service life for the beam cross-section in Figure E12.1. Assuming a surface chloride concentration at all faces of 5%, the concrete has a mean compressive strength of 40 MPa, a critical chloride concentration of 2.4% and an effective diffusion coefficient of 1.98×10^{-11}.

Taking the minimum cover distance in the cross-section to be 30 mm, the service life using Eq. 12.5 is

$$t_{SL} = \frac{1}{12 \times 1.98 \times 10^{-11}}\left[\frac{30/1000}{1 - \sqrt{\dfrac{2.4}{5}}}\right]^2 = 1.27 \text{ years} \tag{E12.1.1}$$

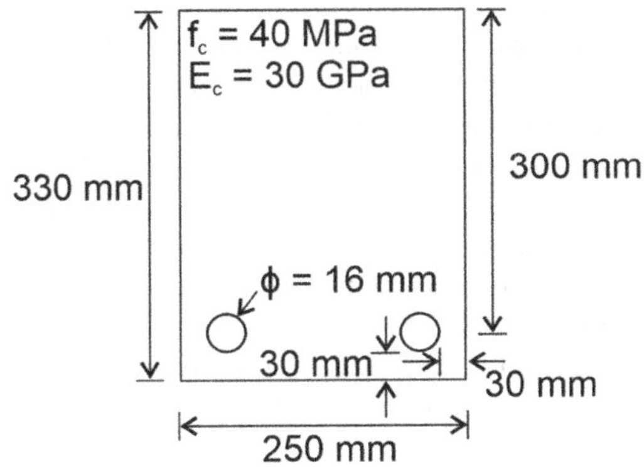

Figure E12.1 Cross section for Example 12.1.

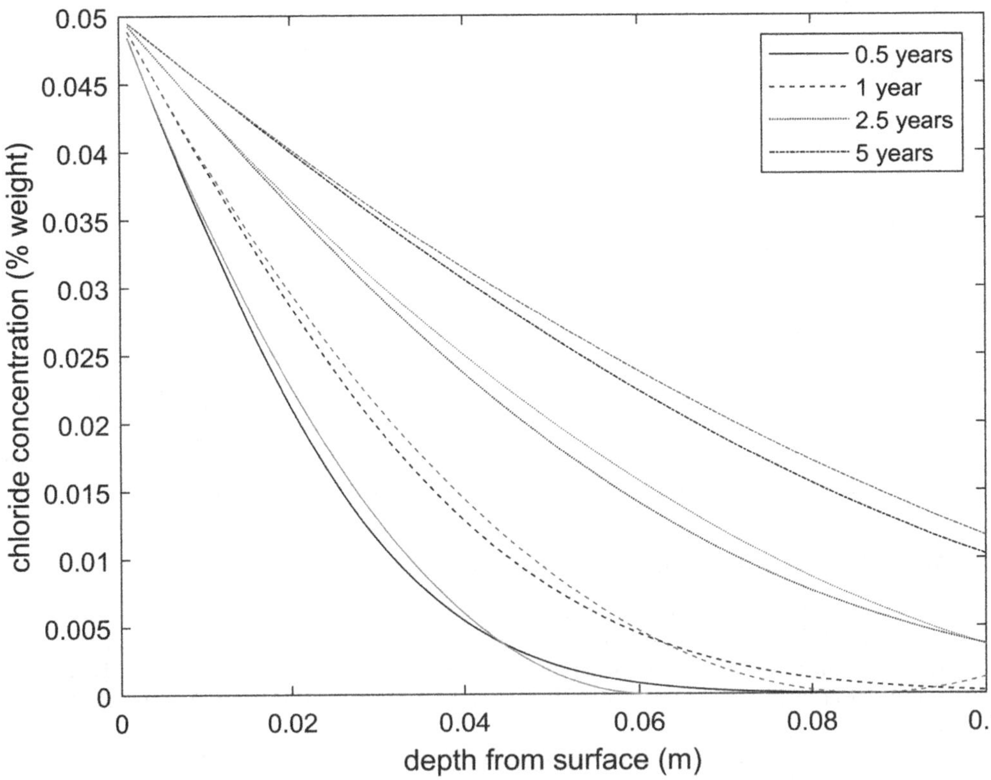

Figure E12.2 Variation in chloride concentration with depth from exposed surface.

As a further example, the variation of chloride concentration throughout the first 100 mm of concrete cover is shown in Figure E12.2, using both Fick's Second Law from Eq. 12.2 as solid lines and the quadratic approximation from Eq. 12.4 as dashed lines.

From this comparison, it can be seen that the analytical solution to Fick's second law is well approximated by the quadratic in Eq. 12.4. It is further observed that the distribution of chlorides varies significantly with time but rapidly stabilises to a relatively linear distribution in space. That is, after 2.5 years, a linear approximation of the variation in chloride concentration with depth would be sufficient.

It is also important to note here that despite extensive testing to measure the chloride diffusion coefficient, fitting a reliable model has been challenging, and models show a very wide scatter from test results. The challenge in fitting models arises from the mechanism of chloride transport; that is, models must capture the impact of concrete porosity at multiple scales and how this changes with ongoing hydration and pozzolanic reactions of the binders, as well as the ongoing binding or release of chlorides from the cement structure (Shafikhani and Chidiac 2019; Razmi et al. 2025).

12.1.2 Carbonation

The carbonation of concrete has a very similar impact to that of chloride ingress in that it leads to the acidification of the concrete pore solution, thereby destroying the passive layer. Carbonation occurs when carbon dioxide from the atmosphere diffuses through the pore space and reacts with calcium hydroxides in the hardened cement to produce calcium carbonates, which lower the concrete pH.

The rate of carbonation is dependent on the diffusion rate of CO_2 through the concrete, which can be modelled using Fick's first law:

$$J = -D_{CO_2}\left(\frac{\partial c}{\partial x}\right)_t \tag{12.6}$$

where at time t, J is the CO_2 flux (the amount of CO_2 that will flow through a unit area in a time interval t), D is the CO_2 gas diffusion coefficient, c is the concentration of CO_2 in the atmosphere and x is the depth of penetration.

Equation 12.8 can be solved assuming steady-state conditions to give the depth of concrete that has undergone carbonation x at any given time t as

$$x^2 = 2D_{CO_2}\frac{c_a}{c_c}t \tag{12.7}$$

in which C_a is the atmospheric concentration of CO_2; C_c is the concentration of CO_2 required to carbonate all hydration products in the area of interest and D_{CO_2} is the CO_2 diffusion coefficient.

Models for the ratio of $\frac{c_a}{c_c}$ for non-carbonated and carbonated concrete have been derived, and those most accepted are presented in FIB (2013). It must, however, be emphasised that this form of modelling is approximate. That is, it is based on assumed conditions, including CO_2 concentration, temperature and humidity, all held constant, which is obviously rarely the case in practice. This is particularly important for concrete that may undergo cycles of wetting and drying because the ingress of CO_2 as a gas is halted when the concrete pore space is saturated. Models also generally do not account for the variation in diffusion that may occur as a result of the inclusion of supplementary cementitious materials, which are known experimentally to impact the rate of carbonation (FIB 2013).

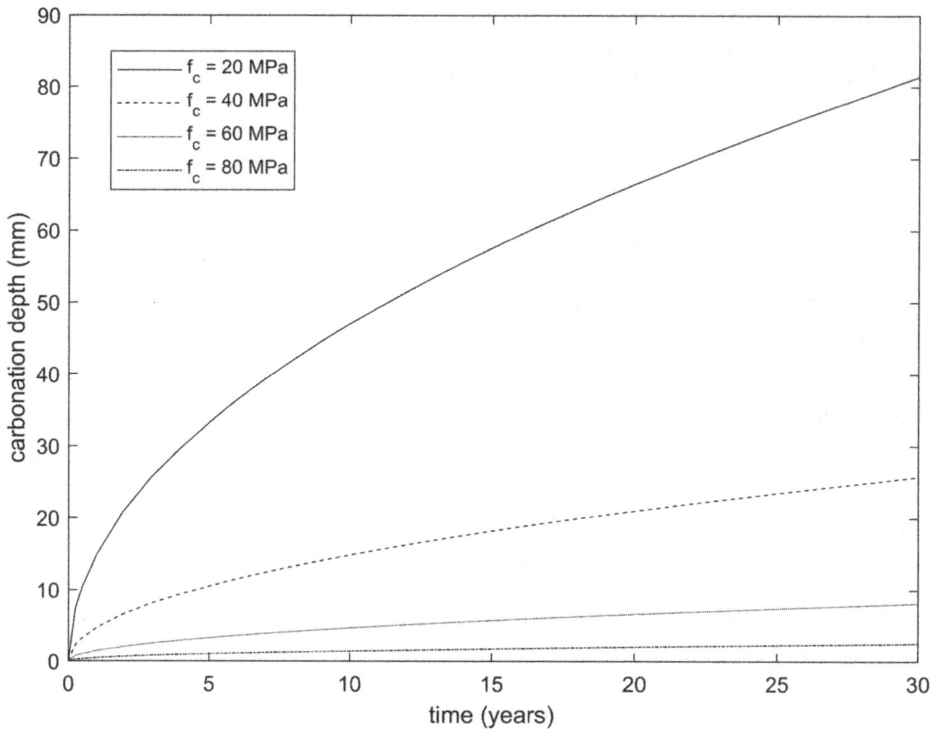

Figure E12.3 Variation in carbonation depth with time and concrete compressive strength.

As a further example of the rate of carbonation, Eq. 12.7 has been solved in Figure E12.3 for concrete compressive strengths ranging from 20 to 80 MPa, taking a typical $\frac{c_a}{c_c}$ value of 8×10^{-6}.

Figure E12.3 shows the significant impact of concrete compressive strength on carbonation depth and, therefore, illustrates the substantial benefit to durability that arises from using higher concrete compressive strengths. For example, if 20 mm of cover is provided to the reinforcement, it would take approximately 2 years for the carbonation front to reach the reinforcement if the concrete compressive strength were 20 MPa, compared to almost 20 years for concrete with a compressive strength of 40 MPa. If the same service life were required for the 20 MPa concrete as for the 40 MPa concrete based on carbonation depth, the concrete cover would need to be more than tripled to approximately 67 mm.

12.2 IMPACT OF CORROSION ON MATERIAL PROPERTIES

12.2.1 Strain-based material properties

Figure 12.2 shows examples of lengths of reinforcing bars that have been subjected to accelerated corrosion by submerging them in a saltwater solution and inducing a current through the bar. Importantly, it can be seen that corrosion is highly non-linear along the length of the bar, with some areas significantly more damaged than others. In some areas, the ribs have been corroded off, and the bar diameter is reduced, while in other areas, the bar still has significant roughness. This is important because the strength and stiffness of rust are negligible compared to those of the remaining uncorroded reinforcement, and as a result, it can be ignored. The strain-based stress/strain relationship can be applied unaltered to the uncorroded core of the reinforcement to predict the residual performance of the bar as follows.

In Figure 12.3, consider an idealisation of the corroded reinforcing bar in Figure 12.2 such that the bar has lost Δ_{m1}% to Δ_{m3}% of its mass in segments 1–3, respectively. This loss of mass Δ_m can be converted to a corresponding average change in diameter ΔD over the bar length L, knowing the density of steel ρ_s=7850 kg/m^3 and the length of the segment that is corroded. This gives

$$\Delta D = \sqrt{\frac{4\Delta_m}{\rho_s \pi L}} \tag{12.8}$$

If a tensile force F is applied to the bar in Figure 12.3, the corresponding stress _σ_1 to σ_3 in each segment is simply the applied force divided by the segment diameter, which can be determined using Eq. 12.8. Hence, the maximum stress within the bar is always defined by the section with the minimum diameter. This means the force in the corroded bar can be determined simply by considering an uncorroded bar with a reduced diameter.

Now consider the extension of the bar when subjected to the load F in Figure 12.3. Knowing the stress in each segment as a function of the change in bar diameter, and using the stress/strain relationship of the uncorroded core in Figure 12.3, the corresponding strain in each segment ε_1 to ε_3 is also known. Taking the average strain ε_{avg} by weighting ε_1 to ε_3 by the segment lengths L_1 to L_3 gives

$$\varepsilon_{avg} = \frac{\varepsilon_1 L_1 + \varepsilon_2 L_2 + \varepsilon_3 L_3}{L_1 + L_2 + L_3} \tag{12.9}$$

It is therefore observed from Eq. 12.9 that the average strain is significantly lower than both the uncorroded material strain and the strain corresponding to the segment with the critical stress, which is the segment with the smallest diameter. While the uncorroded stress/strain relationship can be used to determine the relationship between force and stress in a corroded bar, it cannot account for the loss of ductility that occurs as the bar corrodes.

12.2.2 Bond-slip material properties

As established in the tension stiffening Chapter 4 and shown in the segmental and member Chapters 5–11, the bond between reinforcement and concrete significantly

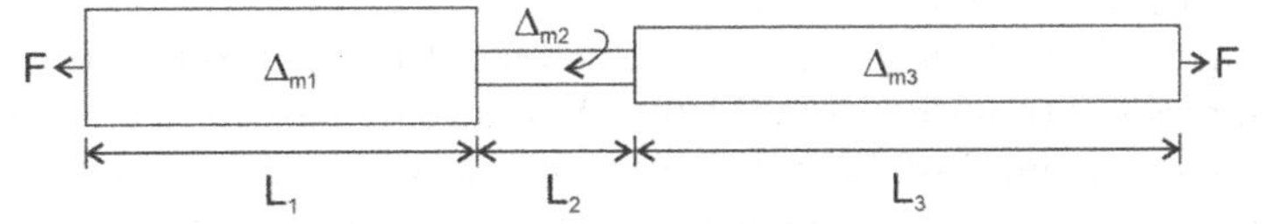

Figure 12.3 Idealisation of bar with varying corrosion.

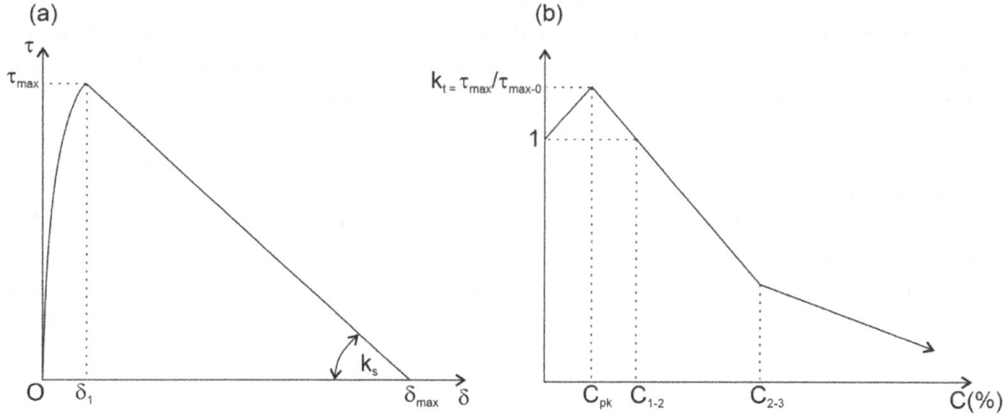

Figure 12.4 Bond model according to Feng et al. (2016) (a) without corrosion; and (b) change to τ_{max} stress with corrosion *C*.

impacts member behaviour at all limit states. At the serviceability limit state, the partial-interaction mechanism controls the formation of cracks and the crack-opening stiffness, and at the ultimate limit state, the bond strength defines whether the maximum force in the reinforcement is controlled by the material strength or by debonding. It is, therefore, essential to quantify the change in bond properties that occur as a result of reinforcement corrosion.

The impact of corrosion on bond is complex. Small amounts of corrosion have been shown to be beneficial, with the expansion caused by corrosion resulting in an increased bond strength due to the generation of local confining stresses $\sigma_{rust.r}$ in Figure 12.1b. However, as the level of corrosion increases, the expansion of the reinforcement generates circumferential stresses $\sigma_{rust.c}$ sufficient to cause the concrete to split, as shown in Figure 12.1b. The formation of splitting cracks leads to a significant reduction in both bond strength and ultimate slip, as explained in Section 3.2.4.

To quantify the change in bond due to corrosion, Feng et al. (2016) calibrated a bond-stress–slip (τ/δ) relationship using the FIB to describe the non-linear rising branch and a linear descending falling branch in Figure 12.4a. That is,

$$\tau = \begin{cases} k_t \tau_{max} \left(\dfrac{\delta}{\delta_1} \right)^{\alpha}; & \delta \le \delta_1 \\ k_t \tau_{max} \left(\delta - \delta_1 \right); & \delta_1 \le \delta < \delta_{max} \end{cases} \tag{12.10}$$

in which τ_{max} is the maximum bond stress that occurs at a slip of δ_1 and δ_{max} is the slip at which the bond stress reduces to zero, which in Feng's model is defined using the slope of the falling branch k_s in Figure 12.4a.

To calibrate Eq. 12.10, Feng et al. (2016) considered 196 pull-out test results covering concrete compressive strengths from 22 to 60 MPa, bar diameters from 10 to 20 mm, and concrete covers from 8 to 70 mm. Based on the collected test data, the maximum bond strength for corroded reinforcement $\tau_{max\text{-}c}$ was normalised by the maximum bond strength without corrosion $\tau_{max\text{-}0}$ to give the piecewise model in Figure 12.4b in which

$$k_t = \begin{cases} C\left(-0.0320\dfrac{c}{d_b}+0.576\right)+1; & 0<C\le C_{1-2} \\ C\left(0.0137\dfrac{c}{d_b}-0.247\right)+1.42+0.0475\dfrac{c}{d_b}-3.94\times10^{-3}\left(\dfrac{c}{d_b}\right)^2; & C_{1-2}<C\le C_{2-3} \\ -0.0016C+0.224; & C>C_{2-3} \end{cases} \tag{12.11}$$

where the critical corrosion levels in Figure 12.4b are given in terms of the ratio of the concrete clear cover to bar diameter ratio $\frac{c}{d_b}$ as

$$C_{pk} = 0.3C_{1-2} \tag{12.12}$$

$$C_{1-2} = 0.288\frac{c}{d_b}+1.72 \tag{12.13}$$

$$C_{2-3} = \frac{-1.20-0.0475\dfrac{c}{d_b}+3.94\times10^{-3}\left(\dfrac{c}{d_b}\right)^2}{0.0137\dfrac{c}{d_b}-0.245} \tag{12.14}$$

Having defined the change in τ_{max}, the change in δ_1 due to corrosion was investigated. Based on statistical analysis, Feng et al. (2016) found that δ_1 reduced in the presence of corrosion due to splitting, but that the most reliable fit to the test data was obtained by holding δ_1 at a constant 0.175 mm. In the calibration data set, δ_1 was found to be constant and equal to 0.175 regardless of the level of corrosion.

Finally, the slope of the falling branch was quantified from each test result and used to calibrate a model for the falling branch slope k_s in Figure 12.4a such that

$$k_s = -0.161\tau_{max\text{-}0} \tag{12.15}$$

and

$$\delta_{max} = 0.175-\frac{k_t\tau_{max\text{-}0}}{k_s} \tag{12.16}$$

Here, it is worth noting that, unlike uncorroded reinforcement, no residual frictional branch is considered, and this is because the combined actions of reducing confinement by splitting the concrete cover and reducing the roughness of the bar surface significantly reduce friction such that it can be ignored.

Worked Example WE-12.2: Bond properties of corroded reinforcement

Let us quantify the bond properties of the tensile reinforcement in Figure E12.1, assuming the concrete has a compressive strength of 40 MPa, the bar has a diameter of 16 mm, and the minimum clear cover to the reinforcing bar is 30 mm, and the bar has corroded such that it has a mass loss C=2%.

Taking the maximum bond stress for a bar with no corrosion as

$$\tau_{max\text{-}0} = 1.25\sqrt{f_c} = 1.25\sqrt{40} = 7.91 \text{ MPa} \quad \text{(E12.2.1)}$$

In order to determine the change in τ_{max} due to corrosion, it is first necessary to identify which branch of Figure 12.4b we are on. Hence, substituting the $\frac{c}{d_b}$ ratio into Eqs. 12.17 and 12.18, the critical corrosion levels can be determined as

$$C_{1\text{-}2} = 0.288\left(\frac{30}{16}\right) + 1.72 = 2.26\% \quad \text{(E12.2.2)}$$

$$C_{pk} = 0.3 \times 2.62 = 0.678\% \quad \text{(E12.2.)}$$

Hence, when C=2%, $k_t=\tau_{max}/\tau_{max\text{-}0}$ is in stage 2 of Figure 12.4b such that from Eq. 12.11

$$k_t = 2\left[0.0137\left(\frac{30}{16}\right) - 0.247\right] + 1.42 + 0.0475\left(\frac{30}{16}\right) - 3.94\times10^{-3}\left(\frac{30}{16}\right)^2 = 2.15 \quad \text{(E12.2.4)}$$

Hence,

$$\tau_{max\text{-}c} = 2.15 \times 7.91 = 17.01 \text{ MPa} \quad \text{(E12.2.5)}$$

and from Eq. 12.15,

$$k_s = -0.161 \times 7.91 = -1.272 \text{ MPa/mm} \quad \text{(E12.2.6)}$$

and from Eq. 12.16

$$\delta_{max} = 0.175 - \frac{2.15 \times 7.91}{-1.272} = 13.54 \text{ mm} \quad \text{(E12.2.7)}$$

In Figure E12.4, the bond properties are compared against those for 0% corrosion. It is essential to note here that the magnitude of the increase in bond for small levels of corrosion is highly dependent on $\tau_{max\text{-}0}$ that can vary by 100% depending on the initial bond conditions described in Chapter 3. That is, here the best-case scenario is presented, where small levels of corrosion lead to a large increase in $\tau_{max\text{-}c}$. By repeating the analysis for a range of corrosion levels, a family of curves is generated, showing that as the level of corrosion further increases, a significant and rapid reduction in τ_{max} and δ_{max} occurs, and by corrosion levels of 6%, there is virtually no change in the bond properties for increased levels of corrosion.

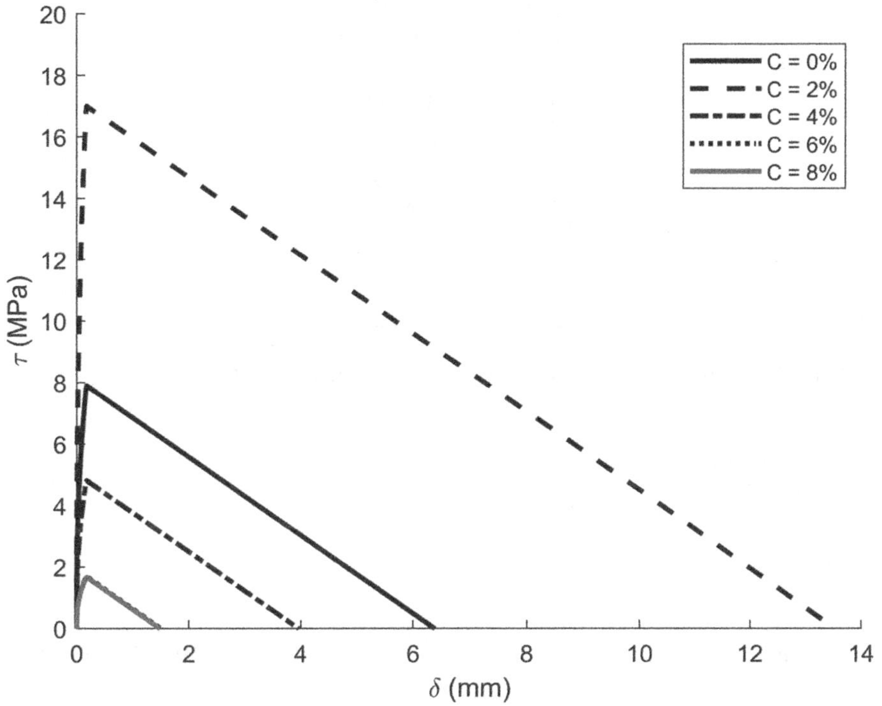

Figure E12.4 Impact of corrosion on local τ/δ bond properties.

12.3 IMPACT OF CORROSION ON THE TENSION STIFFENING MECHANISM

In Section 12.2, the change in strain-based and slip-based material properties was described, and it was observed that while the yield stress of the reinforcement varies proportionally with the mass of the reinforcement lost to corrosion, the reduction in bond properties occurs at a much faster rate. This reduction in bond properties is important, as it not only reduces the crack opening stiffness but also increases the likelihood that the reinforcement will debond. Let us now consider each of these scenarios in turn.

Being mechanics-based, the partial-interaction tension stiffening model presented in Chapter 4 can be applied to corroded reinforcement without modification; hence, it is simply a matter of replacing the material properties with those appropriate for the degree of corrosion present. Let us now consider this impact via an example.

Worked Example WE-12.3: Impact of corrosion on crack opening stiffness

Consider the tension stiffening prism with a cross-section as shown in Figure E12.5, which contains a 16 mm reinforcing bar embedded in 40 MPa concrete with 30 mm clear cover.

If it is assumed that the bar is loaded such that cracking occurs before any corrosion, from Section 4.3.5.2, the crack spacing is given by

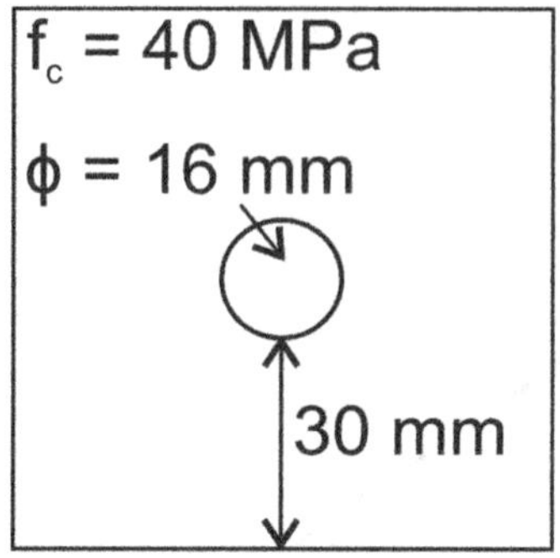

Figure E12.5 Tension stiffening cross-section for Example 12.4.

$$S_p = \left[\frac{2^{\alpha}(1+\alpha)}{\lambda_2(1-\alpha)^{1+\alpha}}\right]^{\frac{1}{1+\alpha}}\left[\frac{f_{ct}-f_{pc}}{E_c}\left(\frac{E_cA_c}{E_rA_r}+1\right)\right]^{\frac{1-\alpha}{1+\alpha}} \quad \text{(E12.3.1)}$$

in which

$$\lambda_2 = \frac{\tau_{max}}{\delta_1^{\alpha}}L_{per}\left(\frac{1}{E_rA_r}+\frac{1}{E_cA_c}\right) = \frac{7.905}{0.175^{0.4}}\times 50.27\times\left(\frac{1}{200{,}000\times 201}+\frac{1}{30{,}000\times 1444}\right)$$
$$= 3.83\times 10^{-5} \quad \text{(E12.3.2)}$$

Substituting into Eq. E12.3.1,

$$S_p = \left[\frac{2^{0.4}(1+0.4)}{3.83\times 10^{-5}\times(1-0.4)^{1+0.4}}\right]^{\frac{1}{1+0.4}}\left[\frac{3.79}{30{,}000}\left(\frac{30{,}000\times 1444}{200{,}000\times 201}+1\right)\right]^{\frac{1-0.4}{1+0.4}} = 107.9\text{ mm} \quad \text{(E12.3.3)}$$

For 2% corrosion, the bond properties are as defined in Figure E12.4, where $\tau_{max\text{-}c}$ = 17.01 MPa at a slip of 0.175 mm, the bond stiffness k_e is

$$k_e = \frac{\tau_{max\text{-}c}}{\delta_{1-c}} = \frac{17.01}{0.175} = 97.22\text{ MPa/mm} \quad \text{(E12.3.)}$$

From Section 4.3.5.4, the partial-interaction crack opening stiffness is given by

$$k_{pi} = E_rA_r\left(\frac{\lambda_1}{\tanh(\lambda_1 S_p)}\right) \quad \text{(E12.3.5)}$$

in which

$$\lambda_1 = \sqrt{k_eL_{per}\left(\frac{1}{E_rA_r}+\frac{1}{E_cA_c}\right)} = \sqrt{97.22\times 49.75\left(\frac{1}{200{,}000\times 394}+\frac{1}{30{,}000\times 1444}\right)}$$
$$= 0.0132 \quad \text{(E12.3.6)}$$

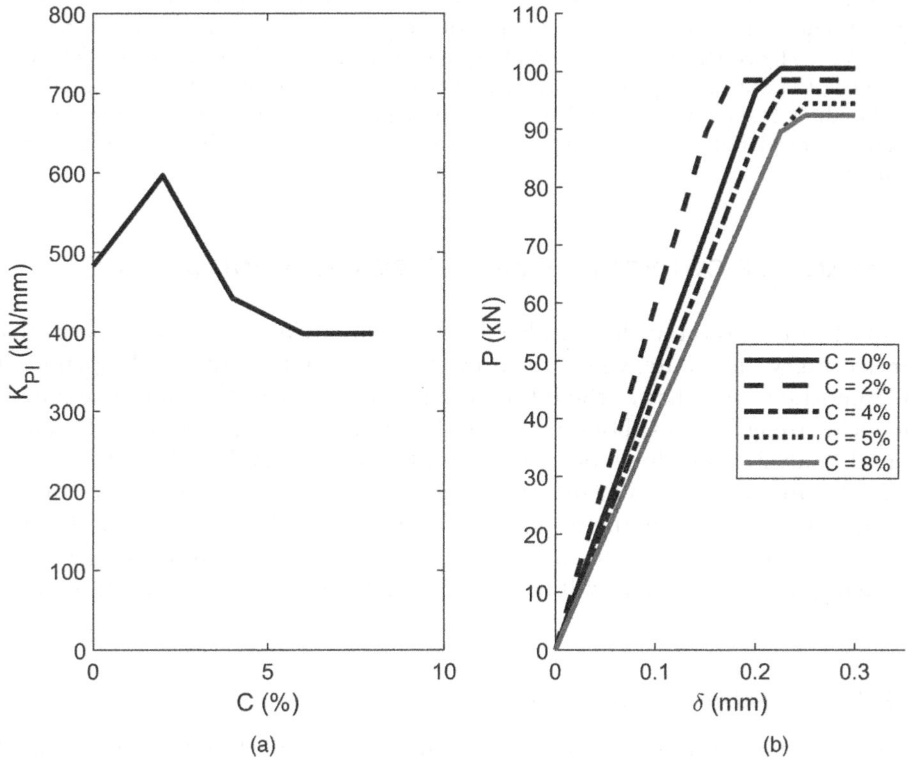

Figure E12.6 Impact of corrosion on (a) crack opening stiffness and (b) *P*/δ and yield force.

and the bar diameter and area have been reduced to allow for the mass lost due to corrosion, giving

$$k_{pi} = 200{,}000 \times 201\left(\frac{0.0132}{\tanh(0.0132 \times 107.9)}\right) = 596.8 \text{ kN/mm} \qquad \text{(E12.3.7)}$$

In Figure E12.6a, the crack opening stiffness is compared for corrosion levels between 0% and 8% mass loss, and in Figure E12.6b, the crack opening stiffness is used to plot the load-slip relationship. The analysis shows that the crack opening stiffness initially increases due to the increased bond at low levels of corrosion before rapidly falling to below that of the uncorroded reinforcement. However, by 8% corrosion, the crack opening stiffness is relatively constant while the yield force continues to decline. This suggests that prior to debonding of the reinforcement, there is an upper limit to the impact of corrosion on serviceability deflections and crack widths, but not on member strength.

Now consider the impact of corrosion on debonding. For reinforcement without corrosion, the bond between the reinforcement and the concrete is provided by the reinforcement ribs and is sufficient such that the reinforcement generally ruptures before debonding takes place. This is not the case for corroded reinforcement. To quantify the force at which debonding occurs, consider the *IC* debonding load that was defined in Section 4.3.3.4 as

$$P_{IC} = \sqrt{\tau_{max}\delta_{max}}\sqrt{L_{per}E_r A_r} \qquad (12.17)$$

in which A_r and L_{per} are taken as the reduced area and bonded perimeter of the reinforcement, and τ_{max} and δ_{max} are taken as that for the corroded reinforcement, $\tau_{max\text{-}c}$ and $\delta_{max\text{-}c}$. By equating the *IC* debonding force with the yield force, it is possible to identify the critical level of corrosion that will cause a transition from yielding to debonding. It should be noted that it is not possible to formulate a simple analytical solution to identify if debonding will control; hence, this must be checked for all analyses.

12.4 IMPACT OF CORROSION ON MEMBER BEHAVIOUR

The impact of corrosion at the member level can be divided into two aspects that at the serviceability limit state, where the impact of corrosion on the crack opening stiffness is important, and that at the ultimate limit state, where the impact of corrosion on the yield strength and debonding strength is important. Significantly, because the segmental analysis procedures developed in Chapters 6 and 7 and used in Chapters 8–11 for flexure and shear are directly dependent on the crack opening stiffness and debonding, and do not contain any additional calibratable factors, they can be applied without modification. Hence, in this section, the analysis procedures will not be presented again; rather, the focus will be on showing the impact of corrosion on member behaviour.

12.4.1 Serviceability limit state

At the serviceability limit state under an instantaneous load, the change in crack opening stiffness can be directly related to a change in flexural rigidity and a change in crack width using the partial-interaction modular ratio defined in Section 4.3.5.3 and applied to a segmental analysis in Chapter 6. In this way, it is possible to directly quantify the change in flexural behaviour that occurs as a result of both the reduction in bar area and the change in bond properties due to corrosion. This differs from standard full-interaction analysis in which the change in bond properties is not considered, and the impact of corrosion cannot be directly assessed; interestingly, this has not been widely investigated.

Worked Example WE-12.4: Impact of corrosion on flexural rigidity and crack width under instantaneous loading

Considering the same cross-section as in Figure E12.1, which is redrawn in Figure E12.7, let us determine the member flexural rigidity and crack width when subjected to a moment of 20 kNm.

The crack spacing has already been determined in Worked Example WE-12.4 to be 107.9 mm, and crack opening stiffness and yield forces for a range of different corrosion levels are extracted from Figure E12.6 and tabulated in Table 12.1.

From Section 4.3.5.4, the partial-interaction modular ratio is given by

$$n_{PI} = \frac{k_{PI} S_p}{E_c A_r} \quad \text{(E12.4.1)}$$

Considering the specific example of 2% corrosion,

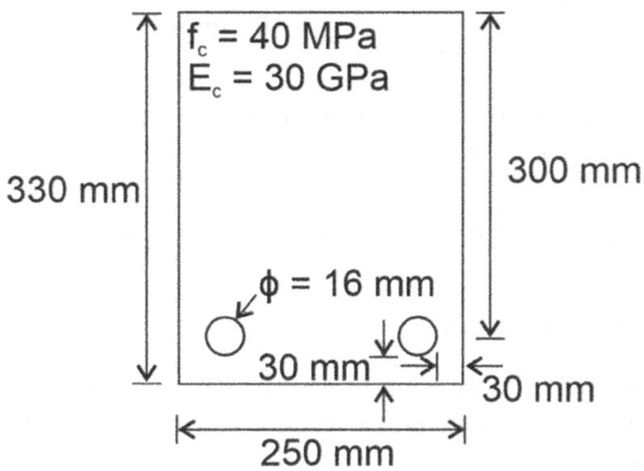

Figure E12.7 Cross section for Example 12.5.

Table 12.1 Impact of corrosion on crack opening parameters and yield force from Example 12.4

C (%)	*kpi (kN/mm)*	P_{yield} *(kN)*	λ_1
0	482.7	100.5	0.00898
2	594.8	98.5	0.01315
4	441.0	96.5	0.00699
6	397.2	94.5	0.00414
8	396.8	92.4	0.00410

$$n_{PI} = \frac{594.8 \times 107.9}{30{,}000 \times 1444} = 5.43 \tag{E12.4.2}$$

The modular ratio can then be used to transform the section using the approach in Section 6.4.1.1. Taking the first moment of area of the transformed section

$$\frac{500 \times d_{NA\text{-}cr}^2}{2} = 5.43 \times 394 \times (300 - d_{NA\text{-}cr}) \tag{E12.4.3}$$

$$d_{NA\text{-}cr} = 78.54 \text{ mm} \tag{E12.4.4}$$

Taking the second moment of area,

$$I_{cr} = \frac{250 \times 78.54^{\ 3}}{3} + 5.43 \times 394 \times (300 - 78.54\ )^2 \ = 167.57 \times 10^6 \text{ mm}^4 \tag{E12.4.5}$$

Repeating the analysis for all corrosion levels in Table 12.1, the variation in flexural rigidity $E_c I_{cr}$ is shown in Figure E12.8a, where it is seen that the 8% mass loss due to corrosion causes an approximately 6% change in flexural rigidity.

Taking an applied moment of 10 kNm, which is approximately 20% of the cross-section flexural strength, the section curvature is given by

$$\chi = \frac{M}{E_c I_{cr}} = \frac{20 \times 10^6}{30{,}000 \times 167.57 \times 10^6} = 1.58 \times 10^{-6}/\text{mm} \tag{E12.4.6}$$

which can be used to determine the slip of the reinforcement relative to the crack face and therefore the crack width

$$w = 2\Delta_{rt} = 2\theta(d - d_{NA}) = \frac{2\chi S_p}{2}(d - d_{NA}) = 2 \times 1.58 \times 10^{-6} \times \frac{107.9}{2}$$
$$\times (300 - 78.54) = 0.051\ \text{mm} \tag{E12.4.7}$$

Repeating the analysis for all corrosion levels, the variation in crack width is shown in Figure E12.8b, where it is seen that the crack width varies by approximately 5% due to the 8% corrosion.

Importantly, neither the reduction in flexural rigidity nor the increase in crack width can be captured in a traditional full-interaction analysis, showing the benefit of the segmental-analysis procedure.

Analysis can similarly be conducted, including concrete time effects (shrinkage and creep), using the procedures outlined in Section 6.5.1.3. For example, assuming a shrinkage strain of 1000με, no concrete creep, an applied moment of 10 kNm, and 2% corrosion

$$d_{NA}^3 - 3e_{eff}d_{NA}^2 + d_{NA}\left[\frac{6}{b}\left(A_1 - e_{eff}A_0\right)\right] - \frac{6}{b}\left(A_2 - e_{eff}A_1\right) = 0 \tag{E12.4.8}$$

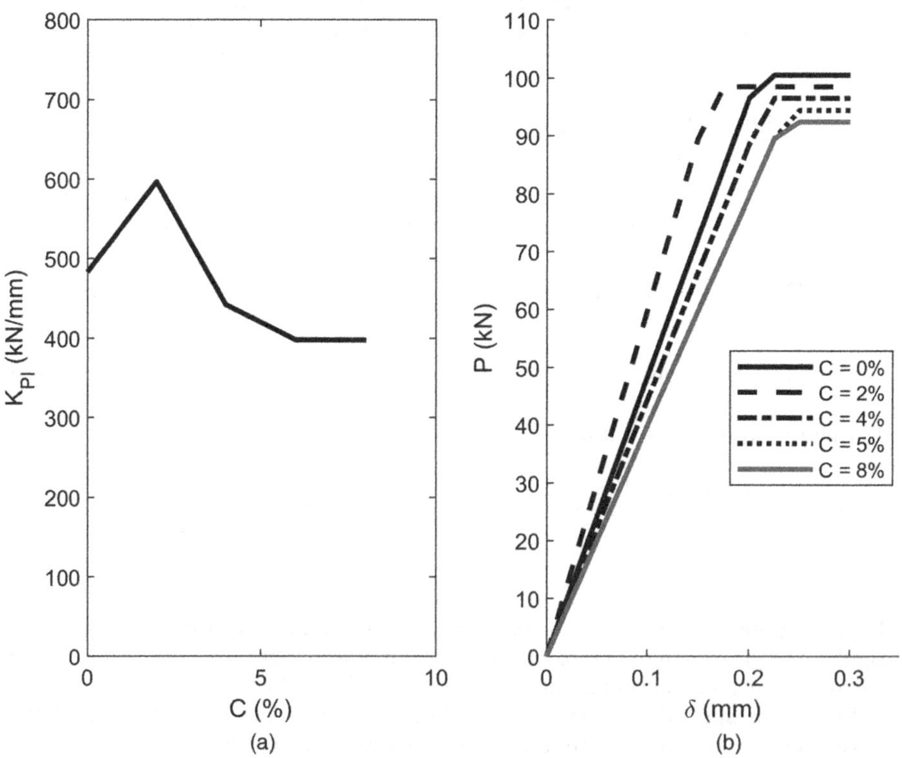

Figure E12.8 Impact of corrosion on (a) flexural rigidity and (b) crack width.

in which

$$e_{eff} = \frac{M + E_r\varepsilon_{sh}\left(A_{rt}d + A_{rc}c_c\right)}{E_r\varepsilon_{sh}\left(A_{rt} + A_{rc}\right)} = \frac{10\times10^6 + 200{,}000\times1000\times10^{-6}\times(394.1\times300)}{200{,}000\times1000\times10^{-6}\times300} = 427 \tag{E12.4.9}$$

$$A_0 = n_{PI}A_{rt} + \left(n_{FI} - 1\right)A_{rc} = 5.43\times394.1 = 2138.6 \tag{E12.4.10}$$

$$A_1 = n_{PI}A_{rt}d + \left(n_{FI} - 1\right)A_{rc}c_c = 5.43\times394.1\times300 = 641.6\times10^3 \tag{E12.4.11}$$

$$A_2 = n_{PI}A_{rt}d^2 + \left(n_{FI} - 1\right)A_{rc}c_c^2 = 5.43\times394.1\times300^2 = 192.5\times10^6 \tag{E12.4.12}$$

Substituting the results of Eqs. E12.4.9 to E12.4.12 into Eq. E12.4.8 and solving yields a neutral axis depth of 37.1 mm.

From Section 6.5.1.3, the corresponding section curvature is

$$\begin{aligned}\chi &= \frac{M}{E_c\left(t,t_0\right)I_{cr}} + \frac{n_{FI}E_c\left(t,t_0\right)\varepsilon_{sh}\left(A_{rt}d + A_{rc}c_c\right)}{E_c\left(t,t_0\right)I_{cr}}\\ &= \frac{10\times10^6}{30{,}000\times167.57\times10^6} + \frac{6.67\times30{,}000\times1000\times10^{-6}\times(394.1\times300)}{30{,}000\times167.57\times10^6}\\ &= 6.20\times10^{-6}/\text{mm}\end{aligned} \tag{E12.4.13}$$

Comparing the curvature to that without shrinkage in Eq. E12.4.6, it can be seen that shrinkage has increased the curvature by approximately four times; however, importantly, the cracked moment of inertia has remained constant.

Now taking the crack width as twice the slip of the reinforcement from the crack face, as given in Section 4.3.5.4:

$$\begin{aligned}w &= 2\left(\Delta_{rt} - \Delta_0\right) = 2\left[\theta\left(d - d_{NA}\right) - \Delta_0\right] = 2\left[\frac{\chi S_p}{2}\left(d - d_{NA}\right) - \frac{\dfrac{\varepsilon_{sh} - \dfrac{f_{ct}}{E_c}}{\lambda_1}}{\tanh\left(\lambda_1 S_p\right)}\right]\\ &= 2\left[\frac{6.20\times10^{-6}\times107.9}{2}(300 - 37.1) - \frac{\dfrac{1000\times10^{-6} - \dfrac{3.79}{30{,}000}}{0.0132}}{\tanh(0.0132\times107.9)}\right] = 0.083\end{aligned} \tag{E12.4.14}$$

The results of the analysis with shrinkage at all corrosion levels are shown in Figure E12.8b, where it can be seen that the general shape of the relationship between corrosion and crack width is similar regardless of the presence of shrinkage, but the addition of shrinkage significantly increases crack widths.

Having shown how the segmental approach can be applied to predict deflections and crack widths under the combined actions of concrete time effects and corrosion, it is interesting to note that the approach may be used as a structural assessment tool. That is, if the crack spacing and crack width were to be measured and the concrete shrinkage strain and creep coefficient could be estimated from existing material models, it is possible to estimate the average level of corrosion between cracks based on the difference in the measured crack opening and the crack opening expected from concrete time effects alone.

12.4.2 Ultimate limit state

In Section 12.4, it was shown that there are two possible failure modes when considering corroded reinforcement: a material failure mode, where the force developed in the reinforcement is a function of the bar strength and crack opening stiffness; and a debonding failure mode, where the bar strength is never reached, but instead, the bar debonds.

In the case of material failure, analysis can be undertaken using an identical approach to that presented in Chapter 6, using either a numerical approach or a closed-form solution. In this case, the only modification required is to change the crack opening stiffness and the material yield force.

In the case where the reinforcement debonds, the analysis presented in Section 4.3.3.4 can be applied without modification.

Worked Example WE-12.5: Change in moment–rotation behaviour due to corrosion

As an example, let us consider the same cross-section as above but producing the entire moment–rotation, moment–curvature, and moment–flexural rigidity relationships. Let us first determine if the reinforcement debonds by considering the case of 8% corrosion, taking the bond properties from Figure E12.4 and using Eq. 12.17:

$$P_{IC} = \sqrt{\tau_{max}\delta_{max}}\sqrt{L_{per}E_r A_r} = \sqrt{1.67\times 1.49}\sqrt{48.19\times 200{,}000\times 369.6} = 94.0\text{ kN} \tag{E12.5.1}$$

which is above the yield force of 92.4 kN; hence, if strain hardening is ignored, as is the case in standard analysis, the capacity of the section is limited by the reinforcement yield force. However, with further increases in corrosion, the debonding force will fall below that of the yield force, and hence the capacity of the member will be reduced.

12.5 NOTATION AND ABBREVIATIONS

A_r	reinforcement cross sectional area
C	mass loss of reinforcement due to corrosion expressed as a %
C_a	atmospheric CO_2 concentration
C_c	concentration of CO_2 required to carbonate all hydration products
c	clear concrete cover

c_{crit}	critical chloride concentration
c_s	chloride concentration at the concrete surface
D	chloride diffusion coefficient
D_{eff}	effective chloride diffusion coefficient
D_{CO_2}	chloride diffusion coefficient
d_b	bar diameter
$d_{NA\text{-}cr}$	neutral axis depth of cracked section
E_c	elastic modulus of concrete
E_r	elastic modulus of reinforcement
f_c	concrete compressive strength
f_{cm}	mean concrete compressive strength
f_{ct}	concrete tensile strength
I_{cr}	cracked moment of inertia
J	CO_2 flux
k_t	proportional change in maximum bond stress due to corrosion
k_e	slope of linear ascending branch of τ/δ relationship
k_{pi}	partial-interaction crack opening stiffness
k_s	slope of falling branch of τ/δ relationship
L	length
M	moment
n_{pi}	partial-interaction modular ratio
PIC	IC debonding strength
P_{yield}	reinforcement yield strength
s	chloride sink coefficient
s_p	crack spacing
t	time
t_{SL}	service life
w	crack width
x	distance
Δ_0	slip of reinforcement at zero load
Δ_D	change in diameter
Δ_m	change in mass
Δ_{rt}	slip of tensile reinforcement relative to crack face
α	coefficient controlling non-linearity of bond stress slip relationship
χ	curvature
δ	bond slip
δ_1	slip at maximum bond stress
δ_{max}	slip at initiation of debonding
ε	axial strain
ε_{avg}	average strain
ε_{sh}	shrinkage strain
θ	rotation
$\sigma_{rust,c}$	circumferential stress due to formation of rust
$\sigma_{rust,r}$	radial stress due to formation of rust
σ	axial stress
τ	bond stress
τ_{max}	maximum bond stress

$\tau_{max\text{-}c}$ maximum bond stress with c% corrosion
$\tau_{max\text{-}0}$ maximum bond stress with 0% corrosion

REFERENCES AND DEVELOPMENT

Feng, Q., Visintin, P. and Oehlers, D.J., 2016. Deterioration of bond–slip due to corrosion of steel reinforcement in reinforced concrete. *Magazine of Concrete Research*, *68*(15), 768–781.

International Federation for Structural Concrete (FIB), 2013. FIB model code for concrete structures 2010. Lausanne, Switzerland.

Khan, M.U., Ahmad, S. and Al-Gahtani, H.J., 2017. Chloride-induced corrosion of steel in concrete: an overview on chloride diffusion and prediction of corrosion initiation time. *International Journal of Corrosion*, *2017*(1), 5819202.

Li, D., Li, L.Y. and Wang, X., 2019. Chloride diffusion model for concrete in marine environment with considering binding effect. *Marine Structures*, *66*, 44–51.

Razmi, A., Bennett, T., Xie, T. and Visintin, P., 2025. A phenomenological model for chloride diffusion coefficient in concretes with traditional and blended binders and alternative fillers. *Construction and Building Materials*, *461*, 139783.

Shafikhani, M. and Chidiac, S.E., 2019. Quantification of concrete chloride diffusion coefficient – a critical review. *Cement and Concrete Composites*, *99*, 225–250.

Chapter 13

Environmental sustainability for structural engineers

Our society is underpinned by the use of concrete; it is an essential component in the construction of not only the structures we live and work in but also the underlying infrastructure that enables our standard of living. At a global level, the importance of concrete is best seen in the fact that, over the past 65 years, the demand for cement has increased approximately 34 times, whilst over the same period the population has only tripled (UN Population Division 2015).

Whilst concrete is clearly an essential enabler of our standard of living, it must be understood that it also has a significant detrimental impact on the environment. The chemical processes involved in the production of Ordinary Portland Cement are highly carbon-intensive, with approximately 6% of all anthropogenic carbon dioxide equivalent ($CO_{2\text{-eq}}$) emissions arising from cement production (Metz 2007). Furthermore, the production of concrete requires large volumes of quarried products such as aggregates and sands. These materials are non-renewable, and their extraction often leads to impacts such as the depletion of natural habitats.

13.1 APPROACHES FOR ENVIRONMENTALLY SUSTAINABLE DESIGN

Given these challenges, structural engineers are increasingly required to explicitly consider the environmental impact of their designs. To assist the construction sector in improving its environmental sustainability, Arnold (2020) and Orr et al. (2021) have developed simple general guidance to inform decision-making throughout the planning, design and construction process. This guidance is summarised in Table 13.1, where the potential $CO_{2\text{-eq}}$ emissions avoided are shown relative to major decisions that are taken at different stages of a project.

In Table 13.1, the largest potential for avoiding carbon emissions arises in the strategic definition of a project, where the decision to build a new structure is made. At this stage, it is clear that if the decision is made to continue using an existing structure in its current state, no additional carbon is expended. Hence, the 'build nothing' solution is the preferred option for minimising environmental impact if all other design constraints can also be met.

If the decision is made that an existing structure is no longer fit for purpose, the next major decision point is in the concept design, where a choice can be made to demolish

DOI: 10.1201/9781003711407-19

Table 13.1 Potential $CO_{2\text{-eq}}$ emissions avoided based on strategic decisions at different stages of a project

Stage of project	*Action*	*Potential avoided emissions*
Strategic definition	Build nothing by exploring alternative solutions	100%
Concept design	Build less by reusing existing structures	50–100%
Technical design	Build using low-carbon materials and lean design approaches	20%–50%
Construction	Use low-carbon construction approaches to build efficiently	Up to 20%

Sources: Based on Arnold (2020) and Orr et al. (2021).

and rebuild a structure or to re-life (reuse) the existing structure. The design for the reuse of a structure can be one of the most challenging aspects of structural engineering, as it is often necessary to increase the loads applied to a structure that was designed for a lower load. A common example is the need to consider an increased seismic load where little to none was considered in the initial design of the structure. In this scenario, there may be a need to significantly increase the lateral stiffness of a structure whilst also avoiding significant additional mass, as this further increases the seismic loads.

The reuse of structures, therefore, requires the structural engineer to develop a sound understanding of the performance of the current structure before devising a unique solution to increase the stiffness or capacity of members or to construct entirely new load paths within the existing structure. In the context of this book, the structural engineer needs to be able to predict the strength, stiffness and remaining durability of the existing structure, and for this, they need to understand both the instantaneous and time-dependent material behaviour presented in Chapters 2, 3 and 12, as well as the member flexural and shear behaviour discussed in Chapters 6 and 7. When designing strengthening options, the ability to reduce the maximum design moment using moment redistribution, described in Chapter 8, can be of significant benefit, as can strengthening with lightweight fibre-reinforced polymers (*FRP*), as described in Chapter 11.

If a new structure must be built, the next best approach for the structural engineer, as outlined in Table 13.1, is to build using low-carbon materials or lean design. The development of low-carbon concretes, or concretes that seek to minimise the usage of natural materials by replacing quarried natural aggregates with various forms of waste materials, has been a significant research focus over the last decade. However, it is unfortunate that very few materials have made the transition from the lab to real-world application. The reason for this is twofold:

i. National codes of practice have often been developed with simplicity in mind and, therefore, contain empirical components that must be recalibrated or shown to be applicable for each new type of concrete. This is often seen as a prohibitively time-consuming and expensive task that requires extensive experimental work at both the material and member level. This book provides an alternative to this expensive and time-consuming experimental work by demonstrating how both the full-interaction and partial-interaction material properties developed in Chapters 2 and 3 can be used to simulate the partial-interaction mechanisms of tension stiffening and concrete softening in Chapters 4 and 5, which, in turn, control all member behaviour in Chapters 6–12. That is, the segmental displacement approaches presented here provide alternative analysis techniques that can significantly increase the speed and reduce the cost of transitioning a new material from the lab into practice.

ii. A second significant issue limiting the transition of new 'low carbon' concretes from the lab to practice is the fact that whilst they claim to be low carbon, they may not be. This unfortunate scenario arises because materials are often produced on the assumption that if they do not use cement or if they incorporate a waste material, they will benefit the environment. This is simply not the case, and it is necessary to conduct careful life cycle assessments to quantify $CO_{2\text{-eq}}$ as a function of the material's performance; this will be discussed in detail in Sections 13.2 and 13.3.

A second major component to consider when constructing a new structure is to conduct lean design. To do this may require the use of more advanced design procedures to help minimise member dimensions. Alternatively, designing lean may involve the use of highly durable materials such as *FRP* bars or ultra-high-performance fibre-reinforced concrete, which are intended to reduce the whole-of-life environmental impact of a structure by increasing its durability and usable service life. In the context of this book, we have deliberately shown how the partial-interaction mechanisms in Chapters 4 and 5 and their use in design procedures in Chapters 6–12 are equally applicable to concrete with fibre reinforcement or *FRP* bars as to traditional concrete with steel reinforcement.

The final aspect presented in Table 13.1 is to build efficiently; however, further consideration of this aspect is outside the scope of this book.

13.2 LIFE CYCLE ASSESSMENT OF CONCRETE MATERIALS AND STRUCTURES

To allow the structural engineer to best select a concrete material for a specific design project, it is necessary to have a means of connecting its environmental impact to its mechanical or durability performance. Life cycle assessment (*LCA*) provides a means of conducting this assessment. It is important to note, however, that the modelling decisions made when setting up an *LCA* can significantly impact the outcomes of the analysis. It is not the purpose of this book to provide the reader with a specific set of modelling decisions, as the most appropriate decisions may vary with location and project type; rather, the focus here is on discussing the impact that different decisions can have on the modelling outcome.

13.2.1 LCA scope

The first step in conducting an *LCA* is to define the scope of the analysis. When dealing with concrete materials, there are two major options to consider:

i. **Cradle-to-gate analysis**: in which the emissions produced for all aspects of concrete production, from extracting raw materials through to the concrete leaving the batching plant, are measured. This cradle-to-gate analysis has the benefit of being easy to define, but is limited in that it cannot consider what happens to concrete throughout its service life and at the end of its life.

ii. The alternative to cradle-to-gate analysis is cradle-to-grave analysis, in which emissions produced for all aspects from concrete production, throughout its service life, and to ultimate disposal are considered. In this form of analysis, the performance of the material in a member may be included, along with emissions associated with construction, consideration of the member's service life and what happens to the concrete after it is demolished. Whilst being more comprehensive, this form of analysis has significant uncertainty, as it is not always clear how long a structure may be in service nor what will be done at the end of its useful life (e.g. demolition or reuse).

As an example of cradle-to-gate *LCA* for concrete, consider the scope of the analysis presented in Figure 13.1, where the concrete may consist of cement, supplementary cementitious materials (SCMs), aggregates (either natural or recycled), and admixtures. To quantify the emissions associated with the production of concrete, the processes used to produce each of the components, including their transport, must be modelled to determine their component emissions and energy consumption.

As an example, consider the concrete *LCA* tool developed by Gursel (2014) and extended by Bennett et al. (2022) and Shobeiri et al. (2024). In this tool, the scope includes

- Fuel pre-combustion, including extraction, processing and transport of fuel that is consumed during all other phases of material production.
- Fuel combustion.
- Electricity generation, including a proportion of emissions associated with the construction and ongoing maintenance of electricity generation plants.
- Quarrying, including drilling, blasting, excavation, crushing, screening and stockpiling of materials required for cement manufacture, as well as for use as coarse and fine aggregates.
- Cement manufacture, including the grinding and blending of raw materials, pyroprocessing, cooling, conveying, milling and grinding, as well as emissions associated with running the plant that are not directly included in these major processes.
- Emissions associated with the production and preparation of *SCMs*.
- Emissions associated with the processing of waste materials for use as aggregates, including emissions related to the running and maintenance of the plant.
- Emissions associated with the production of admixtures.
- Emissions associated with the production and delivery of potable water to the batching plant.
- Emissions associated with concrete manufacture, including the operation of the batching plant.
- Emissions associated with the transport of all materials, including those linked to the initial production and ongoing maintenance of the vehicle.

13.2.2 Emissions factors

It is evident that each of these processes in Figure 13.1 is highly complex, and it is well beyond the role of a structural engineer to model the emissions arising from each process. Modelling is, however, made possible because there are a variety of databases that

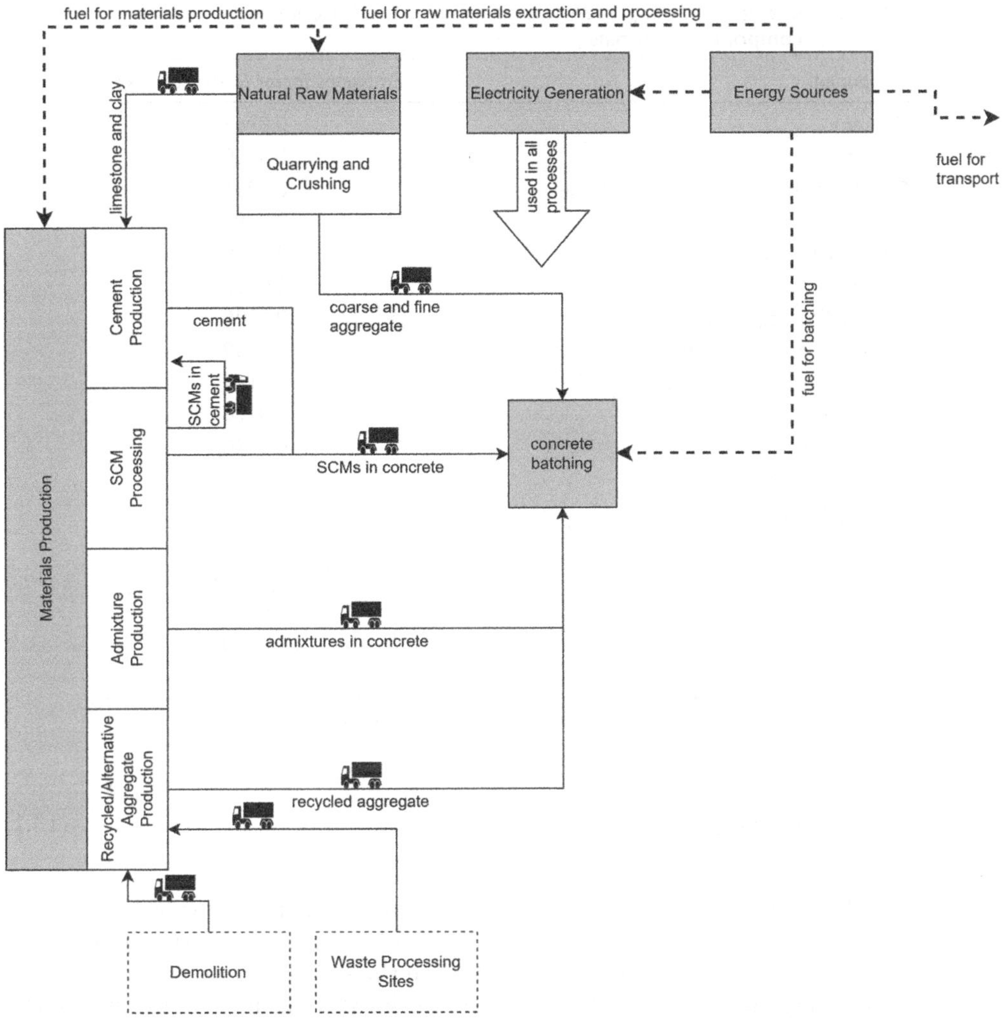

Figure 13.1 Scope of cradle-to-grave LCA for concrete production. (Adapted from Bennett et al. 2022.)

provide emissions factors (the $CO_{2\text{-eq}}$ emissions generated to produce 1 kg of material or to transport 1 kg of material 1 km) and which also include the impact of local energy generation methods, that is, the mix of different types of fossil fuel and renewable energy generation. Examples of emissions factors that have been used in the analyses of Shobeiri et al. (2024) are summarised in Table 13.2 as an indication of magnitudes. These have been sourced from a range of academic studies (Gursel 2014; Roh et al. 2020; Tahanpour Javadabadi 2019; Farina et al. 2017), but it is important to remember that whilst they can be used for comparative analyses, to understand the true magnitude of emissions, appropriate local emissions factors need to be applied.

Knowing the emissions factors of each component of concrete in Table 13.2, the total emissions associated with the production of a unit volume of concrete can be determined using a simple sum product:

Table 13.2 Examples of magnitude of emissions factors for common concrete component materials

Component	*Emissions factor E_i (kg $CO_{2\text{-eq}}$/kg)*
Cement	0.8966
Fly ash	0.0260
Ground granulated blast furnace slag	0.0901
Silica fume	0.0280
Superplasticiser	0.2290
Water	0.0002
Natural fine aggregate	0.0028
Natural coarse aggregate	0.0039
Recycled concrete aggregate	0.0149
Slag aggregate	0.0226
Plastic aggregate	0.1360
Rubber aggregate	0.2450
Glass aggregate	0.2740
Transport mode	*Emissions factor E_{t-i} (kg $CO_{2\text{-eq}}$/kg km)*
Road	12.8×10^{-5}
Rail	2.74×10^{-5}
Sea	1.72×10^{-5}

Sources: Gursel (2014), Roh et al. (2020), Tahanpour Javadabadi (2019), Farina et al. (2017).

$$CO_{2\text{-eq}} = \sum_{i=1}^{n} \left(m_i E_i + m_i d_i E_{t-i} \right) \tag{13.1}$$

where in Eq. 13.1, m_i is the mass of the individual concrete component, E_i is the emissions factor for the specific material, d_i is the distance each material is transported and E_{t-i} is the transport emissions factor associated with the specific mode of transport.

It needs to be emphasised that without considering local conditions, it is only possible to compare the relative performance of different mix designs and not to calculate the absolute magnitude of emissions.

By examining the emissions factors in Table 13.2, it can be concluded that when comparing concretes based on the emissions associated with producing a unit volume, the most sustainable concretes from the perspective of minimising carbon emissions are those that reduce the use of cement. For this reason, significant work has been undertaken to develop mix designs that contain high volumes of SCMs such as fly ash or ground granulated blast furnace slag, which, according to Table 13.2, have emissions an order of magnitude lower than those of cement. The case of SCMs such as fly ash is, however, a significant example of where local interpretation of the use of *LCA* is required. As described in Chapter 2, fly ash is produced as a result of burning coal. If fly ash is considered a waste product, it may therefore be reasonable to set an emissions factor for fly ash that only considers the emissions associated with processing the waste fly ash for use in concrete, as has been done in Table 13.2. However, given that the production and sale of fly ash is a billion-dollar global industry, fly ash may also be considered an industrial by-product. In this case, it might be more appropriate to allocate a proportion of the emissions associated with the burning of coal to the production of concrete. Depending on the approach taken to allocating emissions arising from burning coal, this can increase the emission factor by an order of magnitude (Shobeiri et al. 2021), which obviously has a significant impact on the results of the *LCA*.

Again, it is not the purpose of this book to give an opinion on whether fly ash should be treated as a waste product or a by-product, but rather to explain the factors that may need to be considered when selecting from a range of potential products.

13.2.3 LCA functional unit

Another significant factor that contributes to the interpretation of the results of an *LCA* is the selection of the functional unit, which serves as the basis of comparison. In the past, it has been common practice to compare concrete materials based on the production of a unit volume. However, in a large-scale comparison of more than 5300 concrete mix designs containing conventional and alternative aggregates, as well as a range of SCMs, Shobeiri et al. (2024) demonstrated that the selection of a mix design that minimises $CO_{2\text{-eq}}$ emissions is highly dependent on the choice of functional unit. Specifically, the inclusion of alternative binders and fillers affects the mechanical and durability properties of concrete compared to that produced with only Ordinary Portland Cement. Hence, it is necessary to compare mixes that meet all mechanical and durability performance criteria. The consideration of a functional unit is particularly important when evaluating concretes with alternative aggregates because the incorporation of products such as recycled concrete, crumb rubber, or plastic fragments can negatively impact the mechanical and durability performance of concrete. Therefore, although their inclusion may minimise natural resource usage, $CO_{2\text{-eq}}$ emissions can increase due to the greater amount of binder required to maintain mechanical and durability performance criteria.

If this concept is now elevated to the member level, an appropriate functional unit may be the $CO_{2\text{-eq}}$ emissions and natural resources utilised in the production of a reinforced concrete element that meets a set of strength, serviceability and service life requirements. In this scenario, it may not be the material with the lowest volumetric unit emissions that minimises environmental impact, but rather one with very good strength or durability, which could lead to overall reductions in member dimensions or increases in service life.

13.3 APPLICATION TO DECISION-MAKING

Having now outlined the approach to conducting an *LCA*, including the selection of an appropriate scope and functional unit, it is hopefully clear that it can be a very useful tool to assist the structural engineer in making decisions. With access to appropriate emissions factors for a range of different concrete and reinforcement types, the engineer can design elements using the approaches outlined within this book and determine the environmental impact of their construction. The engineer now has a rigorous means for comparing design options at multiple scales, enabling them to decide whether demolition and rebuild is more appropriate than retrofitting a structure. If a structure is to be rebuilt, this process can also help optimise material selection to minimise overall environmental impact.

13.4 NOTATION AND ABBREVIATIONS

$CO_{2\text{-eq}}$	anthropogenic carbon dioxide equivalent
d_i	distance each material is transported
E_i	emissions factor for the specific material
E_{t-i}	transport emissions factor associated with the specific mode of transport
FRP	fibre reinforced polymer
LCA	life cycle assessment
m_i	mass of the individual concrete component
OPC	Ordinary Portland Cement
SCM	supplementary cementitious material

REFERENCES AND DEVELOPMENT

Arnold, W., 2020. The structural engineer's responsibility in this climate emergency. *The Structural Engineer: Journal of the Institution of Structural Engineer*, *98*(6), 10–11.

Bennett, B., Visintin, P. and Xie, T., 2022. Global warming potential of recycled aggregate concrete with supplementary cementitious materials. *Journal of Building Engineering*, *52*, 104394.

Farina, A., Zanetti, M.C., Santagata, E. and Blengini, G.A., 2017. Life cycle assessment applied to bituminous mixtures containing recycled materials: Crumb rubber and reclaimed asphalt pavement. *Resources, Conservation and Recycling*, *117*, 204–212.

Gursel, A.P., 2014. *Life-Cycle Assessment of Concrete: Decision-Support Tool and Case Study Application*. Berkeley, CA: University of California.

Metz, B., 2007. Intergovernmental Panel on Climate Change, eds., Climate change 2007: mitigation of climate change: contribution of Working Group III to the Fourth assessment report of the Intergovernmental Panel on Climate Change, Cambridge; New York: Cambridge University Press. https://www.ipcc.ch/pdf/assessment-report/ar4/wg3/ ar4_wg3_full_report.pdf.

Orr, J., Ibell, T., Smith, C., Watson, N., Darby, A., Hawkins, W. and Marsh, E., 2021. *Design for Zero*. London: Institution of Structural Engineers.

Roh, S., Kim, R., Park, W.J. and Ban, H., 2020. Environmental evaluation of concrete containing recycled and by-product aggregates based on life cycle assessment. *Applied Sciences*, *10*(21), 7503.

Shobeiri, V., Bennett, B., Xie, T. and Visintin, P., 2021. A comprehensive assessment of the global warming potential of geopolymer concrete. *Journal of Cleaner Production*, *297*, 126669.

Shobeiri, V., Bennett, B., Xie, T. and Visintin, P., 2024. Mix design optimization of waste-based aggregate concrete for natural resource utilization and global warming potential. *Journal of Cleaner Production*, *449*, 141756.

Tahanpour Javadabadi, M., 2019. *Comparative life cycle assessment of incorporating recycled PET aggregates into concrete* (Master's thesis, NTNU).

UN Population Division, 2015. *World Population Prospects: The 2015 Revision*. New York: United Nations Department of Economic and social Affairs.

Part 7

Epilogue

Epilogue

The above failed beam encapsulates virtually all of the mechanisms that control the behaviour of reinforced concrete members. The explanation of the mechanics in this book is not meant to be esoteric or mystifying but was developed to directly simulate what can be seen and measured in practice. Of most importance, we have provided photographs and figures of the mechanism so that the reader can instantly visualise what is happening. The fundamental mechanics are then explained using numerical approaches, because in this way the mechanics can be illustrated with diagrams for further visualisation. Although not complex, these numerical approaches can easily cope with any non-linear material properties and hence they are readily applicable and adaptable to different analysis and design scenarios. Lastly, some closed-form solutions are given where possible. While the closed-form solutions may be more straightforward to implement and are directly relatable to the numerical models, for all but the most mathematically minded, they can often obscure the mechanics in complex and convoluted mathematics. The development of analytical solutions also often requires significant approximations to material behaviour and so they rarely apply to non-linear behaviours and properties. This is not to say that we do not believe in the application of closed-form solutions. Rather, in this book, we have been selective in providing them so as not to distract the reader from the main purpose – the development of a deep understanding of the mechanics of reinforced concrete behaviour (Figure E.1).

In the development of reinforced concrete members, the sequence of stages in the past and also, to some degree, in the present, is to first experimentally test the new type of member so that it can be used in practice within the range of the test parameters, to get the product to market as quickly as possible. The range can then be expanded through further testing, for example, to determine the shear capacity for a range of sections.

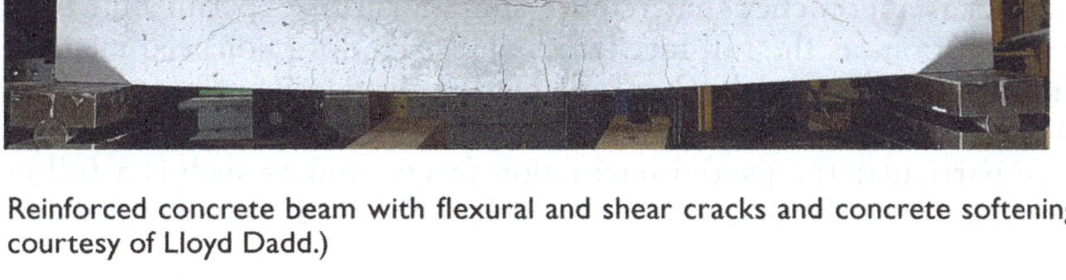

Figure E.1 Reinforced concrete beam with flexural and shear cracks and concrete softening. (Image courtesy of Lloyd Dadd.)

This is referred to as the empirical stage. It can be seen that this is a very expensive way of developing a product, as the tests must encompass all the parameters that the developer envisages, but it does get the product on the market quickly. This approach can also place severe restrictions on the range that can be used. For example, there are not many laboratories that can test a one-metre diameter column. This approach can be used safely when the mechanical behaviour of the member is little understood because the ability of the member to withstand stress resultants is guaranteed by testing.

To further expand the use of the reinforced concrete product, elementary theory is then applied and calibrated through testing. This is referred to as the semi-empirical approach. An example is the derivation of the partial-interaction flexural rigidity required in the design of a member with flexural cracks at discrete positions along its length. This is often tackled by using a combination of the full-interaction uncracked flexural rigidity at a cross-section, which is an upper bound, with the full-interaction cracked flexural rigidity at a cross-section, which is a lower bound, and calibrating this combination of parameters through member testing to allow for partial interaction. A further example is the use of the full-interaction moment-curvature of a cross-section to quantify the partial-interaction moment-rotation of a hinge by calibrating the hinge length empirically; that is, for a given full-interaction curvature at the ultimate moment, the hinge length is determined empirically to give the experimental rotation. It can be seen here that the hinge length determined empirically has nothing to do with the actual hinge in the member but is just a tool for calibration. It may be worth noting that the hinge example is one of the longest-standing problems in RC, and despite a huge amount of testing, an all-encompassing empirical solution is still missing. As with the empirical approach, the semi-empirical approach is only safe to use within the range of parameters tested; that is, interpolation within the range is generally safe while extrapolating outside the range is unsafe.

An alternative to the semi-empirical approach is the semi-mechanical approach, where the elementary theory is now calibrated from the results of numerical models as opposed to experimental testing. The semi-mechanical approach is also limited by the range of the parameters used in the numerical modelling, but as numerical modelling is usually much less expensive than experimental testing, the range can be very large, and so this restriction is relatively minor. The limitation to this approach is really the accuracy of the numerical model. The question is does the numerical model truly simulate all the partial-interaction behaviours of reinforced concrete. If not, then it should only be used within the bounds of its limitations. If the numerical model does simulate all the partial-interaction behaviours, such as the displacement-based approach described in this book, then the semi-mechanical approach is a useful and inexpensive way of promoting the product prior to the development of closed-form solutions.

The next stage, or final stage, is the full-mechanics models that simulate what can be seen and measured in practice, that is, the mechanics. The adjective 'full' has been used here to clearly differentiate this approach from the semi-empirical and semi-mechanical approaches that require some form of experimental or numerical member calibration, as the full-mechanics approach does not need member calibration. It has been shown in this book that using displacement-based approaches, as opposed to strain-based approaches, allows all facets of RC behaviour to be simulated directly, in particular the partial-interaction facets, and as such is a full-mechanics model. To emphasise the point, the full-mechanics-based approach does not require calibration through member testing; all that is needed experimentally are the material properties, which are much less expensive to determine empirically through small-scale tests than large-scale member tests.

The use of full-mechanics-based approaches has many benefits. Developers can concentrate their resources on small-scale material testing of a new product. The material test results can be inserted into the mechanics models for the member behaviour to determine if the material is effective, and if not, the material contents can be changed, retested and reinserted into the member behaviour until a suitable product has been developed for a specific problem. Take, for example, ultra-high-performance concrete, that is concrete with fibres. The optimum fibres to enhance the serviceability behaviour are probably not the same as those required to enhance the ultimate failure, nor those required to enhance durability. The developer can now tackle all three problems at relatively small expense. Knowing the material properties and with the use of the mechanics models, designers can choose a product or a system to solve their specific problem. Empirical researchers can quantify the effect of corrosion or fatigue on the material properties and use the full-mechanics models to determine their effect on the member. Theoretical researchers now have the mechanics to develop closed-form solutions for design. The main advantage of closed-form solutions derived from full-mechanics approaches, over those for semi-empirical or semi-mechanical approaches, is that they clearly show the groups of parameters or variables, that is, equations, that control behaviour and, therefore, give the engineer a real feel for what is happening. A disadvantage of closed-form solutions is that the change in the variation of material properties often has to be simplified to find a closed-form solution, which will lead to reductions in accuracy.

It is important that engineers have a feel for the behaviour of reinforced concrete, which can only be obtained through an understanding of the physical behaviour. For example, why does the engineer need to know about tension-stiffening, that is, the interaction between the reinforcement and the adjacent concrete through the interface bond? Intuition may suggest that increasing the bond strength and stiffness will increase the flexural rigidity of the member. It does not. In fact, it increases the number of flexural cracks, which would appear at first sight to be detrimental. What is important, but not so obvious, is that it significantly reduces the crack width, which is very important in designing for durability. Another example is the collapse of a member or structure through the formation of hinges. The majority of us were brought up to believe that the all-important factor that controls ductility and consequently moment redistribution and collapse is the ability of the reinforcement to yield. This is not the case. The displacement-based approach shows, and it has also been experimentally confirmed, that the ductility of the concrete in compression is a more important factor. In fact, using the displacement-based segmental analysis, it is quite easy to design an FRP-reinforced member that has the same rotational capacity and ability to redistribute moment as a steel-reinforced member. As a final example, when considering adding steel fibres to concrete, one may focus on the change in tensile strength across a flexural crack. What is less obvious, but probably more important, is the increase in ductility of the concrete in compression.

Thankfully, there is no end to the research and development of reinforced concrete while new structures, members and materials are being developed. The availability of a mechanics-based approach will encourage this development by significantly minimising or even eliminating the cost of member testing. Understanding the mechanics of reinforced concrete will help the empirical researcher concentrate on testing where it is most needed, while the theoretical researcher can focus on extending mechanics models to cover the unique properties of each new material. Trying to find closed-form solutions for members with complex material properties is often difficult. So we simplify the mechanics or the material properties to find a simple solution, but by doing this we increase the error. Hence, there is plenty of research required to find a balance between accuracy and ease of application. Notwithstanding the need to further enhance the mechanics models already available and apply them in other areas, such as torsion, stability and frame collapse problems. As such, the mechanics of reinforced concrete behaviour is an interesting, exciting and rewarding area of teaching, research, development and application.

Deric John Oehlers
Phillip Visintin

Index

Note: **Bold** page numbers refer to tables; *italic* page numbers refer to figures and page numbers followed by "n" denote endnotes.

For Product Safety Concerns and Information please contact our EU representative GPSR@taylorandfrancis.com Taylor & Francis Verlag GmbH, Kaufingerstraße 24, 80331 München, Germany

Batch number: 10733701

Printed by Printforce, the Netherlands